MW00838043

INTEGRATION OF DISTRIBUTED GENERATION IN THE POWER SYSTEM

IEEE Press
445 Hoes Lane
Piscataway, NJ 08854

Technical Reviewer
Dr. Bimal K. Bose, Life Fellow, IEEE
Condra Chair of Excellence/Emeritus in Power Electronics
Department of Electrical Engineering and Computer Science
University of Tennessee
Knoxville, TN USA

A complete list of titles in the IEEE Press Series on Power Engineering appears at the end of this book.

INTEGRATION OF DISTRIBUTED GENERATION IN THE POWER SYSTEM

MATH BOLLEN and FAINAN HASSAN

Mohamed E. El-Hawary, *Series Editor*

A JOHN WILEY & SONS, INC., PUBLICATION

Published by John Wiley & Sons, Inc., Hoboken, New Jersey
Published simultaneously in Canada

Library of Congress Cataloging-in-Publication Data:

Bollen, Math H. J., 1960-
Integration of distributed generation in the power system / Math H. Bollen and Fainan Hassan
p. cm. – (IEEE press series on power engineering)
Includes bibliographical references.
ISBN 978-0-470-64337-2 (hardback)
1. Distributed generation of electric power. I. Title.
TK1006.B65 2011
621.31–dc22

2010047229

oBook ISBN: 978-1-118-02903-9
ePDF ISBN: 978-1-118-02901-5
ePub ISBN: 978-1-118-02902-2

10 9 8 7 6 5 4 3 2

CONTENTS

PREFACE

The idea of writing this book first came in February 2008, with its final structure being decided by May 2009 when the main writing work also started. The contents of most chapters were finalized about a year thereafter. In the period of 2.5 years that we worked on this book, there have been a lot of developments in the related area: concerning not only new sources of energy (from biomass to nuclear) but also the power system. For the first time in many years, the power system is on the political agenda, instead of just the electricity production or the electricity market.

Two important concepts of this book, the "hosting capacity" and the use of "risk-based methods" have within the last few months been propagated in important reports by international organizations. The hosting capacity is proposed as a method for quantifying the performance of future electricity networks by both the European energy regulators[1] and by a group of leading European network operators.[2] The latter also recommends the development of risk-based methods for transmission system operation, whereas ENARD,[3] a government-level cooperation within the IEA, makes the same recommendation for the design of distribution networks.

During the last few years, while writing this book, giving presentations about the subject, and listening to other's presentations, we also realized that distributed generation and renewable electricity production are very sensitive areas. It is extremely difficult to keep some middle ground between those in favor and those against the idea. We would, therefore, like to emphasize clearly that this book is not about showing how good or how bad the distributed generation is. This book is about understanding the impact of distributed generation on the power system and about methods for allowing more distributed generation to be integrated into the power system, where the understanding is an essential base.

By writing this book, we hope to help removing some of the technical and nontechnical barriers that the power system poses to a wider use of renewable sources of energy.

June 2011
Math Bollen and Fainan Hassan

[1] European Regulators Group for Electricity and Gas, Position paper on smart grids, June 10, 2010.

[2] ENTSO-E and EDSO. European electricity grid initiative roadmap and implementation plan, May 25, 2010.

[3] J. Sinclair. ENARD Annex III: Infrastructure asset management. Phase 1 final report, March 2010.

ACKNOWLEDGMENTS

The material presented in this book is obtained from different sources. Most of it is work done by the authors themselves, but with important contributions from others. Although some of the ideas presented in this book are much older, the main philosophical thoughts were triggered when André Even introduced the term "hosting capacity" in 2004 during one of the first meetings of the EU-DEEP project. Discussions with other project partners helped in further refining the concepts.

Important contributions, in different forms, were also made by Johan Lundquist (Götene Elförening), Peter Axelberg, Mats Wahlberg (Skellefteå Kraft Elnät), Waterschap Roer en Overmaas, and Emmanouil Styvaktakis. Also, our colleagues and former colleagues Alstom Grid, Chalmers University of Technology, STRI AB (especially Mats Häger, Carl Öhlén and Yongtao Yang, but also many others), Luleå University of Technology, and the Energy Markets Inspectorate should be mentioned for many interesting discussions, which often triggered new ideas.

Of course, we should not forget our families and friends here, having been forced to forget them too often during the past two years.

CHAPTER 1

INTRODUCTION

The electric power system consists of units for electricity production, devices that make use of the electricity, and a power grid that connects them. The aim of the power grid is to enable the transport of electrical energy from the production to the consumption, while maintaining an acceptable reliability and voltage quality for all customers (producers and consumers), and all this for the lowest possible price. The different companies and organizations involved in this have managed to do an excellent job: the reliability and voltage quality are acceptable or better for most customers, and electricity is overall a cheap commodity. There is still a lot of research and other activities going on to make things even better or to improve the situation at specific locations, including work by the authors of this book, but we have to admit that overall the power system performance is excellent.

A sudden change either on the production side or on the consumption side could endanger the situation we have become so accustomed to. Modern society is very much dependent on the availability of cheap and reliable electricity. Several recent blackouts and price peaks have very much shown this. In this book, we will discuss not only the possible impact on the power system of one such change: the shift from large conventional production units to small and/or renewable electricity production. We will discuss not only the problems but also the solutions. Understanding the problems is essential for being able to choose the right solution.

There are different reasons for introducing new types of production into the power system. The open electricity market that has been introduced in many countries since the early 1990s has made it easier for new players to enter the market. In North America and Europe, it is now possible for almost anybody to produce electricity and export this to the power system. The rules for the actual sales of the electricity vary strongly between countries; even the rules for the connection are different between countries. Enabling the introduction of new electricity production is one of the main reasons for the deregulation of the electricity market. More market players will increase competition; together with an increased production capacity, this will result in reduced prices. The price of electricity produced by large conventional installations (fossil fuel, nuclear, hydro) is, however, too low in most countries for small units to be competitive.

The second reason for introducing new types of production is environmental. Several of the conventional types of production result in emission of carbon dioxide with the much-discussed global warming as a very likely consequence. Changing from conventional production based on fossil fuels, such as coal, gas, and oil, to renewable

Integration of Distributed Generation in the Power System, First Edition. Math Bollen and Fainan Hassan.

sources, such as sun and wind, will reduce the emission. Nuclear power stations and large hydropower installations do not increase the carbon dioxide emission as much as fossil fuel does, but they do impact the environment in different ways. There is still carbon dioxide emission due to the building and operation even with these sources, but this is much smaller than that with fossil fuel-based production. The radioactive waste from nuclear power stations is a widely discussed subject as well as the potential impact of an unlikely but nonetheless serious accident. Large hydropower production requires large reservoirs, which impact the environment in other ways. To encourage the use of renewable energy sources as an alternative, several countries have created incentive mechanism to make renewable energy more attractive. The main barrier to the wide-scale use of renewable energy is that it is cheaper to use fossil fuel. Economic incentives are needed to make renewable energy more attractive; alternatively, fossil fuel can be made more expensive by means of taxation or, for example, a trading mechanism for emission rights. Some of the incentive schemes have been very successful (Germany, Denmark, and Spain), others were less successful.

The third reason for introducing new production, of any type, is that the margin between the highest consumption and the likely available production is too small. This is obviously an important driving factor in fast-growing economies such as Brazil, South Africa, and India. In North America and Europe too, the margin is getting rather small for some regions or countries. Building large conventional power stations is not always politically acceptable for, among others, environmental reasons. It also requires large investments and can take 10 years or longer to complete. Small-scale generation based on renewable sources of energy does not suffer from these limitations. The total costs may be higher, but as the investments can be spread over many owners, the financing may actually be easier. The right incentive schemes, economically as well as technically, are also needed here. Instead of building more generation, the recent trend, for example, in Northern Europe, is to build more transmission lines. In this way, the production capacity is shared among transmission system operators. Building transmission lines is often cheaper than building new power stations, so this can be a very attractive solution. Another reason for building new lines instead of new production is that in most countries there is no longer a single entity responsible for ensuring that there is sufficient production capacity available. This means that no single entity can order the building of new production. It is, however, the task of the transmission system operator to ensure that there is sufficient transmission capacity available for the open electricity market. The transmission system operator can decide to build new lines to alleviate bottlenecks that limit the functioning of the open market.

Although growth in electricity consumption has been moderate for many years in many countries, there are reasons to expect a change. More efficient use of energy often requires electricity as an intermediate step. Electric cars are the most discussed example; electrified railways and heat pumps are other examples. Even the introduction of incandescent lamps 100 years ago was an improvement in energy efficiency compared to the candles they were replacing.

No matter what the arguments are behind introducing new electricity production, it will have to be integrated into the electric power system. The integration of large production units, or of many small units, will require investments at different voltage levels. The connection of large production units to the transmission or subtransmission system is in itself nothing remarkable and the investments needed are a normal part of transmission system planning. With new types of production, new types of phenomena occur, which require new types of solutions. Small production units are connected to the low- or medium-voltage distribution system, where traditionally only consumption has been connected. The introduction of large numbers of them will require investments not only at the voltage level where the units are connected but also at higher voltage levels. The variation in production from renewable sources introduces new power quality phenomena, typically at lower voltage levels. The shift from large production units connected at higher voltage levels to small units connected at lower voltage levels will also impact the design and operation of subtransmission and transmission networks. The difficulty in predicting the production impacts the operation of the transmission system.

The terminology used to refer to the new types of production differs: "embedded generation," "distributed generation," "small-scale generation," "renewable energy sources" and "distributed energy resources" are some of the terms that are in use. The different terms often refer to different aspects or properties of the new types of generation. There is strong overlap between the terms, but there are some serious differences as well. In this book, we will use the term "distributed generation" to refer to production units connected to the distribution network as well as large production units based on renewable energy sources. The main emphasis in this book will be on production units connected to the distribution network. Large installations connected to the transmission system will be included mainly when discussing transmission system operation in Chapter 8.

In this book, we will describe some of the ways in which the introduction of distributed generation will impact the power system. This book has been very much written from the viewpoint of the power system, but the network owners are not the only stakeholders being considered. The basic principle used throughout the book is that the introduction of new sources of production should not result in unacceptable performance of the power system. This principle should, however, not be used as a barrier to the introduction of distributed generation. Improvements should be made in the network, on the production side and even on the consumption side, to enable the introduction of distributed generation. Several possible improvements will be discussed throughout this book. We will not discuss the difficult issue of who should pay for these investments, but will merely give alternatives from which the most cost-effective one should be chosen.

The structure of this book is shown in Figure 1.1. The next two chapters introduce the new sources of production (Chapter 2) and the power system (Chapter 3). Chapters 4–8 discuss the impact of distributed generation on one specific aspect of the power system: from losses through transmission system operation.

As already mentioned, Chapter 2 introduces the different sources of energy behind new types of electricity production. The emphasis is on wind power and solar power, the renewable sources that get most attention these days. These are the two sources that will constitute the main part of the new renewable sources of energy in the near future. However, more "classical" sources such as hydropower will also be discussed. The different sources will be described in terms of their variation

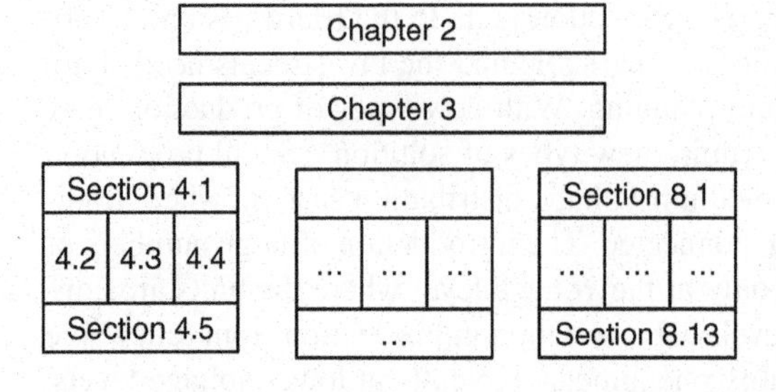

Figure 1.1 Structure of the book: introductory chapters and chapters on specific phenomena.

in production capacity at different timescales, the size of individual units, and the flexibility in choosing locations. These are the properties that play an important role in their integration into the power system.

After a general overview of the power system, Chapter 3 introduces, the "hosting capacity approach." The hosting capacity is the maximum amount of generation that can be integrated into the power system, while still maintaining its performance within acceptable limits. The hosting capacity approach uses the existing power system as a starting point and considers the way in which distributed generation changes the performance of the system when no additional measures are taken. For this, a set of performance indicators is needed. This is a normal procedure in the power quality area, but not yet in other areas of power systems.

Chapters 4–8 discuss in detail various aspects of the integration of distributed generation: the increased risk of overload and increased losses (Chapter 4), increased risk of overvoltages (Chapter 5), increased levels of power quality disturbances (Chapter 6), incorrect operation of the protection (Chapter 7), and the impact on power system stability and operation (Chapter 8).

Chapters 3–8 are structured in the same way, as shown in Figure 1.1. Considering Chapter 5, for example, the first section gives an overview of the impact of increasing amounts of distributed generation on the voltage magnitude as experienced by the end customers. The first section in each chapter is both a summary of the results from the forthcoming sections and an overview of material obtained from the literature. The sections following the first section discuss different details of, in this case, the relation between distributed generation and voltage magnitude. Some of the sections look at the problem from a different perspective or discuss a specific solution. Some of these sections give a general overview, while others go deeper into theoretical models or research results. Most of these sections can be read or studied independent of other sections. The final section of the chapter gives an overview of methods for allowing more distributed generation to be connected without experiencing problems with, in this case, voltage magnitude. The final section of each chapter is again a combination of material from the rest of the chapter and material obtained from the literature. The different solutions presented here include those that are currently referred to as "smart grids." This term has received a huge amount of interest, all the way from fundamental research to politics and newspapers, but it remains unclear what should be included in the term. We will not distinguish here between "smart grid solutions" and "classical solutions," but instead present all the available options.

It is not possible to cover all aspects of the integration of distributed generation in one book. The economic aspects of the different impacts of distributed generation and of the different methods for increasing the hosting capacity are not treated here at all. This is not because economics are not important, they are in fact often the main deciding factor, it is just that we had to stop somewhere. Besides, the economics are very much location and time dependent. The book does not include many detailed simulation studies, but mainly simplified models of the power system and of the distributed generation. There are a number of reasons for this. We would like to propagate the use of such simplified models as a tool to be used during initial discussions; it is our experience that important conclusions can often be drawn from these simplified models. We are also of the opinion that the use of simplified models has a great educational value. The impact of different parameters is much better understood when simplified models rather than detailed simulations are used. Such detailed calculations are, however, needed in many cases before connecting distributed generation to the power system. The structure of the power system is different across the world and the details are very much location dependent. The simplified models of the type presented in this book can be easily adapted to a local situation, whereas simulation studies have to be repeated for each location.

CHAPTER 2

SOURCES OF ENERGY

In this chapter, we will discuss the different sources of energy used for electricity production. We will concentrate on the main renewable sources used for distributed generation—wind power in Section 2.1 and solar power in Section 2.2. Another type of distributed generation, combined heat-and-power (CHP), will be discussed in Section 2.3. We will also discuss the two main sources in use today: hydropower in Section 2.4 and thermal power stations in Section 2.8. Some of the other sources will also be discussed: tidal power in Section 2.5, wave power in Section 2.6, and geothermal power in Section 2.7.

For each of the sources, we will give a brief overview of the status and the prospects, based on the information available to the authors today, for it to become a major source of electric power. Furthermore, an overview will be given of the properties of the source seen from a power system viewpoint. For the major sources, we will concentrate on the variation in the source with time, which is the main difference between renewable sources like the sun, water, and wind, and the thermal power stations. We will not go into details of the way in which the primary energy is transformed into electricity. For further details, the reader is referred to some of the many books on this subject. An excellent overview of energy consumption and production possibilities for the United Kingdom is given in Ref. 286. The analysis can also be easily applied to other countries and hence the book is highly recommended to those interested in energy supply. Another good overview of the different energy sources is given in Refs. 60, 81 and 389. The latter two give an excellent detailed description of the origin and application of some of the sources. A lot of information on wind energy can be found in Refs. 71 and 292. Both books discuss in detail the whole chain from the aerodynamics to the connection with the grid. For solar power, refer to Ref. 337. Besides, for the power system aspects of wind power and other sources of renewable energy, refer to among others Refs. 5, 56, 157, 167, 200, 232, 296, 392, and 458.

There have been many developments in many countries concerning the future energy sources. The reader should realize, when reading this chapter, that it mainly describes the status as of the first months of 2010. Although we have tried to be as objective as possible, we are quite aware that some parts of this chapter may be outdated within a few years. Hence, the reader is encouraged to also refer to more recent sources of information.

Integration of Distributed Generation in the Power System, First Edition. Math Bollen and Fainan Hassan.

2.1 WIND POWER

2.1.1 Status

The kinetic energy from the horizontal displacement of air (i.e., wind) is transformed into kinetic energy of the rotation of a turbine by means of a number of blades connected to an axis. This rotational energy is then transformed into electrical energy using an electrical generator. Different technologies have been proposed and used during the years to produce electricity from wind power. Currently, the main technology on the mechanical side is a two-or three-blade turbine with a horizontal axis. Three competing technologies are in use for the transformation into electrical energy and the connection to the power system: the directly connected induction generation; the double-fed induction generator (DFIG) (more correctly named "double-fed asynchronous generator"); and the generator with a power electronics converter.

Wind power is the most visible new source of electrical energy. It started off as small installations connected to the low-or medium-voltage networks. The last several years have seen a huge expansion of wind power in many countries, with the current emphasis being on large wind parks connected to the subtransmission or transmission system. Single wind turbines of 2 MW size have become the typical size and turbines of 5–6 MW are available, although they are not yet widely used. Developments are going fast, so these values could well have become outdated by the time you read this book.

Single wind turbines in Europe are now mainly being connected to medium-voltage networks; but in more and more cases, groups of turbines are connected together into a wind park. Smaller wind parks, 3–10 turbines is a typical range, can still be connected to the medium-voltage network, but the larger ones require connection points at subtransmission or transmission level. Several parks larger than 500 MW are in operation or under construction in the United States, with Texas and California taking the lead. The biggest wind park at the moment is the Horse Hollow Wind Energy Center in Texas. It consists of 291 turbines of 1.5 MW and 130 turbines of 2.3 MW giving it a total capacity of 735 MW. However, even more larger ones are already under construction or planned. For example, a very large wind park is planned near the north Swedish town of Piteå, with 1100 turbines of 2–3 MW each and an expected annual production between 8 and 12 TWh, that is, between 5 and 8% of the total consumption in Sweden. News items about large wind parks appear almost continuously and the wind power production is growing fast in many countries. Several countries have planning targets of 20% electrical energy from wind power by 2020. Some European countries already produce more than 10% of their electrical energy from wind power; Denmark being on top with 20% of its electrical energy produced by wind. Of the large European countries, Germany (with 7% of electricity coming from wind) and Spain (12%) are the main wind power producing countries.

2.1.2 Properties

Wind power shows variations in production capacity over a range of timescales, from less than 1 s through seasonal variations. There remains difference of opinion about which timescale shows the biggest variation. This depends strongly on the application. We will discuss variations at different timescales in this chapter and in some of the other chapters.

The term intermittent generation is often used to refer to the strong variation with time of wind and solar power. What matters to the power system is however not just the variation with time but also the extent to which the variations can be predicted. For the distribution system, it is merely the actual variations that matter; while for the transmission system, it is both the variations and their predictability that are of importance.

The wind speed and thus the wind power production are difficult to predict longer than a few hours ahead of time. Over larger geographical areas, which is what matters at transmission level, predictions of total production become somewhat better. But even here large prediction errors are not uncommon. Details about this is given in Section 8.4.

The accuracy of the prediction is also important for individual turbines and for wind parks to be able to participate in the electricity market. The details of this strongly depend on the local market rules. We will not further discuss this in this book.

The amount of energy that can be extracted from the wind is not everywhere the same: some locations are more windy than others. What matters for this are the properties of the wind speed over a longer period of time. The average wind speed is an important factor; however, as we will see later, the distribution of the wind speed also matters. The amount of energy that can be produced per year by a wind turbine strongly depends on the location of this wind turbine. The most favorable areas are coastal areas (like the European west coast or the north coast of Africa) and large flat areas (like the American Midwest or Inner Mongolia in China). Mountaintops also often provide good wind conditions. The wind conditions in built-up areas, such as towns or industrial areas, are often not very good, because the buildings are obstacles to wind, taking away energy from the wind and turning it into turbulence. Only the horizontal movement, not the turbulence, is transformed into electrical energy by the wind turbine. Also, in most cases, it is not possible to get permission to build wind power installations close to buildings. As a result, the wind power is often located at a significant distance from the consumption: tens to hundreds of kilometers is not uncommon. Some of the future plans, like for Inner Mongolia, would result in wind power being located thousands of kilometers away from the consumption areas.

Europe is in a rather good position in this sense because the areas with the best wind conditions (mainly around the North Sea), are not too far from the main consumption areas. The fact that the North Sea is very shallow also makes it relatively easy to build large wind parks there.

2.1.3 Variations in Wind Speed

One of the most discussed properties of wind power is its so-called "intermittent" character—the wind speed and thus the power production vary strongly with time over a range of timescales. The variability of the wind is often presented as a power spectrum. This is discussed among others in Ref. 389, according to which the wind speed variance spectrum shows 5-day variations probably related to major weather patterns, daily variations becoming less with altitude, and variations in the range between 10 s and 1 min. The so-called "spectral gap" is present (in the example shown in Ref. 389) between 2 h and 3 min. It is said to be confirmed by many measurements that such a gap appears at almost all locations. It provided a clear distinction between large-scale motion (at timescales above 2 h) and small-scale motion (at timescales less than 10 min).

As stated in Ref. 386, the wind power production "varies very little in the time frame of seconds, more in the time frame of minutes and most in the time frame of hours." Typical standard deviations are as follows:

- 0.1% at 1 s timescale,
- 3% at 10 min timescale, and
- 10% at 1 h timescale.

The analysis of wind speed records from Brookhaven, New York ([418], quoted in Ref. 71 and many others), in the 1950s showed a power spectrum with three distinctive peaks:

- "turbulence peak" between 30 s and 3 min,
- "diurnal peak" around 12 h, and
- "synoptic peak" between 2 and 10 days.

The measurements showed that there is very little energy in the region between 10 min and 2 h. This region is often referred to as the "spectral gap." The presence of this has been confirmed by measurements performed at other locations, for example, Ref. 169 and is widely mentioned in the literature. From a power system operation viewpoint, this is good news. The turbulence peak is a local phenomenon and will average out when many turbines are connected over a wider geographical area. The result is that wind power production will be in general rather constant for timescales up to a few hours. From this it should not be concluded that there are no changes in this range of timescales. For transmission system planning and operation, it is often the worst-case situations that matter. The fact that they are rare does not always matter. We will come back to this topic in Chapter 8. Not all measurements do however show the presence of the spectral gap, nor the diurnal or synoptic peak. This may depend on local conditions, which will especially impact the turbulence part of the spectrum. Measurements presented in Ref. 15 show, for example, that the output power from two wind parks (with 6 and 10 turbines) follows the so-called "Kolmogorov spectrum" (proportional to frequency to the power of $-5/3$) over the time range of 30 s–2.6 days.

It should also be noted here that the "power spectrum" in this context is not the spectrum of the power production but (for a deterministic signal) the square of the magnitude of the spectrum (Fourier series) of the wind speed. For a stochastic signal, the power spectral density is the Fourier transform of the autocovariance function [30].

Turbulence, leading to power fluctuations in the timescale from less than 1 min to about 1 h, is discussed in detail in, among others, Refs. 71 and 292. A distinction thereby has to be made between "turbulence" and "gusts." Turbulence is a continuous phenomenon, present all the time, whereas a wind gust is an occasional high value of the wind speed superimposed upon the turbulent wind. Readers familiar with power quality will recognize the similarity in the distinction between "power quality variations" and "power quality events" (see Section 3.4.1 for more about variations and events).

Turbulence is a complicated process, which is very difficult if not impossible to quantify. It depends not only on local geographical features (like hills and rivers) but also on the presence of trees and buildings. Also, vertical movement of the air due to heating of the earth surface by the sun results in turbulence. As turbulence is a surface-related phenomenon, it will reduce with increasing height. The higher a wind turbine, the less affected by turbulence. In terms of the power density spectrum, the turbulence peak diminishes with increasing height. The shift from individual small turbines to wind parks consisting of large turbines implies that turbulence has become less of an issue for the power system. It remains an issue for the mechanical design of the turbine, but this is beyond the scope of this book.

The level of turbulence is quantified by the so-called "turbulence intensity." To obtain this, the wind speed is sampled with a high sampling frequency (one sample per second or higher) over an interval between 10 min and 1 h. The turbulence intensity is the ratio of the standard deviation to the average over this interval. According to Ref. 292, the turbulence intensity is typically between 0.1 and 0.4, with the highest values obtained during low wind speed. However, some of the standard models for turbulence ([71], Section 2.6.3) recommend the use of a turbulence intensity independent of the wind speed. Reference 292 also states that the Gaussian distribution is an appropriate one to describe turbulence. The distinction between turbulence and gusts is very important here: the probability of a wind gust exceeding a certain value is not found from the Gaussian distribution for turbulence. Of more interest from a power system viewpoint is the spectrum of the turbulence, that is, which frequencies occur most commonly among the fluctuations in wind speed. Several such spectra are shown in Section 2.6.4 of Ref. 71, with their peak between 1 and 10 min. The 50% value of the turbulence peak is between 10 s and 1 h. As mentioned before, the actual spectrum strongly depends on location and time, for example, depending on the wind direction. When the turbulence peak exceeds beyond 1 h, the above-mentioned spectral gap will disappear and the turbulence peak will go over into the diurnal and synoptic peaks.

From a power system viewpoint, what matters are not the variations in wind speed but the variations in power production. The power production is a nonlinear function of the wind speed; variations in wind speed have the strongest impact on the power production when the wind is between (about) 5 and 10 m per second (see Section 2.1.7). Further, the interest is not in the relative variations in power production (in percent) but in the absolute variations (in kilowatt). It is the absolute variations that cause variations in voltage magnitude, that result for example in the kind of fluctuations in light intensity that give complaints from nearby customers about light flicker. More about this in Section 6.2.

2.1.4 Variations in Production Capacity

In this section, we will discuss the variations in production capacity for wind power over a range of timescales, starting at the shortest timescales. The current way of operating wind power implies that variations in production capacity in almost all cases result in the same variations in actual production. In other words, the production is always equal to the capacity. It is possible to reduce the amount of production below the capacity, but that would result in "spilled wind." Any reduction in wind power production will have to be compensated by other sources, in most cases fossil fuel.

At the shortest timescale, seconds and lower, the production of an individual turbine varies mainly due to the impact of the tower on the flow of air around the turbine. This has been studied in detail because the fluctuations in power production cause variations in voltage magnitude that could result in observable variations in light intensity of certain types of lighting. We will discuss this further in Section 6.2. Another source of power fluctuations at the shortest timescale is the fact that the wind speed is higher at higher altitudes. When a blade is pointed upward, it will catch more energy from the wind than when it is pointed downward; the total amount of energy for the three blades will depend on their position. Also, mechanical oscillations in the turbine and the tower as well as the gearbox cause some fast variations in power. All this holds for individual turbines. For large wind parks and for larger regions, these variations will add randomly and become less and less important. Using power electronics techniques and a small amount of storage, variations at a timescale of seconds can be limited even for individual turbines. Energy storage can be present in the form of capacitors connected to a DC bus or by letting the rotational speed of the turbines vary somewhat.

At a longer timescale (minutes), turbulence is the main cause of variations in produced power. The level of turbulence strongly depends on local conditions (landscape as well as weather) and is very difficult to predict. Recommended values for turbulence given in standards and in the literature are mainly used as input in the mechanical design of the installation. In practice, the level of turbulence is not constant at all even at a single location.

A measurement of the power fluctuations for a 600 kW turbine with full power converter, over two different periods of 1 h, is shown in Figure 2.1. Active power values were obtained every second. The two 1 h periods were about 20 h apart. The level of turbulence varies a lot between these two 1 h periods.

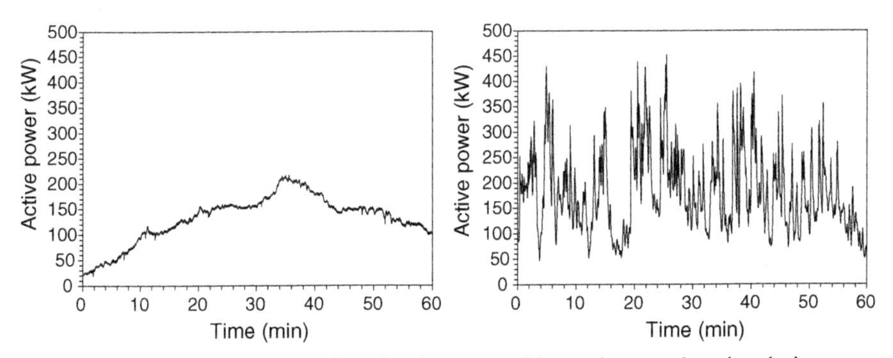

Figure 2.1 Active power fluctuations for the same turbine at the same location during two different 1 h periods.

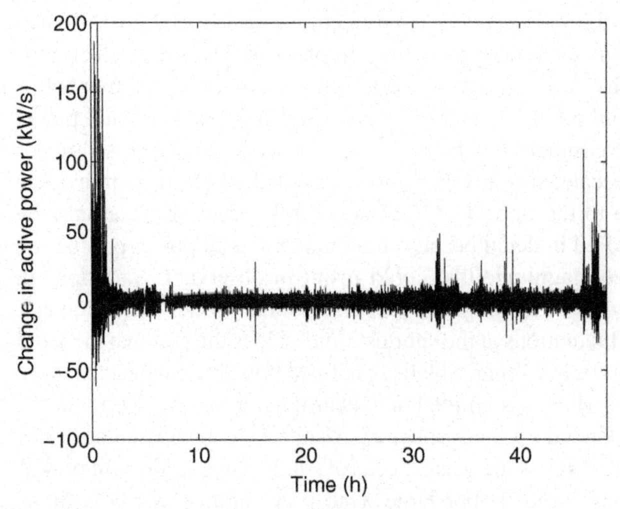

Figure 2.2 Second-by-second change in active power for a 600-kW wind turbine during a 48 h period.

Strictly speaking, note that these figures do not present the turbulence as turbulence is a property of the wind speed. There is, however, no equivalent parameter in use on the power system side of the turbine to quantify the variations in production at this timescale. One option would be to use the ratio between standard deviation and average over a period of 10 min–1 h, in the same way as the definition of turbulence intensity for the wind speed. Closer to existing methods for quantifying voltage quality would be to define the "very short variations" in active and reactive powers in the same way as the very short variations in voltage magnitude [44, 48]. We will discuss this further in Chapter 6.

From the measured values of active power for the 600 kW turbine, the second-by-second variations in power are calculated. The curve in Figure 2.2 is calculated as the difference between two consecutive 1 s averages. Most of the time two consecutive values do not differ more than 20 kW; however, extreme values up to 200 kW can be seen. The fast changes occur only during certain periods, and the largest changes are positive (i.e., a fast rise in power).

The active power produced by the 600 kW turbine during a 48 h period is shown in Figure 2.3. Figure 2.3a shows the measured 1 s averages. From these the 10 min averages have been calculated, shown in Figure 2.3b. The 10 min average shows a much smoother curve; this indicates that the majority of the variations occur at a timescale of less than 10 min.

The standard deviation of the amount of power produced by a 600 kW turbine has been calculated over each 10 min interval (from 600 1 s values). The resulting standard deviation as a function of time is shown in Figure 2.4. Both the absolute value of the standard deviation (in kilowatt) and the value relative to the average power over the same 10 min interval (in percent) are shown. The relative value is up to 50% most of the time. During some periods, the relative standard deviation is very high; this is due to the low values of the average production and does not indicate

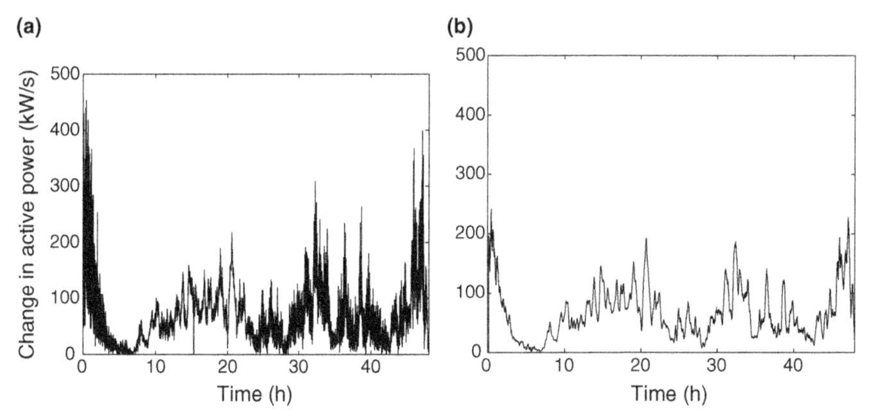

Figure 2.3 1 s (a) and 10 min (b) averages of the active power production for a 600 kW wind turbine during a 48 h period.

any actual high level of variations. As was mentioned before, what matters to the power system is the absolute level of variations. For this measurement, the standard deviation is at most about 100 kW, which is less than 20% of the rating of the turbine.

Measurements of variations in power production at a timescale of seconds to minutes typically require dedicated equipment and are hard to obtain. Performing such measurements over a longer period (for example, 1 year) would also result in large amounts of data. Hourly measurements are more common because they are typically used for metering purposes. In some countries, the electricity market is based on 15 or 30 min intervals, which would typically require measurements of the average production to be available over these intervals.

As an example, Figure 2.5 shows the hourly average power produced by an 850 kW wind turbine located in the middle of Sweden over a period of about 4 months (January–April 2005). Not only does the power vary significantly, it also is

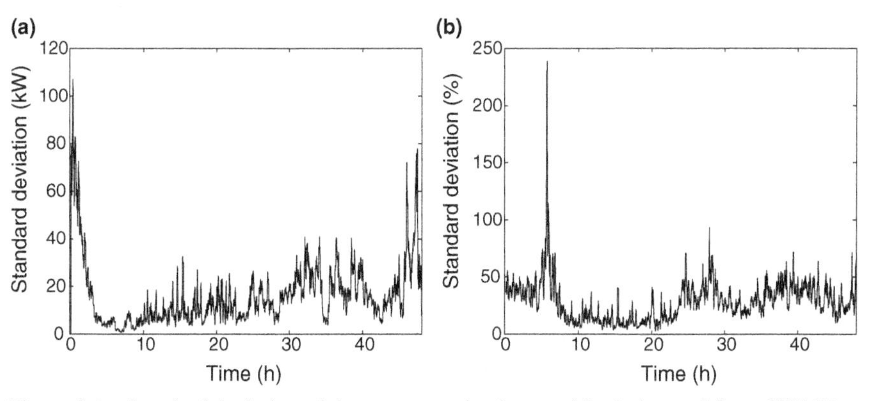

Figure 2.4 Standard deviation of the power production per 10 min interval for a 600 kW turbine during a 48 h period. (a) Absolute values. (b) Relative values.

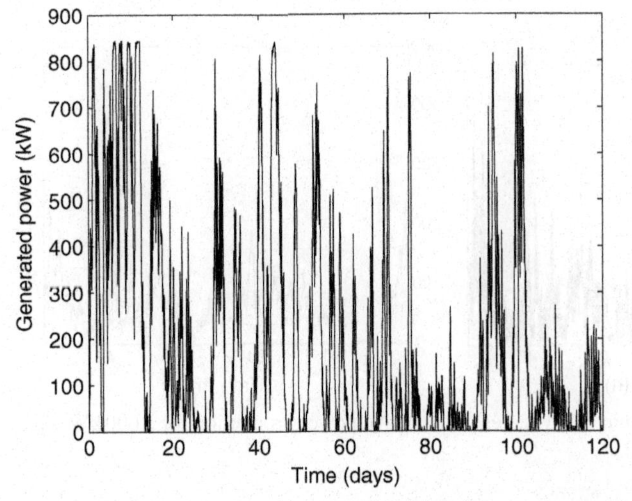

Figure 2.5 Power produced by a wind turbine during a 4-month period.

close to zero a large part of the time. The maximum power production, for which the installation is dimensioned, is reached only during a small fraction of the time.

The variation in production can be presented by using the probability distribution function, as shown in Figure 2.6. This function is obtained from the 1 h averages obtained over a 4.5-year period. The power production was zero during about 14% of the time and less than 100 MW during about 50% of the time. Low production during a significant amount of time is a general phenomenon observed with all wind power installations. This is strongly related to the distribution of the wind speed as will be

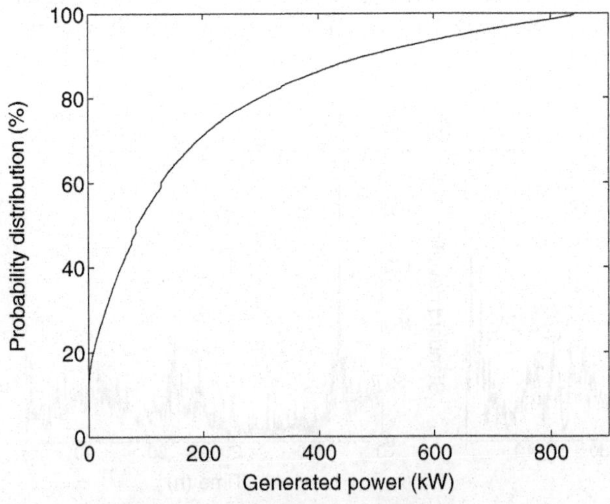

Figure 2.6 Probability distribution function of the hourly power produced by a 850 kW wind turbine over a 4.5-year period.

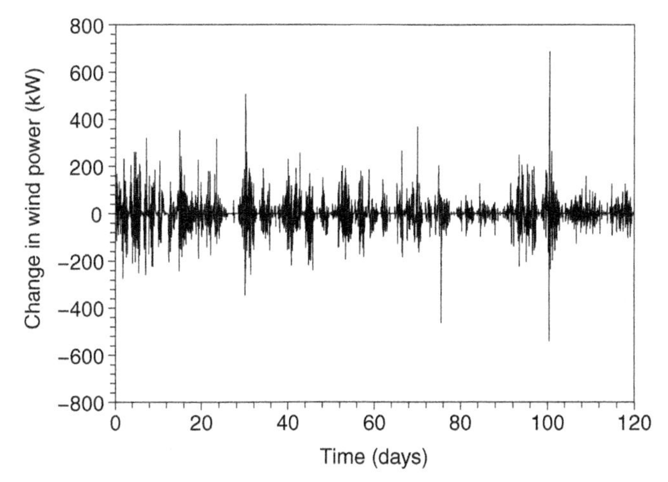

Figure 2.7 Hour-by-hour changes in wind power over a 4-month period.

discussed later. The probability distribution function is defined in Section 2.1.5. The reader may refer to this section for the definition.

Of importance to the operation of the transmission system is not only the fact that wind power varies with time but also the amount of variation at a timescale up to a few hours. The difference between two consecutive hourly averages is shown in Figure 2.7 for the same 4-month period as in Figure 2.5. Although most of the time the changes are less than 200 kW, values up to 600 kW occur.

The probability distribution function of the hour-by-hour changes has been calculated for the 4.5 years over which data were available. The tails of this distribution, that is, the extreme values, are shown in Figure 2.8. Changes of up to 700 kW (almost the rated power of the turbine) occur in both directions. For extreme events, the time between events is sometimes easier to interpret than the probability of the event. In this case, the 1% and 99% lines correspond to events that occur on average once every 4 days. A once-a-month event is found with the 0.15% and 99.85% points, while a

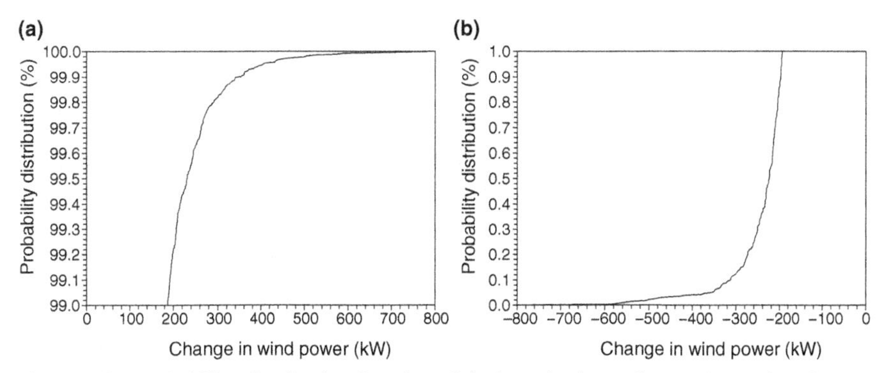

Figure 2.8 Probability distribution function of the hour-by-hour changes in produced power. (a) Large reduction in power (b) Large increase in power.

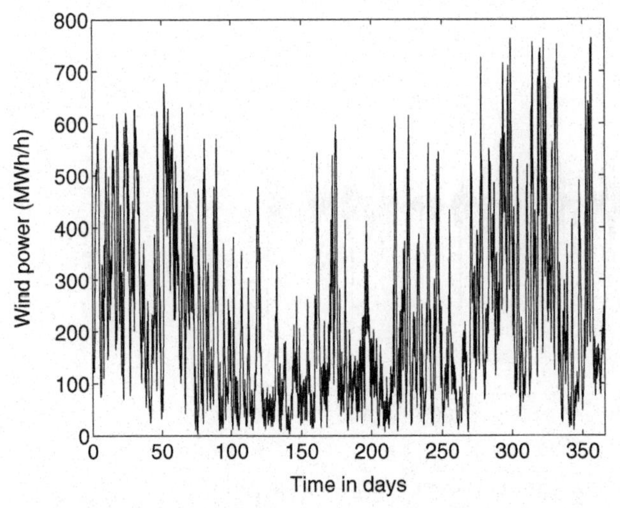

Figure 2.9 Hourly production of wind power in Sweden during the year 2008.

once-a-year event with the 0.01% and 99.99% points. For individual turbines, large changes in production are of less importance. There could be some economic consequences for the power producer depending on the market rules, but rare events have a small impact on the total costs.

The Swedish transmission system operator reports hourly on the electricity production and consumption. The production due to different sources (hydro, nuclear, wind, CHP, etc.) is given in kilowatt hour for every hour during the year [332]. The hourly wind power production during 2008 is shown in Figure 2.9. The total installed capacity was 831 MW at the start of the year and 1067 MW at the end of the year [461]. Wind power in Sweden is (2008 status) mainly found along the west coast and around the Swedish big lakes [132]. The main wind power production is spread over distances up to about 400 km.

The probability distribution of the production over the whole country is shown in Figure 2.10. The production is given as a percentage of the installed capacity. It has been assumed that the installed capacity increased linearly from 831 MW on January 1 to 1067 MW on December 31 [461]. For a comparison with the probability distribution for a single turbine, see Figure 2.6. The total production for the whole of Sweden is never completely zero (the lowest production was 0.4% of the installed capacity); but the production was also never higher than 78% of the total installed capacity. During about 25% of time, the national production is less than 10% of the installed capacity. Thus, even at a national level, one cannot rely much on wind power for energy security. Wind power, however, does have a certain "capacity credit," that is adding wind power to an otherwise unchanged energy mix will increase the availability of the generation capacity. As we will see later, the maximum wind power production in Sweden occurs in winter, when the peak consumption also occurs. (Electric heating is a major consumer of electric energy, which obviously has its peak in winter.) There is at the moment, however, insufficient information available to know what the conditional

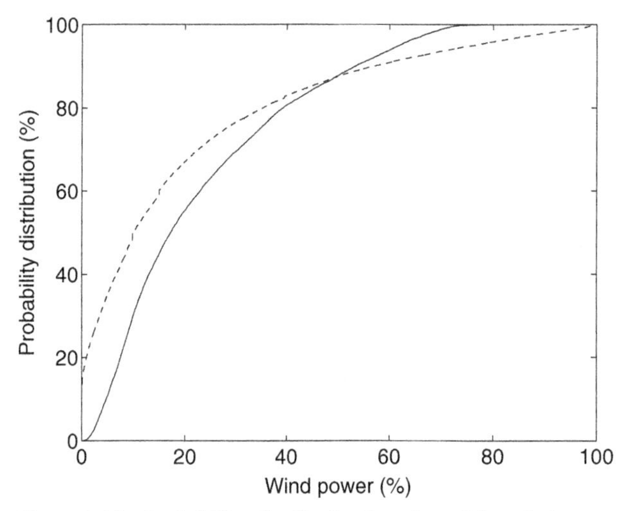

Figure 2.10 Probability distribution function of the wind power production for a single turbine (dashed line) and for the whole of Sweden (solid line).

probability distribution function is for the available wind power production around the time of high consumption. The peak consumption strongly depends on weather in Sweden, which makes a correlation or anticorrelation with wind power production possible.

The distribution for one turbine and the distribution for many turbines spread over a large area have a similar shape, but the spread becomes less for many turbines. For the single turbine, the average is 21% with a standard deviation of 29%. For the country as a whole, the average is 23% with a standard deviation of 18%. The reduction in standard deviation is not as big as expected. This is probably due to most of the turbines over the whole country being exposed to the same weather pattern.

The correlation between the production by one turbine and that by a large number of turbines spread over a large area is shown in Figure 2.11. Each dot corresponds to a 1 h average during the period January 1, 2008–March 31, 2008. There is a weak positive correlation between the two, with a correlation coefficient of 81%.

The hour-by-hour changes in wind power production for the whole of Sweden are shown in Figure 2.12. The figure shows the difference in wind power production for two consecutive 1 h periods. When comparing this figure with the changes for one turbine (in Figure 2.7), one should consider the difference in rated power. The rating of the individual turbine is 850 kW; the highest changes in Figure 2.7 are more than half of the rating. The rating of all wind power in the country is about 1000 MW; the highest changes in Figure 2.12 are only about 10% of rating. For the single turbine, changes of 40% or more of rated occurred nine times per year. For the large area, the highest change was of 12.7%. Changes above 10% of rated occurred only seven times during the 1-year period.

The probability distribution function of the hour-by-hour changes is shown in Figure 2.13. Changes of up to 50 MW upward or downward are very common. Changes of more than 100 MW occur about once a month. Note that 50 and 100 MW

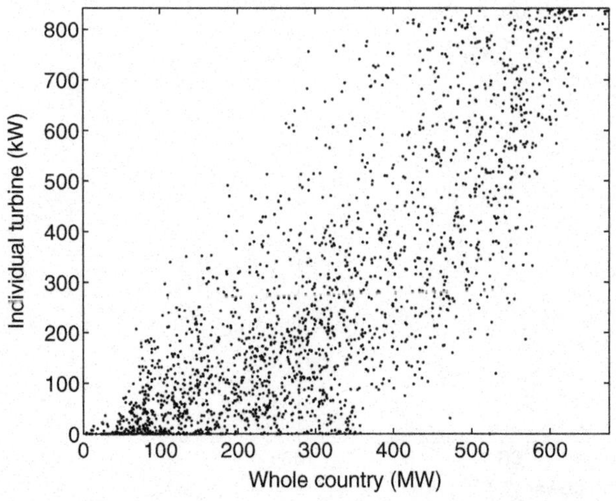

Figure 2.11 Correlation between the wind power production for one turbine and that for the whole of Sweden.

correspond to about 5% and 10%, respectively, of the installed capacity. With higher wind power penetration in Sweden, the variations as a percentage of installed capacity will remain about the same, assuming that the geographical spread of the production units will not significantly change. When discussing operational reserves in Chapter 8, we will see that changes of up to about 1000 MW can be accepted in the existing system. From this we can, as a very rough approximation, conclude that the hour-by-hour changes in wind power production remain acceptable as long as the amount of installed power is less than 10 times the current value of 1000 MW.

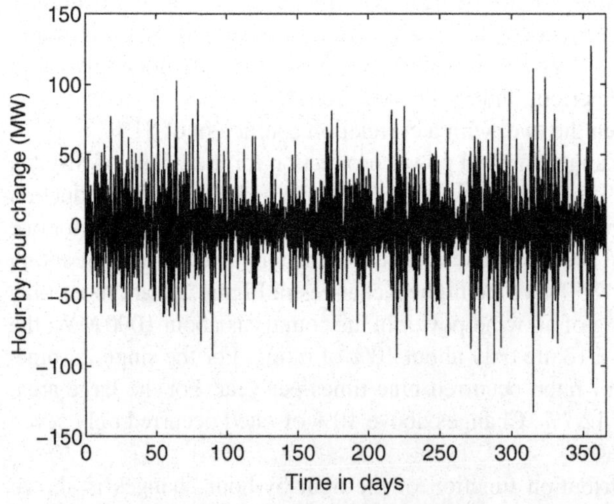

Figure 2.12 Hour-by-hour changes in wind power production for the whole of Sweden during 2008.

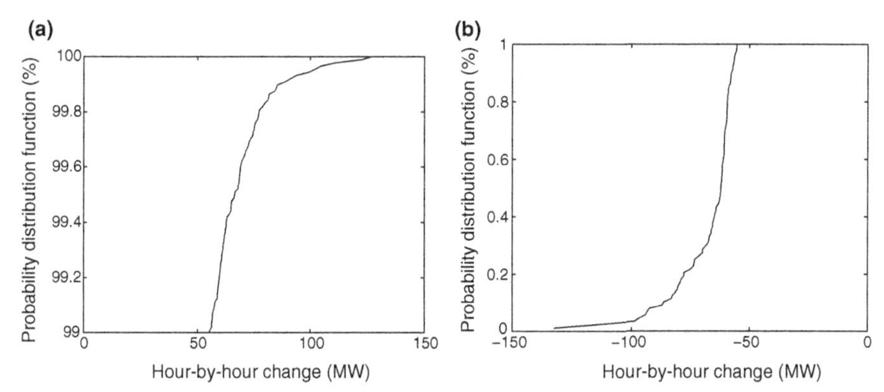

Figure 2.13 Probability distribution function of the hour-by-hour changes in produced wind power for the whole of Sweden during 2008. (a) Large reduction in power (b) Large increase in power.

To illustrate the seasonal variations in wind power production, the average production has been calculated for each 1-week period and merged for 3 years over which complete data were available. The results are shown in Figure 2.14. The highest wind power production occurs in winter, while the lowest during the summer period. A similar picture is obtained when the production for the whole of Sweden is considered, as shown in Figure 2.15 for the year 2008. A correction has been made for the growth in installed production during the year, where a linear growth has been assumed.

As we will see in Section 2.4, a disadvantage of hydropower is that the amount of precipitation (rain or snow) can vary a lot from year to year. As precipitation and wind are determined by the same weather systems, it is important to know if such variations also exist for wind power. It would especially be important to know if there

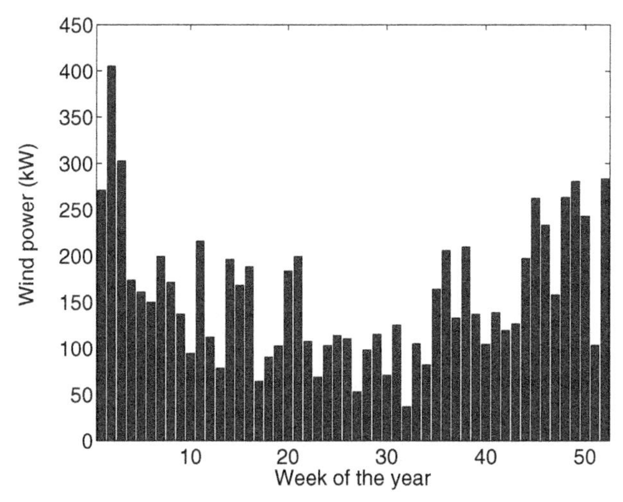

Figure 2.14 Seasonal variations (week-by-week) in wind power production for a single 850 kW turbine.

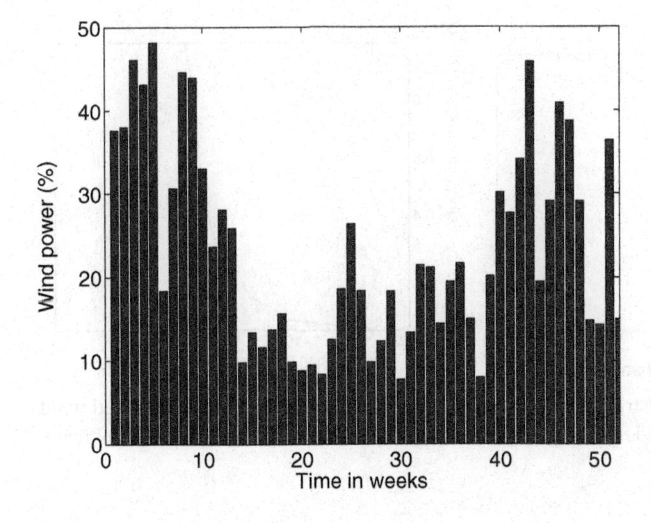

Figure 2.15 Weekly wind power production in Sweden during 2008, as a percentage of the installed capacity.

is any correlation between the annual amounts of energy that can be produced by hydropower and by wind power. Several countries are creating an energy portfolio in which both wind power and hydropower play an important part. Connection of large hydropower and wind power resources throughout large parts of Europe using an HVDC grid is sometimes part of the future vision.

Measurements on the year-by-year variations in available wind energy are presented in Ref. 200 for four regions in Germany. The data were obtained from wind speed measurements at a large number of locations. The statistical results are summarized in Table 2.1. We see at first that the energy production per unit area varies by a factor of 2.5 between wooded lowlands and coastal areas. The same wind park built near the coast will produce 2.5 times as much energy as that produced by one that is built in wooded area away from the coast. The standard deviation of the annual production is 10–20% of the average value.

2.1.5 The Weibull Distribution of Wind Speed

We saw in the previous section that the power production by a wind power unit strongly varies with time. To describe this in a stochastic way, typically the probability distribution function of the wind speed is used. The probability density function $F_X(x)$

TABLE 2.1 Statistics of Annually Available Wind Energy for Four Regions [200]

Region	Average (W/m^2)	Standard deviation (W/m^2)	Ratio
Coastal	178	18.8	0.11
Hills	106	15.5	0.15
Lowlands	88	10.3	0.12
Wooded lowlands	67	12	0.18

gives the probability that a random variable X is less than or equal to x:

$$F_X(x) = \Pr\{X \leq x\} \tag{2.1}$$

From a known relation between wind speed and power production, the distribution of the power production can next be calculated. Instead of a probability distribution function, it is also possible to use measured wind speeds or wind speeds generated from an atmospheric model.

The most commonly used probability distribution to describe the wind speed is the Weibull distribution as, for example, mentioned in Refs. 59, 71, 167, 200, 292, 337, and 456. Most of these references also mention that the Rayleigh distribution (Weibull distribution with a shape factor of 2) is valid for most locations.

The probability distribution function of the Weibull distribution reads as follows [22, 270]:

$$F(t) = 1 - \exp\left(-\left(\frac{t}{\theta}\right)^m\right) \tag{2.2}$$

where θ is the "characteristic value" of the distribution and m is the "shape factor" or "Weibull shape parameter." The expected value of the Weibull distribution is a function of both the characteristic value and the shape factor:

$$\mu = \theta\Gamma\left(1 + \frac{1}{m}\right) \tag{2.3}$$

where $\Gamma(.)$ is the "gamma function" [270]. The probability density function of the Weibull distribution is given by the following expression [270]:

$$f(t) = \frac{m}{\theta}\left(\frac{t}{\theta}\right)^{m-1} \exp\left[-\left(\frac{t}{\theta}\right)^m\right] \tag{2.4}$$

Some examples of shape factor and expected value for the wind speed are given in Table 2.2. The data have been obtained from the web site of the Danish Wind Power Association [106].

The values of shape factor and average wind speed from Table 2.2 are also shown in Figure 2.16 where each star indicates one location. The average wind speed ranges from 4 to 8 m/s with the majority between 6 and 7 m/s. The Weibull shape factor appears to be concentrated around 1.5 and 2.1. In many studies, a Weibull shape factor of 2 is used, resulting in the so-called "Rayleigh distribution." It is interesting to note that the lower values of the shape factor appear mainly for sites in Southern Europe; there appears to be climate impact, and the often made approximation of shape factor of 2 may only hold for the climate of northwestern Europe. For Tera Kora on the island of Curaçao, a shape factor of 4.5 (at an average wind speed of 7.3 m/s) is mentioned in Ref. 175.

The influence of the shape factor on the wind speed distribution is illustrated in Figures 2.17 and 2.18. The former figure plots the probability distribution function of the wind speed for four different values of the Weibull shape factor, in all cases for an average wind speed of 6.5 m/s. The latter figure shows the probability density function.

TABLE 2.2 Shape Factor and Expected Value for the Wind Speed Distribution at a Number of Locations in Europe [106]

Site	Country	Shape factor	Expected value (m/s)
Melsbroek	Belgium	1.91	6.03
Middelkerke	Belgium	2.05	6.73
Berlin	Germany	2.31	5.85
Frankfurt	Germany	1.95	4.92
Helgoland	Germany	2.11	6.60
Beldringe	Denmark	1.97	6.56
Karup	Denmark	2.16	7.62
Kastrup	Denmark	2.34	7.22
Albacete	Spain	1.49	6.73
Menorca	Spain	1.49	6.59
Brest	France	2.02	6.87
Carcassonne	France	2.06	7.62
Bala	Wales	1.58	7.04
Dunstaffnage	Scotland	1.93	7.05
Stornoway	Great Britain	1.86	7.68
Araxos	Greece	1.33	6.07
Heraklion	Greece	1.22	5.79
Brindise	Italy	1.52	6.22
Trapani	Italy	1.37	6.21
Cork	Ireland	1.95	6.96
Malin Head	Ireland	2.03	7.97
Schiphol	The Netherlands	2.18	6.51
Texel Lichtschip	The Netherlands	1.96	6.52
Flores	Açôres	1.55	6.70
Lisboa	Portugal	2.05	6.51

The influence of the shape factor on the distribution is significant. As we will see soon, the power produced by a wind turbine is zero for wind speed below 3–4 m/s and reaches its rated value only for wind speed above 13–15 m/s. The total amount of energy produced by the wind turbine thus strongly depends on the amount of time the wind speed is below 4 and above 13 m/s. This percentage of time varies significantly with the shape factor. It is thus important to consider the shape factor of the wind speed distribution when planning wind power.

2.1.6 Power Distribution as a Function of the Wind Speed

The amount of power generated by a wind turbine is a strongly nonlinear function of the wind speed. The power curve, the relation between wind speed and generated power, is shown in Figure 2.19 for five turbines, all with rated power of 600 kW. The data have been obtained from Ref. 106. We see that the power curves for these five

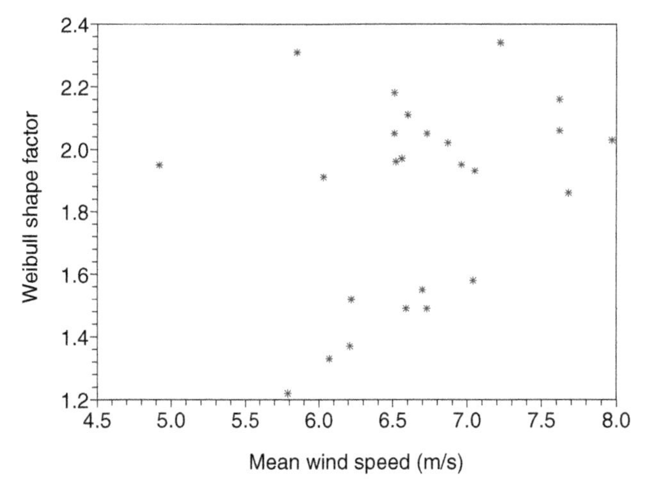

Figure 2.16 Correlation between shape factor and average wind speed for a number of locations in Europe.

turbines are very similar; the differences are mainly in the high wind speed part of the curves.

The power curves are shown in Figure 2.20 for 20 different turbines from 5 different manufacturers, with rated power between 300 and 2750 kW [106]. The generated power is given as a fraction of the rated power.

Although the spread is somewhat bigger than that in the previous figure, still all curves show a very similar behavior. All curves show four distinctive regions:

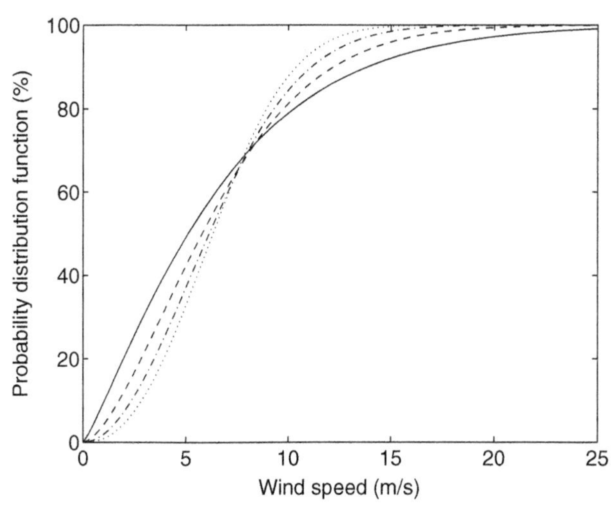

Figure 2.17 Probability distribution function of the wind speed for 6.5 m/s average wind speed and shape factors of 1.2 (solid line), 1.6 (dashed line), 2.0 (dash-dotted line), and 2.4 (dotted line).

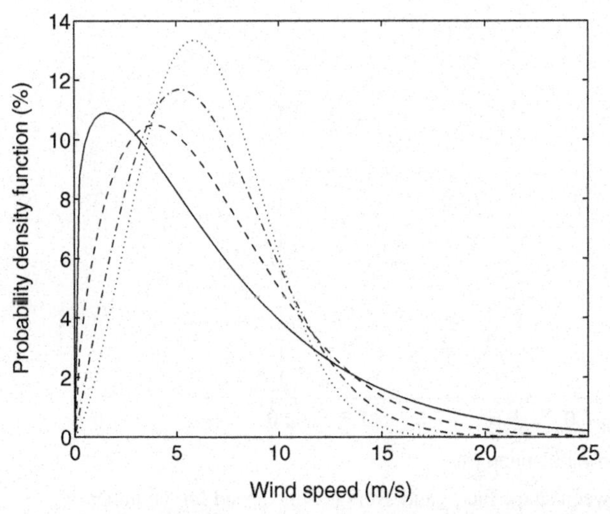

Figure 2.18 Probability density function of the wind speed for 6.5 m/s average wind speed and shape factors of 1.2 (solid line), 1.6 (dashed line), 2.0 (dash-dotted line), and 2.4 (dotted line).

- The power production is zero for small wind speed, below the so-called "cut-in speed." This is where the energy in the wind is insufficient to supply the mechanical and electrical losses. Energy from the grid would be needed to keep the blades moving. The cut-in speed is between 3 and 4 m/s.
- The power production increases fast with the increasing wind speed. In this region, the turbine produces its maximum power for the given wind speed. The amount of energy in the wind is proportional to the cube (the third power) of

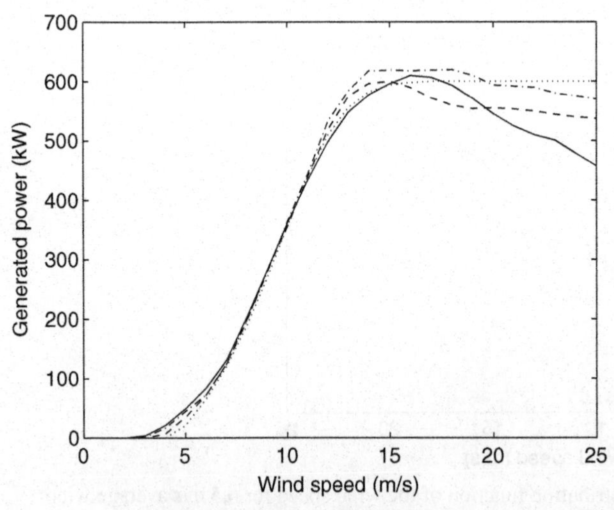

Figure 2.19 Generated power as a function of the wind speed for four different 600 kW wind turbines: type I (solid); type II (dashed); type III (dash-dotted); and type IV (dotted).

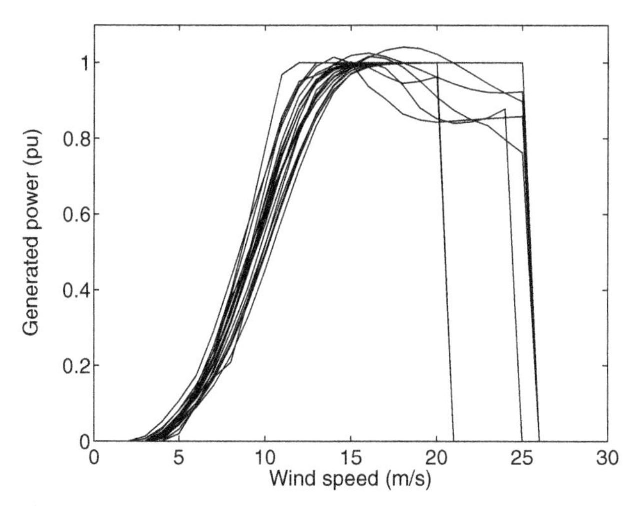

Figure 2.20 Generated power as a function of the wind speed for 20 different turbines.

the wind speed. It is this relation that is mainly responsible for the fast rise in power production. Variations in wind speed in this region will cause large variations in production.

- For further increase in the wind speed, the produced power increases less fast and eventually becomes constant or even decreases somewhat, depending on the type of turbine. The rated power of a wind turbine is the rated power of the electrical installation. This is what sets a limit to the amount of electrical power that can be produced. The power from the blades to the electrical machines should not exceed the machine rating too much, otherwise the machine will be overloaded and fail. There are two methods in use to limit this power: active control by slightly rotating the blades (pitch control), and passive control through the shape of the blades (stall control). With active control, the power is kept constant at its rated value resulting in a flat power curve. With passive control, the power is somewhat higher than rated for medium speed and somewhat lower for high wind speed. The speed at which 90% of rated power is reached ranges from 10.5 to 14 m/s. The wind speed at which the rated power is reached is sometimes referred to as the "rated speed."
- For the very high wind speed, for most turbines from 25 m/s, the blades are fixed and the power production becomes zero to protect the installation against the high mechanical forces associated with high wind speeds. The wind speed at which this occurs is called the "cut-out speed."

Four of the turbines from Figure 2.20 can be compared in more detail in Figure 2.21. All four are from the same manufacturer, but with different size and rating. Two of the turbines have a rating of 2 MW (solid and dashed lines in the figure); for the large wind speed, they produce the same amount of power. This is the region in which the active control system determines the produced power. For the lower wind speed, the 80 m high turbine produces more than the 66 m high turbine:

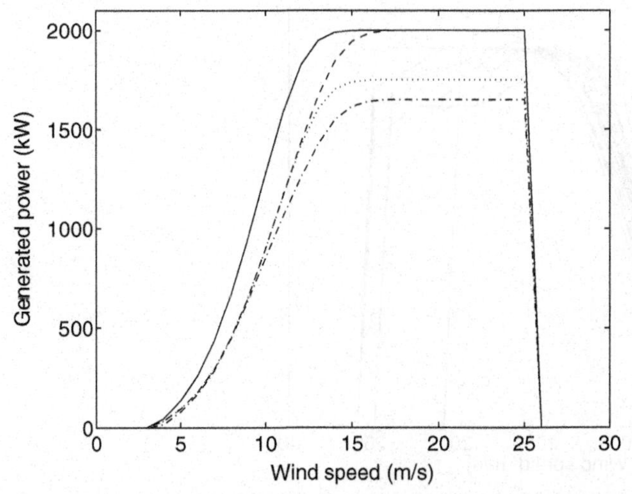

Figure 2.21 Power curves for four different turbines: 80 m, 2 MW (solid); 66 m, 2 MW (dashed); 66 m, 1.75 MW (dotted); and 66 m, 1.65 MW (dash-dotted).

this is where as much power as possible is obtained from the wind. The 80 m turbine (solid curve) is more appropriate for locations with a small average wind speed, whereas the 66 m turbine could be used at location with a high average wind speed. Three of the turbines have a blade length of 66 m, but with different ratings of the electrical installation (1650, 1750, and 2000 kW). For the low wind speed, the three turbines produce the same amount of power. For the higher wind speed, the control system limits the power production.

2.1.7 Distribution of the Power Production

Of importance to the power system is not the distribution of the wind speed but the distribution of the generated power. This distribution is obtained by combining the wind speed distribution with the power curve. A Monte Carlo simulation approach is used here to generate samples from the Weibull distribution for the wind speed. The produced power is next obtained from the power curve of the wind turbine. A sample X from a Weibull distribution with shape factor m and characteristic value θ is obtained from

$$X = \theta(-\ln U)^{1/m} \tag{2.5}$$

where U is a sample from the uniform distribution on the interval [0,1]. The results are shown in Figure 2.22 for an average wind speed of 6.5 m/s and different shape factors. A lower shape factor corresponds to a larger spread in wind speed. In terms of produced power, not only the percentage of time that the production is zero increases but also the percentage of time that the production is close to its maximum increases.

The influence of the average wind speed is illustrated in Figure 2.23. In this case, the shape factor remained constant at 2.0, whereas the average wind speed varied from 6.0 to 7.5 m/s. A higher average wind speed implies that the production is less likely to be zero and more likely to be close to its maximum.

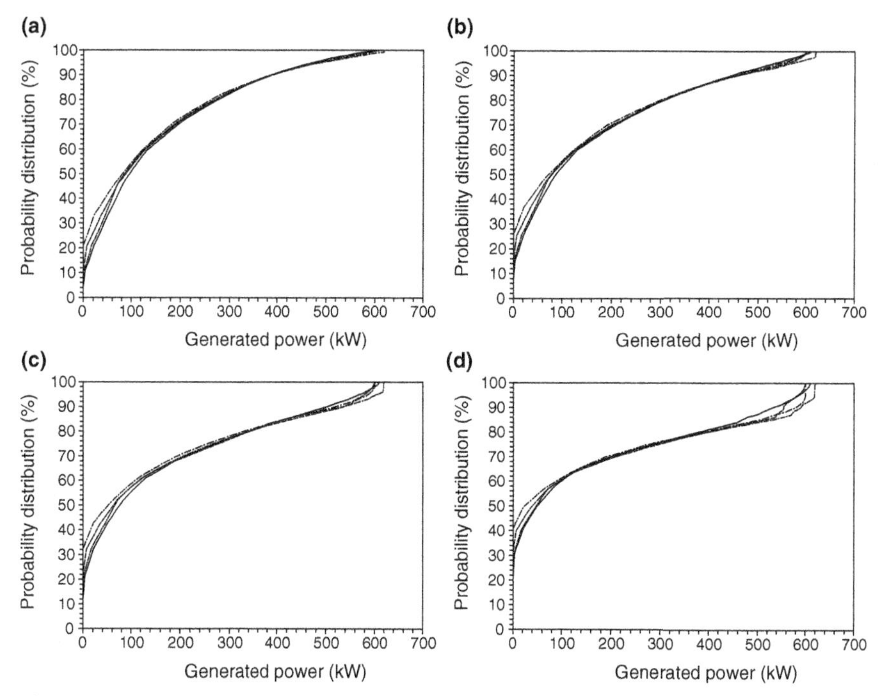

Figure 2.22 Probability distribution function of the generated power with four different 600 kW wind turbines: average wind speed 6.5 m/s; shape factor 2.4 (a), 2.0 (b), 1.6 (c), and 1.2 (d).

The probability distribution of the power produced by a type III wind turbine, for all locations in Table 2.2, is shown in Figure 2.24. The lower the curve, the lower the probability that the production is less than a given value, and thus the better the wind regime. The probability that the production is less than 300 kW (i.e., less than half the rating) varies between 75% and 95%. As the figure shows, the distribution is strongly influenced by the wind regime at the specific location. The earlier figures showed that the difference in production is small for different turbines at the same location. It is thus more important to choose a proper location than the proper turbine.

An alternative way of showing the probability distribution function of the generated power is in the form of a power–duration curve. This curve shows the amount of generated power that is exceeded for more than a certain number of hours per year. The power–duration curves corresponding to the probability distribution functions in Figure 2.24 are presented in Figure 2.25. From the figures one can read that the amount of time the power exceeds half the rated power (300 kW) ranges from about 300 h per year at the worst location to over 2400 h at the best location. Again, we see that the local wind regime has a significant influence on the generated power. It should be noted that the two curves that lay below and to the left of the majority of curves correspond to locations with a not so good wind regime (low-average wind speed) where it is unlikely that large amounts of wind power will be installed.

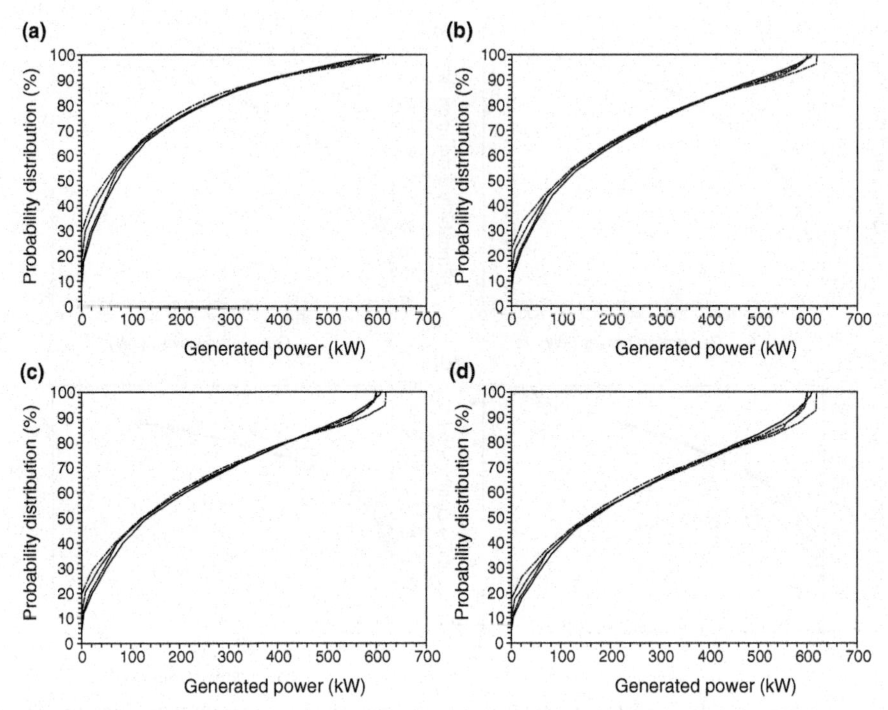

Figure 2.23 Probability distribution function of the generated power with four different 600 kW wind turbines: average wind speed 6.0 m/s (a), 7.0 (b), 7.5 (c), 8.0 (d); shape factor 2.0.

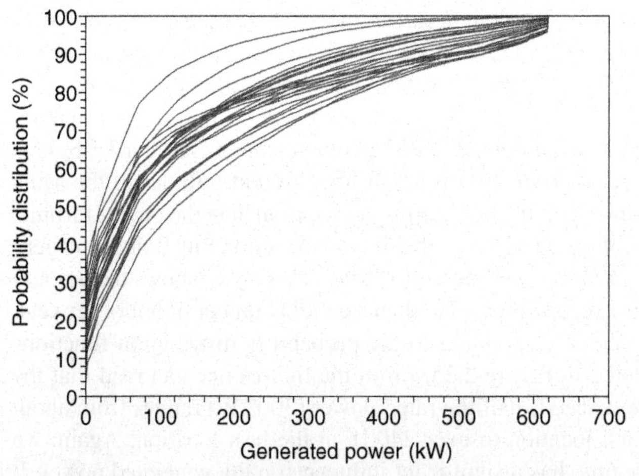

Figure 2.24 Probability distribution function for the power generated by a 600 kW wind turbine (type III) at different locations throughout Europe.

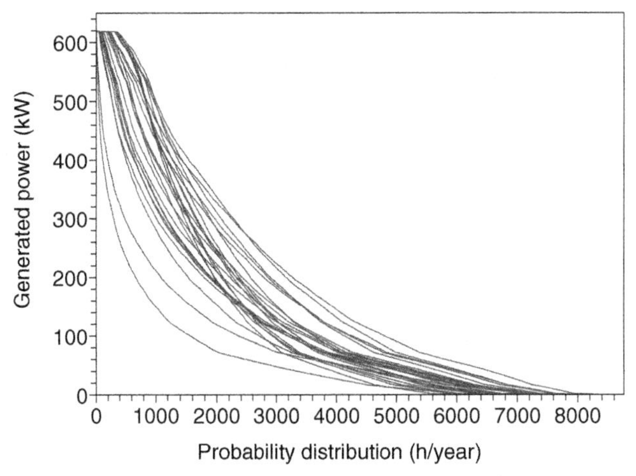

Figure 2.25 Power–duration curve for a wind turbine (type III) at different locations throughout Europe.

2.1.8 Expected Energy Production

The same approach as used before (Figure 2.20 etc.) to obtain the distribution of the produced power has been used to calculate the expected energy production of wind turbines at different places in Europe, using the information on the wind speed distribution from Table 2.2. Samples from the appropriate Weibull distribution were used to generate random wind speed data. These are next used to generate random values for the produced power. The capacity factor is next calculated as the average power production divided by the rated power of the turbine. The calculations performed for a 650 kW machine are shown in Figure 2.26; similar results were obtained for other

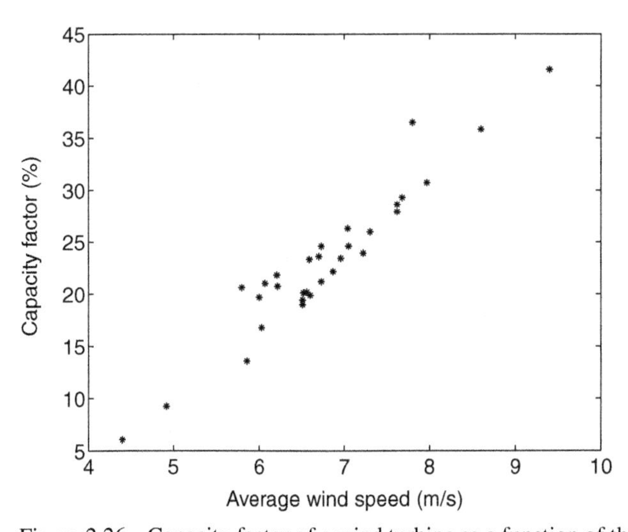

Figure 2.26 Capacity factor of a wind turbine as a function of the average wind speed.

turbines. Most of the locations in the figure correspond to the locations in Table 2.2. Added to this were a location in Sri Lanka and the starting points of IEC wind classes 6 and 7 (8.6 and 9.4 m/s) for which a shape factor of 2 has been assumed.

The capacity factor (and thus the annual energy production) increases linearly with the average wind speed. For an average wind speed of 6 m/s, the capacity factor is less than 20%, that is, the average production of the turbine over a whole year is less than 20% of its rated power. For an average wind speed of 9 m/s, the capacity factor is over 35%. The same turbine at this location produces twice as much energy as one at a location with an average wind speed of 6 m/s. This again confirms that the choice of location is very important. The figure also shows that the shape factor does have a certain impact on the capacity factor; however, the impact of the average wind speed is bigger.

2.2 SOLAR POWER

2.2.1 Status

The amount of energy that 1 m^2 of earth receives from the sun varies strongly between locations. Without clouds, it is highest near the equator and lowest near the poles. Including the impact of clouds, the amount is highest in the deserts. But a solar panel can be tilted toward the sun to compensate for the curvature of the earth. At optimal angle, the amount of energy reaching a solar panel is between 1000 and 2000 kWh/m^2 per year for most locations. In Europe, the best parts are in the south of Spain with insolation above 1900 kWh and the worst parts are in the north of Scandinavia with values somewhat below 1000 kWh.

Building giant solar power installations in desert areas could supply the whole world with energy. Using concentrating solar power, a 1000×1000 km^2 in the desert would be needed [286]. That is not a small size (much more than the "two football fields in the Sahara" that the authors have heard mentioning during informal discussions), but still something that is technically achievable.

The energy from the sun can be used in a number of ways. The most visible use today is in the form of solar panels, often installed on the roof of a building. A big advantage of this type of solar power is that it is produced where it is needed, as electricity consumption is also taking place mainly in buildings. In some countries (like Germany and the Netherlands), this type of installation is getting rather common. A solar panel will only transfer part of the solar energy that falls on it into electrical energy. For the current generation of series-produced solar cells, the efficiency is between 6% and 15%, but individual cells can have higher efficiencies up to 20% [167, 354].

Large photovoltaic (PV) installations have been built recently, including 40 MW installations in Germany and Portugal [452]. The large solar power installations are not based on photovoltaics but on concentrating solar thermal plants, where the energy from the sun is focused using large arrays of mirrors. The concentrated sunlight is used to boil water that in turn powers a turbine like in a conventional thermal power station. The largest of such installations to date is a 350 MW installation in California [451].

Plans for much larger installations exist. A 72 MW photovoltaic plant is planned in northeastern Italy [10], said to be Europe's biggest. The plant was expected to be completed by the end of 2010.

Large installations producing solar power are still rare, but there is a clear trend toward more of these. Large installations can simply be large collections of solar panels; but more often with large installations, the heat from the sun is used to produce steam that in turn powers a turbine to produce electricity. Installations with a capacity of several hundreds of megawatts are planned in Spain, the United States, Germany, and China. All these will be located far away from the consumption of the electricity, so some of the advantages of solar power actually disappear. Large installations reportedly can reach efficiencies between 10% and 15%, similar to those of small installations. However, by using heat storage in molten salt, some of the daily variations in production can be compensated; only for large installations is this economically attractive.

For larger installations, concentrating solar power is the more popular technology at the moment. The efficiency of this technology is lower than that of photovoltaics; in a desert area, concentrating solar gives 15–20 W/m^2, whereas photovoltaics gives about 40 W/m^2. The main reason for choosing concentrating solar is that it is much cheaper per kilowatt hour than photovoltaics. Whenever land is cheap, that is, in remote areas, concentrating solar will appear, whereas photovoltaics is more attractive in built-up areas because it requires less space.

Recently, discussions have started again about very large installations in desert areas. In the United States, the southwest of the country (Nevada and New Mexico) has the space and the sun to make such an installation feasible. In Europe, there are no large open areas, although the southern part of Spain could provide a lot of solar power. However, the attention has instead moved to Northern Africa, where a group of companies have proposed a plan for building huge solar power and wind power installations under the name "DesertTech." The power would be transported to the consumption centers in Europe by means of HVDC lines and cables. For China, the Gobi Desert is often mentioned as a source of large amounts of solar power.

On a smaller and for the time being more practical scale, solar power can play an important role in the electrification of the poorer parts of the world, where the electricity grid is either weak and unreliable or nonexisting. Another important application of solar power is its use to directly heat water, which is referred to as solar thermal. This can be used to cover the hot water need of a household. Such installations are common, for example, in Southern Europe and in China. These installations are not connected to the electricity grid, so there are no integration issues.

2.2.2 Properties

The amount of power produced by a solar power installation depends on the location of the sun in the sky and on the amount of cloud cover. The variations and predictability in cloud cover are similar to that of wind speed. The location of the sun in the sky shows a predictable daily and seasonal variation caused by the rotation of the earth on its axis and around the sun. This will likely make solar power more predictable than wind power, but there is limited experience with prediction for solar power to verify this.

Solar power installations based on photovoltaics have limited economics of scale. The costs per kilowatt remain about the same with increasing size because the cost mainly varies with the solar panels. This makes rooftop installations for domestic or commercial installations an attractive solution. Such small installations are also easy to set up and connect to the grid, although rules on connection vary between countries, and contacts with the local network operator are in almost all countries compulsory. Small rooftop installations are by definition close to the consumption of electricity and they will, therefore, reduce the power transport through both the transmission and the distribution networks.

With thermal solar, large installations are much more attractive than small installations. As a result, they are more likely to be built at remote locations. Very large installations are likely to occur at locations far away from consumption centers, such as Nevada, the Gobi Desert, and Northern Africa.

The daily variation in solar power production indicates that the daily peak falls around noon. This is in most countries not the highest consumption, but certainly a high consumption. It has been suggested to have solar panels tilted and facing southeast or southwest (in the northern hemisphere) so that the peak production corresponds closer to the peak consumption.

The seasonal variation indicates that the annual production peak occurs during summer, where the consumption due to cooling also has its peak in countries with a hot climate. Unfortunately, the daily peak in consumption falls in the afternoon and the annual peaks in consumption falls 1 or 2 months past the date where the sun reaches its highest point in the sky. Despite this, solar power is still expected to offer some contribution to the peak load. Again, an optimal tilt can be chosen to maximize the contribution of solar power to the consumption peak.

The most recent developments point toward storage of molten salt at high temperatures. The salt could be kept sufficiently hot for several hours to still produce steam and thus electricity. This would compensate for any fluctuations in cloud cover on a timescale up to several hours and significantly increase the predictability of the production. It would also allow the energy to be available to cover the evening peak in consumption.

2.2.3 Space Requirements

Both solar power and wind power require a large area to produce significant amounts of energy. As space is limited in many industrial countries, this could become a limit to its use close to the consumption. There remain large amounts of unused space available far away from where most of the population live, but building wind or solar power installations at such locations requires large investments in transmission systems. From Examples 2.1 and 2.2, it follows that solar power can produce several times more energy per area than the wind power. Where space is the limiting factor, solar power is a better choice than wind power. Next, the yield from Example 2.2 is achieved for any available location almost anywhere in Europe. On the other hand, the yield from Example 2.1 is obtained only for areas of several square kilometers with good wind conditions. Note that we did not consider any economic factors in the comparison.

Example 2.1 Consider a large wind park off the western coast of Denmark: Horns Rev A, a site with good wind conditions and with modern turbines. The annual production of the park is about 600 GWh. It takes up about 20 km^2 of space. The yield per area is thus 30 kWh/m^2/year.

Example 2.2 Assume that somebody wants to build a solar power installation in the north of Sweden, not the world's best location for solar power. The insolation here is about 1000 kWh/m^2/year. Assume that the installation has an efficiency of 8%, which is achievable with existing technology. That would result in a yield of 80 kWh/m^2/year.

2.2.4 Photovoltaics

A solar panel uses semiconductor cells (wafers) that produce photocurrent when exposed to the sun, hence referred to as photovoltaic cells. The simplified equivalent circuit in Figure 2.27 represents a PV cell, where the diode "D" with its p–n junction (i.e., its semiconductor material) close to the surface produces the photocurrent I_{ph} as a result of the light falling on it. The current output of the cell I_{pv} is then a DC current with a constant value at different array output voltages, as shown in Figure 2.28 for two different cell temperatures T_1 and T_2 where $T_1 < T_2$. This results in the typical power–voltage characteristics of a PV cell, as shown in Figure 2.29.

A higher amount of irradiation results in higher currents; this will shift the curve both upward and to the left. The operating voltage V_{pv} is adjusted in order to move the operating point at the maximum power point. Maximum power point tracking (MPPT) algorithms are incorporated within the interfacing technology of a PV unit

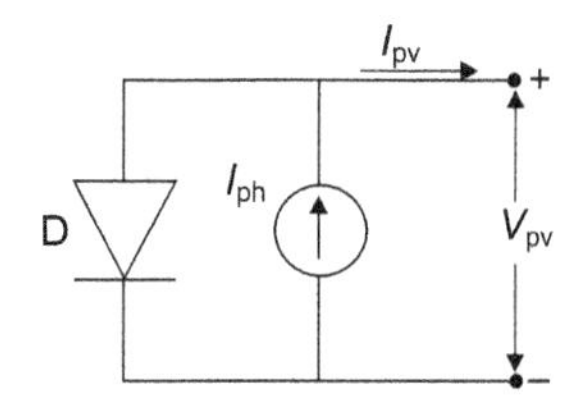

Figure 2.27 Simplified equivalent circuit of a PV cell.

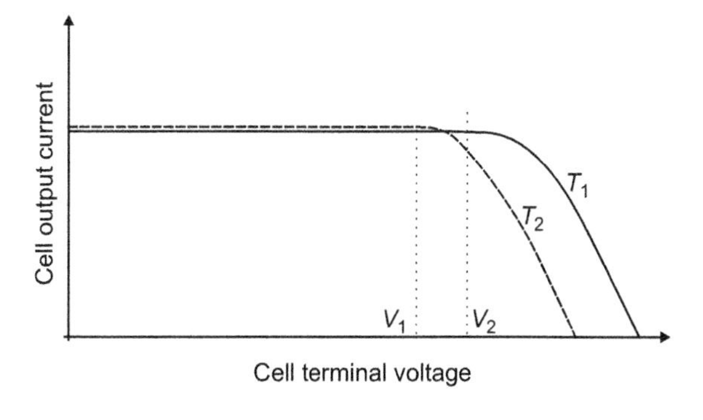

Figure 2.28 Typical current–voltage characteristics of a PV cell at two temperatures $T_1 < T_2$.

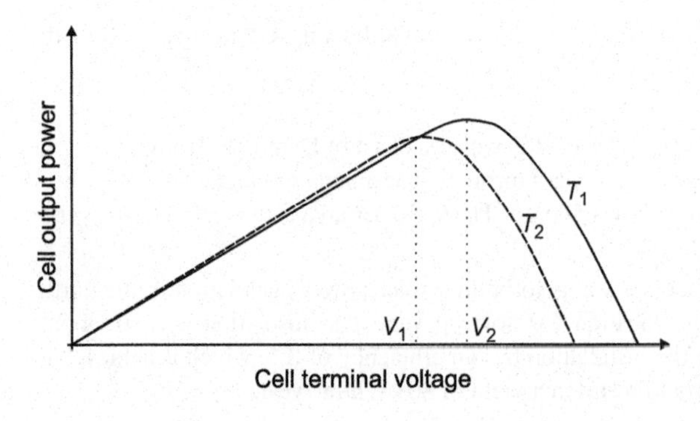

Figure 2.29 Typical power–voltage characteristics of a PV cell at two temperatures $T_1 < T_2$.

in order to adjust the operating voltage V_{pv} so that the maximum power is extracted. There are many MPPT algorithms that could be implemented, but generally three techniques are commonly used [79, 158, 342, 464, 465]:

1. *Perturb and observe* The algorithm perturbs the operating point in a certain direction and samples d*P*/d*V*, if positive indicates the right direction toward the maximum power point (MPP) and vice versa. The algorithm keeps adjusting the required operating voltage through using extra electronic hardware referred to as a chopper or DC/DC converter, as will be explained in Section 2.9.2. This method is easy to implement; however, it has slow response time and may track in the wrong direction in case of rapidly changing atmospheric conditions. Moreover, oscillations around the MPP may result during steady-state operation.
2. *Incremental conductance* This algorithm uses the incremental conductance d*I*/d*V* to calculate the sign of d*P*/d*V*:

$$\frac{dP}{dV} = I + V \times \frac{dI}{dV} \tag{2.6}$$

   Then, the voltage is adjusted accordingly as with the previous algorithm. This method provides faster response than the perturb and observe method; however, it can produce oscillations in case of rapidly changing atmospheric conditions.
3. *Constant voltage method* By assuming that the ratio of cell voltage at maximum power V_m to its corresponding open-circuit voltage V_c is relatively constant throughout the normal operating range,

$$\frac{V_m}{V_c} = K \tag{2.7}$$

   The constant K assumes a value of 0.76, which is an approximate value that is related to the material used for the solar panel [464]. The open-circuit voltage is obtained using an unloaded pilot cell near the array. This method is fast and simple; however, the accuracy could be questioned since the atmospheric conditions could be different for the pilot cell and the array.

2.2.5 Location of the Sun in the Sky

The amount of electrical energy produced by a photovoltaic installation (a "solar panel") is proportional to the amount of radiation that reaches the panel. This amount of radiation is referred to as "insolation" or "irradiation." The former is a somewhat outdated term, according to some authors, that is a shortening of "incident solar radiation." The irradiation consists of three different terms:

- Direct irradiation is the radiation that reaches the panel directly from the sun. This amount depends on the angle between the panel and its direction to the sun. On a clear day, when the panel is directed to the sun, the direct irradiation is about 1000 W/m^2. Even on a clear day, the direct irradiation decreases with the sun being lower in the sky. The direct irradiation fluctuates strongly when clouds pass in front of the sky. Its average value decreases with increasing cloud coverage. The annual amount of direct irradiation can be maximized by tilting the panel with the right angle toward the south. The maximum production is obtained when the tilt angle is around the latitude of the location, when the cloud coverage is assumed the same throughout the year.
- Indirect irradiation is also referred to as "scattered irradiation" or "diffuse irradiation." Indirect irradiation coming from the sky varies only slowly with time, as the fraction of the sky covered by clouds varies. We talk about timescales or hours here. The indirect irradiation coming from the sky increases somewhat with increasing cloud cover, because a cloud is brighter than the clear blue sky. The amount of indirect irradiation from the sky decreases when the panel is tilted because the panel is exposed to a smaller part of the sky.
- Indirect irradiation coming from the earth depends on the amount of direct irradiation at nearby locations. Therefore, this irradiation component decreases with increasing cloud cover. It will further show fast fluctuations because of clouds passing the sun and changing reflections in the neighborhood. The amount of indirect irradiation coming from the earth very strongly depends on location and is difficult to predict. For a horizontal panel, this component is close to zero; it increases for tilted panels.

An important and much discussed property of solar power is its variation with the time of the day and with the time of the year. The variation with the time of the day is due to the rotation of the earth on its axis. The variation with the time of the year is due to the revolution of the earth around the sun together with the fact that the rotational axis of the earth is not perpendicular to the plane in which the earth revolves around the sun. Another annual variation in solar power is due to the elliptical nature of the earth orbit around the sun. The eccentricity of the earth orbit is 1.67%, so the distance between the earth and the sun varies by 1.67% around its average value. The earth is closest to the sun in early January and farthest away in early July. The amount of energy reaching the earth, being inversely proportional to the square of the distance, varies by about 3.3% around its average value during the course of the year. The other variations are, however, much bigger; so the eccentricity of the earth orbit is often not considered in the calculations.

At the top of the atmosphere, an average power of 1353 W is passing through every square meter of the plane perpendicular to the direction of the sun [167, 389]. This value is referred to as the "solar constant." The eccentricity of the earth orbit, as was mentioned before, implies that the amount of power passing through every square meter varies between about 1300 and 1400 W ($1353 \pm 3.3\%$) during the course of 1 year. About 25% of this power is absorbed in the clear atmosphere when the sun is in the zenith; so on a clear day near the equator around noon, about 1 kW of this power reaches the earth.

Both the revolution of the earth around the sun and the rotation of the earth on its axis can be predicted with high accuracy. It can, thus, be predicted very well how much solar power would reach, at any moment in time, to any location on the earth, assuming a clear sky. Accurate equations for the position of the sun can be found in the astronomical literature, with Ref. 300 being the main reference source. In most solar power studies, some simplifications are made, for example, assuming a constant distance between the earth and the sun and a constant speed of the earth around the sun. The errors made by these simplifications are much smaller than the impact of the uncertainties in weather.

The position of the sun compared to the stars is given in terms of its "declination" δ and its "right ascension." The declination is the angle between the sun and the celestial equator; the right ascension is the angle between the sun and the plane through the celestial poles and the equinoxes. (The spring equinox has right ascension zero.) Declination and right ascension are used to indicate the position of a star on the celestial sphere in the same way as "latitude" and "longitude" are used to indicate a position on the earth. The celestial coordinates need to be translated into the horizontal coordinates "elevation" (angle above the horizon) and "azimuth" (angle compared to the north). To translate from celestial to horizontal coordinates, the so-called "hour angle" is introduced, being the difference between the right ascension and the local time. When considering the sun, the local time is strongly related to the position of the sun; so the hour angle can be calculated directly with sufficient accuracy.

The following approximate expressions for the declination and hour angle ω are typically used in solar power studies [128, 389]:

$$\delta = 0.4093 \cos\left(2\pi \frac{(d-173)}{365}\right) \tag{2.8}$$

$$\omega = 2\pi \frac{T_{\mathrm{UC}}}{24} - \lambda_{\mathrm{e}} \tag{2.9}$$

where d is the day of the year (January 1 being day 1 and July 21 being day 202 in a non-leap year), T_{UC} is the coordinated universal time (which is very close to "Greenwich Mean Time"), and λ_{e} is the longitude of the location. Note that all angles are expressed in radians. Knowing the declination and hour angle, the position of the sun compared to the horizon can be calculated using the following expressions:

$$\sin\psi = \sin\phi \sin\delta - \cos\phi \cos\delta \cos\omega \tag{2.10}$$

$$\cos A = \frac{\cos\phi \sin\delta + \sin\phi \cos\delta \cos\omega}{\cos\psi} \tag{2.11}$$

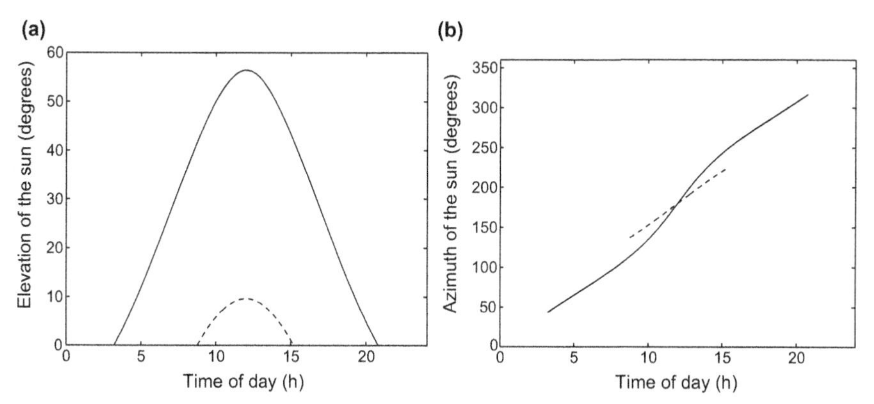

Figure 2.30 Elevation (a) and azimuth (b) of the sun for a location at 57 °N on June 21 (solid line) and on December 21 (dashed line).

where ψ and A are elevation and azimuth, respectively, and ϕ is the latitude of the location. Note that the hour angle has been defined here as being zero at midnight so that the hour angle corresponds to the local solar time, whereas in several other publications, the hour angle is taken zero at noon. Taking the latter convention would give a plus sign in (2.10) and a minus sign in (2.11).

The resulting position of the sun is shown in Figures 2.30–2.32 for three locations and for two days of the year. The location at 57 °N corresponds to Gothenburg, Sweden; the location at 30 °N corresponds to Cairo, Egypt; and the location at 17 °N corresponds to Timbuktu in Mali. Note that the midday sun in Timbuktu is in the north (but close to the zenith) during midsummer.

Once the location of the sun is known in horizon coordinates, the angle of incidence of the solar radiation on the panel can be calculated. Let ψ_S and A_S be the position of the sun, and ψ_P and A_P be the direction of the perpendicular on the solar panel. The angle of incidence of the radiation α is obtained from the following

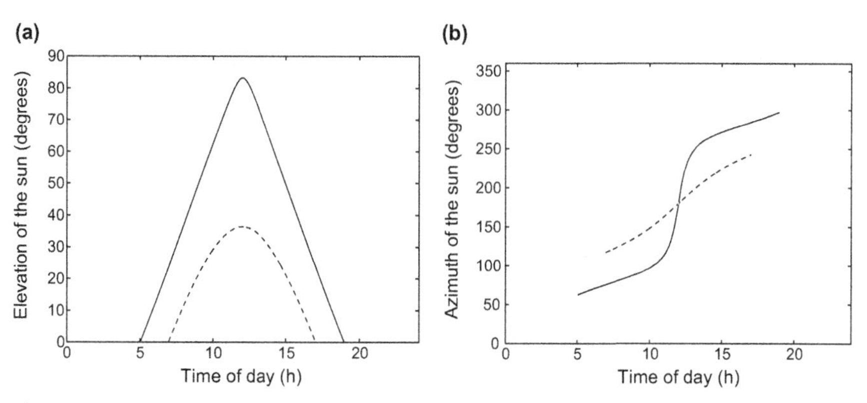

Figure 2.31 Elevation (a) and azimuth (b) of the sun for a location at 30 °N on June 21 (solid line) and on December 21 (dashed line).

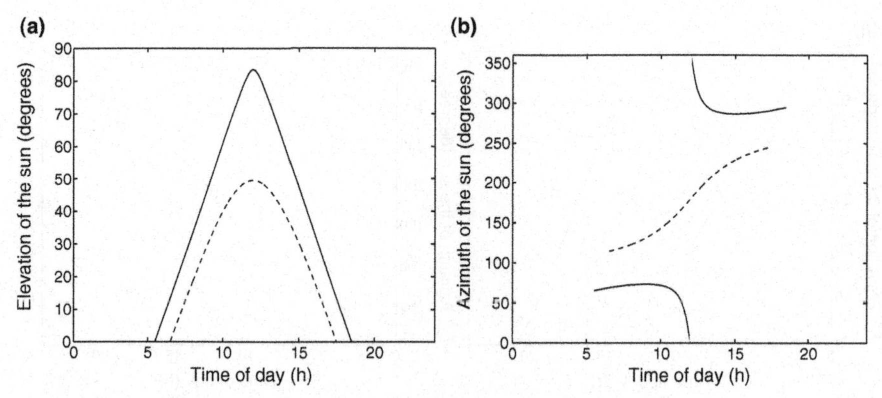

Figure 2.32 Elevation (a) and azimuth (b) of the sun for a location at 17 °N on June 21 (solid line) and on December 21 (dashed line).

expression:

$$\cos\alpha = \sin\psi_P \sin\psi_S + \cos\psi_P \cos\psi_S \cos(A_S - A_P) \tag{2.12}$$

A significant number of publications discuss the different irradiation components [26, 222, 389, 408]. The direct irradiation is easiest to calculate, although there remains some uncertainty on the impact of cloud cover. We will discuss this in the forthcoming section. The indirect irradiation is much more difficult and several models have been proposed. An interesting and thorough overview of the different models available is given in Ref. 325. A total of 22 models relating total and indirect irradiation are compared with measurements at a French Mediterranean site. Most of the models are shown to give large errors, up the 30% rms error. The best models have rms errors between 7% and 10%. It should be noted here that the work on irradiation is very much driven by the need to predict and optimize the annual energy production of a solar panel. This means that the models are acceptable as long as they predict average values. From a power system viewpoint, the short-term predictions and the short-term variation are of more interest. However, not much work has been published on this subject.

From the comparison in Ref. 325, it follows that two models give a reasonable accuracy, with rms errors around 7%. The so-called "Hay model" [198] assumes that the indirect irradiation is composed of a circumsolar component coming from the direction of the sun and an isotropically distributed component coming equally from all directions. In the "Perez model" [344–346] the horizon brightening is added as a third term. The rms error in the latter model is somewhat less, but not significantly. Also, the model proposed by Ma and Iqbal [285] has a performance similar to these models.

Measurements of the irradiation have been made at many locations. Just two examples are shown here. The daily energy received from the sun on a horizontal surface is shown in Figure 2.33 for Mississippi State University (33.50 °N, 90.33 °W). The data were obtained from Ref. 102. Clouds appear every few days, and during the

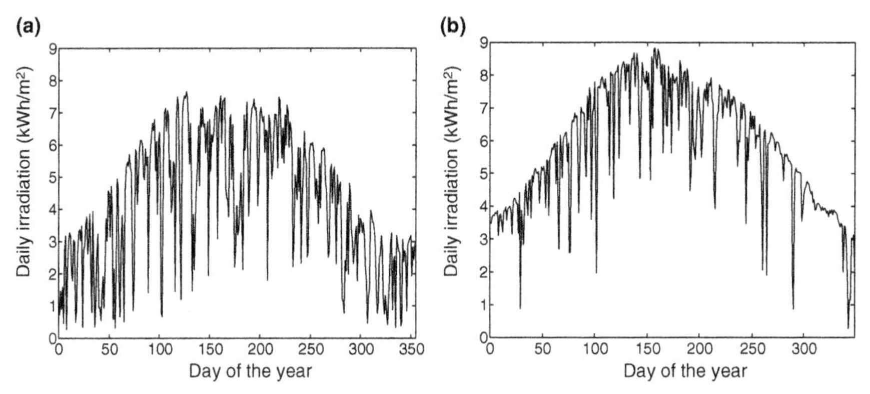

Figure 2.33 Total irradiance for each day throughout 2004 for a site in Mississippi (a) and throughout 1999 for a site in Texas (b).

summer months, there hardly ever seems to be a day completely without clouds. The total energy received per square meter during 2004 was 1497 kWh. The town of El Paso, Texas (31.80 °N, 106.40 °W) has clearly less cloudy days, as shown in Figure 2.33b. The total energy production during 1999 was 2006 kWh. The two locations are at about the same latitude; the difference in energy received on the horizontal surface depends only on the local climate.

2.2.6 Cloud Coverage

The amount of solar radiation that reaches a solar panel not only depends on the position of the sun in the sky but also on the amount of clouds between the sun and the solar panel. Although the position of the sun in the sky can be very accurately predicted many years ahead of time, cloud cover is very difficult to predict even several hours ahead of time. Measurements of cloud coverage have been part of the standard measurements done by meteorologists and are as such recorded over many years for many locations. The cloud coverage is recorded in terms of "oktas," giving the fraction of the sky that is covered by clouds. Zero okta corresponds to a cloudless sky, whereas 8 okta corresponds to a fully clouded sky. A cloud coverage equal to 3 okta means that 3/8 of the sky is covered by clouds.

Measurements of cloud cover over the period 1973–1999 in Gothenburg, Sweden, were used in Ref. 128 to obtain a Markov model describing the changes in cloud cover. A Markov model defines the possible states in which a system can exist (in this case, the nine levels of oktas) and the probability that the system changes to another state from any given state, within a certain time (in this case, 3 h). Measurements of cloud cover made every 3 h were used as input to estimate the transition rates. The transition rate from state i to state j is the probability that the system is in state j in 3 h, knowing that it is in state i now. The resulting transition matrix (with transition rates in percent) is as follows:

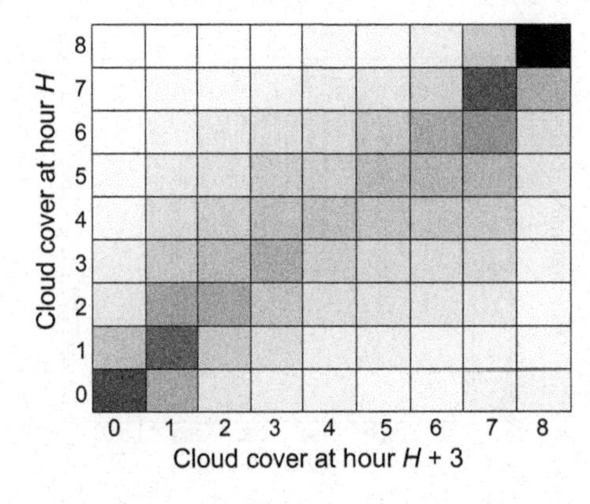

Figure 2.34 Changes in cloud cover obtained from measurements over a 25-year period; averages over the whole year.

$$\Lambda = \begin{bmatrix} 53.8 & 22.5 & 7.1 & 4.7 & 2.7 & 2.3 & 1.7 & 2.6 & 2.6 \\ 15.5 & 45.5 & 14.0 & 9.1 & 4.3 & 3.7 & 3.2 & 3.0 & 1.5 \\ 7.0 & 24.5 & 23.4 & 15.3 & 8.8 & 7.2 & 6.2 & 5.4 & 2.2 \\ 3.8 & 13.4 & 17.7 & 20.3 & 12.6 & 10.6 & 9.0 & 9.1 & 3.4 \\ 2.2 & 8.5 & 12.1 & 15.9 & 16.2 & 14.4 & 13.4 & 13.2 & 4.2 \\ 1.5 & 5.1 & 8.1 & 12.2 & 12.6 & 17.3 & 18.7 & 18.3 & 6.2 \\ 1.0 & 3.0 & 5.2 & 7.4 & 9.5 & 14.2 & 22.2 & 28.0 & 9.5 \\ 0.6 & 2.0 & 2.3 & 3.0 & 3.9 & 6.3 & 11.3 & 50.3 & 20.4 \\ 0.5 & 0.7 & 0.8 & 1.1 & 1.3 & 2.0 & 3.8 & 13.5 & 76.3 \end{bmatrix} \tag{2.13}$$

This matrix is represented graphically in Figure 2.34. The darker the cell, the higher the transition rate. The transition rate is highest along the diagonal, which means that it is most likely that the cloud cover does not change within 3 h. However, the cloudless (0 okta) and the fully clouded (8 okta) states are most stable, with 54% and 76% probability that the cloud cover will be the same 3 h later. The data also show that changes of more than 2 okta within a 3 h period are not very likely.

More detailed calculations are presented in Ref. 127 where the transition rates are shown on a monthly basis. It is shown among others that the transition matrices are different for summer and winter periods.

Measurements performed in Denmark confirmed the strong correlation between cloud coverage and average irradiation at the surface [128, 322]. Based on this correlation, a number of empirical curves were derived. The irradiation reaching the solar panel, which is assumed to be horizontally placed, has been compared with the irradiation for a fully transparent atmosphere:

$$G_0 = 1353 \times \sin \psi \tag{2.14}$$

The results are shown in Figure 2.35: the upper dashed curve represents the fully transparent atmosphere; the three solid curves represent (top to bottom) 0, 4, and 8 okta of cloud cover, respectively.

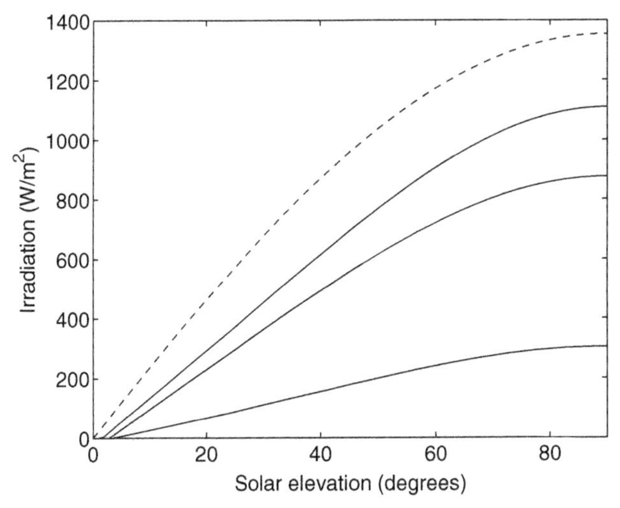

Figure 2.35 Irradiation reaching the solar panel for 0, 4, and 8 okta cloud cover (solid lines) and for the transparent atmosphere (dashed line).

The ratio between the actual irradiation reaching the solar panel and the irradiation for the transparent atmosphere is shown in Figure 2.36. For a completely cloudless sky, at most about 80% of the radiation at the top of the atmosphere reaches the earth surface; for a half-clouded sky, this goes down to about 65%, and for a fully clouded sky, it is less than 25%.

The figure also shows how the atmospheric damping increases fast when the sun gets lower in the sky, that is, smaller elevation angle. When the sun comes closer than 10° to the horizon, the amount of radiation reaching the earth surface drops quickly. It should be noted that these are average values; the actual amount of atmospheric damping may vary strongly from day to day, even for the same amount of cloud cover.

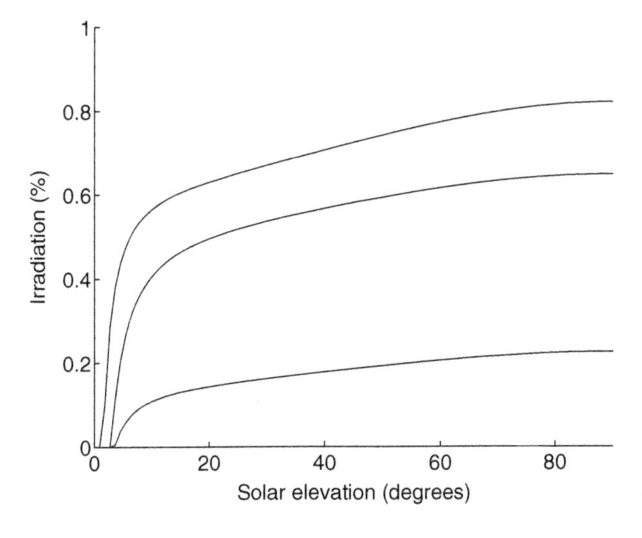

Figure 2.36 Actual irradiation as a function of the ideal irradiation for 0, 4, and 8 okta cloud cover.

TABLE 2.3 Reduction in Irradiation Due to the Clouded Atmosphere

Cloud cover (okta)	Denmark [322]	Australia [329]
0	0.72	0.74
1	0.73	0.72
2	0.69	0.69
3	0.64	0.66
4	0.58	0.63
5	0.54	0.60
6	0.48	0.55
7	0.33	0.44
8	0.18	0.25

The average value of the ratio between actual and ideal irradiation, for elevation angle between 30° and 60°, has been calculated and the results are shown in Table 2.3. The reason for choosing this range is to allow comparison with data obtained for Australia.

The column labeled "Australia" in Table 2.3 is based on data used by the Queensland Department of Natural Resources and Mines to provide irradiation data for sites at which no directed irradiation measurements were made [329]. The average irradiation over a whole day has been compared with the cloud coverage values measured at 9 a.m. and at 3 p.m. For each combination, the most common reduction in irradiation was calculated and placed in a table. For Table 2.3, only the values with equal cloud coverage at 9 a.m. and 3 p.m. were taken. The correlation with the Danish data is very good. The somewhat higher values obtained from the Australian data for clouded skies can be explained from the fact that a fully clouded sky at 9 a.m. and at 3 p.m. does not mean a fully clouded sky all day.

The importance of the correlation between cloud coverage and irradiation is to allow the use of cloud cover data to estimate the performance of solar power at certain locations. Cloud cover data are available for much more locations and over many more years than irradiation data. In this way, the yield and even the optimal tilt of solar power installations can be estimated without the need for doing long-term measurements.

2.2.7 Seasonal Variations in Production Capacity

The amount of radiation reaching a solar panel depends on the following factors:

- The position (Ψ, A) of the sun in the sky. This can be calculated from (2.10) and (2.11).
- The damping $D(\psi, O)$ of the radiation by the atmosphere. This depends on the cloud coverage O and is obtained from Figure 2.35.

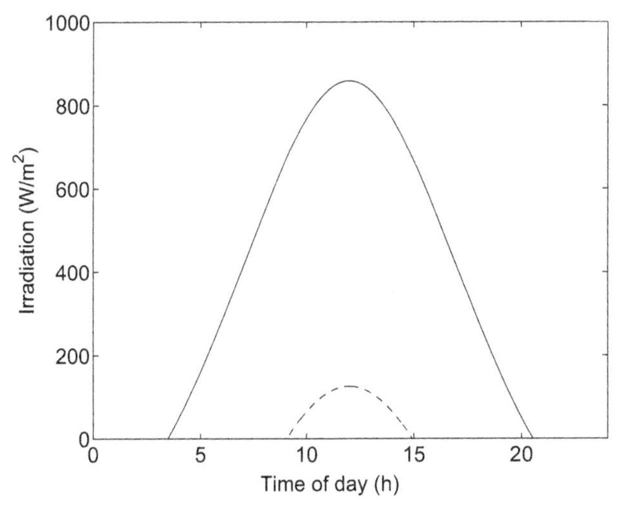

Figure 2.37 Radiation reaching a horizontal panel during a typical cloudless day for a location at 57 °N on June 21 (solid line) and on December 21 (dashed line).

- The angle α between the position of the sun in the sky and the perpendicular of the solar panel. This angle can be calculated from (2.12).

The amount of radiation reaching a horizontal solar panel, in watts per square meter of panel, is obtained from

$$I = 1353 \times D \times \cos\alpha \tag{2.15}$$

According to Ref. 322, the irradiation reaching a horizontal panel, for 0 okta (no or almost no cloud cover), is given by the following expression:

$$I = -24.1 + 894.8 \sin\psi + 238.4 \sin^3\psi \tag{2.16}$$

Combining (2.15) and (2.16) gives the following estimation for the damping D in the atmosphere:

$$D = 0.6613 - \frac{0.0178}{\sin\psi} + 0.1762 \sin^2\psi \tag{2.17}$$

The amount of radiation has been calculated for a horizontal surface at the same location and days of the year as in Figure 2.30, resulting in Figure 2.37. The incident radiation shows a similar behavior as the solar elevation.

The amount of energy obtained by a solar panel depends on the tilt angle of the panel. The effect of tilting on the annual amount of energy received from the sun is shown in Figure 2.38 for a site near London at 52 °N [60].

The annual irradiation shows a small but noticeable variation with tilt angle. Therefore, most solar panels are somewhat tilted upward and toward the south. The daily and annual variations in power production will thus follow the irradiation on a tilted surface. The irradiation reaching a tilted panel has been calculated for Gothenburg (at 57 °N) on June 21 and on December 21. The results are shown in Figure 2.39.

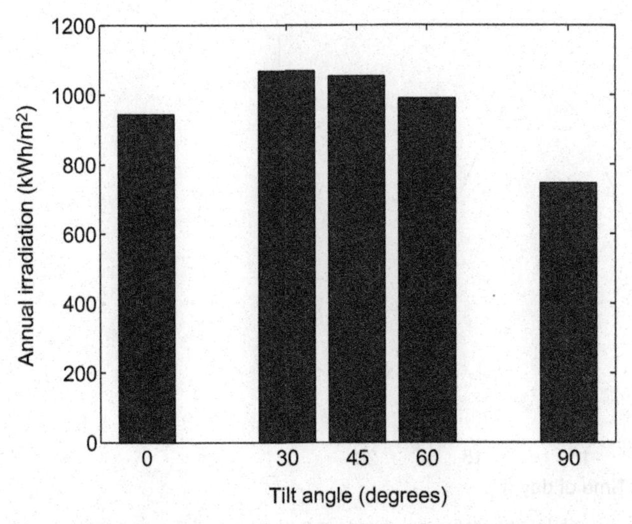

Figure 2.38 Annual irradiation as a function of the tilt angle.

During summer, the highest production is obtained for 30° tilt, whereas in winter the highest production is obtained for 60° tilt.

The daily irradiation has been calculated using the above equations by integrating the irradiation over a 24 h period. Next, the daily irradiation has been calculated as function of the day of the year for different tilt angles. The results are shown in Figure 2.40 for Gothenburg (57 °N), where zero tilt angle corresponds to a horizontal panel. For a panel with a tilt angle of 60°, the daily production is constant during a large part of the year; the total annual production equals 2125 kWh in this case. For a tilt angle of 30°, the summer production is higher but the winter production is lower, with an annual total of 2047 kWh. For tilt angles of 0° (horizontal) and 90° (vertical), the annual production is 1459 and 1663 kWh. Note that the vertical panel gives higher annual production than the horizontal panel.

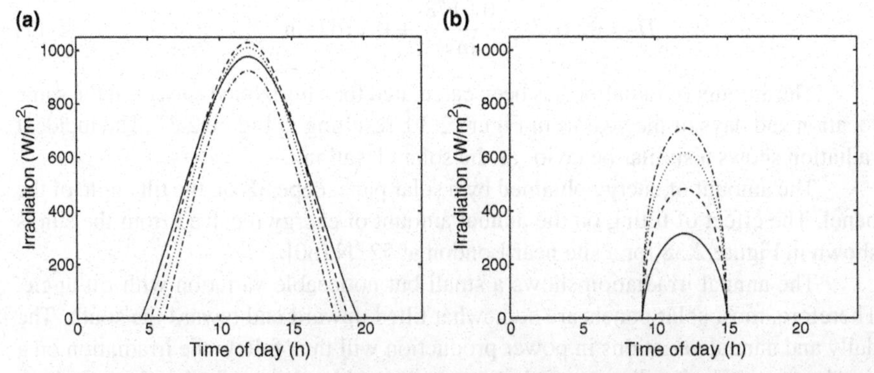

Figure 2.39 Irradiation reaching a panel tilted at 15° (solid), 30° (dashed), 45° (dotted), and 60° (dash-dotted) for a location at 57 °N on June 21 (a) and December 21 (b).

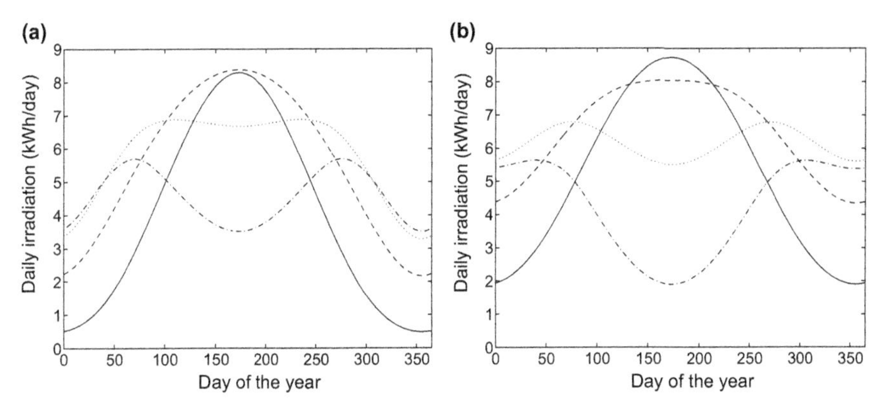

Figure 2.40 Variation in irradiation on a tilted panel through the year for a location at 57 °N (a) and 43 °N (b); tilt angle 0° (solid), 30° (dashed), 60° (dotted), and 90° (dash-dotted).

The same figure shows the seasonal variations in solar production for a location at 43 °N (upstate New York). The flattest curve is obtained for 30°tilt, whereas the difference between summer and winter is smallest for 60° tilt. Finally, the results are presented for Cairo (30 °N) and Timbuktu (17 °N) in Figure 2.41.

The impact of the tilt angle on the annual irradiation is shown in Figure 2.42 for four locations. The optimal angle shifts to lower values for locations closer to the equator and the maximum power production increases.

Note that these calculations assume a cloudless sky. The actual energy production will be lower due to cloud cover. The cloud cover will typically be different for the summer and winter seasons, which will also impact the optimal tilt angle. Two other phenomena have also not been considered in the calculations either: the amount of energy produced by a solar panel is not exactly proportional to the amount of energy that reaches its surface. Instead, the photoconversion efficiency becomes somewhat lower for lower irradiation. The efficiency for 500 W/m^2 irradiation is about 95% of the efficiency for 1000 W/m^2 irradiation. For an irradiation equal to 100 W/m^2, the

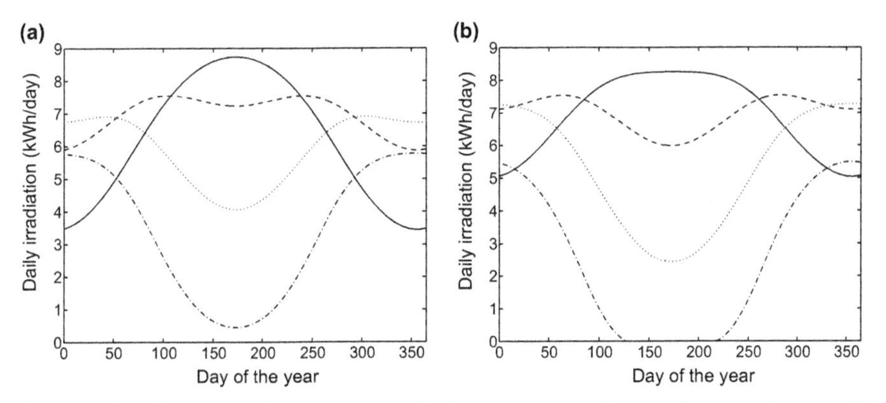

Figure 2.41 Variation in irradiation on a tilted panel through the year for a location at 30 °N (a) and 17 °N (b); tilt angle 0° (solid), 30° (dashed), 60° (dotted), and 90° (dash-dotted).

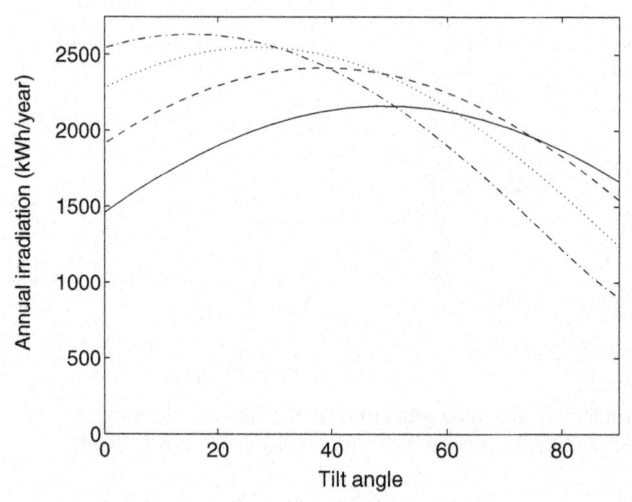

Figure 2.42 Variation in annual irradiation with tilt angle 57° (solid), 43° (dashed), 30° (dotted), and 17° (dash-dotted) North.

efficiency has gone down to about 65%. Also does the efficiency go down somewhat for low angles of incidence. For angles of incidence below 50°, the effect is negligible; for an angle of incidence of 60°, the reduction is about 10%; for an angle of incidence of 80°, the reduction is about 45%; and for angles of incidence above 85°, the power production becomes 0 [337].

2.2.8 Fast Variations with Time

Variations in irradiation at timescales of seconds and minutes are due to shading. The main effect is the passing of clouds in front of the sun. But also persons, animals, or objects passing between the sun and the panel will result in changes in irradiation. The impact of partial shading on the production depends on the connection of the individual cells. In most cases, the cells are connected in series to obtain a voltage suitable to transport the energy to the grid, in the range of 50–200 V, depending on the size of the installation. If one cell in a series is shaded, this will reduce the production of the whole panel by a fraction much more than the size of the part that is shaded. In Ref. 275, it is shown that shading one cell in a chain of 24 can reduce the total output by almost 50%.

Passing clouds are obviously not a concern when the sky is fully clouded or completely cloudless. It is with partially clouded skies that large variations in irradiation, and thus in power production, can occur. The largest changes in irradiation occur when the sky is almost cloudless. One single cloud can reduce the direct irradiation to zero, but will not contribute much to the indirect irradiation. With increasing cloud coverage, the number of rapid changes in production will increase, but the size of the steps will decrease because the clouds result in higher background radiation. With cloud coverage above 50% (4 okta), the number of rapid changes in production will also decrease.

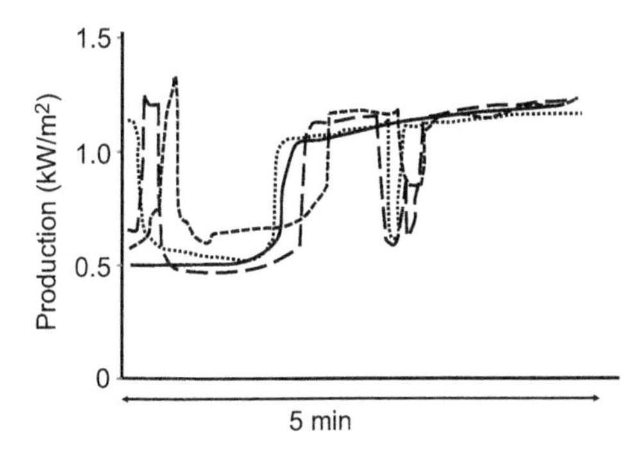

Figure 2.43 Variation in power production with four individual houses during the passage of a cloud [241].

According to Ref. 167, variations in irradiance due to cloud cover can take place as fast as seconds. But Ref. 408 states that a 1 min resolution in time is sufficient to include the effects of passing clouds. Measurements of the variations in irradiance and power production are shown in Refs. 241 and 242. The beam irradiance (direct irradiance measured perpendicular to the sun) and total irradiance are measured with a 1 s resolution. The direct irradiance is shown to drop from 900 W/m^2 to 0 in about 6 s. Calculations using typical altitude and speed of clouds [242] show that this transition time is between 2 and 20 s if we assume the cloud to have a sharp edge. While the beam irradiance drops from 900 W/m^2 to 0, the total irradiance on a horizontal surface drops from about 700 to 100 W/m^2. In Ref. 241, measurements are shown of an installation consisting of 28 panels of 2 kW rated power each, spread over an area of $500 \times 350\,\text{m}^2$. The power production shows changes of about 40 kW (70% of rated power) within half a minute.

The measurements of the changes in power production with four individual houses [241] have been reproduced in Figure 2.43. From these measurements and the calculations shown in Ref. 242, we conclude that changes in power production can take place at timescales down to a few seconds. For a single system, excursions as large as 75% of rated power take place with a rate of change up to 10% per second. For the complete installation, covering 28 systems, excursions as large as 60% can take place, up to 3% per second. The size of the installation damps the speed of the changes, but the impact on the total change is not so big.

The impact of changes in cloud cover on power production is studied in Ref. 234 by means of a physical-based model of the shape and movement of clouds. Some of the results of that study have been reproduced in Tables 2.4 and 2.5. Two different cases have been studied in Ref. 234: a transition from cloudless to fully clouded skies (the so-called "squall line"). The result of the passing of such a squall line is the reduction of production from a high value to a low value. The time it takes for the line to pass the whole service area is given in Table 2.4. For a small service area, the production drops in less than 2 min, whereas it may take more than 2 h for a large service area.

The other weather situation studied is the passing of cumulus clouds, where the sky is partly clouded. The result is that the production of each solar panel shows strong fluctuations. The impact on the production in the whole service area is given in Table 2.5. The largest drop in production has been calculated over different time

TABLE 2.4 Transition Time for Loss of All Solar Power Due to a Passing Squall Line [234]

Service area (km^2)	Transition time (min)
10	1.8
100	5.5
1000	17.6
10,000	55.3
100,000	175.7

TABLE 2.5 Loss in Solar Power Due to Passing Cumulus Clouds [234].

Service area (km^2)	Maximum drop during the interval (% of total capacity)			
	1 min	2 min	3 min	4 min
10	15.9	19.1	19.6	19.6
100	5.5	7.5	7.5	7.5
1000	2.8	3.1	3.1	3.1
10,000	2.7	2.7	2.7	2.7
100,000	2.7	2.7	2.7	2.7

windows, ranging from 1 to 4 min. For small service areas, changes up to 20% are possible within a few minutes. But even for the largest area, the total production may drop by almost 3% within 1 min.

The study presented in Ref. 462 is based on measurements of 1 min irradiance during a 1-month period. The highest variations in irradiance occur for medium values of the average, corresponding to cloud cover between 3 and 5 okta. For a single location, the standard deviation of the 1 min values over a 2 h period around noon is about 25% of the maximum value. The average of five locations at distances between 10 and 20 km shows a standard deviation of about 10% of its maximum value. For fully clouded skies (7 and 8 okta) and for cloudless skies (0 okta), the standard deviation is close to zero.

Measurements of total irradiance have been obtained with a 5 s resolution in Ref. 460. The amount of fluctuations in irradiance at different timescales has been determined by means of a wavelet transform. It is also shown here that the fluctuations are longest for partly clouded skies. For fully clouded and cloudless skies, the fluctuations are small. The fluctuations are strongest for timescales between 5 and 10 min, corresponding to the passing of individual clouds.

The impact of cloud coverage at longer timescales is illustrated in Figure 2.44. The figure shows measurements of the global horizontal irradiance with 5 min time resolution during three consecutive days in January 2004. The measurements were

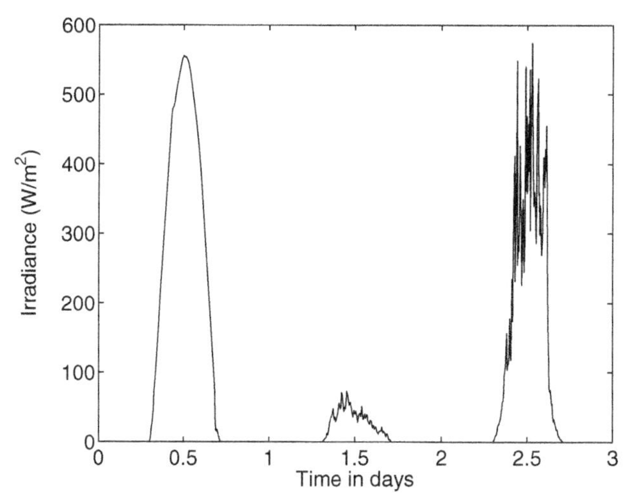

Figure 2.44 Variation in irradiance for three consecutive days.

performed at Mississippi State University (33.50 °N; 90.33 °W). The data were obtained from Ref. 102.

During the first day, the irradiance follows the path of the sun along the sky. The second day was fully or largely clouded and the irradiance was small. The third day was partly clouded, resulting in large fluctuations in irradiance due to clouds passing in front of the sun. Note also that the peak values on day 3 are higher than that on day 1. The presence of clouds increases the diffuse irradiance. The difference in irradiance from one 5 min period to another is shown in Figure 2.45 for the same

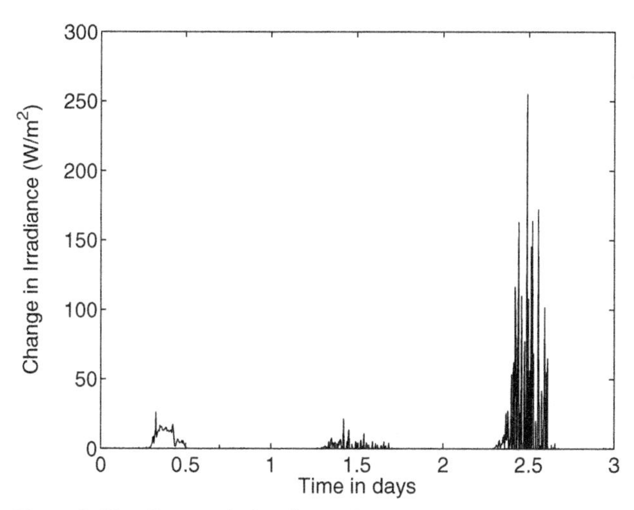

Figure 2.45 Changes in irradiance from one 5 min period to another during three consecutive days.

TABLE 2.6 Cases Per Year with the Change in Irradiance Exceeding a Certain Value

Value being exceeded (W/m^2)	Cases per year
400	462
450	290
500	173
550	89
600	51
650	22
700	4

3-day period. During the first and second day, the changes are small. During the third day, the changes are up to half of the maximum irradiance.

The change in irradiance from one 5 min period to the next has been calculated for the whole of the year 2004. The highest value measured was 737 W/m^2. The number of times per year that high values were recorded is shown in Table 2.6: 462 times per year the change was more than 400 W/m^2, and so on. The table shows that high values are not a rare event: they do occur regularly.

2.3 COMBINED HEAT-AND-POWER

2.3.1 Status

Combined heat-and-power, also known as "cogeneration," refers to the use of recovered exhaust heat of any production unit for another process requirement. This in turn results in improvement in the energy utilization of the unit. By so doing, the overall thermal efficiency of generation may be raised from 40–50% to 70–90%. The upper limit of 90% holds for large installations with a very well-defined and constant heat demand.

Combined heat-and-power does not have to be a renewable source of energy; in fact, many CHP installations use natural gas as a source. The use of biomass as a source is the only renewable form of CHP. The direct combustion of organic matter to produce steam or electricity is the most advanced of the different CHP processes and, when carried out under controlled conditions, is probably the most efficient.

Large CHP installations are used for production of steam in industrial installations, for space heating in the agriculture, and for district heating. Agricultural CHP is very common in the Netherlands and in Denmark, where about 25% of electricity comes from CHP. Recently, incentives toward smaller generation units have resulted in a growth in CHP in some countries.

An application that receives a lot of attention in the literature is the so-called "micro-CHP": small installations that produce both electricity and heat for local space heating. Possible applications, next to domestic houses, would be hotels, shopping centers, and offices. For example, a somewhat larger unit produces 105 kW electricity and 172 kW heat [296]. A market for much smaller units may be emerging, intended for heating of domestic premises. An example is a unit that produces 1 kW electricity together with 7.5–12 kW heat [442].

Example 2.3 The following examples of heat and electricity production are given in Ref. 105:

- A thermal power station needs 63.6 units of primary energy to produce 35 units of electricity.
- A high-efficiency (condensing) boiler needs 55.6 units of primary energy to produce 50 units of heat.
- A CHP installation needs 100 units of primary energy to produce 35 units of electricity and 50 units of heat.

Comparing these shows that 19.2 units of primary energy can be saved by using CHP, a saving of about 20%.

This first example shows that CHP is a good idea when electricity uses the same primary energy as heat production. What matters here is not the average source of primary energy, but the marginal source: what is saved by producing heat and electricity at the same time. When the marginal source is fossil fuel, CHP will reduce the emission. The next example considers the situation when the marginal source is nonfossil, like hydro or nuclear. In that case, CHP will actually increase the emission.

Example 2.4 Using the same values as in Example 2.3, assume that electricity comes from clean sources, but that CHP requires natural gas as a primary source of energy. Producing electricity and heat separately requires 63.6 units of clean sources and 55.6 units of natural gas. Producing electricity and heat together requires 100 units of natural gas. Next, 63.6 units of clean sources are replaced by 44.4 units of natural gas.

The marginal source is determined by the electricity market and by the operational security concerns in the transmission system. It will vary with time, for example, being coal at night and hydropower at daytime. For an optimal reduction of emission, the scheduling of CHP will have to depend on the marginal fuel being used. An intelligent pricing system for electricity could achieve this, which is again beyond the scope of this book. Such a pricing system could, however, result in the CHP units switching on and off based on the marginal fuel (i.e., the electricity price) of the hour. A situation might occur where the units are switched off during peak consumption. This will have an adverse impact on the operation of the power system.

2.3.2 Options for Space Heating

Combined heat-and-power is presented as a very efficient way of producing both electricity and heat. This is true when the comparison is made with a thermal power plant not using heat recovery. In industrial installations with a large heat demand (often in the form of steam), it may indeed be a good idea to use the waste heat of the power generation to produce steam. Whether CHP for space heating is a good idea depends on the alternatives. In Examples 2.5 and 2.6, CHP is compared with electricity from a thermal power station to supply a heat pump.

Example 2.5 Consider a CHP installation with an efficiency of 75%, producing electricity and heat at the ratio of 1:1.7 (data for a 100 kW installation according to Ref. 296). Every unit of primary energy produces 0.278 unit of electricity and 0.472 unit of heat. Assume that the local heat demand is 10 kW; with this installation, this requires 21.2 kW of primary energy. Next to the required heat, the installation produces 5.9 kW of electricity.

A heat pump or "ground-source heat" transports ambient heat (typically from underground) into buildings. The efficiency of heat pumps is a factor of 3–4, that is, 1 kW of electricity can transport 3–4 kW of heat [286, 458]. Note that this efficiency higher than 1 is not against any physical law; the heat is not produced out of nothing, it was already there. The heat is simply transported against the temperature gradient. The electricity is not used to produce heat, it is only used to transport already existing heat.

Example 2.6 Consider a heat pump that requires 1 kW of electricity to produce 3 kW of heat. (the lower limit of the range given in Refs. 286 and 458). To produce the same 10 kW of heat as in Example 2.5 requires 3.3 kW of electricity. The CHP installation produced an additional 5.9 kW of electricity, so we have to produce 9.2 kW of electricity to make a fair comparison. Using a thermal power station with an efficiency of 45% will require 20.4 kW of primary energy. This is slightly less than the 21.2 kW needed for the CHP installation in Example 2.5.

Exercise 2.1 Repeat the calculations when direct electric heating is used for space heating. Assume an efficiency of 90% for the electric heating, which is on the low side. Consider the same electricity production as in Example 2.6.

Exercise 2.2 Repeat the calculations when the primary energy is directly used for space heating, whereas the electricity is again produced from a thermal power station. Assume an efficiency of 90% for the heating, which is achievable by using modern condensing boilers.

From these examples it follows that CHP for space heating is not more efficient than using heat pumps with the electricity coming from a thermal power station. This does not mean that CHP will not be used for space heating. An individual customer often makes a decision based on economic criteria. Producing heat and electricity locally may simply be cheaper than buying electricity and using some of this to power a heat pump. An important factor to consider in this is that the price of electricity for the end customer is only partly determined by the costs of the primary fuel. The use-of-system fee and the local taxes have big influence on the price. By producing locally, the customer not only saves the costs of the primary fuel but also the other contributions to the price.

2.3.3 Properties

The production capacity of a CHP unit depends directly on the heat demand. The higher the heat demand, the higher the production capacity. The heat is produced as a by-product of the electricity production, so the heat demand sets a minimum limit to the electricity production. When the electricity production is higher than needed, the excess heat will have to be cooled away, which reduces the efficiency of the installation. In the same way, when the electricity production is less than needed, heat has to be produced in another way. This is technically possible, but most installations use a fixed ratio between electricity and heat production, where the local heat demand determines the production.

With industrial installations, the heat demand is very constant and predictable. When CHP is used for room heating, either industrial, in offices, or domestic, the heat demand depends on several weather parameters: the difference between the indoor and outdoor temperatures, the wind speed, and the irradiance. Each of these parameters is difficult to predict, so the amount of production cannot be accurately predicted.

If we consider the prediction 1 day ahead of time, that is, upon closure of the day-ahead market, errors of a few degrees in temperature prediction are not uncommon. This will result in prediction errors in production up to about 10% (see Example 2.7). This appears an extreme case, although no data have been found on actual prediction errors in CHP production.

Example 2.7 Assume a system with 6000 MW CHP. The rated capacity is obtained for an indoor temperature 30° above the outdoor temperature. If we assume that the heat demand is directly proportional to the temperature difference, each degree corresponds to 200 MW production. A prediction error of 5°, average over the whole geographical area, would correspond to an error in production of 1000 MW.

Small-scale CHP, often referred to as micro-CHP, like these for office buildings or domestic heating, often uses an on–off control where the production of heat and electricity is either zero or at its maximum. As a result, the production will continuously vary between zero and its maximum value. This will result in large power fluctuations close to the point of connection, but at transmission level, the variations will average out and the predictability will not be impacted.

Combined heat-and-power can be located only where there is a heat demand, as heat cannot be transported over large distances. As heat demand goes together with electricity demand, CHP is always located close to the electricity consumption. This is an obvious advantage of CHP for the power system.

Another advantage of CHP is its close correlation with the electricity consumption. For industrial CHP, the correlation is obvious. Combined heat-and-power for space heating is most likely to occur in countries with a cold climate, where the electricity consumption is highest during winter. Also, the daily load curve of CHP follows the consumption rather well. We will see some examples of this in Section 2.3.5.

2.3.4 Variation in Production with Time

In Sweden, during 2008, the energy production by CHP was 5.2% of the total consumption. CHP consists in part of industrial installations where the heat demand depends on the industrial process. This production is constant throughout the year,

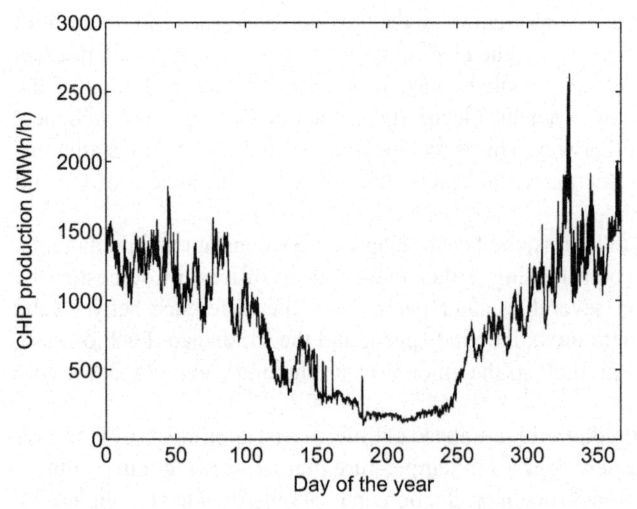

Figure 2.46 Energy production by combined heat-and-power in Sweden during 2008.

with a possible reduction during holiday periods. The production of CHP by space heating depends on the outside temperature and will thus show strong daily and annual variations. The hourly energy production by CHP in Sweden, for 2008, is shown in Figure 2.46. The annual variation is clearly visible.

In Ref. 192, a comparison is made between micro-CHP with and without thermal storage. Thermal storage allows longer on–off cycles for the generator. This will reduce the maintenance requirements of the generator unit, because it experiences less starts and stops. It will also change the impact on the power system. In fact, the reduced starting and stopping is also advantageous for the power quality. Another reason for having long heating cycles is the electricity consumption during start-up and shutdown of the unit. This consumption is independent of the duration of the heating period, so the short heating periods are very inefficient.

For the system with heat storage, the generator was running at high power for 1–2 h, followed by 1–2 h during which the generator was not producing any power. An example of electricity production during a number of heating cycles is shown in Figure 2.47, which is reproduced from Ref. 23. The figure is based on measurements that were obtained from a domestic installation that included a heating storage tank. Just before and just after the heating period, the unit consumes electricity, especially

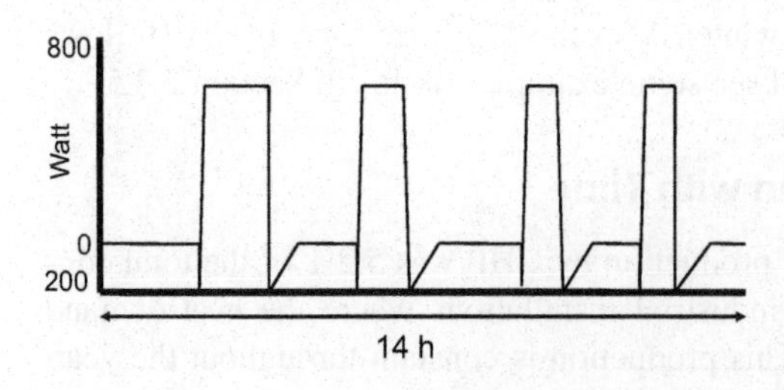

Figure 2.47 Example of the electricity production by a micro-CHP unit with heating storage.

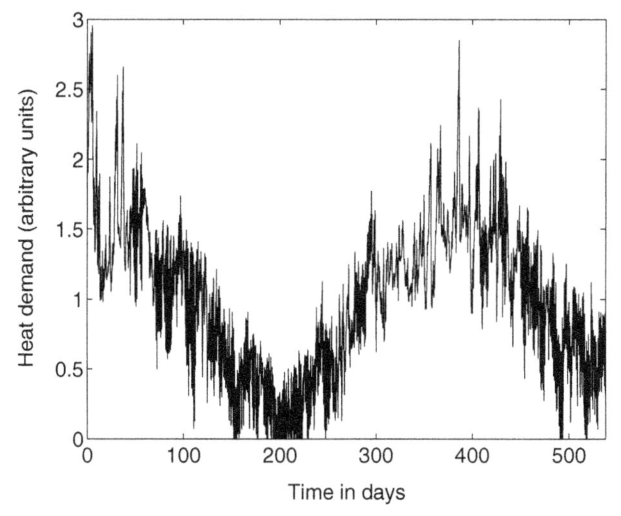

Figure 2.48 Variation in heat demand during a 1.5-year period.

at the end of the heating period. The duration of a heating period varies between 1 and 2 h.

The variation in heat demand for space heating has been estimated from a 1.5-year record of temperature values for the town of Ludvika, in the middle of Sweden. Temperature measurements were available for each 10 min. The hourly heat demand was calculated by adding the six differences between the assumed inner temperature and the measured outer temperature over each 1 h period. The results are shown in Figure 2.48, where one unit of heat demand is the average value during the whole measurement period. The horizontal axis (day 0) starts at midnight on January 1, 2003. As expected, the heat demand is lowest during summer and highest during winter.

The heat demand in the previous figure has been calculated assuming a room temperate of 23°C. The heat demand strongly depends on the room temperature (i.e., on the thermostat setting), as shown in Table 2.7. The heat demand, and thus the energy consumption, changes by about 5% for each degree change in room temperature. Energy saving is beyond the scope of this book, but it is still worth to point out the

TABLE 2.7 Impact of Room Temperature on Heat Demand for Space Heating

Temperature (°C)	Annual heat demand (%)
20	83.0
21	88.6
22	94.3
23	100
24	105.8
25	111.6

potential. This example holds for space heating; the same holds for cooling but the other way around.

Measurements of the electricity production of small CHP units were performed as part of an integrated European project (EU-DEEP) and as part of a number of experiments toward the integration of distributed generation in the electricity grid and market. Some of the measurement results are presented in Ref. 151. The gas consumption of a group of university buildings in Greece, where gas is only used for heating, on a typical winter day shows a peak at 7 a.m. and a second peak at 5 p.m. After the morning peak, the gas consumption drops linearly to zero in about 4 h. After the evening peak, the demand drops to zero in about 2 h. At another group of buildings in the same campus, the gas consumption shows a similar pattern, with peaks in the early morning and in the late afternoon. The gas consumption does, however, not drop to zero in this case.

Experience with a 1 kW Stirling engine for CHP [151] showed that it was running five to eight times per day for 35–60 min. Measurements were made in November, the number of cycles and the duration of each cycle depend on heat demand. The reactive power is close to zero when the active power production is at rated power, which is reached about 20 min after the start of the engine. During these 20 min, the unit produces active as well as reactive power. Also, after the heat production had stopped, the unit produces and consumes reactive power.

Experiments with an 80 kW (electrical) CHP installations show five periods of production during 1 day; three of them last only 10 min and come within 1 h, while the other two last about 2.5 h each with half an hour in between. In all cases, the power production did go from zero to maximum within several minutes; the return to zero production takes place even faster.

2.3.5 Correlation Between CHP and Consumption

The power system of the western part of Denmark is connected to Sweden and Norway through HVDC and with Germany through a synchronous connection. The maximum consumption (not considering the transmission system losses) during 2008 was about 3700 MW. The electricity production in this part of Denmark consist of conventional generation dispatchable by the transmission system operator, combined heat-and-power, and wind power. The variation in consumption and production during 2008 is summarized in Table 2.8. In terms of energy, 59.3% of the consumption is supplied by conventional generation, 23.3% by CHP, and 24.6% by wind power. The sum of

TABLE 2.8 Production and Consumption in Western Denmark During 2008

	Maximum (MW)	Mean (MW)	Minimum (MW)	Maximum percentage	Minimum percentage
Consumption	3692	2403	1270	–	–
Conventional generation	3166	1426	415	141	17
CHP	1292	561	146	42	6
Wind power	2177	591	0	121	0

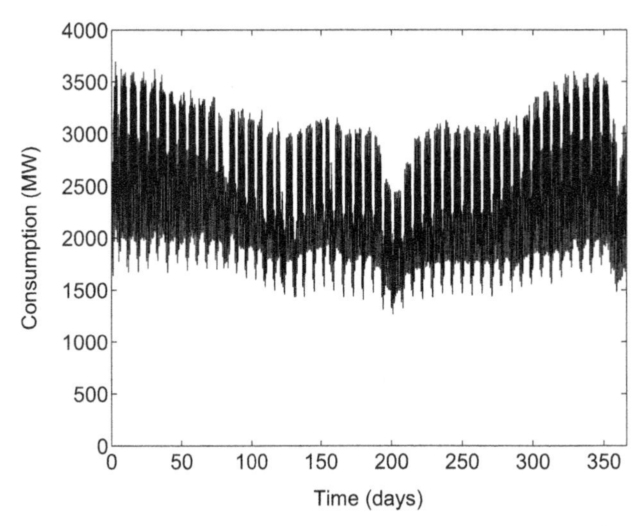

Figure 2.49 Net consumption in western Denmark during 2008.

these three contributions is more than 100% because of the losses in the transmission system (2.4% on average) and the exchange with the neighboring countries. The data in Table 2.8 and the forthcoming figures are obtained from Ref. 142.

The consumption for the western part of Denmark is shown in Figure 2.49 as a function of time. The total available production capacity, including import from neighboring countries, should, at any moment in time, exceed the consumption. What matters for the planning of production capacity is the maximum consumption and the amount of time that the consumption is above certain levels. A curve like the one in Figure 2.50 is used for generation planning.

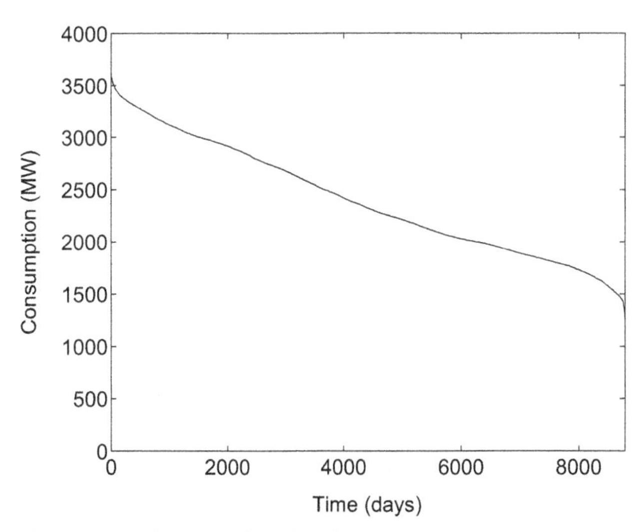

Figure 2.50 Consumption–duration curve.

All the sources of production available to the system operator should be sufficient to cover the maximum consumption plus a safety margin to accommodate for unavailability of the production. Either deterministic methods, such as a fixed percentage margin, or probabilistic methods, such as a fixed loss-of-load expectancy, can be used. Details of these can be found in the literature on power system reliability [29].

The area under the curve is the total electrical energy that is consumed during the year. This is filled with different sources of production, including distributed generation. From the viewpoint of the transmission system operator, international connections can be treated as production. The availability of the international connections, however, has to be considered very carefully. The peak consumption might well occur at the same time in the neighboring countries. In that case, a low availability of the international connections should be considered in the calculations.

It should also be noted that under the open electricity market, there is no longer anybody responsible for the planning of the production facilities. A shortage of production will result in a high electricity price, which in turn will make it attractive for investors to build new production facilities.

As already mentioned, combined heat-and-power stands for about one quarter of the annual production of electrical energy in western Denmark. This production is not under control of the system operator and is, therefore, often referred to as a "negative load." The same holds for wind power. The amount of production that has to be scheduled by the system operator is equal to the consumption minus the production by CHP minus the production by wind power. The different duration curves are shown in Figure 2.51.

The highest consumption is equal to 3692 MW. The highest value of consumption minus CHP is equal to 2742 MW, a reduction of the peak by 25.7%. For the planning of production facilities, CHP is clearly a positive factor, especially if we

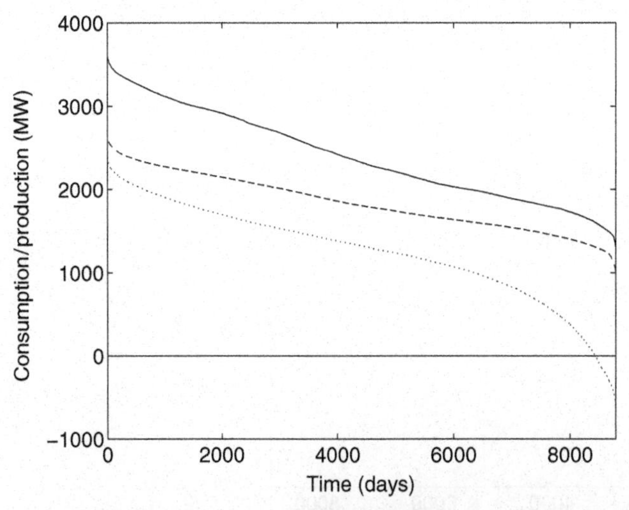

Figure 2.51 Duration curves for the consumption (solid), consumption minus combined heat-and-power (dashed), and consumption minus combined heat-and-power and wind power (dotted).

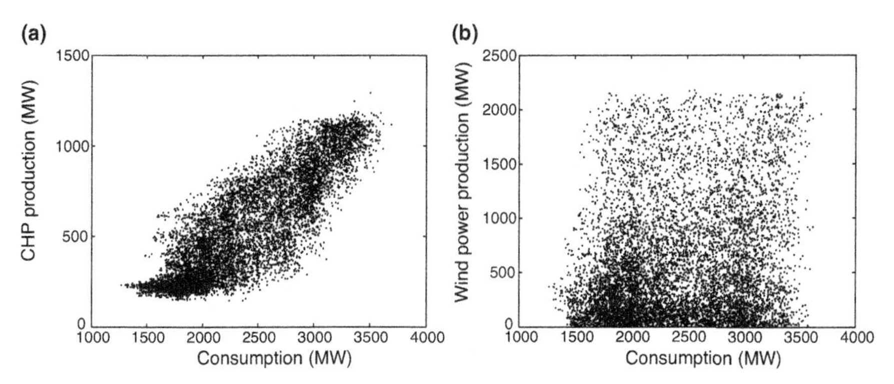

Figure 2.52 Correlation between consumption and combined heat-and-power (a) and wind power (b).

consider that the energy production from CHP is 23.3% of the consumption. The use of distributed generation increases the generation reliability or reduces the need for conventional production capacity. This is not the case however with wind power. Although wind power stands for 24.6% of the energy production, it reduces the need for conventional production facilities by only 6.3%. The term "capacity credit" is sometimes used to indicate how much the amount of conventional production facilities can be reduced because of the introduction of distributed generation. Calculating a capacity credit however requires that some kind of measure for the reliability of the system is defined. The capacity credit is defined as the amount with which the conventional generation can be reduced, after introduction of distributed generation, while maintaining the same value for the reliability measure as before introduction of the distributed generation. It is obvious that the capacity credit depends on the choice of the reliability measure.

The different impact of CHP and wind power can also be seen in Figure 2.52. The correlation between consumption and CHP is much stronger than the correlation between consumption and wind power.

2.4 HYDROPOWER

Hydropower uses the flow of water toward the sea to generate electricity. The most visible form is by using large dams to create reservoirs. The difference in water levels upstream and downstream of the dam allows the use of the potential energy in water to be used effectively. Such installations range in size from a few tens of megawatts to several hundreds of megawatts, with some very large installations in operation as well. The power station with the Three Gorges Dam in China is the biggest in the world with a total capacity of 22,500 MW. Other very large installations are the Itaipu plant at the border of Brazil and Paraguay (12,600 MW), the La Grande Plant in Quebec, Canada (15,000 MW), and the Cahora Bassa plant in Northern Mozambique (2060 MW). Some countries rely to a large extent on hydropower for their electricity production: for example, Brazil (83%), Argentina (25%), Canada (60%), Chile (50%), Norway (99%), and Sweden (50%). In absolute numbers, Canada and China are the largest hydropower producers, both with around 330 TWh production per year. For China, this however only amounts to 15.8% of its total electricity production [136].

Although the amount of electricity produced from hydro has been steadily rising, its share of the total has been going down. In 1973, 1400 TWh electricity was produced from hydro globally; this was 21% of the total electricity production. In 2007, the corresponding values were 3100 TWh and 15.6% [224].

The current trend in hydropower is toward very large installations in Asia, Africa, and South America. Plans are being discussed for a 40,000 MW installation called "Grand Inga" in the Democratic Republic of Congo, with the electricity being transported by HVDC links over whole of Africa. The current installation, known as "Inga," has a capacity of 1775 MW [143, 390]. The use of these resources will require huge investments in the transmission grid. In Africa, international connections are required that could be difficult for political reasons. In Asia and South America, this is much easier to achieve.

No significant extension of large hydropower is expected in Europe or North America. This is not only partly due to the public resistance against large dams but also due to most of the attractive locations are already in use. According to Ref. 167, 20–25% of the global hydro potential has been developed already. It remains unclear how much of the remaining part can be actually developed. Remote location are in itself no longer a barrier; experience with other projects (Itaipu, Three Gorges Dam, and Cahora Bassa) has shown that large amounts of power can be transported over long distances using HVDC technology. The main barriers are the social and environmental impacts of large artificial lakes in remote locations, as well as the extremely high costs of building such installations. The (actual or perceived) safety of the dam is another issue that could become a barrier to building very large hydropower installations.

Small hydropower may be easier to build, but there are at the moment no indications of any growth in this area. This could be not only due to the lack of proper incentive schemes but also due to the lack of potential. According to Ref. 167, the global potential for small hydro is about 500 GW, where an upper limit of 5 MW is used for small hydro. Assuming a utilization of 2500 h per year, this would result in 1250 TWh per year of electricity of about 6% of the global consumption in 2007. An example of incentives against building small hydropower installations is the property tax in Sweden for power stations: 0.2% for wind power; 0.5% for CHP; and 2.2% for hydropower.

The 2009 energy outlook by the U.S. Department of Energy sees hydropower as providing the main growth in electricity from renewable energy. Between 2010 and 2030, hydropower is expected to grow globally from 3381 to 4773 TWh—an increase by 1392 TWh. Wind power on the other hand, is expected to grow from 312 to 1214 TWh; an increase by 902 TWh [144]. Where wind power is expected to show a much bigger percentage growth (290 versus 41%), the growth is still less in absolute terms.

2.4.1 Properties of Large Hydro

Hydropower based on large reservoirs is by definition located at remote areas. Large-scale hydropower always requires a strong transmission system. Extending the amount of hydropower requires further strengthening of the transmission system. Three examples are the 400 kV grid in Sweden, the 750 kV grid in Quebec, and the long HVDC links, for example, in Brazil and China.

Hydropower shows especially the long-term variations at timescales that range from days through years. This makes hydropower very attractive from the points of view of electricity market and transmission system operation. Both for the electricity market and for the transmission system operation, timescales longer than a few days are normally not important.

These long-term variations are, however, important for the planning of the generation. Any system with a large amount of hydropower needs to consider the occurrence of dry years. This can be either by overdimensioning of the reservoir size (so that water from a wet year can be kept for a dry year), by adding thermal generation to the energy mix, or through sufficiently strong links with other regions. The latter has also clear economic advantages to sell the surplus of energy during wet years.

2.4.2 Properties of Small Hydro

Flow of river and small hydropower could be located closer to the customer than the large hydro, but it remains restricted to mountainous areas. Small hydropower schemes could be used to strengthen the distribution grid in such mountainous areas.

Small hydropower refers to units varying from 10 kW to about 30 MW. Sometimes they are classified as microhydro (10–100 kW), minihydro (100–500 kW), and small hydro (500 kw–30 MW).

The People's Republic of China has nearly 100,000 small hydropower plants constructed in the last 20 years. The distinctive features of small hydropower plants include being usually run-of-river type, construction in a relatively short period of time, using well-developed technology with an overall efficiency of over 80%, and having automatic operating systems with low operation and maintenance costs. Without significant storage capacity, large variations in available water flow may be experienced. In the United Kingdom, the capacity factor for hydro, that is, the ratio of actual annual energy generated to energy produced at rated output over 12 months, is approximately 30% that is nearly the same as for wind energy [310].

2.4.3 Variation with Time

The variation in hydropower with time depends on the flow of water into the reservoir or on the actual flow through the river. To illustrate this, data were used of the flow speed of the river Maas (known as Meuse in French) near the city of Borgharen in the Netherlands. This is a typical "rain river"; the sources of this river are the mountain ranges in the east of Belgium and in the northeast of France. The flow of river at this location was obtained from Ref. 366 for midnight every day from 1975 through 2008. The average flow is 239 m^3/s. The highest flow measured during this period was 2959 m^3/s. The probability distribution function of the flow is shown in Figure 2.53.

The variations in flow during 2008 are shown in Figure 2.54a. The flow is highest during the winter months (January and February) and lowest during the summer months. A similar pattern is presented in Ref. 167 for a small river in England (with

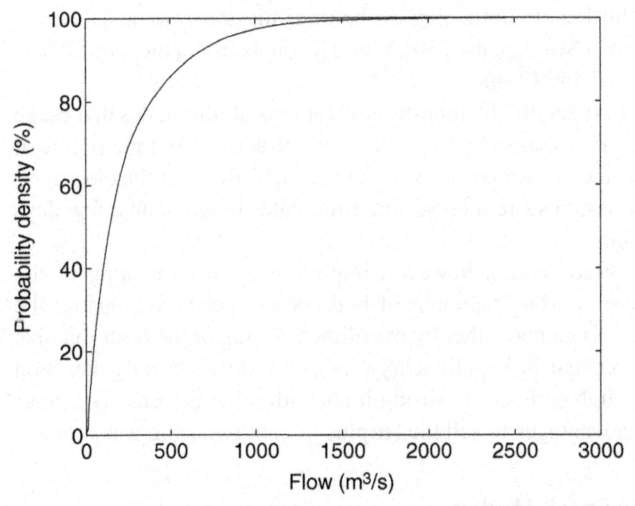

Figure 2.53 Probability distribution function of the flow of river.

an average flow of about 10 m^3/s. As is shown in Figure 2.54b, this is a normal pattern, returning in most years. With some exceptions (the large peak in month 7 corresponds to the notorious flooding in the summer of 1980), periods with high flow occur during the winter months.

Most hydropower installations use a reservoir to collect the water, so the electricity production is no longer directly linked to the flow. What matters in that case is the integrated flow over a longer period of time. In Figure 2.55, the average flow over each 30-day period is shown. The yearly pattern is clearly visible, but with large differences between the years.

One of the well-known properties of hydropower is its year-by-year variation. The amount of water flowing into a reservoir can vary significantly between a wet year and a dry year. This is illustrated in Figure 2.56, showing the total flow for each of the 34 years for which measurements were available. The average flow is about

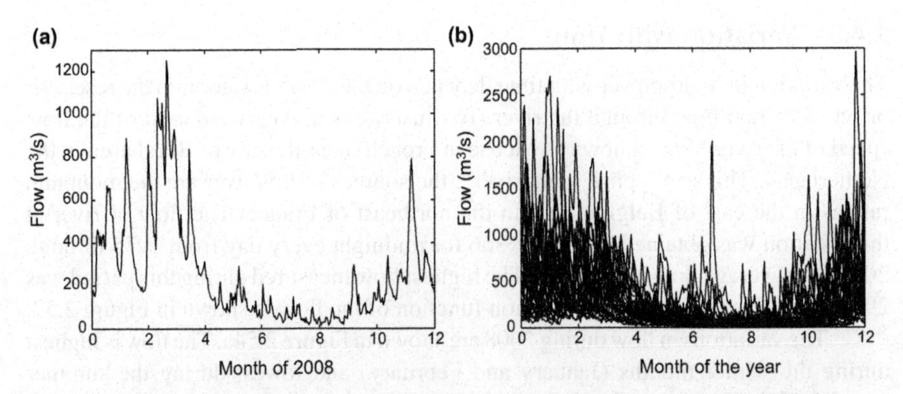

Figure 2.54 Variation in flow of river for 2008 (a) and during the years 1975–2008 (b).

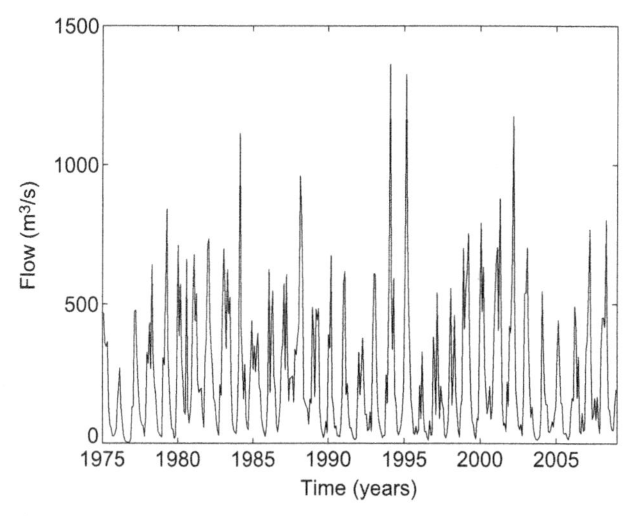

Figure 2.55 Variation in 30-day average of flow of river.

7.5 billion m^3/year with a standard deviation of 2.3 billion m^3/year. The two highest values were both somewhat more than 11 billion m^3/year, whereas the two lowest values were 2.3 and 3.8 billion m^3/year.

As a comparison, Figure 2.57 shows the variations in electricity production from hydropower for the whole of Spain [312], roughly during the same period. The variations are somewhat smaller: the standard deviation is 23% of the mean for Spain, whereas it is 30% of the mean for the single river in Figure 2.56. These rather big variations in longer term availability of hydropower is one of the main drawbacks of hydropower and an important reason why some countries have diversified their energy portfolio so that it is no longer dominated by hydropower.

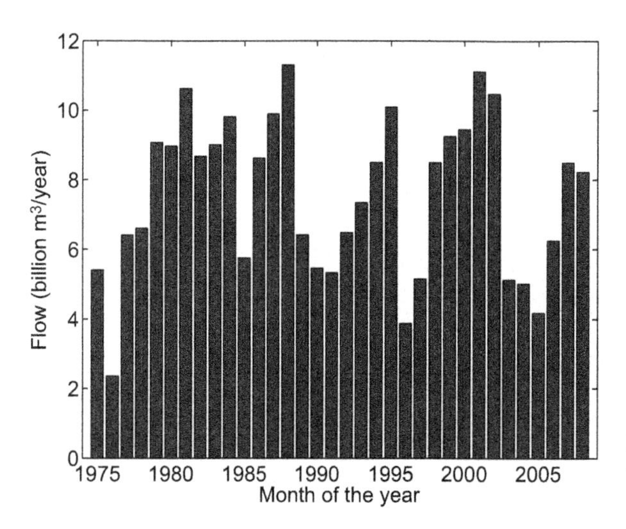

Figure 2.56 Year-by-year variations in flow: 1975–2008.

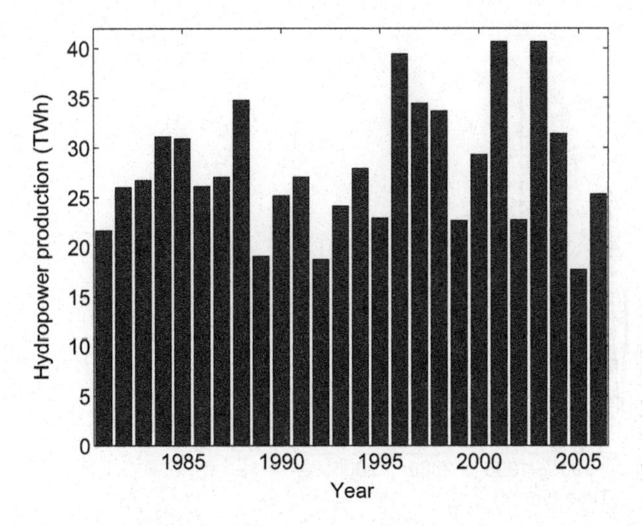

Figure 2.57 Year-by-year variations in electricity production from hydro in Spain: 1981–2006.

The year-by-year variations in hydro production are not as large for most other countries: it depends strongly on the size of the country and on the difference in climate between different parts of the country. However, even a country like Chile (with a huge spread in climate) has experienced some years with a clearly reduced production, superimposed on a generally increasing trend due to building more production capacity, as is illustrated in Figure 2.58 [311].

The day-by-day changes in flow of river are presented in Table 2.9 for the same location as before. The difference has been calculated between the measured flow and the value measured 24 h before. In 90% of the cases, this difference is less than 100 m^3/s. The flow in most cases does not see large changes between consecutive days.

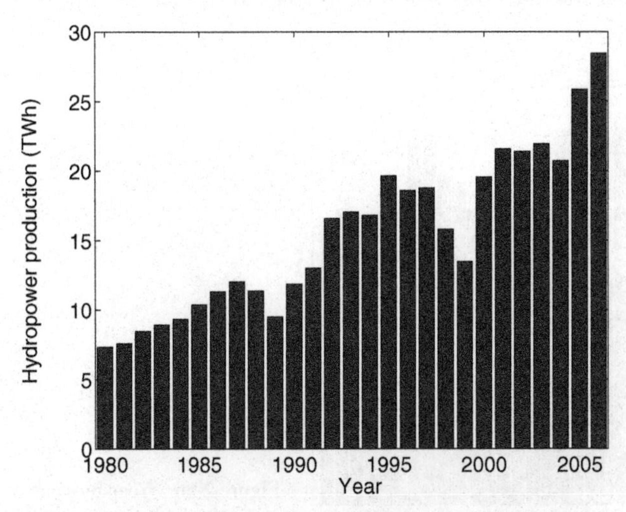

Figure 2.58 Year-by-year variations in electricity production from hydro in Chile 1980–2006.

TABLE 2.9 One-Day Changes in Flow of River

	Increase (m^3/s)	Decrease (m^3/s)
Max	+973	−660
1–99%	+288	−165
5–95%	+100	−85
10–90%	+45	−54

The hour-by-hour variations will be even smaller. We also see that large increases are somewhat more likely than large decreases.

2.5 TIDAL POWER

Tidal power makes use of the difference between high tide and low tide or the water flow between high tide and low tide. One of the possible schemes traps the water at high tide and lets it flow back only at low tide, thus making use of the difference in sea level between high tide and low tide. Other schemes use the water flow between the high and low tides to generate electricity. The former scheme (known as "tidal lagoon" or "tidal barrage") can be used at locations with a high difference between the low and high tides; the latter scheme (known as "tidal current") can be used at locations with a strong flow. Tidal lagoon schemes will produce power around low tide, that is, about once every 12 h. Tidal current schemes will produce power in between high and low tides, that is, about once every 6 h. A combination of the two schemes is able to produce power six times a day. The so-called "hydrokinetic" schemes can be used with turbines under the water level being powered by the water flow. As water has a much higher density than air, much smaller turbines can be used than for wind power. A scheme with 875 submerged turbines in the Niagara river is mentioned in Ref. 9. A tidal lagoon creates a storage behind the dam and can be combined with pumped storage to add extra power.

The tides are due to the gravitational pull of the sun and the moon (more precisely, it is the gradient in the gravitational field that causes the tides). The rotation of the earth causes a 24 h period in the tides. The magnitude of the tides, that is, the difference between the high and low tides, varies with a period of about 1 month due to the rotation of the moon around the earth. The power produced by a tidal scheme is thus strongly fluctuating with time, but very predictable.

There are only a limited number of locations at which tidal power is possible. According to one of the companies involved in tidal power [17], about 15% of the electricity needs of the United States can be covered by tidal power. A similar figure is mentioned for the United Kingdom. The potential varies, however, strongly between countries. Large potential for tidal power production is in India and Australia. Worldwide the economic potential of tidal power is stated to be around 180 TWh per year [17] or about 1% of the electricity consumption in 2007 [224]. Estimations of the potential vary significantly, with Ref. 76 mentioning a potential of 2000–3000 TWh per year (10–15% of the global consumption in 2007).

According to Ref. 167, the potential for tidal barrages in Europe is 105 TWh per year with an installed capacity of 64,000 MW (giving an utilization of 1640 h per year). This corresponds to 3% of the 2007 electricity consumption in Europe. The potential is further very strongly concentrated in Britain (48%), France (42%), and Ireland (7%).

A 240 MW tidal power station has been in operation since 1966 at the mouth of the river Rance in Brittany, France [167, 364, 450]. The annual production is 600 GWh, equal to 2500 h of peak production per year or 29%. A 20 MW installation has been in operation since 1984 in the Anapolis River in Canada [444]. Although the potential is regularly stated to be large, no other installations of similar size have emerged. Plans to build a huge tidal power installations in the Severn Barrage between England and Wales have been discussed for many years now. A peak power of 8000 MW is stated for such an installation. Assuming the same 29% utilization as for the French station, this would produce 20 TWh of electricity per year or 5% of the UK electricity consumption for 2008. Environmental objections against blocking the mouth of the Severn, as well as the huge investment costs, make it very unlikely that anything will be built soon.

Tidal current installations have less environmental impact because the turbines are located below the water level and there are almost no parts of the installations visible above the water. Several installations are being discussed. A 30 MW tidal current installation is planned for Pentland Firth in Scotland [17]. A 300 MW tidal current installation is planned for South Korea where 300 turbines of 1 MW each are positioned on the sea floor [363]. A similar 200 MW installation is planned for New Zealand [362]. According to Ref. 167, the estimated UK potential for tidal current schemes is 36 TWh per year, which would be 9% of the 2007 electricity consumption.

Tidal power installations, if they emerge, will likely be large installations with sizes of tens of megawatts and up. As these installations will be located where the power is available, which is typically not where the production is, enforcement of the transmission system will be required.

2.6 WAVE POWER

According to Ref. 389, the average energy in the wave in the northern Atlantic Ocean is about 10 kJ/m^2. This is said to correspond to the accumulated wind energy up to a height of about 200 m. Referring to Ref. 389 again, "on a medium time scale, the characteristics of the creation and dissipation mechanism may make wave energy a more [constant] energy source than wind energy, but on a seasonal scale, the variations in wind and wave energy are expected to follow each other."

The seasonal variation in wave power in the United Kingdom is shown in Ref. 167 (with reference to Ref. 384). The annual capacity factor is around 28% (2500 h per year), but varies from almost 50% in January and February to 10% in July. According to Ref. 375, the average power density of the waves is 5–10 kW/m on the Swedish west coast and 20–80 kW/m on the Norwegian coast. The utilization of a wave power installation would be 3000–4000 h per year for Sweden and 4000–5000 h per year for Norway.

The global potential for wave power is, according to Ref. 167, 200–5000 GW, most of which would be offshore installations. The potential for shoreline installations is a factor of 100 less. The upper value of the estimation would, assuming 2500 h of utilization per year (as in the above data for the United Kingdom), result in 12,500 TWh of electricity per year being produced from wave power, which would be over 60% of the global electricity consumption in 2007. It is this value that explains the enthusiasm among proponents of wave power. The lower estimate would, however, only be able to cover only 2.5% of the consumption. Reference 105 states a potential of 600 TWh/year for the European Atlantic coast, which would be about 25% of the electricity consumption.

Many competing technologies are under development, but none of them has come any further than prototypes or installations not connected to the grid. The energy in waves varies from 5 kW/m during summer up to 1000 kW/m during a winter storm [75]. It is very difficult to manufacture generators that are able to produce electricity even during summer, without being damaged by a storm.

Plans for several pilot installations connected to the grid, however, do exist [453]. A 2 MW installation on the Portuguese coast was taken out of operation after being connected to the grid for about 1 year. A pilot installation with 400–500 turbines and a total rated power of 10 MW is planned in Smögen on the Swedish west coast. This is stated to be the world's biggest wave park [375]. The power is generated by rather small units, 10–50 kW, that are located on the sea bottom. Parks can, however, consists of many such units with total powers of tens of MW. The generators produce power in the form of pulses whenever a wave passes above the generator. The power production is smoothed by adding many units together into one park and by using a limited amount of local energy storage. The larger the park, the smaller the need for storage. Plans for 250 MW of wave power, using turbines of about 500 kW rating, have been presented for Ireland [439].

Compared to tidal power, the production by wave power is less predictable. It is, however, more constant at a timescale of hours and it is available at more locations than tidal power. As many cities and towns are located near the coast, offshore installations could feed directly into the local distribution grid, thus reducing the loading of the transmission grid.

It remains unclear what wave power will look like, as seen from the power system. Near-shore installations may feed into the distribution system of town and villages close to the coast. Such installations may be rather small. Installations further out to sea, or even on the ocean, would have to be of large capacity (hundreds of megawatts or more) to be economically attractive. Connection to the grid will have to take place by undersea cable, using HVDC technology for installations more than a few tens of kilometers from the shore. The transmission grid would likely require enforcement at the place where the wave power is connected.

2.7 GEOTHERMAL POWER

Geothermal energy is, after hydropower and biomass, the third most exploited renewable energy source with more than 6000 MW installed generation capacity in

21 different countries [94]. Geothermal energy extracts heat from the earth's crust through an aquifer or inject water flow in the case of hot dry rock technology. Energy conversion can be either via mechanical means or the heat can be used directly. The main country that uses geothermal power for a large part of its energy production is Iceland. Geothermal provided 66% of the country's energy, mainly in the form of space heating. Hydropower provided for 15%. About 500 MW of electricity is produced in geothermal plants. The untapped potential is said to be as high as 50 TWh, which is more than sufficient to provide for the whole of the country, but only a small part of the global energy consumption (about 20,000 TWh in 2007).

An overview of installed geothermal power worldwide is given in Chapter 6 of Ref. 81. The global total has grown from 3900 MW in 1980 to 9000 MW in 1995. Of the 1995 production, 35% was in the United States and 24% in the Philippines. Enormous amounts of energy are present in the core of the earth, but only at a limited number of locations (Iceland being one of them) can this energy be practically used. At these locations, steam is produced in a natural way and it is the steam that can easily be used. It is further mentioned in Refs. 81 and 286 that the production of some geothermal plants has dropped significantly over time because the geothermal heat (i.e., the hot steam) is used faster than it can be recovered by the heat in the center of the earth.

Studies have been done toward the feasibility of directly using the hot magma deep underground. A study done in 2005 concluded that the whole of the U.S. energy production could be covered for at least 500 years by drilling holes as deep as 15 km throughout the country [286].

Assuming that geothermal energy will be utilized at a large scale (which is far from certain), it will be at remote areas, thus again requiring enforcement of the transmission system.

2.8 THERMAL POWER PLANTS

The bulk production of electricity in the world today takes place in thermal power stations. The majority of power comes from large units, with sizes starting somewhere around 100 MW up to more than 1000 MW. Often, several units are located at the same site, with up to 5000 MW per site. Especially such large installations can only be connected to the transmission or subtransmission system.

The fuel used consists of fossil fuels (oil, gas, coal, and lignite) and uranium. In almost all cases, the fuel is transported, sometimes from the other side of the world. The actual location of the production unit is not restricted by the presence of the fuel and they can be built close to the consumption centers (i.e., the cities). The location of the fuel, however, does impact the location in some cases. Two examples for Great Britain are the large coal-fired power stations that were built in the 1970s in the north of England close to the coal mines and the gas-fired power stations that were built in the 1990s near the coast, closest to the North Sea gas. This is only possible due to the presence of a strong transmission grid.

TABLE 2.10 Global Electricity Production by Fuel [224]

Fuel type	1973	2007
Gas	12.1%	20.9%
Oil	24.7%	5.6%
Coal	38.3%	41.5%
Nuclear	3.3%	13.8%
Hydro	21.0%	15.6%
Total	6 661 TWh	19 771 TWh

Thermal power stations always require access to cooling water and the largest units are, therefore, located near the coast or near main rivers. This also limits the choice of locations, but still makes it possible to have them located not too far from the consumption. The need for large amounts of cooling water also implies that the production can still be impacted by the weather. For units located near a river, a limit on production is sometimes imposed during hot and dry periods when the access to cooling water is limited. These are in many countries exactly the periods when electricity consumption is highest.

The global electricity production in 2006 was 18,930 TWh, of which 66.9% was produced from fossil fuel and 14.8% from nuclear power. Thus, over 80% of electricity produced around the world comes from thermal power stations. The remaining is mainly hydropower, with 16% of the total production.

The global electricity production in 1973 and 2007, split up by the type of fuel used, is shown in Table 2.10 according to the International Energy Agency [224]. Overall, the electricity consumption has grown by a factor of ~ 3. Oil is the only fuel that has reduced in absolute numbers (from about 1500 to 1000 TWh). All other sources have increased. In terms of percentage share, there has been a shift away from oil toward nuclear and gas. The share of hydro has somewhat reduced and the share of coal has somewhat increased.

When considering the whole energy consumption (i.e., not only electricity), gas, coal, and oil make up 80.3% and waste and biomass (also a thermal source) 10.6%. The remainder is nuclear (6.5%), large hydro (2.2%), and other renewables (0.4%) [167]. Comparing this with the global electricity production for 2007 (Table 2.10), we see that nuclear and hydro have about the same share of electricity production, but in terms of primary energy, nuclear is three times as large as hydro. Both hydro and nuclear are almost exclusively used for electricity production. The reason for the difference is the inefficiency of thermal power stations. To produce 1 MWh of electricity requires 2–2.5 MWh of primary energy. Electricity production in a thermal power station is a Carnot process where the efficiency is limited by the laws of thermodynamics.

Many of the new power stations being built or scheduled are thermal power stations: gas, coal, and nuclear. Gas-and coal-fired power stations have been built for years now, with gas taking over oil as shown in Table 2.10. The last several years have also seen the reemergence of nuclear power as an acceptable option. Data given in Ref. 459 quantify this. As of April 1, 2010, there were, globally,

- 438 nuclear reactors in operation, with a total capacity of 374 GW;
- 52 units being built (51 GW);
- 143 units planning or on order (156 GW); and
- 344 units proposed (363 GW).

We will stay out of the discussion surrounding nuclear power, and will concentrate only on the possible impact of this on the power system. Nuclear power units are always large installations. The list above gives the average size of 850 MW for existing units and 1050 MW for future units. This will in most cases require an enforcement of the transmission system, especially when considering that in most cases a number of units are built close together. In Chapter 8, we will discuss the different types of reserves needed for the secure operation of the transmission system. The amount of reserves, in a system dominated by thermal units, is determined by the size of the largest production unit. Building large nuclear power generators (Finland is building a 1600 MW unit and France a 1630 MW unit [459]) will require additional reserves to be available. Nuclear power units are further known for long and often unpredictable maintenance times associated with the strict safety requirements in place. This implies that reserve capacity needs to be in place for this. Note that the average time under maintenance for nuclear power plants is similar to that for other plants. It is the occasional long maintenance time that does not significantly impact the availability statistics, but that requires some kind of reserves to secure the electricity supply. It is important to note here that in the deregulated electricity market, there is no longer any stakeholder responsible for guaranteeing the long-term availability of electricity, although some countries have allocated this task to a government agency. Nuclear power is often presented as a more reliable alternative to solar or wind power, but the reliability issues associated with nuclear power are different, not absent. From a power system operation viewpoint, nuclear like all types of thermal generation has however less impact than large amounts of wind or solar.

Other thermal power stations (gas- and coal-fired) are normally smaller than nuclear power stations. Especially modern gas-fired combined cycle units have moderate size up to a few hundred megawatts. This is most advantageous from the power system operation point of view. Coal-fired stations have somewhat the tendency of being larger and more often with several units together. One of the reasons for this is the difficulty in finding locations and the transport of the fuel. Both gas- and coal-fired power stations fit well in the existing transmission and substransmission infrastructure, but several coal-fired units together could require investments in the transmission system. New coal-fired power stations are more likely to be built close to the sea because transport of coal over land is expensive. Gas-fired power stations can be built more freely: the local environmental impact is much smaller, and the transport of natural gas via underground pipeline is easy.

Future developments may see carbon capture and storage, which would make coal-fired power plants more attractive again. According to Ref. 445, the carbon dioxide emission can be reduced by 80–90%. Some "clean coal" projects are being planned, but it remains unclear what the outlook is. According to Ref. 167 (with reference to the "Stern report" [393]), in 2025 and 2050, about 20% and 40% of the reduction in carbon dioxide emission in the United Kingdom will come from carbon capture and storage. A 30 MW pilot installation is in operation since September 2008, with a lignite-fired power plant in Germany [420]. A pilot for precombustion capture of carbon dioxide will be started in August 2010 with the coal gasification plant in Buggenum in the Netherlands. Several other installations are being built or planned.

The longer term future may see the emergence of nuclear fusion plants. This may become the unlimited source of energy of the future. At the time of writing this, two competing projects (one European, one American) are expected to reach the breakeven point (where the energy production is the same as the amount of energy needed to create the nuclear reaction) within a few years. There remains, however, great doubt among other researchers if this point will actually be reached [286, 376]. It also remains unclear how a nuclear fusion power station will look like, as many engineering challenges remain unsolved [307]. If these challenges are all solved however, there will be enough fuel in the oceans (in the form of Deuterium) to "supply every person in a ten-fold increased world population with 100 times the average US consumption for one million years" [286]. The possible impact of nuclear fusion on the power system is as yet unknown. Large-scale use of nuclear fusion for power generation remains at least 25 years away.

2.9 INTERFACE WITH THE GRID

The connection point of an energy source to the grid is usually referred to as the point of connection (PCC). Its definition depends on the ownership and utility interconnect requirements. Two different possible definitions are shown in Figure 2.59, where typically the interconnection relay (protection) is installed at the PCC [310]. It will be assumed here that the PCC is the connection point before the interconnection

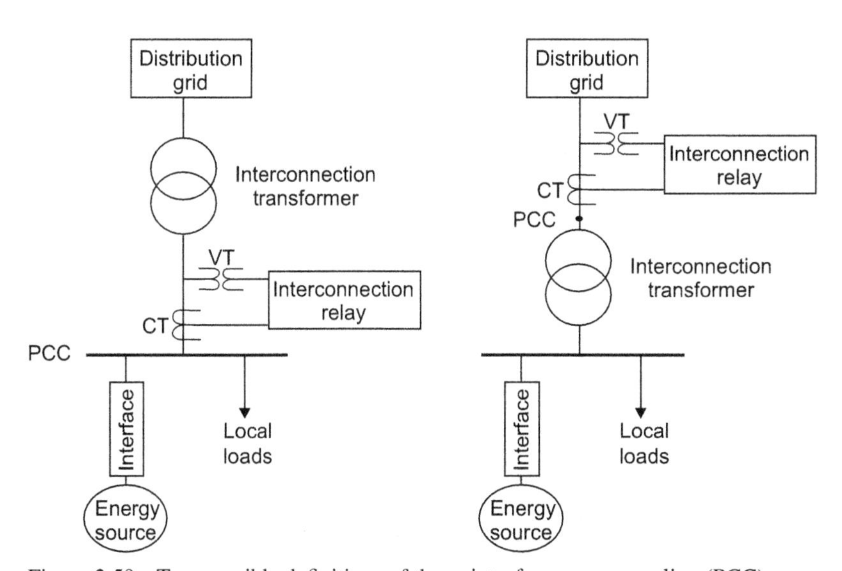

Figure 2.59 Two possible definitions of the point-of-common coupling (PCC).

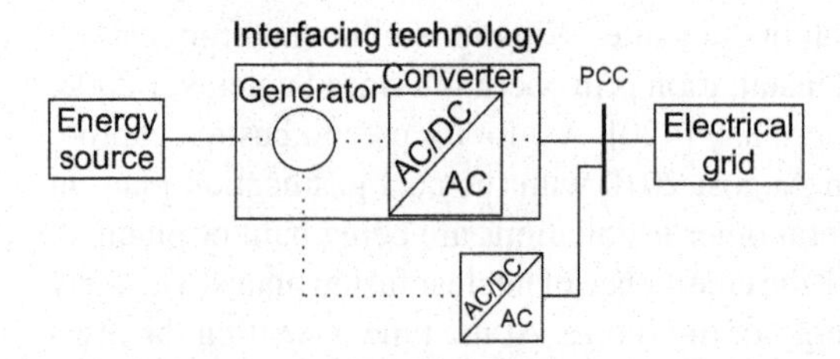

Figure 2.60 Interfacing of energy sources with the grid.

transformer (the left-hand schematic of the figure). The technology that connects an energy source to the PCC, which is referred to as the interfacing technology, may comprise another transformer.

There are different interfacing technologies used; in common is the use of generators and power electronics converters, as shown in Figure 2.60. The main goal of the interfacing technology is to accommodate the energy produced to the grid requirements. Depending on the nature of their produced power, different energy sources use different technologies to interface/couple with the grid, as reported in Table 2.11.

The interfacing technologies are classified here into four categories: direct machine coupling, full power electronics coupling, partial power electronics coupling, and modular or distributed power electronics coupling. The different grid couplings are explained below in more detail.

2.9.1 Direct Machine Coupling with the Grid

In general, the sources that could be directly coupled to the grid are those that originally produce mechanical power, such as wind power, small hydropower, and some CHP. It is then more efficient to transfer the mechanical power into electrical power through direct machine coupling to the grid without any intermediate stage.

TABLE 2.11 Interfacing Technologies for Different Energy Sources

Energy source type	Source of energy	Electrical generator	Power electronics
Wind power	Wind	SG, PMSG, IG, DFIG	Optional, AC/AC
Hydropower	Water	SG	N/A
Fuel cell (CHP)	Hydrogen	N/A	DC/AC
Biomass (CHP)	Biomass	SG, IG	N/A
Microturbines (CHP)	Diesel or gas	SG, IG	Optional, AC/AC
Photovoltaic (solar power)	Sun	N/A	DC/AC
Solar thermal (solar power)	Sun	IG	N/A
Wave power	Ocean	LSG	AC/AC
Flow of river (small hydro)	Rivers	PMSG	AC/AC
Geothermal	Earth temperature	SG, IG	No

SG, synchronous generator; PMSG, permanent magnet synchronous generator; IG, induction generator; DFIG, double-fed induction generator; N/A; not applicable; LSG, linear synchronous generator.

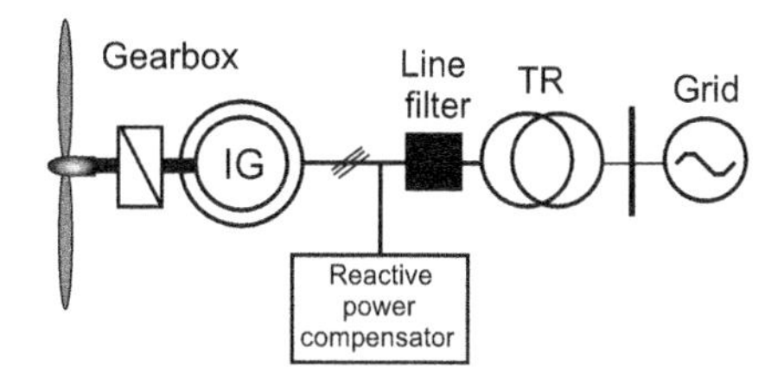

Figure 2.61 Direct induction generator coupling for a wind turbine.

The choice of the type of the machine depends on the nature of the mechanical power supplied. For a constant mechanical power, resulting in constant rotating shaft speed, the synchronous machine is a proper candidate, whereas for strongly variable power, the induction machine is more suitable. This is mainly because of the inherent oscillations damping that the induction generator provide through the slip speed difference between the rotor and stator. Permanent magnet synchronous machines are utilized for slow rotational speed turbines. A linear synchronous machine is used to transfer vibrational energy into electrical energy.

The synchronous machine is used to interface small hydropower and some CHP, where the mechanical power is controlled through controlling the speed of the shaft. Synchronous generators can be a source of both active and reactive powers to the electrical system. They typically require synchronizing equipment to be paralleled with the grid.

The induction machine is used to interface variable speed energy sources with the grid. The clear example of this is wind energy sources, connected as shown in Figure 2.61. But they generally suffer from the disadvantage of drawing reactive power from the system and also from drawing high starting currents that may cause local flicker. Connection to the network hence requires an extra reactive power compensator, as shown in the figure, which could be basically a capacitor bank. Hence, induction machines for direct connection of DGs are generally limited in size [264, 310]. They are restricted because of their excitation that requires reactive power supplied basically from the grid. On the other hand, induction generators are like induction motors, requiring no synchronizing equipment. They are less costly than synchronous generators and the protection required is only over/undervoltage and frequency relaying. For wind power and low-head run-of-river turbines (small hydro), the rotational speed of the turbine is usually slow and hence the interfacing need either a gearbox or a large multiple pole generator for energy conversion, for example a permenant magnet synchronous generator (PMSG).

2.9.2 Full Power Electronics Coupling with the Grid

The main task of the power electronics interface is to condition the energy supplied by the DG to match the grid requirements and to improve the performance of the energy source. Power electronics equipment has the capability to convert power from one form to another by using controlled electronic switches, and hence called power electronics converters (or converters in short). For instance, they are used to convert direct current (DC) power to alternating current (AC) power that matches the grid requirements. Such converters are referred to as DC/AC converters. With DC energy sources, the power electronics interface may consist of one DC/AC converter or

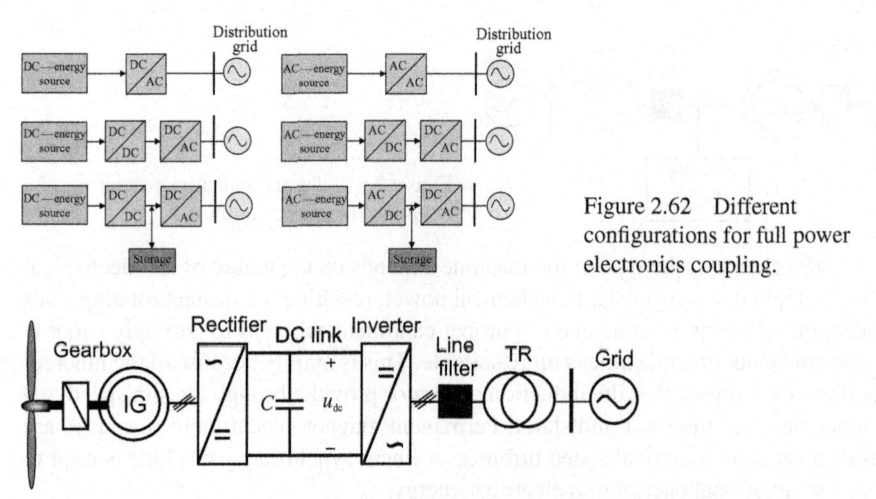

Figure 2.62 Different configurations for full power electronics coupling.

Figure 2.63 Full power electronics interfaced wind turbine with induction generator.

an intermediate DC/DC conversion stage can be added to achieve a specific goal, for example, in order to regulate the output voltage so that the maximum available power is extracted as in the case of PV systems. The same is valid for AC sources, where they have to be adjusted to match the grid requirements by a DC intermediate stage (see Figure 2.62). Energy storage could also be connected in the DC stage, called the DC link, to adjust the energy injected into the utility grid in all operating conditions.

The arrangement that consists of AC/DC and DC/AC converters is also referred to as frequency converter, since it connects two different AC systems with possibly two different frequencies together. The frequency converter is in a back-to-back configuration, when the DC link is directly connecting the two converters together as shown in Figure 2.63, or an HVDC/MVDC system configuration when there is a transmission DC cable in the DC link for either HV or MV applications, as shown in Figure 2.64.

The back-to-back configuration in Figure 2.63 is used to interface a variable speed wind turbine with the grid. The aerodynamic power of the turbine is set through the pitch, stall, or active stall control. The pitch control is the most popular type for the recent generations of wind turbines, where the pitch angle is dynamically adjusted in order to capture the maximum wind energy below rated wind speed. The aerodynamic power is then transformed through a gearbox and an induction generator (IG) to electrical power. This electrical power is then adjusted using the frequency converter

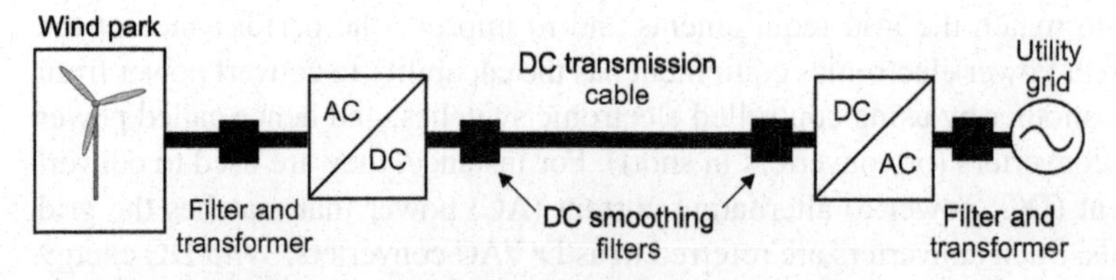

Figure 2.64 DC power transmission for a wind farm.

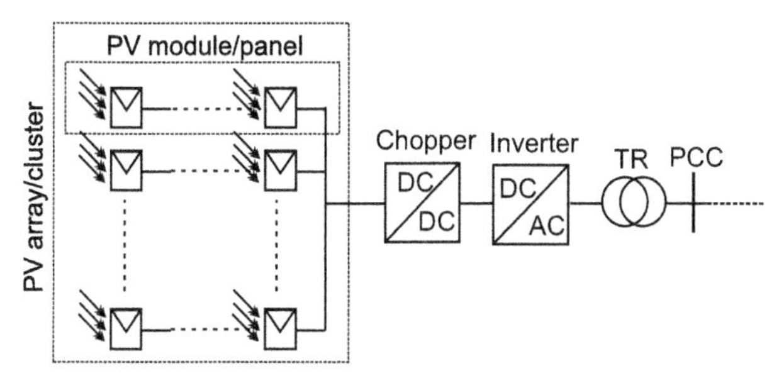

Figure 2.65 Conventional photovoltaic array interface to the grid through central conversion.

to match the grid requirements. An equivalent setup can also be found in literature, where a multipole permanent magnet synchronous generator (PMSG) replaces the gearbox and IG set [190, 191, 434]. This gearless setup results in reduced acoustic noise and mechanical losses. Also, the PMSG inherently has higher efficiency than the IG and does not require an external excitation. Moreover, the PMSG can be driven over a wider range of operating speeds [434]. Still, different design aspects are under research for such setup [173, 190, 191, 343, 434].

Another example of the full power electronics interface is the photovoltaic distributed generation shown in Figure 2.65, where central conversion is applied to extract the maximum power of the photovoltaic array and inject it to the grid. The task of the DC/DC converter, as mentioned before, is to adjust the operating voltage of the PV panel so that maximum power is extracted at different temperatures (or irradiation levels) as has been described previously in Section 2.2.4.

2.9.3 Partial Power Electronics Coupling to the Grid

Other arrangements of power electronics coupling to the grid may implement smaller sizes of the converter, where the converter is rated for a certain percentage of the DG apparent power—hence, referred to as partial power electronics coupling. Two such systems are shown in Figures 2.66 and 2.67.

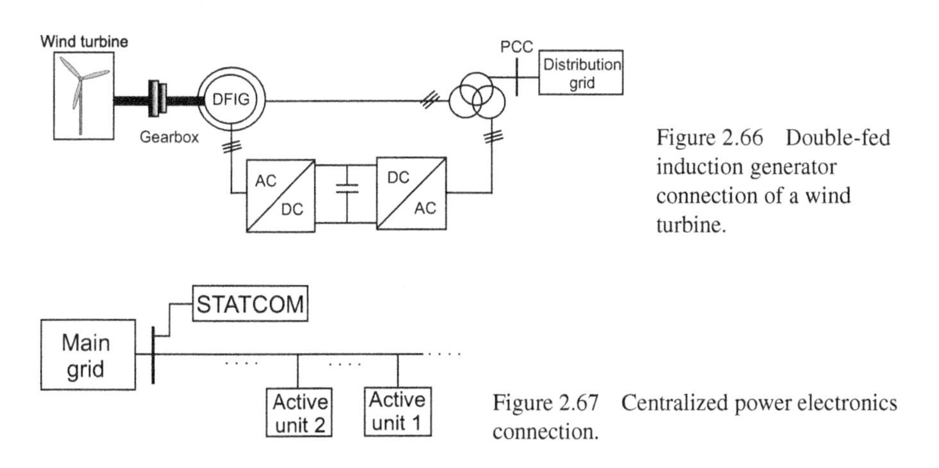

Figure 2.66 Double-fed induction generator connection of a wind turbine.

Figure 2.67 Centralized power electronics connection.

In a variable speed wind turbine with double-fed induction generator, the converter feeds the rotor winding, while the stator winding is connected directly to the grid. This converter setup, through decoupling mechanical and electrical frequency and making variable speed operation possible, can vary the electrical rotor frequency. This turbine cannot operate in the full range from zero to the rated speed, but the speed range is quite sufficient. This limited speed range is caused by the fact that a converter considerably smaller than the rated power of the machine is used. In principle, the ratio between the size of the converter and the wind turbine rating is half of the rotor speed span [264]. In addition to the fact that the converter is smaller, the losses are also lower and the control possibilities of the reactive power are similar to the full power converter system.

A partially rated power electronics converter at the connection point of a wind farm (or any other aggregate of sources) is usually needed to mainly provide a voltage dip ride-through capability, which is a required feature regarding different grid codes, and possible reactive power support. Usually a STATCOM (static var compensator), which is mainly a voltage source converter (VSC), is used for this purpose as shown in Figure 2.67.

The voltage source converter is the main enabling technology to interface energy sources at the front end to the grid for either full or partial power electronics interfaces. Its basic structural feature is a capacitor at the DC link and the use of self-commutating switches to transfer the energy to the grid in the appropriate form. To get a closer idea about how a VSC works, a simple single-phase transistor bridge is considered, as shown in Figure 2.68. The switches s1–s4 are represented using insulated gate bipolar transistors (IGBTs) that are commonly used in medium-voltage applications. An IGBT is a self-commutated switch; meaning that it can turn on and off on full current, which facilitates the use of high-frequency switching that produces better waveforms regarding the harmonics content. In the figure, the switches s1 and s2 are turned on at the same time, while switches s3 and s4 are off resulting in positive DC voltage V_{dc} appearing on the AC side. After this the switching states are reversed, meaning that s_1 and s_2 are off while s_3 and s_4 are on, resulting in a negative DC voltage $-V_{dc}$ appearing on the AC side. The output voltage v_o is then continuously varied between the positive and negative of the magnitude of the DC voltage.

The VSC is then connected to the grid through an inductance filter (L-filter) and a transformer, as shown in Figure 2.69 where L refers to the summation of the L-filter and leakage inductance of the transformer. To transfer the energy from the energy source to the grid, the VSC is controlled in such a way so as to produce a phase

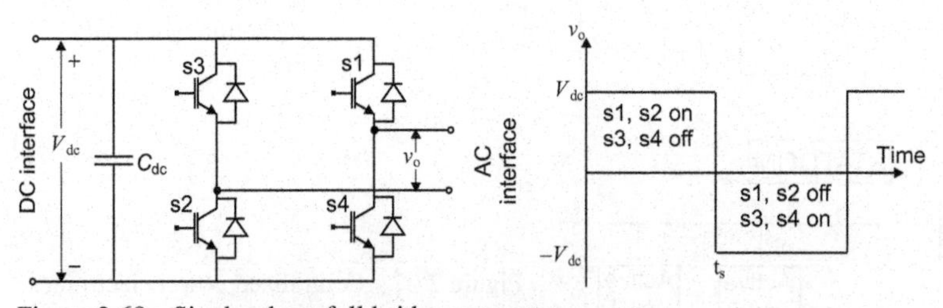

Figure 2.68 Single-phase full bridge converter.

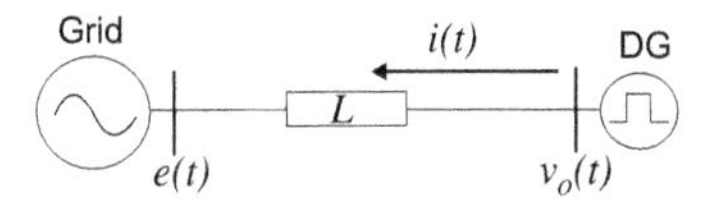

Figure 2.69 Connection of VSC DG to grid.

shift between its output voltage and the grid voltage. Since the VSC output voltage amplitude is generally maintained constant, the active power transfer is controlled through the injection of current into the grid. Assume the grid voltage is purely sinusoidal with an amplitude E:

$$e(t) = E\sin(\omega t) \tag{2.18}$$

where ω is the angular frequency of the grid voltage. The output voltage of the VSC is

$$v_{\mathrm{o}}(t) = V_{\mathrm{o}}\sin(\omega t + \delta) + \Sigma v_{\mathrm{h}} \tag{2.19}$$

where δ is the phase difference between the converter output voltage and the grid voltage and Σv_{h} is the summation of the output voltage harmonic components. The current flowing from the DG to the grid can be expressed as a summation of a steady-state component and a transient component:

$$i(t) = \frac{\int v_{\mathrm{o}}(t) - e(t)}{L}\mathrm{d}t = \frac{\int V_{\mathrm{o}}\sin(\omega t + \delta) - E\sin(\omega t)}{L} + \frac{\int \Sigma v_h}{L} = i_1(t) + i_h(t) \tag{2.20}$$

where $i_1(t)$ is the current fundamental component and $i_{\mathrm{h}}(t)$ is the current higher harmonic components.

From the above equations, and assuming $V_{\mathrm{o}} = E = 1$ per unit, the fundamental component of the current is zero if $\delta = 0$. However, the harmonic components of the current will still exist since they depend on the converter voltage harmonics. When δ assumes a value, a current will start to flow into the grid having a fundamental component with superimposed harmonic components. This is shown in Figure 2.70; the current (or active power) reference to the converter controller has a step from zero to a positive value at $t1$. The controller reacts by setting a small phase shift between the converter output voltage, solid line in the figure, and the grid voltage, dashed line in the figure, that results in the required current step. Important to observe is that the phase angle between the DG voltage and the current is zero, indicating unity power factor operation. This is generally how the available output power of the energy source is injected into the grid.

The power electronics interface also has the capability of injecting reactive power into the grid, and hence compensating for different power quality locally at the PCC. An example is shown in Figure 2.71, where a reactive power reference is given to the controller at $t2$. In this case, a small decrease in the voltage amplitude will be achieved resulting in a change of the phase and amplitude of the DG injected current. This change is directly related to the reactive power injected into the grid. Note that this operation is not a unity power factor operation any more, and is generally not allowed by most of the DSOs. However, it could be considered in the future in order to offer some ancillary services to the grid, as will be discussed in next chapters.

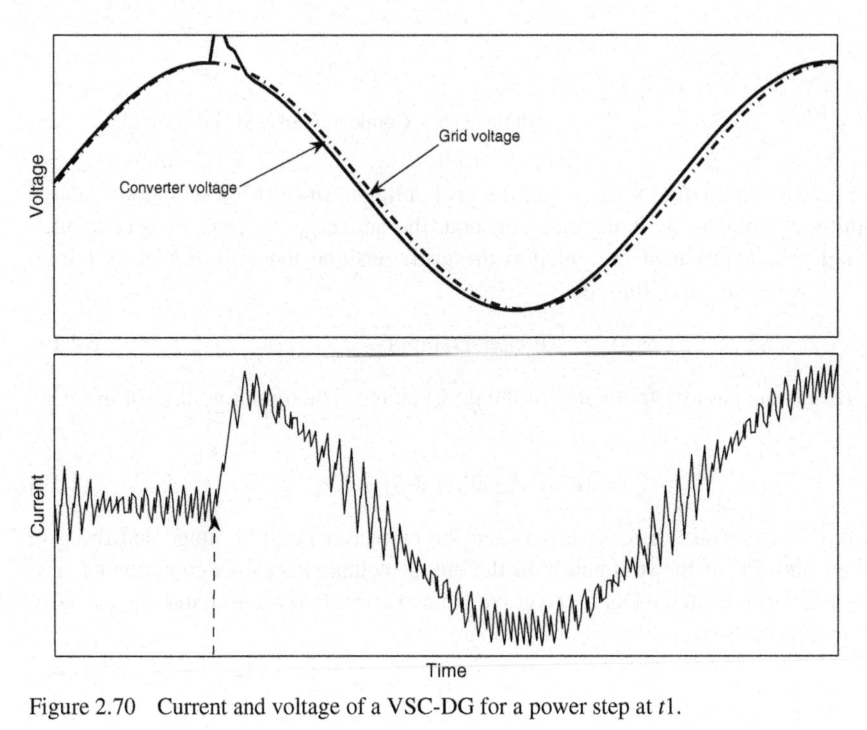

Figure 2.70 Current and voltage of a VSC-DG for a power step at $t1$.

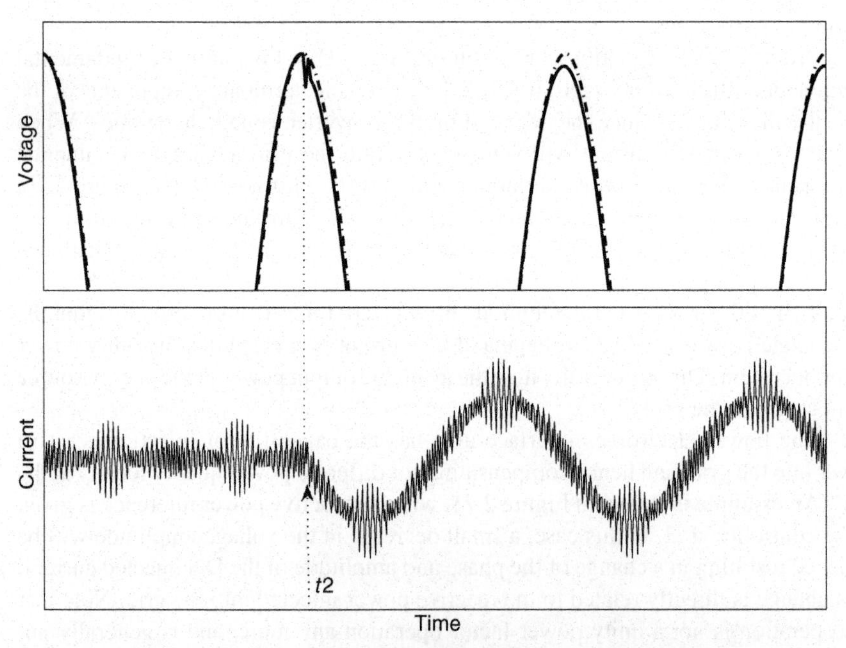

Figure 2.71 Current and voltage-tops of a VSC-DG for a reactive power step at $t2$.

Two important points are of interest for a proper operation of full power electronics-interfaced DGs: the detection of the grid voltage angle and the minimization of the injected current harmonics. The first point is important for the controller to produce the proper switching commands that would result in the required transfer of power. It is done through adopting proper control algorithms (referred to as phase-locked loop algorithms), which use the grid voltage measurement to extract the instantaneous voltage-phase angle. The second point is important in order to provide a good power quality. Mainly, there are two ways to achieve this: either to use a higher order filter (for instance an LC-filter) or to implement control techniques and converter topologies that produce output voltages that are more closer to the sine wave shape (for instance, using multilevel converters). There are different control techniques that are developed for this purpose, yet they are out of the scope of this book and hence will not be further discussed.

2.9.4 Distributed Power Electronics Interface

Distributed power electronics interfaces refer in this context to a number of distributed generation that are connected to the same local grid through power electronics converters. If such units belong to the same owner, their operation can be coordinated in a way to achieve certain benefits such as regulating the local voltage as will be discussed in more detail in Chapter 4.

The module-integrated photovoltaic system, shown in Figure 2.72, is also a type of distributed active interface structure that has been basically developed in order to increase the efficiency and reliability of the solar power cells. This is possible since different solar cells in an array or a cluster are exposed to different irradiation. Hence, by operating each integrated converter at a different point that is related to the MPP results in better reliability compared to using a central conversion. It has been found that using this structure increases the power efficiency to about 96% in addition to providing a better power quality [431].

Another example is a wind farm with full power electronics-interfaced wind turbines. Implementing a proper control architecture that makes use of the distributed controllability of the converters makes such setup reliable, reconfigurable, and self-healing. A possible distributed control architecture is shown in Figure 2.73. As shown

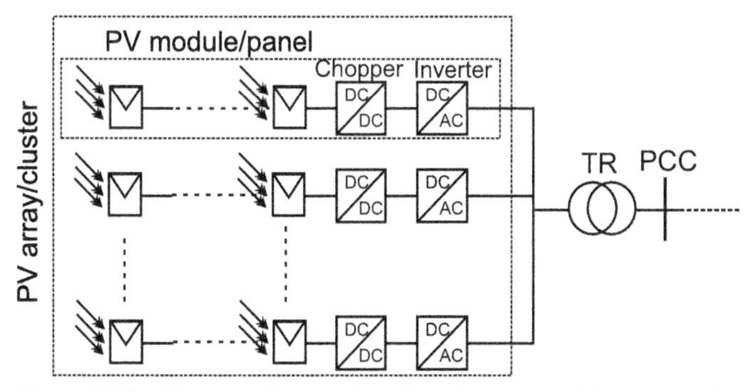

Figure 2.72 Photovoltaic array interface through modular conversion.

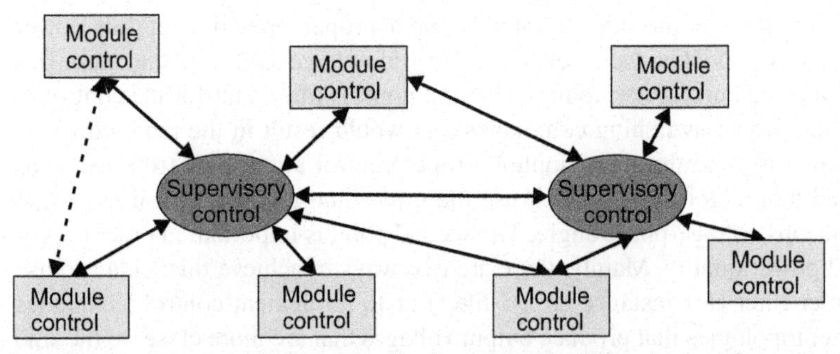

Figure 2.73 Distributed control for distributed power electronics interfaces. A module refers to a converter. The dashed line refers to possible communication between modules.

in the figure, the supervisory control can be divided into a number of controllers. A module control is connected to two or more supervisory controllers so that if one fails, it will be directed to the other one. Using this concept, the module control may have its primary signal from a supervisory controller and a backup signal from another supervisory controller in a scenario that provides a reliable, self-organizing, and self-healing control structure. This control structure additionally supports the plug and play operation, where the setup can be upgraded through plugging in a new module that instantly communicates with a supervisory controller and other modules in order to contribute to the overall control function.

2.9.5 Impact of the Type of Interface on the Power System

The interfacing technology of distributed generation to the grid determines its impact on the distribution system operation and also the impact of different upstream contingencies on the energy sources. The effect of the slow changes on the power production of the DGs as well as of the fast changes produced, for instance, when using power electronics interfaces is discussed in more detail in the following chapters. Briefly, the interfacing technology impacts the system in different ways:

- It determines the quality of the power supplied by the energy source. Usually the grid operator (DSO) sets certain requirements over the power quality injected by the DG. It is the DG owner responsibility to ensure that the DG technology meets the DSO requirements.
- It determines the impact of different power quality on the energy source. Different interfaces have different sensitivities to different power quality events at the grid.
- It determines the efficiency of the energy source. For example, using modular power electronics for solar power interfacing results in more efficient systems compared to central full power electronics interfacing.
- It determines the availability of the energy source. This is again determined by the sensitivity of different interfaces to the different power quality phenomena originated from the upstream network.

TABLE 2.12 General Comparison of Interfacing Technology

Interfacing technology	Controllability	Robustness	Efficiency	Cost
Induction generator	−	−	+	−
Synchronous generator	+	+	++	+
Partial power electronics	++	−	+	++
Full-power electronics	+++	−	−	+++
Modular or distributed power electronics	++++	+	+++	++

"−" for Less and "+" for More.

- It impacts the cost of the DG installation.
- It determines the complexity of the DG application.
- It determines the controllability of the energy source. For instance, full power electronics interfaces introduce the most controllability of all interfacing technologies since they provide independent control over active and reactive powers injected into the grid.

Different interfacing technologies are compared in Table 2.12 regarding their controllability. As a rule of thumb, a trade-off must always be made between the high performance of the interface and the overall cost of the installation.

The impact of distributed generation, in general, will be treated in more detail regarding different phenomena at the grid in Chapters 4–9.

2.9.6 Local Control of Distributed Generation

Defining the local controllability of DGs is an important aspect that needs to be considered to facilitate the integration of new generation units and optimize the operation of the distribution system. If their local controllability is available for the grid, this will optimize the number of the grid assets and increase the operational efficiency of the distribution grid. Different distributed generation technologies provide different control capabilities over both the active and reactive powers. These capabilities can be utilized in many ways in order to increase the amount of DG integration into the distribution system, referred to as the hosting capacity as explained in the next chapter, and to facilitate the implementation of plug and produce. Generally, the active power output of a DG can be controlled in four different ways [410]:

1. *To extract maximum power* Wind and solar power units are usually operated under this criterion in order to increase their efficiencies. In the periods where the outputs of such units exceed the grid demand, the use of energy storage would be necessary. The storage unit operation can also be ruled by the energy prices; for instance, charges when prices are low and discharges when prices are high.
2. *To limit maximum power injected* The curtailment of the maximum power injected is required by some grid codes in contingency situations. This could

be met, for instance, as above, by using storage or through the control of the generation unit. Big generation units can be operated using this criterion, which requires these units to be dispatchable. This means that they receive a control command from a centralized grid controller, also referred to as supervisory control.

3. *To keep a production threshold* Beyond a certain available production power, a local control can be set in order to reduce the actual production with a certain value. The difference between the actual production and the available power can then be considered as a spinning reserve for network management.
4. *To set the production profile according to the balancing requirements of the grid* The supervisory control can also be used to set the production profile of the dispatchable (big) production units for better grid management. CHP units are more likely to be dispatchable, since they have the highest running cost due to the use of gas as an energy source.

The control of reactive power is also possible, though a fixed reactive power operation is desirable regarding current grid codes. However, this might change in the future, where a source may be utilized mainly for controlling the reactive power instead of producing active power. Power electronics-interfaced sources are enabling technologies for such a scenario.

At present, virtually most of the distributed generation are connected to low-voltage distribution networks and not to high-voltage transmission grids. Individually most of these units are not subject to central system control, partly because of the high cost of conventional telemetry, meaning that in the United Kingdom, for example, system control and data acquisition (SCADA) has only low penetration at voltages below 132 kV [264]. At these levels, therefore, distribution networks operate largely with incomplete and uncertain information. Generally, any network problems caused by renewable DGs arise because of the intermittent nature of their output or the uncertainty of information at the PCC, and not particularly because they are renewable energy sources.

In order to better utilize the DGs capabilities, facilitate their integration, and lessen their negative impact on the grid, introducing different operational scenarios is vital. In one scenario, different units with different control criteria can be combined into one vital source (also referred to as a cell) where the total power exchange command between this energy source and the distribution system is set through a supervisory controller. An internal controller is then used to manage the energy within the cell according to certain operation and economic cost functions in order to optimize the cell operation. Adding a new generation unit to a cell would change/update its operational cost function and may add some controllability to the cell to be utilized by the supervisory controller.

In order to automate this process, the use of communication and information technology between DGs and the energy management system becomes necessary. This indeed is an important aspect envisioning the future grid, which is referred to as the smart grid [314]. There are currently a standardization process for setting information models for different distributed generation units. The following standardization efforts are acknowledged:

- **IEEE 1451** Standards for plug-and-play sensors, where communication protocols, and information models and scenarios are discussed.
- **IEC 61850-7-420** Includes information models for four DER types: photovoltaic power systems, fuel cell power plants, reciprocating engine power systems, and CHP systems.
- **IEC 61400-25** Sets information models for monitoring and control of wind power plants.
- **IEC 61970** Sets common information models for energy management systems.

Such standardization efforts are important in order to provide information infrastructures that facilitate the connection of any sources regardless of its vendors or type, which would eventually facilitate a plug-and-produce scenario.

Such operational scenario has the potential to increase the amount of distributed generation units connected to the distribution grids or the hosting capacity, which is explained in the next chapter.

CHAPTER 3

POWER SYSTEM PERFORMANCE

3.1 IMPACT OF DISTRIBUTED GENERATION ON THE POWER SYSTEM

3.1.1 Changes Taking Place

The introduction of distributed generation is part of a number of changes that take place in the power system at the same time. Most countries in the world are impacted by this, although the different changes do not occur everywhere at the same time. The main changes that are taking place, that have taken place, or that will most probably be taking place are the following ones:

- Electricity generated from fossil fuel is replaced by electricity generated from renewable sources of energy. Even when new thermal generators or large hydropower units are built, their location is rarely decided by what is best for the power system. These shifts are discussed in Chapter 2.
- Large generator units connected to the transmission system are replaced by small units connected to the distribution system. The terms "distributed generation" and "embedded generation" are both used. Neither of the terms fully describes the specific differences with "conventional generation." The term "distributed energy resources" has also appeared a few years ago.
- The "vertically integrated utility" (with generation, transmission, and distribution having one owner) is split up. The split is different in different countries, but the typical separation is between generation, transmission, and distribution. The generation market is open to competition. The transmission and distribution operators have a natural monopoly, with oversight by a regulatory body.
- The generation assets are no longer owned by one or a few owners, instead many new players have entered the electricity market. The technology is available for domestic customers to generate their own electricity using rooftop solar panels or a small combined heat-and-power unit in the basement. The investment will not always be profitable but the risk is affordable for a large part of the domestic customers. Legal and administrative barriers exist in many countries but even these are being removed.

Integration of Distributed Generation in the Power System, First Edition. Math Bollen and Fainan Hassan.

3.1.2 Impact of the Changes

All these changes will have their impact on the power system: the perceived and actual impacts are being discussed at many platforms, not only in scientific papers and books (like this one) but also on the Internet, in newspapers, and at a range of meetings. It is difficult to get an overview of all the impacts because the generation, the network, and even the consumption change all the time. Further, will the different changes impact each other. Examples of changes in the network are the replacement of overhead lines by underground cables and the building of HVDC lines. Examples of reduction in consumption are the replacement of incandescent lamps by compact fluorescent lamps and the replacement of electric heating by heat pumps. Examples of increase in electricity consumption are electric heating (instead of oil), high-speed trains (instead of airplanes), and electric cars (instead of gasoline-fired cars).

Probably the most important impact of the deregulation of the electricity market is that there is no longer any single organization that has control over the location and production of the individual generator stations. Even when we consider only the large conventional power units, their production is determined primarily by the market. The system operator is only allowed to intervene in the market when the resulting power flows would endanger the operational security of the market. Also, the location of new power stations is determined by the owners of those stations themselves. The network operator has to be informed and can set limits to the connection, but it cannot control the developments.

The introduction of smaller generator units makes the situation even more complicated. Those units are often completely out of control of the system operator. With large generator units, the system operator is at least aware of their existence and of the intended amount of production. The system operator is also, in most cases, able to change the production of large generation units when their production level would otherwise endanger the operational security of the system. We say that those units are "dispatchable": they can be turned on or off on demand. Smaller units, distributed generation, cannot be dispatched in this way. The system operator will simply have to accept their production whatever it is. This type of generation is sometimes referred to as "negative load," but this is again an oversimplification of reality. With renewable sources of energy, the production is further strongly fluctuating with time and difficult to predict. This makes the planning and operation of the power system even more difficult. The fluctuations and the difficulty to predict occur at different timescales for different types of energy sources.

The large conventional power stations are connected to the transmission or subtransmission network. These networks are designed for power transport in both directions. Distribution systems are, however, designed for power transport in only one direction: toward the consumers. As we will see in the forthcoming chapters, this will cause a number of problems, some more severe than others.

The cost of protection and control equipment is rather independent of the size of a generator unit: an underfrequency relay for a 1 kW microturbine costs the same as an underfrequency relay for a 1000 MW nuclear power station. In the latter case, the costs are negligible, whereas in the former case they can make the difference between profit and loss for the owner of the unit. The result is that the owners of small units will be less willing to install anything that is not absolutely necessary for producing electrical energy. The owner of a small unit will do an economic optimization for his own pocket. Some network operators introduce strict connection rules, including requirements on protection and metering. This can, however, easily result in an unnecessary barrier to an increased use of renewable energy.

3.1.3 How Severe Is This?

Now all this should be put into perspective. The introduction of distributed generation will not immediately result in a complete collapse of the power supply. Fluctuations and uncertainty also occur in the consumption of electricity. The consumption at national level can differ by a factor of 2–3 between the highest and lowest annual values. The general consumption pattern can be easily predicted with its daily, weekly, and seasonal variations. But prediction errors of hundreds of MW are not uncommon. Both winter peaks (in a cold climate) and summer peaks (in a hot climate) are strongly weather dependent and thus equally difficult to predict as the production by solar or wind power.

Hydropower, a source that has been integrated into the power system very successfully, is prone to variations and prediction errors as well, but at a timescale completely different from that for other renewables. The differences between a dry year and a wet year can be huge, in terms of TWh production available. The power system has to be designed to cope with this because nobody can predict the amount of rain or snowfall months into the future. There has to be, for example, thermal capacity available to cover the consumption in case there is not sufficient water in the reservoirs.

The power system further has to be designed and operated in such a way that it can cope with the loss of any large production unit. The design and operation of the transmission system is based on the so-called "dimensioning failure," which typically refers to the loss of the largest production unit or the loss of two large units shortly after each other. The size of this dimensional failure is typically between 1000 and 2000 MW. This is in fact an uncertainty the system operators have managed to cope with for many years: the amount of available generation might suddenly drop by up to 2000 MW for large transmission system. Such a system is therefore also able to cope with uncertainties in production from new sources up to 2000 MW.

The presence of new sources of generation at remote locations is often mentioned as a problem, for example, with wind power. But the existing power stations are also being built further and further away from the main areas of consumption. In a future energy scenario based on coal, large power stations in Europe will more likely be built near the coast where coal can be transported more easily and where cooling water is most easily available. In the same way, gas-fired power stations in the United Kingdom were built mainly near the coast, where the gas was available, but far away from the main consumption areas. New nuclear power stations will be built near the coast or near large rivers, but certainly away from the population centers.

It is difficult to decide what the main challenges are with the connection of large amounts of distributed generation to the power system. It strongly depends on the local properties of the distribution system, on the properties of the energy source, and on the kind of interface used. The potential adverse impacts at distribution level include the following:

- Overload of feeders and transformers could occur due to large amounts of production during periods of low consumption. This is discussed in Chapter 4.
- The risk of overvoltages increases due to production at remote parts of a distribution feeder. This could in theory already occur before the production exceeds the consumption. In some cases, but those are likely to be rare, undervoltages may occur due to production on another feeder. Voltage magnitude variations are discussed in Chapter 5.
- Due to the introduction of distributed generation, the level of power quality disturbances may increase beyond what is acceptable for other customers. The impact of distributed generation on power quality is discussed in detail in Chapter 6.
- Incorrect operation of the protection is another potential consequence of large amounts of distributed generation. This could be failure to operate or unwanted operation. The potential failure to detect uncontrolled island operation is the most discussed impact on the protection. Protection is discussed in Chapter 7.

For the transmission system operator, the increase in uncertainty due to fluctuating and unpredictable power sources is one of the main concerns. Although there is already a serious level of uncertainty in the existing power system, the introduction of new sources of generation will introduce more and new types of uncertainty. Uncertainties occur over a range of timescales: from the short notice time for the connection of new installations to prediction errors at a timescale of minutes or even seconds. Changes in weather, important for solar power and wind power, occur at a timescale of hours to days. With large penetration of wind power, a passing weather front could result in the loss of thousand of MW of production within a few hours. For a system that is not designed to cope with a loss of more than, for example, 1500 MW, this could be a serious problem. Finally, the behavior of small or novel generators during large disturbances is very much unknown. This could make it difficult for the network operator to operate the system in a secure way. The impact of distributed generation on the operation of the transmission system is discussed in detail in Chapter 8.

3.2 AIMS OF THE POWER SYSTEM

From the previous sections, we can conclude that the introduction of sufficiently large amounts of distributed generation will reduce the performance of the power system. A serious discussion on this is possible only when there are ways to quantify the performance of the power system. To quantify the performance of the power system, it is essential to go back to the basics: what is the actual aim of the power system?

When considering the aims of the power system, it is very important to distinguish between what we call "primary aims" and the "secondary aims." The primary aims are related to the customer: reliability of supply, voltage quality, and the tariffs. The secondary aims are internal aims that are set by the network operator to achieve its primary aims. Examples are preventing of component overload, correct operation of the protection, current quality (i.e., limiting the emission of disturbances by equipment or customers), operational security, and costs. Both primary and secondary aims are important and should be considered when studying the impact of new sources of generation.

It is not to say that primary aims are more important than the secondary aims. In fact, the system and network operators will mainly consider the secondary aims. In a well-designed system, fulfilling the secondary aims will automatically result in the primary aims being fulfilled. But what matters in the end to the customers are the primary aims of the system: reliability, quality, and tariffs. All the rest is "just engineering." This distinction is very important because the introduction of new sources of generation may require a change in secondary aims; the primary aims will, however, always have to be fulfilled.

When we start with the primary aims, we can see that they cannot be always fulfilled. The tariff is always higher than the customers would like, interruptions do occur, and perfect voltage quality is not feasible either. Instead, so-called "performance indicators" have to be introduced and the performance of the power system has to be measured against these indicators. As far as the tariffs are concerned, the actual tariffs are the indicators. A lot of work has been done on the introduction of suitable indicators for quality and reliability. For reliability, it is the number and duration of interruptions that matters. Several reliability indices are defined in IEEE Standard 1366 [219]. For voltage quality a whole range of indicators have been defined, some of which we will come across in the forthcoming chapters. An overview of voltage quality indices is also given in Ref. 78.

We can define "performance indicators" for secondary aims as well, for example, the maximum current through a transformer or the probability of fail-to-trip for a protection relay. We will come across them in the forthcoming chapters.

3.3 HOSTING CAPACITY APPROACH

Introduction of distributed generation will impact the performance of the power system. In fact, every change in the generation or in the load will impact the performance of the power system. For a fair assessment of the impact, which in turn is a basis for discussion on measures to be taken, the above-mentioned performance indicators are needed. The impact of distributed generation can be quantified only by using a set of performance indicators.

In some cases, the performance of the system will improve after the connection of distributed generation; in other cases, it will deteriorate. It is the latter cases that have received the most attention. A deterioration of the supply is not directly a concern, as long as the resulting quality is within an acceptable range. In other words, for every performance index there exists a limit that should not be exceeded. This is shown schematically in Figure 3.1.

The performance index decreases with increasing amount of distributed generation, in this example. For small amounts of distributed generation, the index remains above the limit and the system performance is still acceptable. But for larger amounts of distributed generation, the performance becomes unacceptable. The "hosting ca-

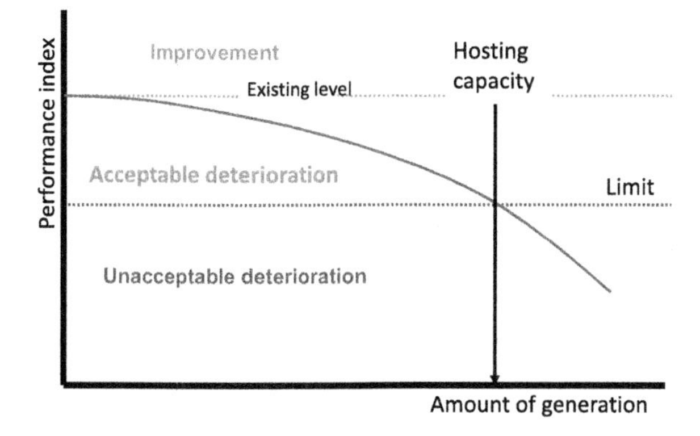

Figure 3.1 Hosting capacity approach; performance index is high in ideal case.

pacity" is defined as the amount of distributed generation for which the performance becomes unacceptable.

In Figure 3.1, the performance index was defined such that a higher value corresponds to a better performance. In practice, many performance indices are defined such that a lower value corresponds to a better performance. This is illustrated in Figure 3.2. The deteriorating performance with increasing amount of distributed generation corresponds to the rising level of the performance index.

An example is the maximum voltage magnitude experienced by any of the costumers in a distribution network. Adding generation to the feeder will result in an increase for this maximum voltage magnitude. For small amounts of generation the increase will be small, but for larger amounts of generation the voltage could become unacceptably high. In that case, the hosting capacity has been exceeded. The increase in voltage magnitude due to distributed generation will be discussed in detail in Chapter 5.

There are also cases in which the introduction of distributed generation will initially result in an increase in performance of the power system. But for large amount of distributed generation, the performance will deteriorate. This is illustrated in Figure 3.3. Two hosting capacities can be distinguished in this case: the first hosting capacity above which the system performance deteriorates compared to the no-generation case and the second hosting capacity above which the system performance becomes unacceptable. Examples of this are the risk of overload and the losses in the

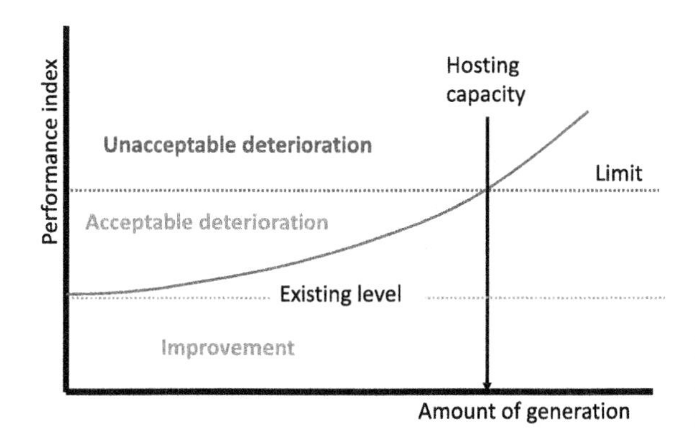

Figure 3.2 Hosting capacity approach; performance index is low in ideal case.

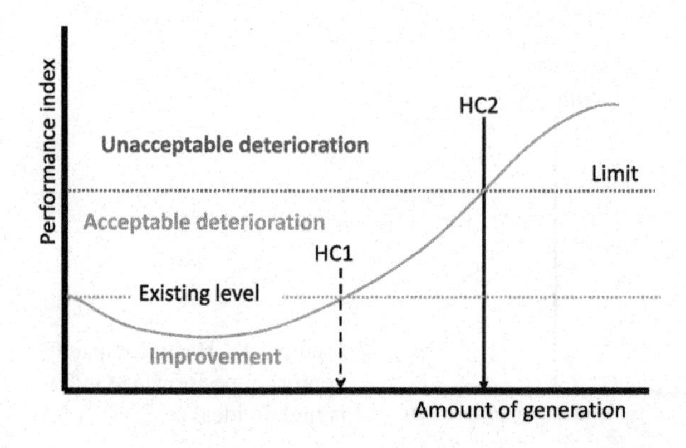

Figure 3.3 Hosting capacity approach; initial improvement in performance.

grid. Both will initially reduce but will increase when large numbers of distributed generation are connected. More details are provided in Chapter 4.

The first hosting capacity, in Figure 3.3, is often referred to implicitly in discussions on reliability. The connection of distributed generation should, according to some, not result in a deterioration of the supply reliability. Apart from the mistake that reliability is quantified by a number of indices, the mistake made in such a statement is that a certain deterioration of the supply should be acceptable for most customers.

In a simplified way, the hosting capacity approach proceeds as follows:

- Choose a phenomenon and one or more performance indices
- Determine a suitable limit or limits
- Calculate the performance index or indices as a function of the amount of generation
- Obtain the hosting capacity

To know how much distributed generation can be connected, it is important to define appropriate performance indicators. The hosting capacity is based on this. The choice of index and limit will have a big influence on the amount of distributed generation that can be accepted. We will see several examples of this in the forthcoming chapters.

The above approach gives a hosting capacity for each phenomenon (voltage magnitude, overload, protection mal-trip) or even for each index (number of interruptions, duration of interruptions, etc.). It is the lowest of these values that determines how much distributed generation can be connected before changes in the system (i.e., investments) are needed.

The hosting capacity is a tool that allows a fair and open discussion between the different stakeholders and a transparent balancing between their interests, for example,

- Acceptable reliability and voltage quality for all customers
- No unreasonable barriers against new generation
- Acceptable costs for the network operator

The impact of distributed generation is different for the different stakeholders. An attempt to give a complete overview of all stakeholders and the way in which they are impacted by distributed generation is given in Ref. 43. This is a complex job among others because new stakeholders will certainly appear, like the "aggregators" mentioned in Refs. 43 and 113. In this book, we will try to keep the discussions technical and as much as possible "stakeholder neutral." In some cases, the point of view of the different stakeholders is important and the ones that are most directly involved are the following:

- The network operators, at distribution and transmission levels
- The owners of the distributed generators
- The owners of the existing large production units
- The other customers

The network operators have an important task of electrically connecting the production and consumption. So, they must maintain acceptable reliability and voltage quality at a reasonable price. At the same time, the owners of the distributed generation units want to connect to the grid for a reasonable price, without excessive curtailment of production. Both small and large units want to make profit by selling electrical energy, which makes them competitors. From a purely economic viewpoint, distributed generation is not in the interest of the owners of the large production units. This is, however, a normal property of any open market and should not be seen as an adverse impact of distributed generation.

Higher performance requirements with the connection of distributed generation will result in better reliability and quality for other customers, but it will make it more expensive for the owners of distributed generation to connect to the grid. This in turn could slow down the introduction of renewable sources of energy.

3.4 POWER QUALITY

Power quality concerns the electrical interaction between the network and its customers. It consists of two parts: the voltage quality concerns the way in which the supply voltage impacts equipment and the current quality concerns the way in which the equipment current impacts the system [42]. Most of the recent emphasis is on voltage quality, with voltage dips and interruptions being dominant. The underlying philosophy behind power, voltage, and current quality is the view of the power system as in Figure 3.4. The "customers" may be consumers ("end users"), producers, or both. In our case, the customers that will be primarily considered are the distributed generators.

When considering distributed generation, the situation becomes somewhat more complicated. It makes sense to consider three different power quality aspects for distributed generation instead of two. The third aspect, next to voltage and current quality, is the tripping of generator units due to voltage disturbances. We will come back to this later, but first we will provide a general discussion on voltage quality and current quality as it concerns distributed generation.

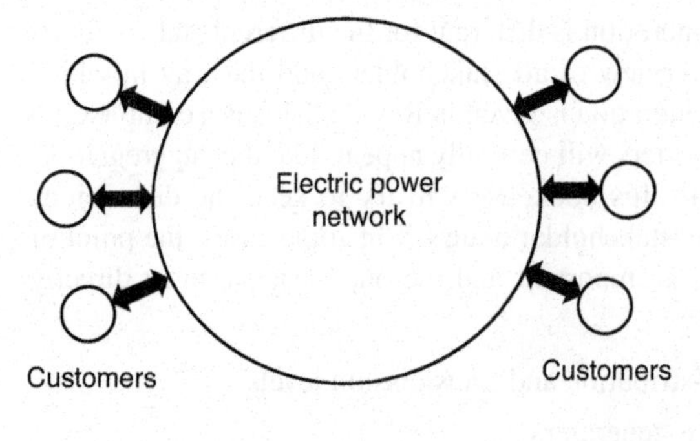

Figure 3.4 Modern look at electric power systems: network and customers.

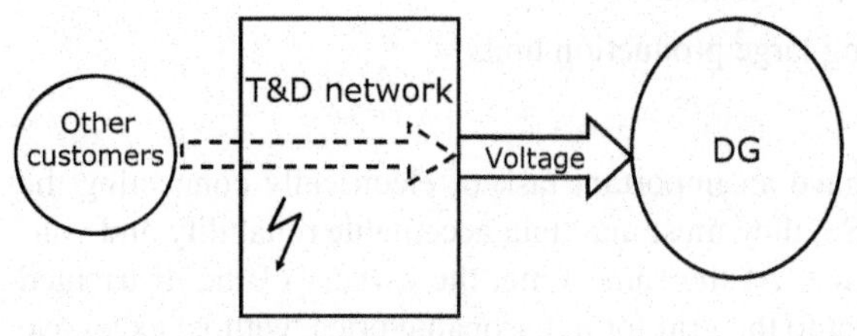

Figure 3.5 Power quality and distributed generation: voltage quality.

3.4.1 Voltage Quality

The first aspect of power quality, the so-called "voltage quality," is illustrated in Figure 3.5. The voltage quality at the terminals of the generator is determined by other equipment connected to the grid and by events in the grid.

Any generator unit is affected by the voltage quality in the same way as all other equipment. The impact of voltage disturbances includes a reduction of the lifetime of equipment, erroneous tripping of the equipment, and damage to equipment. This issue should be addressed in the same way as for end-user equipment. However, an important difference between generator units and other equipment connected to the grid is that the erroneous tripping of generator units may pose a safety risk: the energy flow is interrupted potentially leading to machine overspeed and large overvoltages with electronic equipment. To give guidelines for the immunity of generator units against voltage disturbances, a distinction between variations, normal events, and abnormal events is useful, as will be discussed further in Section 3.5.

3.4.2 Current Quality

The second aspect of power quality, the so-called "current quality," is illustrated in Figure 3.6. The generator current impacts the network and through the network other

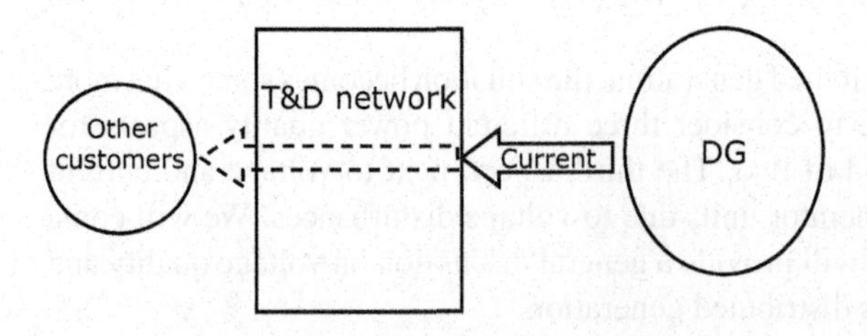

Figure 3.6 Power quality and distributed generation: current quality.

customers. The impact on other customers should remain within acceptable limits and so should the impact on the network. This is the basis of the "hosting capacity approach" discussed in Section 3.3.

Distributed generation can also be used to improve voltage quality in a controlled way. This requires a means to control the current magnitude, the phase angle of the current compared to the voltage, and the current waveform. Some of this is possible with a synchronous machine, but the main research activity concerns the control of distributed generation with power electronics interface. The impact of distributed generation on current and voltage quality is discussed in detail in Chapter 6.

3.4.3 Multiple Generator Tripping

The third aspect of power quality is illustrated in Figure 3.7. It in fact concerns an extension of the "voltage quality." As we will see in Section 3.5.3, the immunity of generator units against abnormal events is a trade-off between the consequences of the unit tripping (both economical and safety) and the costs of achieving immunity. The tripping of multiple units may, however, have an impact on the network that will be unacceptable for other customers.

The tripping of one single generator unit, for whatever reason, may adversely impact the network or other customers. This is, however, seen as a current quality issue and would be considered with the connection requirements of the unit. What is a separate and more serious issue is the simultaneous tripping of multiple units due to a common cause. The common cause would in this case be a voltage quality disturbance.

Tripping of multiple generators could occur when the level of a voltage quality variation, for example, the unbalance, reaches exceptionally high values. It could also occur for voltage quality events such as voltage dips. As will be explained in Section 3.5, the generators should be designed such that they can cope with all normal voltage quality variations and with normal (i.e., commonly occurring) voltage quality events. The immunity of distributed generators is, however, a matter of economic optimization by the operator of the unit. It is not economically attractive to invest in additional immunity against events that rarely occur. The same kind of economic trade-off is made when choosing the immunity of equipment in industrial installations [96, 97]. It can thus be expected that distributed generation will trip for extreme levels of voltage quality variations and for severe events. As all generators connected to the same part of the network will experience a similar disturbance, the tripping of multiple units may occur. At distribution level, such a situation may be accepted occasionally, for example, once a year or less. From a customer viewpoint, it would not be more

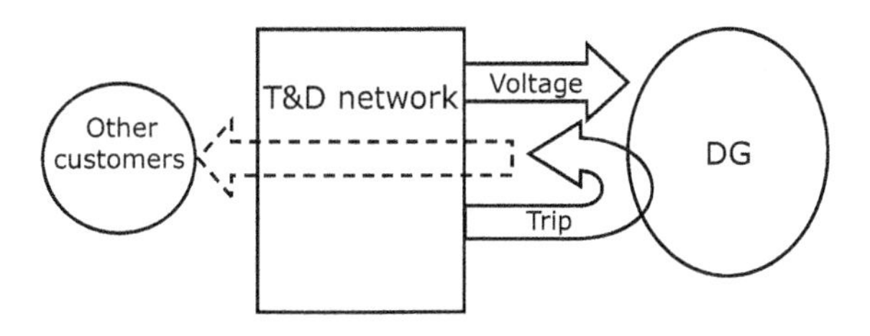

Figure 3.7 Power quality and distributed generation: tripping of generator units.

severe than a complete interruption of the supply; several interruptions per year are seen as acceptable in distribution networks [39, 53]. (Note that we refer to the number of interruptions experienced by the worst-served customers; the average number of interruptions per customer, as expressed by SAIFI [219], should be less.) However, it should be verified that no extreme events appear due to tripping of multiple generators with a large risk of damage to end-user equipment.

The case that has achieved most attention is the potential mass tripping of distributed generation, or large wind parks connected to the transmission system, due to events originating in the transmission system. Such a disturbance could spread over a large geographical area, resulting in the loss of large amounts of generation. The resulting changes in power flow at transmission level or even the lack of spinning reserve could cause a large-scale blackout. As such blackouts have become all but unacceptable, the transmission system operators have started to look seriously into this issue. A first result of this closer attention is the setting of strict requirements on the immunity of generation against voltage and frequency variations, including voltage dips and short-duration frequency swings. This is often referred to as "fault ride through."

The situation is made more complicated by protection requirements set by the operator of the distribution network to which a generator is connected. The network operator often requires the generator unit to disconnect from the network when a disturbance occurs in the distribution network. There are two reasons for this. The first is that instantaneous tripping of distributed generation during a fault will prevent any interference with the protection of the distribution feeder. The second reason for tripping distribution units is to prevent uncontrolled islanding of parts of the distribution system. Without such protection, the situation could occur that one or more generators supply part of the load. This could result in dangerous situations for maintenance personnel. Apart from that, the island could operate with a voltage magnitude and frequency far away from the nominal values. The danger of equipment damage due to this will be obvious. To prevent this, many distributed generators are equipped with sensitive protection that disconnects the unit upon detecting an unusual voltage magnitude or frequency. The so-called ROCOF relay ("rate of change of frequency") is one of the elements in such protection. Also minimum frequency, maximum frequency, minimum voltage–time, and maximum voltage–time settings are in use. Typical values for minimum and maximum frequency are 49 and 50 Hz, respectively. For more details on this, the reader is referred to Chapter 7.

The tripping of distributed generation has been reported for many years in Britain. Tripping of the HVDC link with France would often result in the shutdown of industrial combined heat-and-power plants because the ROCOF relay would disconnect the generator from the grid. The amount of distributed generation was, however, small enough to not endanger the secure operation of the transmission system. In the European transmission system, large-scale tripping of distributed generation was first observed during the Italian Blackout on September 28, 2003 [414]. When the frequency dropped to 49 Hz, 1700 MW of generation connected to the distribution grid tripped on minimum frequency [104, 419]. This added to the shortage of generation, but keeping all distributed generation connected would probably not have prevented the blackout. A less well-known fact that also contributed to the shortage of generation was that about 4000 MW of main generation tripped due to a number of reasons before the frequency reached 47.5 Hz. It should also be noted that the tripping of the distributed generation, when the frequency reached 49 Hz, was not a spurious trip. The disconnection was exactly as intended. The tripping of the main generators before the frequency reached 47.5 Hz was, however, not as intended; the basic philosophy is that large units remain connected down to a frequency of 47.5 Hz.

During the next large disturbance in the UCTE network, on November 4, 2006, the unpredictable impact of distributed generation during severe disturbances became even more clear. Due to overloading of a number of transmission lines in Germany, the interconnected grid split in three parts, one with a shortage of generation (the western part), one with a surplus of generation (the northeastern part), and one being close to balance (the southeastern part). The western part had, immediately after system splitting, a shortage of 9500 MW resulting in a sharp drop in the frequency. This frequency drop resulted in the disconnection of about 10,800 MW of generation that made the situation more severe [415]. About 4200 MW of wind power was tripped, about 60% of all the wind power connected to this part of the grid. Of the combined heat-and-power units about 30% reportedly tripped due to the underfrequency. The loss of the distributed generation contributed to it in that about 17, 000 MW of load was disconnected.

In the northeastern part of the grid, which showed a fast increase in frequency, the tripping of large amounts of distributed generation was clearly of assistance to the system. However, the automatic, but noncontrolled, reconnection resulted in severe stress to that part of the system and according to the UCTE report on the blackout [415] a further splitting of the system was not far away at certain moments.

More recently, a case has been reported in Great Britain where the loss of a 1000 MW production unit resulted in the loss of several hundred MW of distributed generation due to tripping of the ROCOF relay. According to unconfirmed information, the total loss of production exceeded the size of the dimensioning failure. But as far as the information available at the moment, the amount of primary reserve was sufficient to prevent underfrequency load shedding.

These issues will be discussed in detail in Section 8.10.

3.5 VOLTAGE QUALITY AND DESIGN OF DISTRIBUTED GENERATION

The magnitude, frequency, and waveform of the voltage impact the performance of all equipment connected to the grid, including distributed generation. For end-user equipment, a set of standard documents exists to ensure the compatibility between equipment and supply. For distributed generation, no such set of standards exists yet (February 2009) but an early draft of an IEC standard has been distributed recently [214].

In the absence of standards and as a general guidance, some thoughts will be presented here on the immunity of distributed generation against voltage quality disturbances. A distinction is thereby made between variations, normal events, and abnormal events. The distinction between variations and events is common in the power quality literature (although not all authors use the same terminology) and explained among others in Ref. 42 and 45.

It should be noted that the discussion in this section is solely based on the viewpoint of the distributed generation. As mentioned before, the network operator is likely to impose requirements on the immunity of distributed generation in the future.

3.5.1 Normal Operation; Variations

Power-quality variations are the small disturbances in voltage and current that occur during normal operation of the power system; examples of power quality variations are harmonics and unbalance. The design of any generator unit should be such that existing levels of voltage variations do not lead to premature failure or disconnection of the unit. A higher level of disturbance will typically lead to faster component aging. A very high level may lead to immediate disconnection of the generator from the grid or even to equipment damage.

For large installations, the local level of variations may be used in the design. Such an approach runs the risk, however, that the disturbance level might in future increase beyond the design criterion. For smaller, off-the-shelf equipment, a global criterion should be used. The design should be such that it can tolerate the disturbance level at any location. Even for larger distributed generators, it is recommended to use a global criterion, to prevent unwelcome surprises in the future. It is, however, not easy to obtain information on global disturbance levels. The only available guides on this are the European voltage characteristics standard EN 50160 [134] and the compatibility levels given in IEC 61000-2-2 [210]. An additional margin might be considered to cope with the well-known fact that 95% levels are given for most variations instead of 100% levels. However, the levels for voltage quality variations given in EN 50160 and IEC 61000-2-2 are high values. At the vast majority of locations, the disturbance levels are lower most of the time. In case it is expensive to add additional immunity beyond the levels in international standards, the generator unit, protection equipment might be added to the generator to prevent damage due to high disturbance levels. The setting of such protection should obviously be such that tripping is a rare event.

The only variation for which EN 50160 does not give useful values is voltage fluctuation. A long-term flicker severity equal to 1.0 is given as a limit. But this level is exceeded at several locations in the system, as the authors know from experience. The good news is that the fast voltage fluctuations that lead to light flicker rarely have an adverse impact on equipment. There remain a number of variations that are not sufficiently covered in standard documents. Waveform distortion above 2 kHz remains an uncovered territory [51, 260]. Another example is voltage magnitude variations on a timescale between a few seconds and a few minutes [44, 48].

3.5.2 Normal Events

Power quality events are large and sudden deviations from the normal voltage magnitude or waveform. Examples of voltage magnitude events are voltage dips and transients. Also, rapid voltage changes can be treated as events. According to Ref. 45, the distinction between variations and events is in the way they are measured. Variations are recorded at predefined instants or intervals, whereas events require a triggering algorithm. In Ref. 47, a further distinction is made between normal and abnormal events, based on their frequency of occurrence.

Normal events are mainly due to switching actions in the power system or within customer installations. Examples of normal events that may lead to problems with equipment are voltage dips due to motor starting or due to transformer energizing and capacitor energizing transients. It is important that generator units are immune against all normal events to prevent frequent tripping of the installation.

Whereas EN 50160 gives reasonable guidelines for variations, no useful information is available for normal events. The best design rule available is that the equipment should tolerate the worst normal event. Similar design rules are needed for end-user equipment such as adjustable-speed drives. The regular tripping of such drives due to capacitor energizing transients [7, 117, 299, 429] shows that it is not straightforward to implement these rules.

Another normal event that may lead to tripping of small generator units is transformer energizing. The second harmonic could be a problem for some equipment (leading to DC current [333]) and it may interfere with some protection and fast control algorithms. The interface between the generator and the grid should be designed such that it can tolerate voltage dips due to transformer energizing. The main concern will be the tolerance of power electronics interfaces against the high levels of even harmonic components.

The design issues are the same with generator units as with end-user equipment, but the consequences of tripping are different. A difference from the viewpoint of the generator operator is that tripping could pose a safety issue. Tripping implies that the energy flow from the unit to the grid is interrupted immediately. The energy input on the chemical or mechanical side of the unit may take some time to stop. The result is a surplus of energy in the unit that expresses itself in the form of high overvoltages or overspeed in case of rotational machines. Safety measures are obviously needed especially for mass-produced equipment. But as any safety measure has a finite failure risk, the number of trips should be limited in any case. As normal events may have a high frequency of occurrence, it is of utmost importantance that the units are immune against any commonly occurring events.

3.5.3 Abnormal Events

Abnormal events occur more rarely and are more severe. The border between normal and abnormal events is rather arbitrary as this example will show. Voltage dips due to faults may be seen as abnormal events because faults are unwanted situations that are removed from the system as soon as possible. But one may equally well classify voltage dips due to faults cleared by the primary protection as normal events and voltage dips due to faults cleared by the backup protection as abnormal events. A similar classification is made in Ref. 181 between normal and abnormal transients.

The immunity requirements of small generators against abnormal events such as frequency swings and voltage dips will depend on the severity of those events. The consequences of abnormal events may be very severe and it is not feasible to design generators that can cope with any such event. The design should be such that the unit is not damaged, but it may have to be disconnected to prevent damage. In that case, the unit will no longer supply electrical energy to the network. This will give a loss of revenue to the operator of the generator.

The loss of a large (conventional) generation unit will lead to a frequency swing in the whole system. The problem is more severe in Scandinavia and in the United Kingdom than in continental Europe because of the size of the interconnected systems. The problem is most severe on small islands. Frequency swings due to loss of a large generator unit are common and may occur several times a week. Large frequency deviations are much more rare and should not occur at all in a large interconnected system. In reality, they do occur occasionally, with intervals of several years or more in between.

A second group of abnormal events of importance to distributed generation are short circuits and earth faults. Their main immediate impact on generator units is the voltage dip at their terminals. Voltage dips at the terminals of generator units may have a range of impacts. This holds for rotating machines as well as for power electronic converters. The impact of voltage dips on end-user equipment is discussed in detail in a range of publications [42, 97]. The impact on generating equipment is ruled by the same phenomena but the impact is often in the opposite direction. Whereas induction motors show a drop in speed during a dip, induction generators show an increase in speed. But both take large amounts of reactive power when the voltage recovers after the fault. Whereas rectifiers trip on undervoltage at the DC bus, inverters trip on overvoltage because the energy flow into the grid gets reduced. But both inverters and rectifiers experience overcurrents during unbalanced dips and when the voltage recovers after a fault.

No document exists at the moment that indicates the expected number of voltage dips or frequency swings at a location in the power system. Surveys have been done for some countries, but the published results in most cases give only average values, if any results are published at all. Even when measurement results are available, these have a limited predictive value. Stochastic prediction methods are needed, similar to the methods used to estimate system reliability. Several publications discuss such methods for voltage dips and examples have been shown for both transmission and distribution systems [5, 8, 9]. The method presented in Ref. 97 for selecting equipment immunity against voltage dips can be adapted for distributed generation. Work toward a global voltage dip database has started and some early but very interesting results are presented in Ref. 97.

No such methods have been applied to predict the number of frequency swings, but their development would be straightforward as the whole system can be modeled as one node for frequency calculations. It will also be easier to obtain appropriate statistics for frequency swings than for voltage dips.

3.6 HOSTING CAPACITY APPROACH FOR EVENTS

Where it concerns power quality events, such as interruptions and dips, the quality of supply is characterized through the number of events per year with certain characteristics. It is important to realize that there exist no general objectives for most

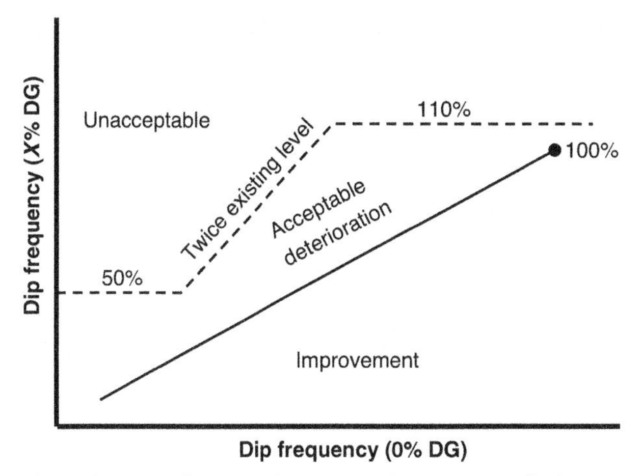

Figure 3.8 Hosting capacity concept for (power quality) events.

power quality events. Some benchmarking studies have been performed, mainly for interruptions, but there are no generally accepted limits. Some countries use national requirements on the maximum number of interruptions and on the maximum duration of an interruption. These could be used as indices and limits to determine the hosting capacity for networks in those countries. To apply the hosting capacity approach for events requires the development of new limit values.

Instead of a fixed limit for all networks and all customers, the approach proposed here uses a limit that depends on the existing event frequency. The proposed limits are presented in Figure 3.8. The horizontal axis gives the existing event frequency (e.g., the number of dips per year) for a given site; the vertical axis gives the event frequency after introduction of a certain amount of distributed generation. For points at the diagonal (the solid curve), both event frequencies are the same and the introduction of distributed generation does not have any impact. For sites below the diagonal, the quality of supply improves; for sites above the diagonal, it deteriorates.

The setting of limits corresponds to subdividing the part above the diagonal into "acceptable deterioration" and "unacceptable deterioration." For example, one could use a combination of three limits, according to the dashed line in Figure 3.8.

- The basic limit is a maximum percentage increase in event frequency, 100% increase in the figure (to twice the existing level): no customer should see more than this percentage increase in event frequency.
- Next to this relative limit, an absolute limit is introduced to prevent an excessively high event frequency for those sites with the highest event frequencies in the existing system, 110% of the maximum existing event frequency in the figure.
- For sites with a very low event frequency in the existing situation, a small absolute increase will translate into a large relative increase. To prevent this from forming barriers to the introduction of generation, an absolute limit is also

used for the sites with the lowest existing event frequency. The example used in the figure is half of the highest event frequency for the existing sites.

Note that this approach is fundamentally different from the approach proposed for power quality variations. For power quality variations, only an absolute limit is used. The limits for power quality variations are chosen such that a customer will not experience any equipment problem (in EMC terms, "no interference") as long as the index value remains below the objective. With events, the situation is different. Every single event may cause interference (depending on the equipment immunity) so that any increase in the event frequency will be observed by the customer. Having the same maximum limit for all customers could lead to a large number of complaints from those customers that have a good supply quality in the existing system.

When comparing existing and future event frequency, it is essential that they be obtained by using the same methods. As measurements of future event frequency are not possible, it is proposed that simulations be used for both scenarios. Obviously, the simulations have to be such that they result in realistic values. The limits in Figure 3.8 refer to individual customers or locations in the system ("site indices" in power quality terminology). Next to those it may also be needed to consider the impact of distributed generation on system indices.

3.7 INCREASING THE HOSTING CAPACITY

In the forthcoming chapters, same examples of methods to allow the connection of more distributed generation to the power system will be discussed. In all cases, the primary aims of the power system should still be fulfilled, although the actual requirements may change due to the changing circumstances. In general, one can distinguish between four different approaches to prevent distributed generation from interfering with the ability of the power system to fulfill its primary aims:

- Do not allow any distributed generation to be connected to the power system. This may appear a nonsolution, but it would be the preferred solution when, for example, distributed generation would cause much more environmental impact than the existing sources of generation. The reason for allowing distributed generation to be connected to the power system is that it has a number of important advantages, for example, environmental advantages. These are considered to outweigh its potential disadvantages for the power system.
- Allow limited amounts of distributed generation. This limit is the amount that does not result in any unacceptable adverse impact on the reliability or quality for other customers. When this cannot be achieved with the existing system, the costs for network enforcement (or other solutions) have to be covered by the generator as part of the connection fee. A discussion on what is acceptable quality and reliability is important and may result in a higher hosting capacity. Also, a trade-off may be made between the need for renewable sources of generation and the quality and reliability requirements of the existing customers.

- Allow unlimited amounts of distributed generation, but place strict requirements on the generator units. This will include not only protection and control but also the ability to quickly reduce or increase production on request. This approach accepts the need for the integration of new sources of generation, but it shifts the costs to the generator owners. The gain that can be obtained from additional protection requirements will be limited because the amount of loss production (e.g., "spilled wind") will increase quickly with increasing amounts of distributed generation. New control concepts, starting with power electronics control, will have to be developed. Also, the development of new communication-based concept such as "smart grids" and "microgrids" is very much driven by the expected wide-scale penetration of renewable sources of generation.
- Allow unlimited amounts of distributed generation, without any requirements beyond a small number of basic ones (e.g., safety and overcurrent protection). Instead, it becomes the task of the network operator to ensure that quality and reliability remain within their acceptable limits. Also, here a discussion is needed on what is acceptable quality and reliability, as well as possibly a new trade-off between the interests of the different stakeholders. This approach will require significant investments in the power system with large penetration of distributed generation. The costs of this cannot be bore by the network operator, but have to be covered, for example, through an increase in the use-of-system fee, as a surcharge on the energy costs, or covered by the state through an increase in taxes. Especially the latter solution would underline the importance given to renewable energy. Similar technological solutions as under the previous bullet point will most likely be developed. The difference is that the developments will be driven and paid for by the network operators instead of by the generators. An ancillary services market could be a good tool to encourage the generator units and even ordinary customers to contribute to increasing the hosting capacity.

CHAPTER 4

OVERLOADING AND LOSSES

The risk of component overload and the losses in the grid are both related to the rms value of the current. The risk of overload is related to the highest values of the current, whereas the losses depend on all values but with higher values contributing more than smaller values. But, as we will see in this chapter, they are impacted in a similar way by distributed generation. For small amount of DG, losses and risk of overload decrease; for larger amounts, both increase.

In Section 4.1, the impact of distributed generation on losses and overload is summarized. Sections 4.2 and 4.3 discuss the impact of DG on overload in detail, with the former treating radial networks and the latter meshed networks. The impact of DG on losses in discussed in detail in Section 4.4. Finally, Section 4.5 presents some of the methods for allowing more distributed generation to be connected without a high risk of overload and without excessive losses (i.e., increasing the hosting capacity). Both more classical methods and advanced methods proposed in the technical literature will be presented.

4.1 IMPACT OF DISTRIBUTED GENERATION

Distributed generation is connected closer to the consumption than conventional large-scale generation. The power therefore is transported a shorter distance, which reduces the losses. Also, the power flow from higher voltage levels toward the load is reduced. This decreases the risk of overloading at the higher voltage levels. Thus, distributed generation is in general advantageous for overloading and losses. While at a specific location the impact on overloading can be estimated using the general knowledge about the size of the installed generation and the maximum and minimum loads, the impact on losses needs more detailed knowledge about the behavior of both generation and load over time.

The biggest reduction in losses is obtained when the generation is located at the same premises as the consumption. This is the case for domestic CHP and for rooftop solar panels. The impact of this is quantified in Ref. 408, where the losses are calculated for a feeder in the center of Leicester, UK, without distributed generation and when 50% of the houses are equipped with rooftop solar panels. The results are summarized in Table 4.1. The introduction of distributed generation does somewhat reduce the losses in the low-voltage network, but the impact is rather limited compared to the amount of distributed generation we talk about here. Note that the losses are

Integration of Distributed Generation in the Power System, First Edition. Math Bollen and Fainan Hassan.

TABLE 4.1 Impact of Rooftop Solar Power on the Losses in the Low-Voltage Network

DG penetration	Winter losses	Summer losses
0%	67 kW	34 kW
30%	60 kW	29 kW
50%	58 kW	27 kW

expressed in kW, not in kWh, in the tables. This is to emphasize that only a specific operational state has been considered in Ref. 408. The duration of this operational state should be known before the annual energy savings can be estimated.

Reference 221 shows the change in losses with different levels of solar power and micro-CHP on the same network as in Ref. 408. Some of the results from that paper are reproduced in Table 4.2.

We see that a combination of solar power and domestic CHP reduces the losses by up to 50%. But even here the reduction in losses is only a small percentage of the total amount of energy produced by the distributed generation. The impact of a single photovoltaic installation on the losses is studied in Ref. 424. For a 55 kW solar power installation connected 350 m from the distribution transformer, the losses are shown to drop by about 25%.

In Section 4.4, it is shown that connection of wind power to a medium-voltage feeder can reduce the losses by up to 40% when the optimal amount of generation capacity is installed. For a distributed generator that is operating continuously, the losses can be reduced by up to 80%.

With further increase in the amount of distributed generation, the losses during reverse power flow will start to dominate. It is shown in Section 4.4 that the losses will become larger than those without generation when the average production becomes more than twice the average consumption. This is a significant amount of distributed generation and even in that case the increase in losses will just be a small percentage of the amount of environment-friendly energy produced. According to Ref. 424, the losses become larger than the initial losses when a solar power installation larger than 110 kW is connected to a low-voltage feeder.

As shown in Ref. 186 for the Swedish island of Gotland, wind power could result in an increase in the feeder losses locally, but a decrease in losses for the whole system.

TABLE 4.2 Distribution System Losses for Different Penetration of CHP and Solar Power [221]

Distributed generation	Winter	Summer
None	67 kW	34 kW
100% CHP	36 kW	28 kW
50% solar	58 kW	27 kW
100% CHP and 50% solar	33 kW	24 kW

The expected increase in losses due to distributed generation, for the year 2020, is estimated in Ref. 280. The increase in losses is shown to be about 1.6% of the production by distributed generation, or about 0.22% of the total consumption. The increase in losses is mainly due to dedicated feeders for wind power installations. A small increase in losses is not any concern, as long as the increase in losses does not become a significant part of the production due to distributed generation, which is not the case. There does not seem to be any need for considering the increase in losses due to distributed generation as a limiting factor.

Although distributed generation does impact the losses, this should not be used as an argument for or against DG. The importance of DG is in the use of renewable sources of energy. Any impact on the losses will just be a few percent of the amount of energy injected. The losses will not significantly change the global energy balance or any decision in favor or against DG. What could become an issue is the involvement of different stakeholders. The costs of the losses are at first carried by the network operator and next charged to its customers as part of the use-of-system tariff. Even if the total losses drop, the losses per kWh transported could still increase, resulting in an increase in the use-of-system tariff.

The impact of distributed generation on the risk of overloading is more direct and the limit for the amount of distributed generation that can be connected is also a hard limit. Once the loadability of a component is exceeded, the overload protection will trip the component; alternatively, the component will fail, typically resulting in a short circuit. In either case, this will lead to an interruption for one or more customers. It is shown in Section 4.2.1 that the risk of overloading increases when the maximum production becomes more than the sum of maximum and minimum consumption. The feeder will become overloaded when the maximum production becomes more than the sum of the loadability and the minimum consumption. Smaller amounts of generation will reduce the loading of the feeder. According to Ref. 435, CHP is seen by the Shanghai Municipal Electric Power Company as a suitable way of reducing peak load in distribution and subtransmission networks. As is shown in Sections 4.2.3–4.2.6, the reduction in peak load requires generation with a high availability, such as industrial CHP or hydropower. Domestic CHP will not produce any electricity in Shanghai during summer because of the zero heat demand. It was shown in Ref. 408 that rooftop solar panels reduce the daytime loading of the distribution transformer but not the evening peak. The maximum load was shown to remain the same.

As shown in Ref. 457, overloading is the limiting factor for connecting distributed generation to urban networks in Toronto, Canada. The hosting capacity of a 4.16 kV feeder was shown to be about 6 MW. The risk of reverse power flows through the HV/MV transformer is the subject of Ref. 111. The following criterion is used for determining the hosting capacity: the reverse power flow should not occur during more than 5% of the hours in 1 year. This criterion was applied to a set of 256 existing medium-voltage networks in Italy. The hosting capacity was shown to be on average 4.5 MW per medium-voltage network. Increasing the allowed percentage of time with reverse power flow to 20% only moderately increased this amount to 5.4 MW. Looking at individual medium-voltage networks, the hosting capacity was shown to vary between 2 and 7 MW for 90% of the networks. It was also shown, however, that a simplified assessment procedure could significantly reduce the hosting capacity.

In order to estimate the amount of distributed generation that will cause overloading and increased losses, the hosting capacity for each phenomenon is identified. It is shown in this chapter that two hosting capacities regarding overloading, at which actions are to be taken, can be identified. The first hosting capacity is when the maximum generation is more than the sum of maximum load and minimum load. When this level is exceeded for the load and generation downstream of a feeder section, a more detailed study is needed. Note that in most cases no immediate mitigation methods are needed after there is a certain margin between the maximum consumption and the ampacity of the feeder section. The second hosting capacity is reached when the sum of generation and minimum load exceeds the ampacity of the feeder section. In this case, mitigation actions are needed, such as enforcement of the feeder section or curtailment of the generation during periods with low consumption, as discussed in the last section of this chapter.

In meshed networks, like the ones common in transmission and distribution grids, the situation can become more complicated. The connection of production to the distribution network will reduce the load. The same holds for the connection of a large wind park to a substation to which mainly consumption is connected. In general, the reduction in loading for one or more individual buses will result in a reduction in the power flows through the network. As shown in Section 4.3.1, there are situations where the reduction in minimum load at certain buses directly limits the ability of the subtransmission or transmission grid to transfer power. When power transfer through the region is important, it may not be possible to connect any distributed generation to those buses. The same may be the case when already a large amount of generation is connected to a subtransmission bus. That generation may have taken up all hosting capacity, leaving no further space for distributed generation.

4.2 OVERLOADING: RADIAL DISTRIBUTION NETWORKS

4.2.1 Active Power Flow Only

Consider a location along a long distribution feeder operated radially, with total downstream consumption equal to $P_{\text{cons}}(t)$ and total downstream generation $P_{\text{gen}}(t)$. The reactive power flows are neglected, so the total power flow at this location is

$$P(t) = P_{\text{cons}}(t) - P_{\text{gen}}(t) \tag{4.1}$$

We assume that no overloading situation exists before the introduction of distributed generation. As long as the maximum power flow after connection is less than that before, there will be no overload. The condition to be fulfilled is

$$P_{\text{max}} < P_{\text{cons,max}} \tag{4.2}$$

With large amounts of generation, the maximum power flow occurs for maximum generation and minimum consumption:

$$P_{\text{max}} = P_{\text{gen,max}} - P_{\text{cons,min}} \tag{4.3}$$

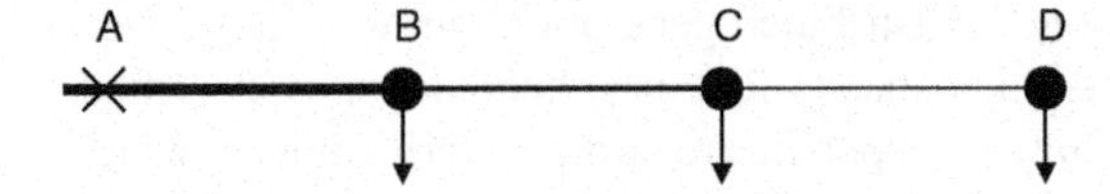

Figure 4.1 Example feeder.

so that the condition for guaranteeing no overload is

$$P_{\text{gen,max}} < P_{\text{cons,max}} + P_{\text{cons,min}} \tag{4.4}$$

This criterion should be fulfilled for every location on the feeder. This is a sufficient requirement that could be used as a first hosting capacity level. As long as the maximum generation is less than the sum of minimum and maximum consumption, no overloading will occur. When the maximum generation exceeds this first hosting capacity level for any location along a feeder, additional studies are needed to determine a second hosting capacity level. The second hosting capacity is reached when the maximum current becomes equal to the ampacity of the feeder section. The maximum power through the feeder section is again given in (4.3). This should be less than the maximum permissible power:

$$P_{\text{gen,max}} - P_{\text{cons,min}} < P_{\text{max,limit}} \tag{4.5}$$

resulting in the following condition:

$$P_{\text{gen,max}} < P_{\text{max,limit}} + P_{\text{cons,min}} \tag{4.6}$$

This approach for finding the hosting capacity is illustrated in Examples 4.1 and 4.2. In the examples, it is assumed that the maximum and minimum loads are well known. In reality, information about the minimum load is often not available. This could easily result in an underestimation of the hosting capacity. This will be discussed in detail in Section 4.2.7.

Example 4.1 Consider the simple feeder shown in Figure 4.1 with conductor data as in Table 4.3 and consumption data as in Table 4.4. The ampacity values were taken from Ref. 1 for underground cables. A nominal voltage equal to 15 kV has been used to calculate the maximum permissible (apparent) power from the ampacity. The same feeder will be used in forthcoming examples; the tables therefore contain more data than needed for this example. The first hosting capacity, according to (4.4), for feeder section CD is

$$P_{\text{gen,max}} = 2.5\,\text{MW} + 700\,\text{kW} = 3.2\,\text{MW}$$

TABLE 4.3 Conductor Data for the Example Feeder in Figure 4.1

Feeder section	AB	BC	CD
Conductor size	185 mm^2	95 mm^2	35 mm^2
Ampacity	388 A	268 A	151 A
Maximum permissible power	10 MVA	7 MVA	3.9 MVA

TABLE 4.4 Consumption Data for the Example Feeder in Figure 4.1

	B	C	D
Maximum active power	2 MW	3.5 MW	2.5 MW
Maximum reactive power	1.3 Mvar	2 Mvar	1.3 Mvar
Minimum active power	500 kW	900 kW	700 kW
Minimum reactive power	300 kvar	500 kvar	500 kvar

For feeder section BC, the whole downstream load should be considered, that is, load C plus load D. The first hosting capacity is

$$P_{\text{gen,max}} = (2.5 + 3.5\,\text{MW}) + (700 + 900\,\text{kvar}) = 7.6\,\text{MW}$$

For feeder section AB, we get

$$P_{\text{gen,max}} = (2.5 + 3.5 + 2\,\text{MW}) + (700 + 900 + 500\,\text{kW}) = 10.1\,\text{MW}$$

Example 4.1 gave the amount of generation that can be safely connected downstream of each feeder section. When limits are needed for the amount of generation in each load point, different combinations are possible. The maximum amount of generation supplied from load point C is equal to 7.6 MW minus the amount supplied from D. A possible allocation of limits for the amount supplied from each load point is

- Load point B: 2.5 MW.
- Load point C: 4.4 MW.
- Load point D: 3.2 MW.

Example 4.2 For the example feeder in Figure 4.1, the second hosting capacity (when the actual overload starts to occur) for feeder section CD is

$$P_{\text{gen,max}} = 3.9\,\text{MW} + 700\,\text{kW} = 4.6\,\text{MW}$$

For feeder section BC, the sum of the maximum power of this section and the minimum of all downstream consumption gives the second hosting capacity:

$$P_{\text{gen,max}} = 7\,\text{MW} + (700 + 900\,\text{kW}) = 8.6\,\text{MW}$$

For feeder section AB, the result is

$$P_{\text{gen,max}} = 10\,\text{MW} + (700 + 900 + 500\,\text{kW}) = 12.1\,\text{MW}$$

When the second hosting capacity for the first feeder section is exceeded, the circuit breaker at location A will trip when maximum generation coincides with minimum consumption. One may say that the overload is "self-revealing" and there is no risk of equipment damage. Instead, the overload will result in an interruption, causing a reduction in the supply reliability.

The situation becomes different when the second hosting capacity is exceeded for feeder section BC or CD. As there is no specific protection for these feeder sections, they may become overloaded without the feeder protection (at location A) observing an overload. The result could be damage to equipment. This requires enforcement of the feeder section.

Exercise 4.1 In Example 4.2, a second feeder section (CE) is connected to point C in Figure 4.1, with 35 mm^2 conductor cross section. The minimum and maximum consumption at point E are 900 and 500 kW, respectively. Calculate the first and the second hosting capacity for the different feeder sections. Compare the results with the hosting capacity found in the above example. Explain the differences.

4.2.2 Active and Reactive Power Flow

When including the contribution of the reactive power to the current, the expressions become more complicated although the methodology remains the same. The maximum apparent power for a feeder section without generation is equal to

$$S_{\max,1} = \sqrt{P_{\text{cons,max}}^2 + Q_{\text{cons,max}}^2} \tag{4.7}$$

where $P_{\text{cons,max}}$ and $Q_{\text{cons,max}}$ are maximum active and reactive power consumption, respectively. Here it is assumed that the time variations in active and reactive power demand are the same, so the maximum reactive power is reached when both active and reactive power consumption are at their maximum. When active and reactive power consumption show significantly different variations with time, this should be included in the study. In this case, a time-domain simulation might be more appropriate. Some examples are given below.

With a substantial amount of generation connected to the feeder, the maximum apparent power is reached for minimum consumption and maximum generation:

$$S_{\max,2} = \sqrt{(P_{\text{gen,max}} - P_{\text{cons,min}})^2 + Q_{\text{cons,min}}^2} \tag{4.8}$$

where it is assumed that the generation neither produces nor consumes reactive power. The maximum apparent power with generation should be less than the maximum apparent power without generation.

$$S_{\max,2} < S_{\max,1} \tag{4.9}$$

This results in the following expression for the first hosting capacity:

$$P_{\text{gen,max}} < P_{\text{cons,min}} + \sqrt{P_{\text{cons,max}}^2 + Q_{\text{cons,max}}^2 - Q_{\text{cons,min}}^2} \tag{4.10}$$

The second hosting capacity is reached when the current through the feeder section reaches the ampacity of the feeder. The maximum current, or apparent power, is found from (4.4). This should be less than the maximum permissible apparent power (ampacity times nominal voltage) resulting in the following expression:

$$P_{\text{max,gen}} < P_{\text{cons,min}} + \sqrt{S_{\text{max,limit}}^2 - Q_{\text{cons,min}}^2} \tag{4.11}$$

Example 4.3 Consider again the example feeder in Figure 4.1. The first hosting capacity, using (4.10), for section CD is

$$P_{\text{gen,max}} = 700 + \sqrt{2500^2 + 1300^2 - 500^2} = 3400\,\text{kW}$$

For feeder section BC, the first hosting capacity is

$$P_{\text{gen,max}} = 1600 + \sqrt{6000^2 + 3300^2 - 1000^2}$$

In both cases, the estimation is somewhat higher than that calculated in Example 4.1, where the reactive power has been neglected. The approximated expression thus gives a safe limit.

Exercise 4.2 Calculate the first hosting capacity for feeder section AB in Example 4.3.

Example 4.4 For the same example as in Figure 4.1 and in the previous examples, the second hosting capacity, using (4.11), for feeder section CD is

$$P_{\text{gen,max}} = 700 + \sqrt{3900^2 - 500^2} = 4500\,\text{kW}$$

Again, the estimation is somewhat higher than that calculated in Example 4.1.

Exercise 4.3 Repeat the calculations in Example 4.4 for feeder sections BC and AB.

4.2.3 Case Study 1, Constant Production

To illustrate the impact of distributed generation on the loading in a realistic way, a measurement of the active and reactive parts of the consumption has been used, and combined with a model of the power injected by distributed generation. The power through a 130/10 kV transformer supplying a region with mainly domestic load without any distributed generation was measured every minute during a 4-week period. For the simulations, it was assumed that a constant amount of generation was added on the secondary side of the transformer. The maximum apparent power was calculated as a function of the amount of generation, resulting in the curves shown in Figure 4.2. For each of the curves, the ratio between generated active power and reactive power consumed by the generator was assumed to be constant. For the lower curve, the reactive power was assumed zero (unity power factor). For the upper curve, the reactive power was equal to the active power (power factor 0.71). For the intermediate curves, the ratio between reactive and active power was (top to bottom) 0.8, 0.6, 0.4, and 0.2. This corresponds to power factor values of (top curve to bottom curve) 0.71, 0.79, 0.85, 0.95, and 1.00.

For generation at unity power factor, the first hosting capacity (crossing with the horizontal line) is about 25 MW. With increasing amount of reactive power, the hosting capacity decreases, becoming about 13 MW for a power factor of 0.7. Note that even in the latter case, the installation of significant amounts of generation (i.e., up to 13 MW, or 70% of maximum consumption) gives a reduction in feeder loading.

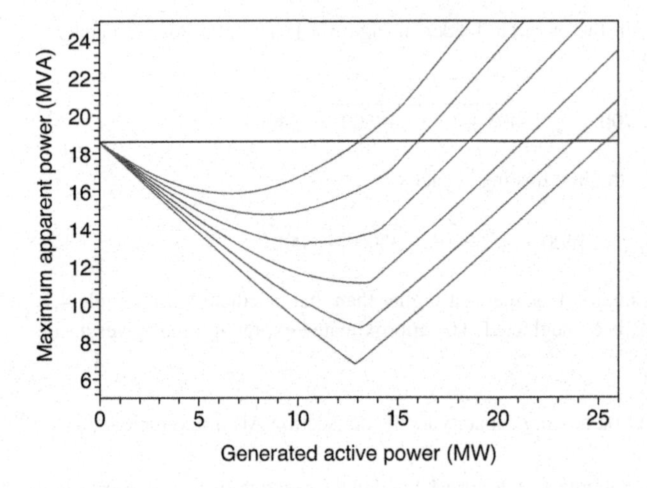

Figure 4.2 Maximum apparent power as a function of the generated active power for a medium-voltage network. The different curves correspond to different amounts of reactive power consumed by the generators.

4.2.4 Case Study 2: Wind Power

When the distributed generation is present in the form of wind power, the stochastical aspects of the generation will have to be considered. As the wind power is not available a large part of the time and is rarely equal to its nominal value, the reduction in peak load will not be as high as with generation that is available a large fraction of the time. To illustrate this, the same transformer currents as in Section 4.2.3 have been used. Random values of the production have been obtained by assuming Weibull distributed wind with a mean value of 7 m/s and a shape factor of 2. A scaled version of a 600 kW wind turbine was used to translate the values of the wind speed into values of the active power production. The reactive power production is assumed to be zero. See Chapter 2 for more details about stochastic models of wind power production.

For a number of values of the nominal wind power, from zero to 30 MW, the 99%, 99.9%, and 99.99% values of the apparent power through the transformer have been calculated. The results are shown in Figure 4.3. For wind power values up to about 25 MW, the risk of overloading only marginally decreases. The risk of overloading increases fast when more than 25 MW is connected on the secondary side of the transformer.

Compare Figure 4.3 with Figure 4.2 where constant production has been assumed. The hosting capacity is the same, around 25 MW, but for constant production the maximum apparent power shows a huge decrease (down to less than half) for moderate amounts of distributed generation. For random sources such as wind power, moderate amounts of penetration will not give any significant reduction in the risk of overload.

It is also interesting to compare the values found with some properties of the original load profile. The "hosting capacity" found was 25 MW, which is 138% of the maximum consumption. The sum of maximum and minimum consumption is 25.2 MW, which is a good approximation of the hosting capacity, both for constant and for random production.

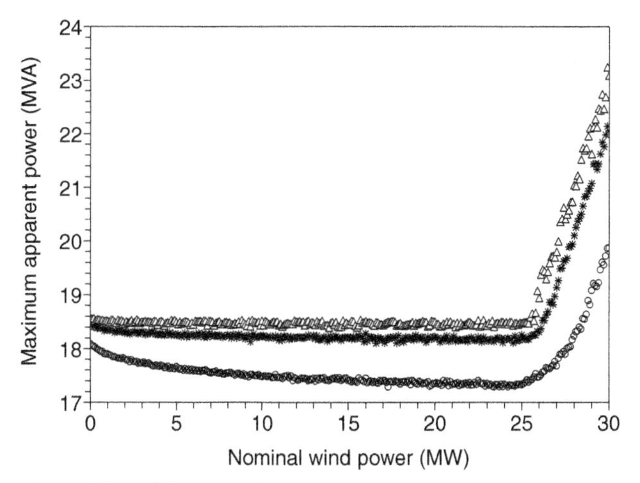

Figure 4.3 High-percentile values of apparent power through a transformer with increasing amount of wind power on the secondary side: 99.99% (triangle); 99.9% (star); 99% (circle).

4.2.5 Case Study 3: Wind Power with Induction Generators

Wind power installations based on induction generators consume reactive power that causes additional loading on the feeder. To model the impact of this on the risk of overloading, it is assumed that the reactive power is consumed only by a fixed series reactance. This series reactance represents mainly the leakage reactance of the induction generator and the reactance of the turbine transformer. This reactance varies between 0.20 and 0.40 per unit with the turbine rating as a base. The above-mentioned indicators for the risk of overload have been recalculated for reactance values of 0.25 and 0.40 per unit, with the results shown in Figures 4.4 and 4.5, respectively.

The main impact of the consumption of reactive power is that the first hosting capacity, where the overload risk is the same as that without wind power, shifts to lower values. The first hosting capacity shifts from about 26 MW to about 23 MW for reactive power consumption of 25% and to about 21 MW for reactive power consumption of 40% of the active power. When the amount of wind power is less than the first hosting capacity, the risk of overload is not impacted by the reactive power consumption.

4.2.6 Case Study 4: Solar Power with a Hotel

To illustrate the impact of solar power on the loading, the measured consumption for a large hotel located at 38°N has been used. Like before, measurements of consumption have been combined with a model of the production. The average active and reactive power consumption per phase have been measured every minute during a 7-day period in summer. The total active power consumption (sum of three phases) is shown in Figure 4.6 as a function of the time of day. The maximum consumption is about 590 kW; the minimum about 230 kW. The sum of maximum and minimum consumption is 820 kW, which is the first approximation for the hosting capacity.

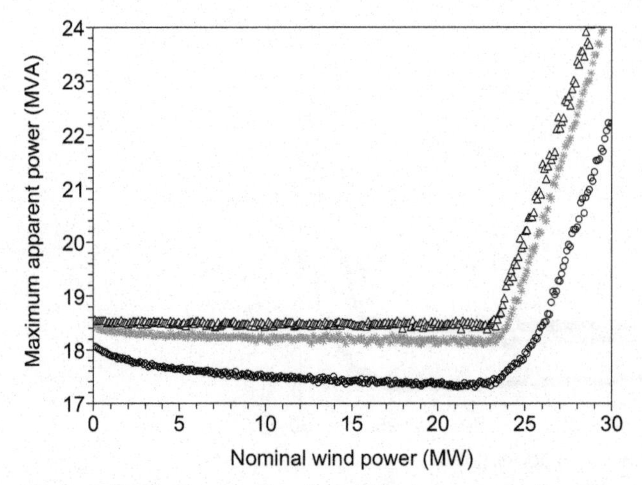

Figure 4.4 High-percentile values of apparent power through a transformer with increasing amount of wind power on the secondary side: 99.99% (triangle); 99.9% (star); 99% (circle). The reactive power consumption at maximum power is 25% of the active power.

Next 300 kW solar power is installed, resulting in a reduction in active power flow. It is assumed that the sun is at its most northern position in the sky (i.e., June 21) and that the sun reaches its highest at 13:30 local time. (The hotel is about 7°W of the center of its time zone, which gives about 30 min delay; also there is 1 h additional delay due to summer time.) The production of 300 kW is reached when the sun is in the zenith, but during this time of year the maximum production is about 97% of that value. The production as a function of time has been calculated using the expressions for a horizontal panel in Section 2.2. The active power consumption as a function of

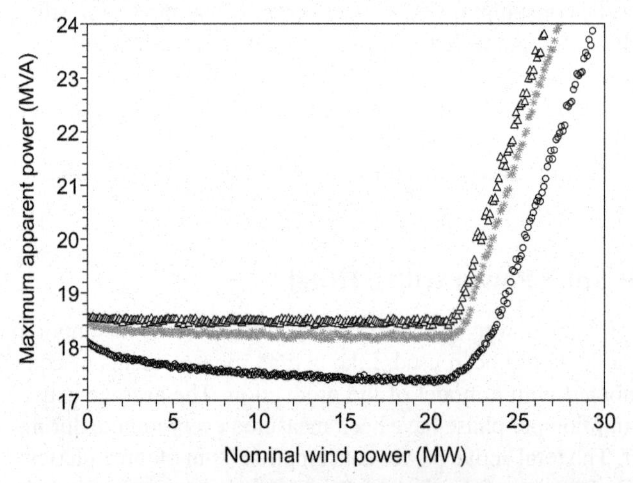

Figure 4.5 High-percentile values of apparent power through a transformer with increasing amount of wind power on the secondary side: 99.99% (triangle); 99.9% (star); 99% (circle). The reactive power consumption at maximum power is 40% of the active power.

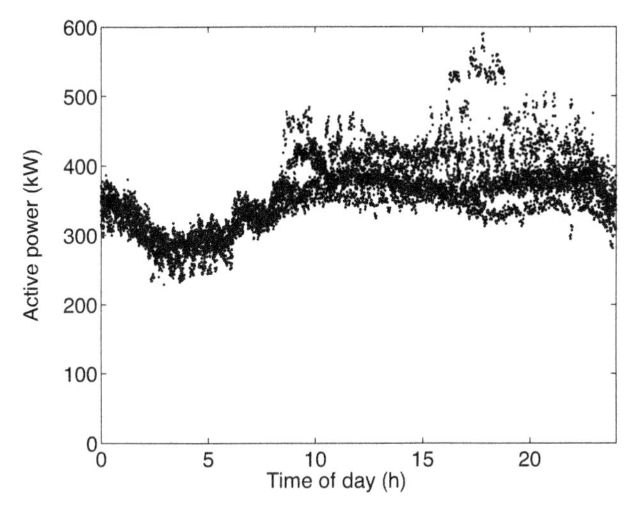

Figure 4.6 Active power consumption for a large hotel during seven consecutive days in summer.

time of day is shown in Figure 4.7. The consumption around noon is reduced a lot, but in the evening and night the production is obviously not reduced. The maximum load is about 500 kW (only a small reduction), whereas the minimum load is reduced to about 40 kW.

The active power flow for an even larger amount of solar power (with 750 kW maximum production) is shown in Figure 4.8. During several hours around noon, the hotel becomes a net producer of power, with a maximum of around 400 kW at noon.

For even higher amounts of solar power, the production around noon becomes higher than the consumption in the evening. The loading of the supply would exceed the loading without solar power. The highest value of the apparent power in any of

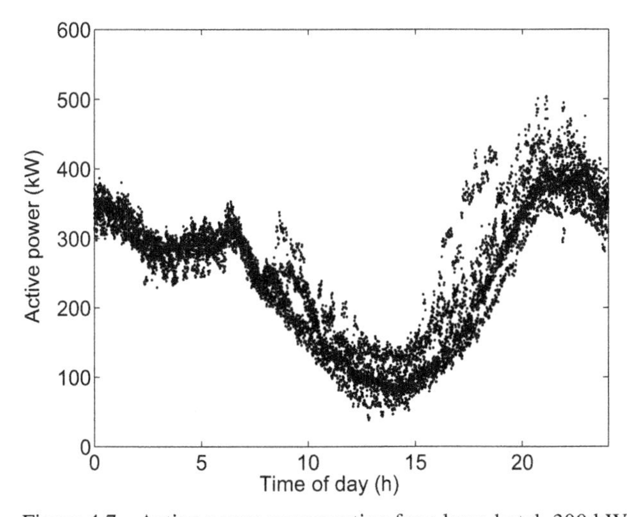

Figure 4.7 Active power consumption for a large hotel, 300 kW solar power.

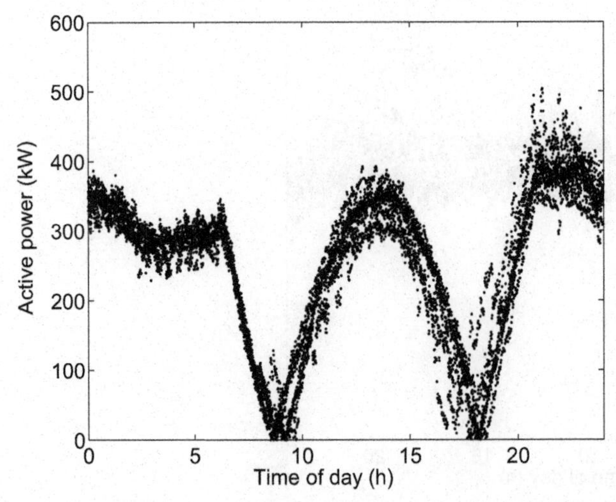

Figure 4.8 Active power consumption for a large hotel, 750 kW solar power.

the three phases has been calculated as a function of the amount of solar power. The solar power is assumed to be equally distributed over the three phases. The results are shown in Figure 4.9. The system loading (i.e., the maximum apparent power) slightly decreases up to about 300 kW solar power. For higher amounts of solar power, the maximum occurs when the sun is below the horizon. When more than 1000 kW solar power is installed, the maximum loading occurs at noon and will increase linearly with the amount of solar power. The value of 300 kVA per phase (slightly above the original maximum) is reached for 1140 kW installed solar power. For more solar power, the risk of overload is increased and additional investments are needed, like a larger transformer.

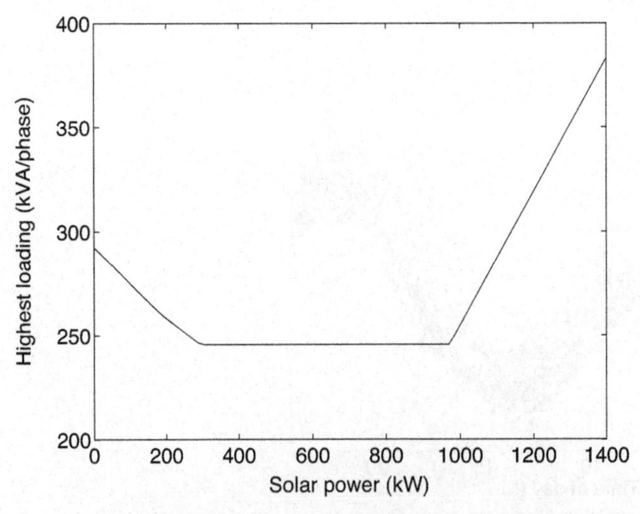

Figure 4.9 Maximum apparent power per phase as a function of the amount of solar power.

This example shows that the hosting capacity for solar power, as far as overload is concerned, can be higher than that according to the first approximation (minimum plus maximum loads): in this case 1140 kW instead of 820 kW. The reason for this is that the maximum production never occurs at the same time of day as minimum consumption. This example also shows that solar power, at least in this case, does not contribute much to the maximum loading of the system. This is because the consumption remains high in the early hours of the evening, after the sun has set.

4.2.7 Minimum Consumption

It was shown at a few places in the previous sections that the amount of generation that can be connected to a distribution feeder (the "hosting capacity") is directly dependent on the minimum consumption. Every watt reduction in minimum consumption gives 1 W reduction in the hosting capacity. The minimum consumption in a distribution network is for most network operators not well known. Before the introduction of distributed generation, the minimum consumption was not of importance, neither during the design nor during the operation of the network. The maximum consumption matters a lot and is an important input to the design of a distribution feeder. The so-called "after-diversity maximum consumption" per customer plays an important role in determining the required rating of the components. When distributed generation will start to become an important part in the future, concepts such as "after-diversity maximum production," "after-diversity minimum consumption," and "after-diversity minimum production" will play an equally important role. Detailed measurements and simulations will be needed to obtain typical values of these design inputs.

In the mean time, the minimum consumption still needs to be estimated to determine the hosting capacity. In low-voltage networks, the diversity between customers is large because of the low number of customers. The ratio between maximum and minimum consumption is large at low voltage. The additional hosting capacity due to the minimum consumption is small. Further, the minimum consumption is uncertain, so a low value needs to be chosen to prevent a too high risk of feeder overload. Choosing the minimum consumption equal to zero is an acceptable decision for low-voltage feeders. Obviously, when a more accurate minimum value is available, this one should be used.

In medium-voltage networks, the minimum consumption is still a significant part of the maximum consumption. Setting the minimum consumption to zero would result in a serious underestimation of the hosting capacity. It is therefore important to have an accurate estimate of the minimum consumption. Modern metering equipment is able to collect consumption patterns with a much higher accuracy than the classical meters. These consumption patterns can be used for an accurate estimation of the minimum consumption.

It is further important to consider any correlation between minimum consumption and maximum production. As shown in Section 4.2.6, this can have a significant influence on the hosting capacity. This matters most for sources that are strongly dependent on the time of day, such as solar power and some kinds of combined heat and power. What matter with those sources is not the minimum consumption over the whole day but the minimum consumption when maximum production can be expected (around noon for solar power).

4.3 OVERLOADING: REDUNDANCY AND MESHED OPERATION

4.3.1 Redundancy in Distribution Networks

Many distribution networks are meshed in structure, but operated radially. This allows for interruptions to be limited in duration by means of switching actions. The principles are discussed, for example, in Ref. 42 and 258. To ensure that backup operation is possible, the network should also in that case not be overloaded. Consider the simple network in Figure 4.10.

During backup operation, the network is also operated radially, so expression (4.4) still holds, but with generation and load being the sum of the two feeders:

$$(P_{\text{gen},1} + P_{\text{gen},2})_{\max} < (P_{\text{cons},1} + P_{\text{cons},2})_{\max} + (P_{\text{cons},1} + P_{\text{cons},2})_{\min} \quad (4.12)$$

During normal operation, the following relation should hold for the two feeders to prevent an increased risk of overloading:

$$\begin{aligned} (P_{\text{gen},1})_{\max} &< (P_{\text{cons},1})_{\max} + (P_{\text{cons},1})_{\min} \\ (P_{\text{gen},2})_{\max} &< (P_{\text{cons},2})_{\max} + (P_{\text{cons},2})_{\min} \end{aligned} \quad (4.13)$$

In theory, situations could occur where conditions (4.13) are fulfilled, but condition (4.12) is not. The load maxima could occur at different instances in time, so that $(P_{\text{cons},1} + P_{\text{cons},2})_{\max} < (P_{\text{cons},1})_{\max} + P_{\text{cons},2})_{\max}$, where the load minima and the generation maxima could occur at the same time. Although this is in theory possible, it does not appear a very likely situation to occur. It should further be noted that (4.12) gives only the "first hosting capacity," beyond which the risk of overload becomes higher than that in the network without distributed generation. Of more concern is the "second hosting capacity" beyond which the risk of overload becomes unacceptable.

Extreme cases might still occur, mainly when large units are connected toward the end of a feeder. There are a number of solutions to cope with this:

- Limit the generator size
- Enforce the distribution network
- Accept that backup operation is not possible in all cases
- Curtail the amount of generation during backup operation

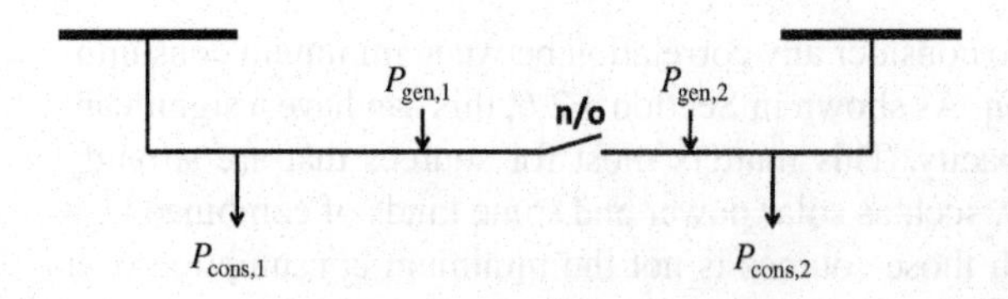

Figure 4.10 Radially operated meshed network: distribution feeder with possibility for backup supply.

The latter is the most reasonable solution; curtailment is needed only during periods of high production, low consumption, and when at the same time the main feeder is not available. The probability of this will be small. The main costs will be the more detailed instructions that have to be followed by the distribution network operator to prevent overloading during backup operation.

More advanced solutions could be considered where the hosting capacity is calculated after every switching action. The production from distributed generation is curtailed once it exceeds the hosting capacity for the actual operational state of the distribution system. In that case, there will not be any limit on the installed capacity, but only on the actually produced power.

4.3.2 Meshed Operation

Transmission and distribution networks are operated meshed, so that the loss of a component does normally not result in an interruption for any of the customers. The meshed operation has some potential impacts for the risk of overloading. In a radial network, distributed generation will always result in a reduction in the network loading, as long as the amount of distributed generation is less than the "first hosting capacity." When the network is operated meshed, this is no longer always the case. An example is shown in Figure 4.11: a large power flow takes places between two regions in an interconnected network. The result is a large power flow in the connection A–B as well as a lesser flow in the loop A–C–D–B.

The load connected to bus C will increase the flow in the connection A–C but at the same time reduce the flow in the connection D–B. Adding (distributed) generation to node C will initially result in a reduction of the load, and thus an increase in the power flow between D and B. If the connection D–B is close to its loading limit, already a small amount of generation can result in overloading.

Situations like this are not likely to occur due to small-size distributed generation. The reduction in loading will occur for all buses in a similar way. The concern is, however, especially with the connection of large wind parks to the subtransmission networks. Long-distance power transport could result in parallel flows through the subtransmission network. The connection of already a small amount of generation could result in some possible overloads. Very small hosting capacity values could result when all possible operational states have to be considered. We will come back to this in Section 4.3.3.

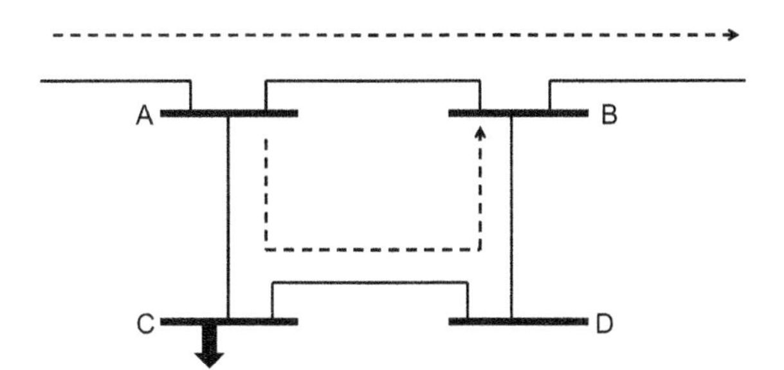

Figure 4.11 Example of a large power flow in a meshed network.

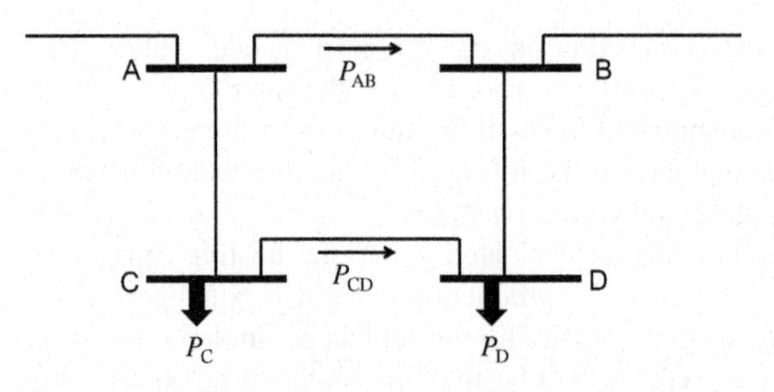

Figure 4.12 Example of a meshed subtransmission network.

Consider the network shown in Figure 4.12. A linearized DC loadflow is used to obtain an expression for the power flow through connection C–D due to a large power transfer through the region.

Neglecting the resistance, the active power transferred over a line is a function of the angular difference between the voltages on both sides of the line:

$$P_{ij} = \frac{\sin\theta_{ij}}{z_{ij}} \tag{4.14}$$

with z_{ij} the impedance of the line. For not too large angles, we can linearize this expression into the following one:

$$P_{ij} = \frac{\theta_{ij}}{z_{ij}} \tag{4.15}$$

For connection C–D in Figure 4.12, this reads as

$$P_{\mathrm{CD}} = \frac{\theta_{\mathrm{CD}}}{z_{\mathrm{CD}}} \tag{4.16}$$

For the other connections, similar expressions are obtained, where it has been used that the flow through A–C is the sum of the flow through C–D and the load at bus C, and that the flow through D–B is the difference between the flow through C–D and the load at bus D:

$$P_{\mathrm{CD}} + P_{\mathrm{C}} = \frac{\theta_{\mathrm{AC}}}{z_{\mathrm{AC}}} \tag{4.17}$$

$$P_{\mathrm{CD}} - P_{\mathrm{D}} = \frac{\theta_{\mathrm{DB}}}{z_{\mathrm{DB}}} \tag{4.18}$$

$$P_{\mathrm{AB}} = \frac{\theta_{\mathrm{AB}}}{z_{\mathrm{AB}}} \tag{4.19}$$

Combining (4.16)–(4.19) and using that $\theta_{\mathrm{AB}} = \theta_{\mathrm{AC}} + \theta_{\mathrm{CD}} + \theta_{\mathrm{DB}}$ gives the following relation between the flow through A–B and the flow through C–D:

$$P_{\mathrm{AB}} = \frac{(z_{\mathrm{CD}} + z_{\mathrm{AC}} + z_{\mathrm{BD}})P_{\mathrm{CD}} + z_{\mathrm{AC}}P_{\mathrm{C}} - z_{\mathrm{BD}}P_{\mathrm{D}}}{z_{\mathrm{AB}}} \tag{4.20}$$

When the flow through C–D is limiting the power transfer through the region, (4.20) gives the maximum permitted power through A–B that will not result in overloading of connection C–D. Increased load at bus C will increase the transfer capacity; increased load at bus D will decrease the transfer capacity. Connection of generation to bus C will reduce the transfer capacity; connecting generation to bus D will increase the transfer capacity. Power transfer through a region might take place in both directions. The result is that the hosting capacity at both buses can be low, or even zero.

4.3.3 Redundancy in Meshed Networks

In transmission and subtransmission networks, planning and operation are based on the so-called $(N-1)$ principle. Under this principle, the loss of any single component should not lead to an interruption for any of the customers. The details of this principle, and the differences in application between transmission and subtransmission, vary between network operators. We will come back to this when discussing transmission systems in Chapter 8.

A simple example is shown in Figure 4.13: a 50 MW load is supplied through a double-circuit line, with a capacity equal to 60 MW per circuit. When the minimum load is 20 MW, up to 140 MW (120 + 20) of generation can be connected without overloading the line. However, when the $(N-1)$ principle is applied, the hosting capacity is only 80 MW (60 + 20).

The $(N-1)$ principle requires that the loss of any component does not result in an interruption for any of the customers. This principle is used as an operational principle by almost all transmission system operators. At subtransmission level, the principle is less strictly applied during operation, although the details differ between operators. Of importance for the discussion in this chapter is that the $(N-1)$ principle is used during the design of the network. The design criterion to be used reads something like: "For any operational state, the loss of any component should not result in an overload anywhere in the network." Most network operators do not perform calculations for all operational states and for all component losses. Instead, a limited number of operational states are considered, typically including maximum demand and minimum demand. Prior knowledge about the most critical operational states is needed for this. For each operational state, either all single contingencies (loss of one component) or a predetermined list with the most severe contingencies is studied. The calculations are repeated for increasing amounts of generation until one or more network elements become overloaded. The highest amount of generation not resulting in an overload is the hosting capacity.

A solution that allows more generation to be connected, while still maintaining the $(N-1)$ principle, is to use the so-called "intertrip scheme." An intertrip scheme

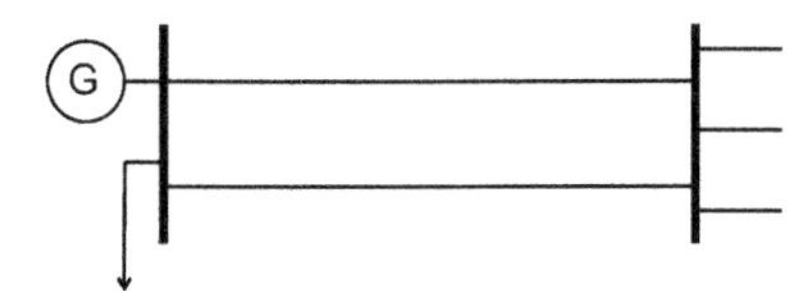

Figure 4.13 Bus with load and generation connected to the main grid via two lines.

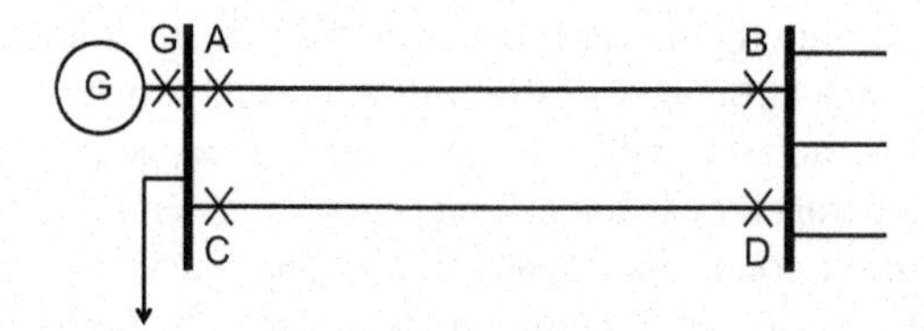

Figure 4.14 Circuit breakers along a double-circuit line for intertrip scheme.

ensures that a certain breaker is opened once another breaker is opened. The most common application is with distance protection of overhead lines, to speed up the opening of the breaker farthest away from the fault [131, Section 17.3; 209, Section 6.6]. An intertrip scheme of use here would be, with reference to Figure 4.14, to trip breaker G whenever one of the breakers A, B, C, or D is tripped.

In that way, the whole transport capacity of the double-circuit line can be used, but without risking line overload when one of the two lines trips. The intertrip scheme allows, in this case, to connect up to 140 MW generation instead of 80 MW. The drawback is that the generation, or part of the generation, is tripped whenever one of the lines is tripped. This, however, occurs only for a limited amount of time, typically a few percent, when maintenance is included.

The choice of operational state and of contingencies is mainly based on experience by the network operator. The most severe operational state and the most severe contingencies should be chosen. The experience is, however, based on the operation of the existing system, that is, without distributed generation. With large amounts of distributed generation, different operational states and different contingencies may become the ones that cause the highest component loading. It is therefore recommended to take as many operational states and contingencies as is practical. Some preconsiderations are also recommended on what might be the operational states and contingencies that result in the highest component loading.

The simplified network shown earlier in Figure 4.11 is reproduced in Figure 4.15, this time with the thermal loadability of the connections indicated. Also, the range of load at buses C and D is indicated in the figure.

We assume that the $(N-1)$ principle is used to determine the maximum amount of generation that can be connected. The 30 MW capacity of connection C–D sets the limit for the amount of generation at buses C and D. During the loss of connection A–C and minimum load at bus C, no more than 40 MW generation can be connected to bus C. During the loss of connection B–D, the amount of generation at bus D

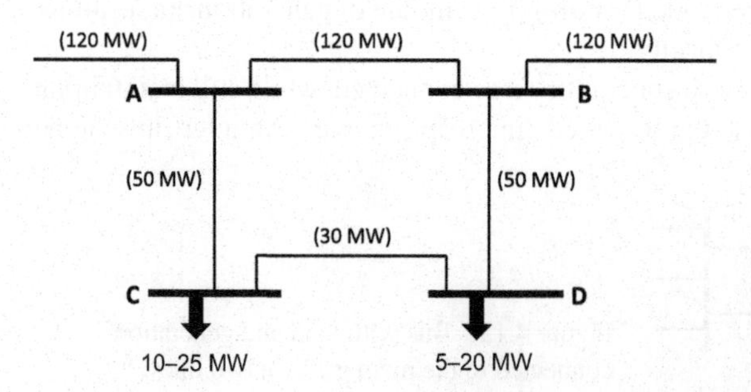

Figure 4.15 Example of subtransmission network with busload and loadability of the connections indicated.

is limited to 35 MW. The total amount of generation is limited by the capacity of connections A–C and B–D at no more than 65 MW (50 + 10 + 5).

The hosting capacity for new sources of generation could be smaller than this for a number of reasons. Two such reasons are the presence of existing generation and large power flows through the region. When existing generation is present, for example, in the form of hydropower, this is included in the hosting capacity as calculated before. Consider a 30 MW hydropower generator already connected to bus C. The hosting capacity for new generation, for example, a wind park, is in that case only 10 MW for buses C and 35 MW for buses C and D together.

As shown in Section 4.3.2, the presence of parallel flows could result in a very small hosting capacity. During a high power transfer through the region, the loadability of line C–D could become the limiting factor. When A–B is a single connection (i.e., its loss is a single contingency), the transported power through the region can never be more than 30 MW. A more likely situation is that A–B is a double-circuit line or a corridor with a number of lines in parallel. The loss of one circuit would increase the power flow through C–D. The maximum amount of power that can be transported through the region is the one that gives 30 MW power through connection C–D after loss of part of the connection A–B. Using (4.20), the maximum power through A–B is

$$P_{\mathrm{AB}} = \frac{(z_{\mathrm{CD}} + z_{\mathrm{AC}} + z_{\mathrm{BD}})}{z_{\mathrm{AB}}} P_{\mathrm{CD}} + \frac{z_{\mathrm{AC}}}{z_{\mathrm{AB}}} P_{\mathrm{C}} - \frac{z_{\mathrm{BD}}}{z_{\mathrm{AB}}} P_{\mathrm{D}} \tag{4.21}$$

with P_{CD} the maximum permitted power through C–D, P_{C} and P_D the loadings (consumption minus generation) at nodes C and D, respectively, and z_{AB} the highest value of the impedance between A and B after the loss of one circuit.

Adding generation to bus C will reduce P_{C} and thus reduce the amount of power that can be transported through the region. The larger the z_{AC}, that is, the longer the line A–C, the more the transfer capacity is impacted by generation connected to bus C. When large power transfers through the region are common, the hosting capacity could be very small or even zero. In case large power transfers occur in both directions, the hosting capacity for new generation at bus D may be equally small.

Also in this case, intertrip schemes might be able to result in a higher hosting capacity. One possibility is to trip connection A–C when one of the circuits between A and B is tripped. The network becomes radial and loop flows are no longer possible. Alternatively, the generation, or part of the generation, at bus C might be tripped.

Intertrip schemes use rather simple logic. The result is often a complete trip of an installation in all cases, even when no overload results and even when a small reduction in generation is enough. Modern protection and communication technology allows for more subtle solutions. For example, the generation connected to bus C could be limited such that the power flowing through connection C–D remains just below its limit. Such limits could be activated upon loss of one of the circuits between A and B or only using the power flowing through connection C–D as an input parameter. With multiple generators feeding into the same bus, or at nearby buses, market principles could be used to determine which generator limits its production first.

An intertrip scheme will also allow more generation to be connected in the case shown in Figure 4.11, where the generation connected to bus C will result in an

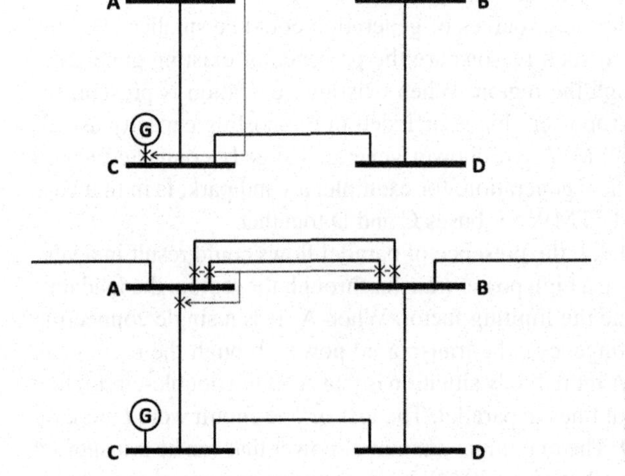

Figure 4.16 Intertrip scheme in a meshed system.

Figure 4.17 Alternative intertrip scheme in a meshed system.

overload of connection C–D during large power flows through the region. As shown before, the resulting hosting capacity could be small or even zero when the $(N-1)$ principle has to be fulfilled. The network is reproduced in Figure 4.16, where the two connections between A and B are shown. The majority of the power will flow between A and B but some of the power will flow through C–D. Upon loss of one of the connections between A and B, connection C–D will become overloaded. Adding generation to bus C will increase the amount of power flowing through C–D for the same flow between A and B. When the transport capacity of C–D is the limiting factor, the presence of generation at bus C will reduce the amount of power that can be transported through the region. The intertrip scheme will trip the generator whenever one of the four breakers in the connections between A and B is opened. In this way, the $(N-1)$ principle is still fulfilled, while at the same time a (much) larger amount of generation can be connected.

An alternative intertrip scheme would be to open the loop upon the loss of one of the connections between A and B, as shown in Figure 4.17. This scheme requires less wiring, especially when the distance protection is already equipped with an intertrip scheme. This scheme also would not require any tripping of the generation. The costs of the scheme would thus be less than those in Figure 4.16. However, the various operational states should be studied carefully to verify that no other overload situation occurs.

4.4 LOSSES

Estimating the increase or decrease in losses due to the introduction of generation to the distribution network requires much more detailed studies than estimating if a possible overload may occur. An increase in losses can be compensated by a reduction at another moment in time or at another location. What matters are the total losses over all feeder sections and integrated over time. Without distributed generation, these total losses are

$$F_0 = \sum_{s=1}^{N_S} \int_0^T \{L_s(t)\}^2 \, dt \tag{4.22}$$

with N_S the number of feeder sections and L_s the load downstream of feeder section s. It has been assumed here that the resistance per unit length is the same throughout the feeder. With generation present, the production G_s downstream of feeder section s should also be included, resulting in the following expression for the total losses:

$$F_{DG} = \sum_{s=1}^{N_S} \int_0^T \{L_s(t) - G_s(t)\}^2 \, dt \tag{4.23}$$

The reduction in the losses due to the introduction of generation is the difference between (4.22) and (4.23), which can be written as

$$\Delta F = \sum_{s=1}^{N_S} \int_0^T G_s(t) \{2L_s(t) - G_s(t)\} \, dt \tag{4.24}$$

As long as this expression is positive, the losses are reduced by the introduction of generation. As long as $2L_s(t) - G_s(t) > 0$, for all t and all s, the expression under the integral is always negative, so the integral and the subsequent summation always give a negative result, and thus a reduction in losses. However, this is an overly strict condition, far below the actual hosting capacity (where the total losses would start to increase).

When the feeder resistance differs with location along the feeder or when the total losses over several feeders are considered, the I^2R per feeder section and per time interval should be added to obtain the total losses.

Exercise 4.4 Reformulate the above expressions when including the difference in feeder resistance with location.

Assuming that load and generation have similar behavior in time for all feeder sections, the following condition can be obtained:

$$G_{\text{mean}} < 2 \times L_{\text{mean}} \tag{4.25}$$

In other words, as long as the average generation is less than twice the average load, no significant increase in losses is expected. The condition that load and generation have the same time behavior may seem rather unrealistic. However, the result will be valid approximately as long as the generation pattern is not opposite to the load pattern. It should also be noted that an exact calculation of the hosting capacity is not needed where the losses are concerned. An increase in losses by some tens of percent (compared to the existing level of losses, for example, from 5% to 6% of the consumed energy) is not really a concern, neither from a cost viewpoint (losses are a minor part of the total cost of electricity supply) nor from an environmental viewpoint (the gain by using renewable or energy-efficient sources is much higher than the increase in losses).

A more detailed study of the losses may, however, be needed to determine the economic impact on the network operator or the specific reduction in losses due to one generator. Such a study would require detailed information on load and generation patterns and their correlation.

4.4.1 Case Study 1, Constant Production

The losses have been calculated for the same consumption pattern as in Section 4.2.3. The losses over the 4-week period for the system without generation are rated as 100%. Next, the losses have been calculated for different amounts of generation with different ratios between generated active power and consumed reactive power. The results are shown in Figure 4.18. For each of the curves, the ratio between generated active power and reactive power consumed by the generator was assumed constant. The lower curve corresponds to zero reactive power (unity power factor). For the upper curve, the reactive power was equal to the active power (power factor 0.7). For the intermediate curves, the ratio was (top to bottom) 0.8, 0.6, 0.4, and 0.2.

The first hosting capacity (where the losses are the same as those without DG) is shown in Table 4.5 as a function of the ratio between generated active power and consumed reactive power, both for the loading and for the losses. In both cases, the hosting capacity is the amount of active power that can be injected without increasing the maximum apparent power or the total losses. In both cases, the hosting capacity decreases significantly when the generator units consume reactive power.

All of the results presented in Table 4.5 have been obtained by combining the measured consumption with a simplified model of the production. The values for

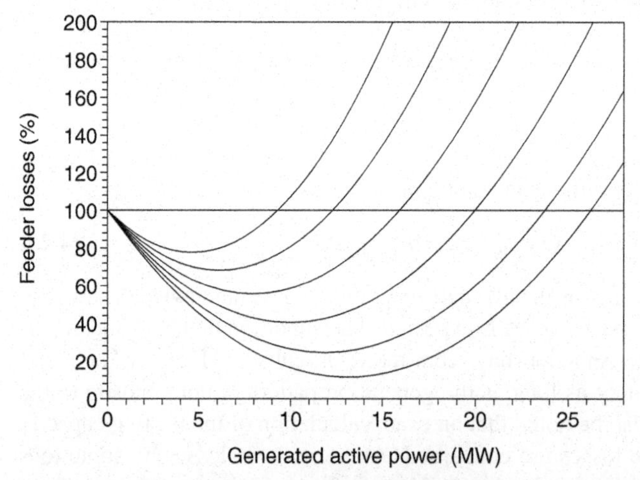

Figure 4.18 Feeder losses as a function of the amount of generated active power. The different curves correspond to different amounts of reactive power consumed by the generators.

TABLE 4.5 First Hosting Capacity for Overload and Losses: Measured Consumption Pattern

Q/*P* ratio	Power factor	Overload	Losses
0.0	1.00	25.3 MW	26.1 MW
0.2	0.98	23.7 MW	23.6 MW
0.4	0.93	21.3 MW	19.7 MW
0.6	0.86	18.5 MW	15.5 MW
0.8	0.78	16.0 MW	11.9 MW
1.0	0.71	13.1 MW	9.0 MW

The first column gives the ratio between consumed reactive power and generated active power for the generator unit.

active and reactive power consumption were obtained from a measurement as 1 min averages over a 4-week period. Whereas the consumption varies continuously in a realistic way, the production is assumed constant, in both active and reactive power, during the whole 4-week period.

4.4.2 Case Study 2: Wind Power

When the distributed generation is not always available, the reduction in losses becomes less. To quantify this, we again assumed an average wind speed of 7 m/s with a shape factor of 2.0. The same calculations as in Figure 4.3 have been performed to obtain time series of apparent power for different amounts of wind power. The losses have been calculated as the sum of the squares of all the apparent power values over the measurement period. The results are shown in Figure 4.19, where the losses for

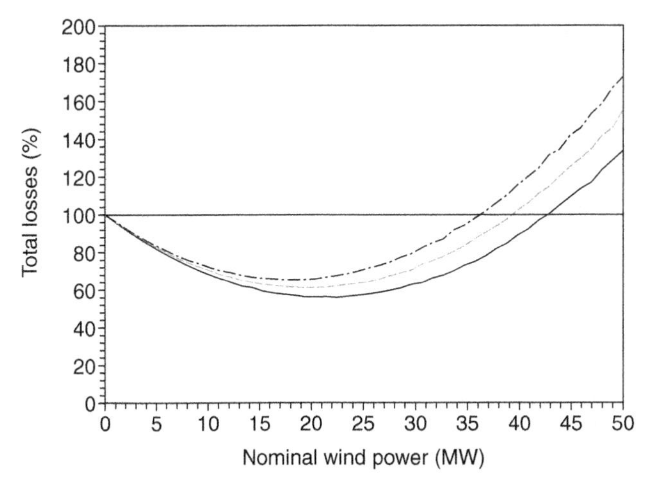

Figure 4.19 Estimated total feeder losses as a function of the amount of wind power: zero reactive power (solid line); 25% reactive power at peak production (dashed); 40% reactive power at peak production (dotted).

the no-wind case have been set to 100%. The different curves correspond to different amounts of reactive power consumed by the generator. The figure shows a decrease in losses up to about 20 MW wind power. The minimum losses are about 55% of the losses for the zero-wind case. For higher amounts of wind power, the losses start to increase again, and for more than 43 MW wind power, the total losses become higher than those without any wind power.

The impact of the consumption of reactive power is that the losses increase: the minimum occurs for a smaller amount of installed wind power and is less deep. Also, the first hosting capacity (where the losses are the same as those without wind power) is reached for a smaller amount of installed capacity.

4.5 INCREASING THE HOSTING CAPACITY

The calculations shown in this chapter indicate that the loading capacity of a distribution feeder does not put a severe limit on the amount of distributed generation that can be connected. The first hosting capacity, beyond which the maximum current becomes higher than that for the no-wind case, is equal to the sum of maximum and minimum consumption. Still the hosting capacity is sometimes exceeded, typically when large amounts of DG are connected to weak parts of the distribution grid. At subtransmission level, situations may, however, occur where the hosting capacity is low or even zero.

There are a number of ways to increase the amount of generation that can be connected to the power system. Some of these solutions are technically simple and cheap, others are simple but expensive (like building new transmission lines), others are complicated and expensive, still others remain a technical challenge and might not be practically implementable yet. In the forthcoming sections, we will discuss some possible solutions to increase the hosting capacity for distributed generation. Here we will mainly consider those cases where the risk of overload is the limiting factor. A minor increase in losses is not a serious concern, compared to a minor overload of a line. In the latter case, the line will trip, often resulting in an interruption for one or more customers.

Most of the solutions to be discussed below will be considered when a large generator unit (larger than the hosting capacity) wants to connect to the network. But a network operator might decide to be proactive by increasing the hosting capacity for smaller distributed generation, including microgeneration.

4.5.1 Increasing the Loadability

The thermal rating of a line or cable can be increased by using wires with a larger cross section. The simple rule holds here that more copper means more power that can be transported. Instead of rewiring the connection, a new line or cable can be placed in parallel. Two identical lines or cables in parallel have twice the transport capacity as one line or cable. Wires with a larger cross section also result in reduced losses.

This solution works for any size of generation, but the costs of rewiring can be rather high when connecting larger generation units to remote parts of the network. With the connection of a large unit, the network operator would typically charge these costs to the generator. When the overload is due to several smaller units, this is not always possible and costs may be charged to all customers.

4.5.2 Building New Connections

New lines and cables may also reduce the loading of existing connection. Dedicated lines or cables could be used for larger installations so as to connect them to parts of the network with sufficient transport capacity. This might be at a higher voltage level, in which case also a new transformer is needed. Again, the costs could become high.

Building new connections will typically also reduce the total losses in the system. However, dedicated feeders will often increase the losses because generation and consumption no longer have the possibility to cancel each other. The situation that typically occurs is that the losses increase at the voltage level where the generation is connected (e.g., 11 kV) but they decrease in the transformer at a higher voltage level (e.g., 33 kV).

4.5.3 Intertrip Schemes

Instead of not allowing new generation to be connected to the system, it can be allowed during those operational states for which it will not result in overloading. By using intertrip schemes, also referred to as "special protection schemes," as discussed in Section 4.3.3, the generator is disconnected once the network configuration changes. In this way, the generator can produce power whenever the network is in its normal state, which is the majority of time. An example is the intertrip scheme installed on a wind farm in New Zealand connected to the 66 kV grid [67]. Upon the tripping of one of the 66 kV lines connected to the local substation, a number of medium-voltage feeders in the collection grid are tripped, reducing the production to a value below 30 MW.

Intertrip schemes make sense only for larger installations, such as the ones connected to subtransmission systems, when the $(N-1)$ principle could otherwise severely restrict the amount of generation that can be connected. Manual disconnection would be too slow to prevent overload. One may argue that intertrip schemes like this do not comply with the $(N-1)$ criterion, after all a customer (the generator unit in this case) experiences an interruption. This would obviously have to be discussed with the generator owner; the alternatives would be a smaller generator or no generator at all.

At distribution level, such schemes might be used to prevent overloading during backup operation. Intertrip schemes could be connected to the normally open switch, disconnecting generation once this switch is closed. But as the closing of this switch is a manual operation (either locally or from a control room), detailed switching instructions could serve the same purpose and be more effective. For example, only generation downstream of the normally open switch has to be tripped, but the downstream direction depends on which of the feeders is faulted. This is not always easy to implement in a standard intertrip scheme.

Developments in protection and communication technology make it possible to implement more complex schemes, for example, considering the downstream direction for a normally open switch. Alternative overload protection schemes could be designed, for example, if line current exceeds threshold and local generation exceeds local consumption, and trip generation. If the current remains above the threshold after the generation is tripped, the line should obviously be tripped to prevent overload damage.

Any intertrip scheme can only be applied to larger generator units. For wind parks connected to the subtransmission level, the costs of an intertrip scheme (even including communication links) are negligible compared to the total costs of the installation. But for small generator units, such as domestic microgeneration, intertrip schemes do not appear practical.

The disadvantage of any intertrip scheme is that it has to be preprogrammed and should cover all situations. A fail-to-trip could easily result in the loss of several connections and an interruption for a larger geographical area, depending on the voltage level concerned. The network operator should avoid this as much as possible, resulting in many unnecessary trips. For example, when the presence of a generator results in overload with maximum generation and low load, the intertrip scheme will also trip the generator at high load and for lower amount of production. More advanced protection schemes, as will be discussed in the next section, can reduce the number of unnecessary trips, but the basic problem remains. It should also be noted that intertrip schemes are proven technology with which many network operators have experience. However, the advanced protection schemes are not in use yet and will probably require further development and pilot installations.

4.5.4 Advanced Protection Schemes

The hosting capacity approach presented in Sections 4.2.1 and 4.2.2 is based on the worst case: maximum production minus minimum consumption is not to exceed the ampacity of the feeder. However, with many types of distributed generation, maximum generation is available only for a small percentage of time. Also, the demand, most of the time, is larger than the minimum demand.

An alternative approach would be to have a kind of "time-dependent hosting capacity," where at every moment in time, the maximum amount of generation that can be injected is calculated. This value is then communicated to the generation and if the production exceeds the hosting capacity of the moment, the production is curtailed. When the generation consists of one or two large units, the communication is straightforward. When the generation consists of many small units, an aggregation scheme is needed. Various aggregation methods can be used. The concept of microgrids represents an aggregation scheme, where enough local power generation along with local loads is looked upon as one controllable cell from the power system viewpoint [92, 113, 151, 263]. Within the microgrid a communication infrastructure, which sets the control reference for all the energy sources, introduces a solution for an optimization problem in order to produce the required power output of the cell

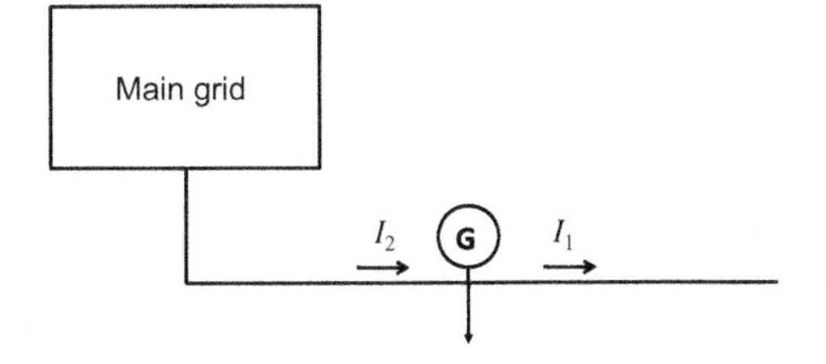

Figure 4.20 Advanced overload protection scheme in a radial distribution feeder.

and at the same time utilize the energy sources in an economical manner [202]. An example of such an aggregation structure is the Danish pilot cell project [282], which aims at developing a new solution for optimal management and active grid utilization of the large amount of distributed generation present in Denmark.

In a radial distribution feeder, an overload can only be mitigated by reducing downstream load. With reference to the situation shown in Figure 4.20, the production by the generator can impact only the current I_2, but not the current I_1. When I_2 exceeds its maximum limit ("overload limit") and the local production exceeds the local consumption, the overload can be mitigated by reducing the production. When the consumption dominates, the overload can be mitigated by increasing production or reducing load. Of interest for the discussion here is the situation where the overload is due to the generation. Instead of tripping the feeder by the overload protection, the "advanced overload protection" reduces the generation until the current becomes less than the feeder capacity. Alternatively, the protection scheme calculates the maximum permissible production (being the sum of the feeder capacity and the local consumption) at regular intervals. This value is communicated to the generator. If the production exceeds the maximum permissible production, the production is curtailed.

The necessary condition for this scheme to function is that the direction of the main source is known. In a radially operated meshed network, the direction to the main source can change. This requires measurement of two currents (I_1 and I_2 in Figure 4.20) as well as information about the network configuration to be available to the protection scheme. Such information may be obtained from a SCADA system, or alternatively, the protection scheme could determine the direction locally by using correlations between changes in production and changes in power flow on both sides of the connection point. In either case, both currents need to be measured. This scheme will allow larger units to be connected, but the connection costs will be higher because of the additional protection scheme and the additional components such as current and voltage transformers.

At subtransmission and transmission level, such an advanced protection scheme gets more complicated as the load at a certain location can influence the current through several connections. Also, it is no longer obvious what the upstream and downstream directions are. When one generator, for example, a large wind park, is responsible for the overload risk, the implementation of an advanced protection scheme is relatively simple. Suppose that the connection of a windpark at location A is limited by the transfer capacity of connection X. To connect a wind park with a higher maximum production, the production should be curtailed whenever the current through connection X exceeds its maximum permissible level. Here it is obviously

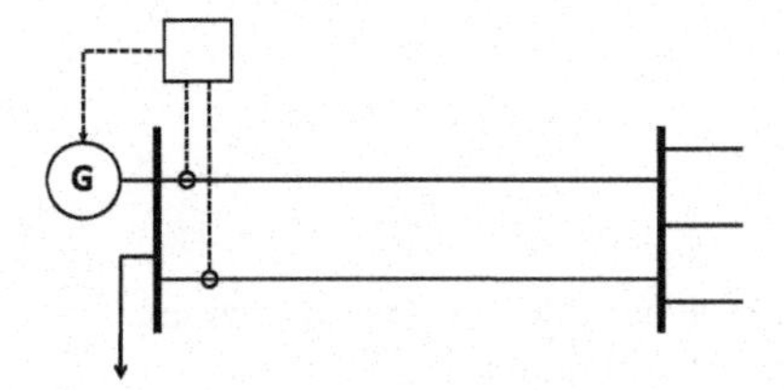

Figure 4.21 Advanced protection scheme for generation connected via a double-circuit line.

assumed that there is no overload risk before the connection of the wind park. In other words, every overload of connection X is due to the wind park.

In theory, it should still be possible to calculate a production ceiling for the wind park from measurement of a number of power flows. However, this may require measurement at several locations. In this case, a reduction of production whenever the loading through connection X exceeds its limit seems the easiest solution. The speed of reduction would have to be a compromise between the risk of the actual overload protection tripping the connection and the risk of excessive but unnecessary reduction of the production.

To be able to comply with the $(N-1)$ principle, the production will have to be curtailed rather quickly. The earlier discussion mainly concerned rather slow changes in loading: due to changes in demand and due to changes in production. This loss of a connection will, however, result in an almost instantaneous, often large, change in loading of the other connections. For example, when one circuit of a double-circuit line trips, the current through the other circuit will roughly double. This situation is shown in Figure 4.21. An intertrip scheme, as discussed before, would lead to excess curtailment, especially when one of the circuits is out of operation regularly.

Implementing the idea of controllable cells with assigned lower curtailment limits seems to be a feasible paradigm with large amounts of small generation. The lower limit will depend on the types of active units within the cell or on its energy management optimization design. For example, if storage and generation units exist, a decision may be made between how much energy to store and how much production to reduce. A communication infrastructure is then needed to communicate the output power flow and the highest and lowest limits of each cell to a central controller. A control hierarchy could then provide a set of commands for each cell, and when the production curtailment limit is reached the protective action will be initiated. The primary control action, that is, curtailing the output power flow of the cell, can be performed very fast using the storage capability of the cell, which is usually connected through power electronics devices that make such a fast response possible. The same fast response is possible with any generation with a power electronics interface such as solar power or micro CHP. For wind turbines with direct connection to the grid and pitch control, a fast response is still possible before the overload protection is initiated. Combined with the microgrid operational possibility, when the cell is disconnected from the grid due to the protection action, the energy sources may continue to operate in islanding mode.

An advanced overload protection scheme could be used that sets a production ceiling based on the momentary load. This ceiling is regularly updated. If the updates occur often enough and if the curtailment of the generation takes place fast enough, the $(N-1)$ criterion can still be guaranteed. For the double-circuit line in the figure, the production ceiling could be calculated as follows:

- IF demand exceeds generation THEN no ceiling.
- IF generation exceeds demand AND both circuits in operation THEN ceiling = 2 times capacity plus demand.
- IF generation exceeds demand AND one circuit in operation THEN ceiling = capacity plus demand.

Demand curtailment could be implemented as well to prevent overload due to excessive demand. Next to a production ceiling (when production exceeds demand), a demand ceiling is needed when demand exceeds production. The ceilings could be calculated using the following algorithm:

- IF demand exceeds generation AND both circuits in operation THEN demand ceiling = 2 times capacity plus production.
- IF demand exceeds generation AND one circuit in operation THEN demand ceiling = capacity plus production.
- IF production exceeds demand AND both circuits in operation THEN production ceiling = 2 times capacity plus demand.
- IF production exceeds demand AND one circuit in operation THEN production ceiling = capacity plus demand.

When the generation consists of many small units, an aggregation scheme could again be implemented. The same would hold for load curtailment when several customers are involved.

In a meshed system, as in Figure 4.11, it is no longer possible to give an easy expression for the amount of generation that can be injected before overload of one of the components occurs. A full load flow would be needed. In case a control algorithm is used, a simplified expression based on the DC load flow could possibly be used.

4.5.5 Energy Management Systems

Establishing an energy management system (EMS) has the advantage of avoiding the need for changes in the infrastructure (or primary components) of the system through the use of the information and communication technology (ICT) and the flexibility of the DG units [410]. In other words, the elements of an energy management system can be retrofitted into the existing system, hence potentially minimizing the cost and engineering efforts to increase the hosting capacity. This is accomplished through distributing the energy assignments of renewable generators over time by making use of their intermittent nature or through the use of storage units. Through synchronized measurements, data processing, and communications, different generators and storage units can be assigned different duties in different timescales. Assuming the renewable energy having a high share of the generation, the use of their intermittent nature, if enough information is available, plays a vital role to manage a particular part of the distribution grid [85].

In order to establish a successful energy management system, there are basically two system studies that have to be carried out:

- **Grid identification**: A control area has to be established. The use of storage devices in the control area is paramount in order to increase (or at least not deteriorate) the efficiency of a control area. Through power system analysis and identification of different active units' technologies and capabilities, a definition of effective control area should be possible. A control area may be defined as an area of the power system where the coordinated control of different active devices is possible in order to reach a global control goal, which in this context is increasing the hosting capacity without overloading the feeders or increasing the losses.
- **Operational scenarios**: All possible operational scenarios have to be studied for a specific control area. Detailed knowledge about all active devices and load patterns is essential [470]. A number of optimum operational solutions to meet, for instance, different cost functions may result. The mutual exclusive optimum solutions may be listed in a sort of database in order to be used for the EMS design.

As has been introduced in Section 2.9.4, the use of distributed control techniques (as opposed to central control of distributed active units) provides more flexibility and control capabilities for DG units connected in close proximity. The use of such distributed control techniques in Open Modular Architecture Control (OMAC) may facilitate the connection of new DG units to the distribution system and its operation as a part of the overall energy management system [18]. One approach of OMAC design is known as the Knowledge Server for Controllers (KSC). It is defined as a server providing the capability of intelligent data processing for other systems. It allows a client system to reach and exploit external intelligent processing resources because it does not have any [316]. The KSC represents a database with hundreds of executable modules that are dedicated to provide the optimum solution for different problems. A data transfer between the KSC and a client control allows for the choice of the proper module and hence solution on request from the client, as shown in Figure 4.22. A KSC can host a number of clients in parallel, which reduces the cost and size of client systems (DG units in this context) and facilitates their connection and contribution to the grid in a smooth way.

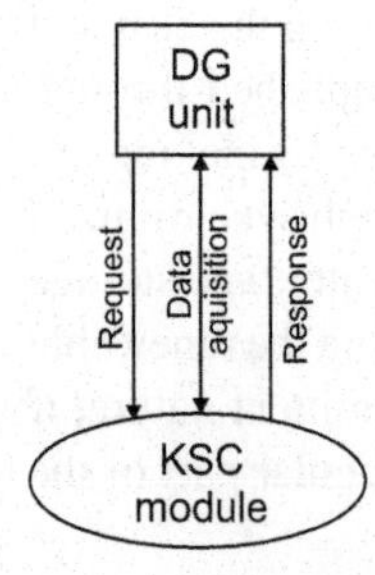

Figure 4.22 KSC approach.

Different approaches are also possible with the use of different automation techniques [410]. Many distribution operators are aware of the benefits of applying automation in distribution in general [385]. Such benefits are, for instance, improving the system reliability, planning, design, and response to outage and power quality issues. If distribution automation is established, the implementation of energy management systems is highly facilitated [269, 340].

4.5.6 Power Electronics Approach

The power electronics approach can be utilized in order to increase the line loadability and balance distribution transformers' loading through power routing and as an enabling technology for energy management systems.

A series or parallel controlled device can be operated in a way to either change the line parameters or direct a portion of the power transmitted by one line to be transmitted by another line. Also, the charging of energy storage devices when the line is overloaded is another tool to achieve energy management and increase the hosting capacity. In all cases, power electronics is the main tool for introducing a fast and enhanced dynamic performance of the system.

4.5.6.1 Dynamic Series Compensation A typical solution using series flexible AC transmission system (FACTS) devices usually introduces a means of manipulating the series impedance of a feeder through the connection of controlled series reactances. This in turn increases the feeder's power transfer capacity and hence increases the second hosting capacity as expressed in (4.6) or (4.11). A typical dynamic series compensator can be modeled as an impedance-dependent voltage source connected in series with the line. For details about the different practical devices that represent a dynamic series compensator, the reader may refer to any textbook on FACTS (for instance Ref. 297).

The dynamic series compensator can basically operate in two modes: tracking mode or regulating mode. In the tracking mode, the virtual impedance of the dynamic compensator changes in a way to increase or decrease the power flow through the feeder reflecting the changes in the generation and load. On the other hand, in the regulating mode the virtual impedance of the series compensator changes in a way to keep the power flow through the feeder constant regardless of the changes in generation and load. This latter case is possible in meshed networks. To better understand the basic operation of dynamic series compensation, the following example is set.

Example 4.5 Consider the simple two parallel lossless feeders in Figure 4.23. In the figure, line 1 (connected to bus 1) is series compensated using a power electronics device. Assume that the summation of local loads and generators results in the power flow direction as indicated in the figure. The line parameters and balanced system data are as follows:

$$X_1 = 0.5\,\text{pu}$$
$$X_2 = 0.5\,\text{pu}$$
$$\sin\delta = 0.5 = s_1$$

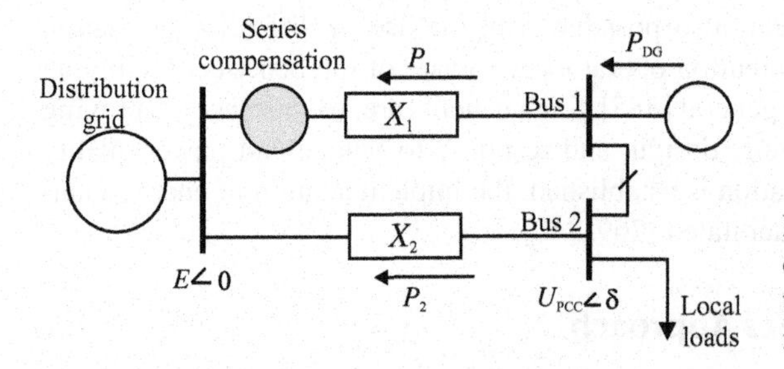

Figure 4.23 Example of dynamic series compensation to control power flow in a meshed network.

The series compensation has a compensation factor

$$\frac{\lambda}{X_1} = -0.3 = s_2 \tag{4.26}$$

where λ is the virtual impedance of the series device and is assumed negative to provide a capacitive compensating effect. Note that λ may also assume positive values for inductive compensating effect. The impedance of the series compensated line is then

$$X_s = X_1 + \lambda = X_1(1 + s_2) = 0.35. \tag{4.27}$$

This results in $P_1 = 1.428$ pu and $P_2 = 1$ pu.

Note that s_1 and s_2 are the two controllable states in this example. It is further assumed that the operational limit of the lines is $P_{\mathrm{ref}} = 1.44$ pu. It is assumed that the generation has suddenly increased 0.12 pu with $t = 0.3$ s. With the series compensation working in a regulating mode, this increase should be forced to flow through line 2 since line 1 is already reaching the operational limit. Using the first state (s_1) as a control signal, a simple proportional–integral (PI) control can be implemented as

$$s_1 = k_{\mathrm{p}}(P_1 - P_{\mathrm{ref}}) + k_{\mathrm{i}} \int (P_1 - P_{\mathrm{ref}} \mathrm{d}t \tag{4.28}$$

where the proportional gain is set $k_{\mathrm{p}} = 0.1$ and the integral constant is set $k_{\mathrm{i}} = 0.1$. As shown in Figure 4.24, the dynamic series compensation has regulated the power flow through line 1. The power flow through line 2 has increased to about 1.12 pu to accommodate the increase in generation. The compensation factor of the dynamic compensator has changed from -0.3 to -0.218.

The emerging trend of series compensation is to implement voltage source converter (VSC) technology. As explained in Section 2.9, the VSC offers the independent control of active and reactive powers flowing through a feeder. A VSC series compensator is referred to as "series-connected synchronous-voltage source compensator" or "SSSC". An SSSC can absorb or inject P and Q independently; however, to control the active power flow, its DC side has to utilize some kind of storage. The amount of active power injected/absorbed consequently depends on the size of the storage utilized at the DC side.

In a way to provide more flexibility, a back-to-back (BtB) converter connection can be implemented. A series-connected BtB converter has been proposed in Ref. 383, as shown in Figure 4.25, in order to achieve load balancing in industrial parks. The two distribution transformers can have common or separate upstream power

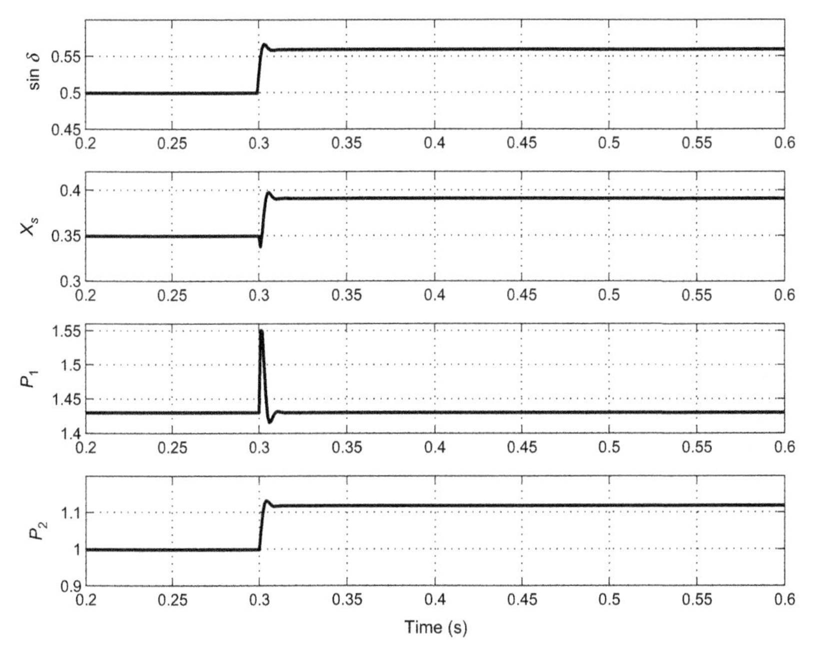

Figure 4.24 Dynamic series compensation of Example 4.5 for a step in the generation of 0.12 pu: δ is the voltage angle at the connection point of the generator; X_s is the series compensated line reactance; P_1 is the active power flow through the series compensated line; P_2 the power flow through the parallel-connected feeder.

source. The operation of such converter is somewhat similar to Example 4.5. The main advantage of such a setup is the controllability of both active and reactive power flows in both feeders at the same time without the need for communication signals. The current and voltage measurements at the connection point should suffice to determine the overloading condition and the required control action. A series-connected compensator, however, is basically connected at the sending end of the line. Hence, using local measurements may result in some sections of the feeder still becoming overloaded.

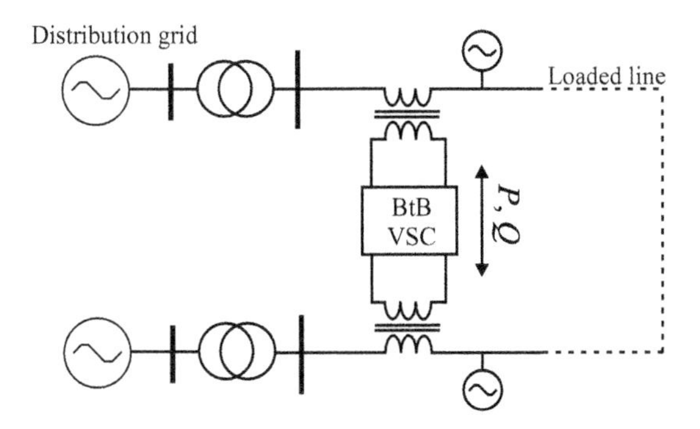

Figure 4.25 Transformer balancing through BtB connection or alternatively an MVDC transmission.

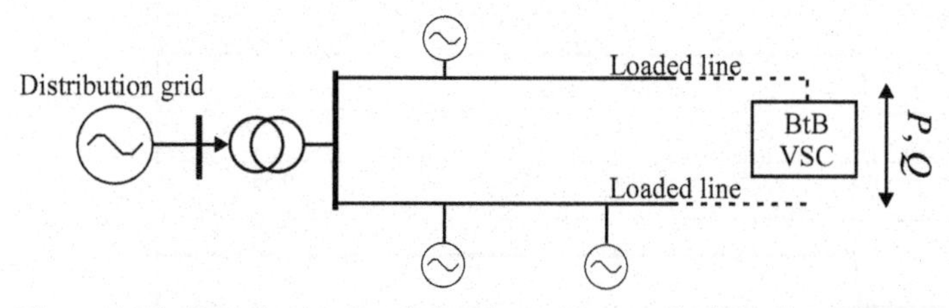

Figure 4.26 Distribution line balancing with an end-of-the-line BtB connection.

4.5.6.2 Dynamic Shunt Compensation Tying two feeders with a shunt-connected BtB system with the use of remote signals for the control is also a possibility. Shunt FACTS are usually aiming at compensating the voltage at the connection point through injecting reactive power. Hence, they are usually utilized at the end of a line.

By connecting two radial lines, a meshed distribution system is created. Through the use of BtB system as shown in Figure 4.26, both the active and reactive power transfers between the two lines are controlled. This solution may also target the first hosting capacity as its control aim.

4.5.6.3 Energy Storage The connection of energy storage devices through power electronics converters provides fast response in order to store the energy when the generation is reaching the hosting capacity, and then releasing the energy when needed. The storage is a key element of an energy management system and hence ICT is a key element for successful operation. The control aim here may target the first hosting capacity.

For the dynamic series and parallel compensation, the protection system should be made aware of their operation in order to avoid nuisance tripping during transient operation and also to provide correct tripping signals in case of compensator failure. Same applies for the storage; hence, the use of ICT is paramount for such solutions even though not used directly within the control of such devices.

All the above solutions assume that the DG units are mostly renewables, hence making the use of their (and also the load) intermittent nature. There may still appear a situation when the lines are overloaded and curtailing generation may still be needed.

4.5.7 Demand Control

Instead of curtailing the amount of production, the consumption can be controlled as well. This is a subject that is currently being discussed a lot. Certain loads could be held available for periods when the production by distributed generation is too high. Increasing the consumption will prevent overloading of the feeder. Such a scheme is discussed in Ref. 246 mainly to compensate fluctuations in wind power production. Charging of plug-in hybrid cars is mentioned as such a load. Also HVAC (heating, ventilation, and air conditioning) can be used to compensate fluctuations in production. The scheme can also be used to prevent overload without having to curtail generation. This requires, however, some preparation on the load side; for example, the cars should not be fully charged and the temperature setting of the HVAC should not be at its lowest or highest value. Some industrial processes are also said to offer opportunities.

One of the issues under discussion is the use of distributed generation and storage for reducing peak load. In Ref. 441, the use of storage systems is discussed. Peak load limitation is mentioned as one of the applications of a storage system. It is also mentioned that the requirements for peak load limitation may be conflicting with the requirements placed by other applications. The use of energy storage to reduce the peak load of a feeder is also mentioned in Ref. 112. According to that paper, a capacity between 500 kW and a few MW per feeder is needed for 2–10 h.

Demand control for large industrial installations has been in use for many years. These are nonautomatic schemes in which a phone call is made from the transmission system operator to the industrial installation to ask for a reduction in consumption. When the industry owns an on-site generator a request can be made to increase the production instead. Such systems require a certain lead time, most ideally during the day-ahead planning, but in some cases emergency calls are made requiring a reduction as fast as possible. Such systems do not function for a large number of small customers. When a system is systematically overloaded, rotating interruptions are sometimes put in place. This is also a method of demand control, but not a very convenient one.

More recently, discussions have started on using automatic methods combined with fast communication channels, to control the demand of large numbers of loads and the production of large numbers of small generator units. The most advanced system that has been tested in practice is the cell controller introduced by the Danish TSO [283]. The aim of the controller is that the distribution grid, seen from the transmission grid through the 150/60 kV transformer, operates in the same way as one single generator. For example, a reference for the reactive power exchange is set by the TSO. From this reference value for the total reactive power exchange, reference values are set for the reactive power exchange at the interface of each of the generators in the distribution grid. Also, static var compensators and synchronous condensers can be part of the scheme. In a similar way, the total current through the transformer and through other network components can be limited.

4.5.8 Risk-Based Approaches

Most of the limits set before, such as "maximum production less than feeder capacity plus minimum consumption," are deterministic limits. The main concern with such deterministic limits is that they are based on a worst-case approach. In this example, the amount of generation that can be connected is determined from the maximum production and the minimum consumption. For renewable sources of generation, the maximum production is reached only for a few percent of time. For many domestic loads, the minimum consumption is reached only for 1 or 2 h during a few days or weeks per year. As a result, the worst case (maximum production and minimum consumption) may only be reached once every few years. By allowing a small probability for the feeder to be overloaded, more distributed generation can be connected.

There are, however, a number of arguments against a risk-based approach and it is important to clearly mention those here. As explained in Chapter 3, one of the primary aims of the power system (i.e., of the network operator) is to ensure a high reliability of supply. Preventing overloads is "only" a secondary aim, but an important one because of the consequences of an overload. When a distribution feeder is overloaded, the overload protection will trip the feeder after a certain time. The higher the overload, the faster the protection will trip the feeder. If the feeder is not equipped with overload protection, excessive currents would significantly reduce the lifetime of the components. Overload situations will thus, either immediately or in due time, result in interruptions for the customers. Using a deterministic criterion based on the worst case is thus in the best interest of the existing customers because it prevents supply interruptions due to overloads. It might, however, also result in a barrier to the introduction of renewable sources of energy.

Before any risk-based approach will be implemented in reality, a discussion is needed about who carries the risk: the customer experiencing the interruption; the network operator; or the generating customer. When the overload protection trips the overloaded component sufficiently fast so that it does not suffer any significant loss of life, the risk for the network operator is low. It is only the customer suffering the interruption that carries the risk. However, when penalties, compensation, or price caps are associated with the number of interruptions, the network operator will likely be reluctant to accept more interruptions. The role of the regulator becomes very important in that case. A scheme could even be introduced where the generating customers are responsible for paying compensation when excessive production results in supply interruptions. In that case, the risk is carried by the generating customers. To remove barriers against the integration of renewable sources of energy, a national government may even decide to spread the compensation costs over all energy consumers or over all tax payers in the country.

To perform a risk analysis, appropriate data should be available on both consumption and production patterns. The presence of advanced metering has made it possible to collect detailed consumption data for many customers without excessive costs. Information on production patterns may be harder to obtain, but for sources such as sun and wind, there are reasonably accurate models available that are being improved continuously. Historic weather data are available for most locations in the industrialized world over decades or longer. These form a good base to generate production data for distributed generation based on solar or wind power. Production data for micro-CHP will require further development, but historic temperature data will certainly be a good base here.

The number of interruptions that are acceptable varies strongly between customers. Some customers do not accept any interruptions at all, whereas others are willing to accept a certain number of interruptions per year. A number of regulators have set requirements on the maximum permissible number of interruptions; an overview of existing regulations in Europe is given in Ref. 53. The maximum permissible number of interruptions originating at medium voltage varies between 2 and 25 per year. Not many countries have requirements for low-voltage and high-voltage levels, so no conclusions can be drawn for those voltage levels. In Refs. 39 and 208, it is concluded from an analysis of customer complaints that up to three interruptions per year are acceptable to rural domestic customers, assuming that none of the interruptions lasts longer than 8 h.

At subtransmission and transmission level, the risk analysis becomes more difficult. A distinction has to be made here between two different risks.

- The risk of a supply interruption due to component overload. The probability of an interruption should be low because an interruption originating at subtransmission or transmission level will in most cases impact many customers. The tripping of a component by the overload protection will not always lead to an interruption because the network is operated meshed. However, tripping of one component due to overload often results in an overload elsewhere. The loss of two or more components often results in an interruption.
- The risk that the $(N-1)$ criterion no longer holds. The $(N-1)$ criterion is a secondary aim, where a rather high probability of noncompliance can be accepted as long as the primary aim (no supply interruptions) is maintained.

Calculation of these risks and probabilities is more complicated than for radial distribution feeders, but still not overly complicated. The data requirements are similar to those at lower voltage levels: data on consumption and production patterns and component availability. The choice of acceptable risk levels (probabilities; expected number of interruptions per year) is not straightforward and will require some further discussion. An important first step would be to estimate what are the risk levels in existing subtransmission and transmission systems. As long as the introduction of new sources of generation does not significantly increase the risk, no further changes in the network are needed.

4.5.9 Prioritizing Renewable Energy

When the hosting capacity has to be shared between different generators, priority could be given to the most environment-friendly sources of energy. This could be during the planning stage, by not allowing polluting sources to be connected, or only during the operational stage. The former would occur more likely at distribution level, whereas the latter is a more typical solution at subtransmission and transmission level. Consider, for example, again the subtransmission network shown in Figure 4.11. An existing coal-fired power station at bus C could severely limit the amount of wind power that could be connected. However, during operation, priority could be given to the wind power: the coal-fired power station would be curtailed or turned off during large interregional power transfer. When only wind power is connected to bus C, the amount of interregional power transfer could be limited to allow for more wind power to be produced.

4.5.10 Dynamic Loadability

The overload protection of a component such as a line or a transformer is based on comparing the current through the component with a current setting. This current setting is the maximum permissible current through the component and is determined such that this current will not cause damage to the component. The actual physical limit is, however, (in almost all cases) not the actual current but the temperature of

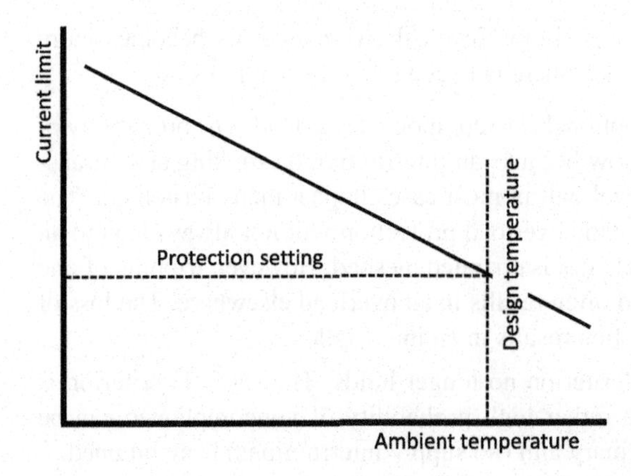

Figure 4.27 Current limit for a component as a function of the ambient temperature.

the conductor or of the insulating material. This temperature depends not only on the current but also on the ambient temperature and on the speed with which the heat is transferred to the environment. The latter is, for example, strongly dependent on the wind speed. During periods with lower ambient temperature and higher wind speed, more current could be allowed through the component for the same risk of damage due to overload. The actual current limit (i.e., where the maximum permissible conductor or insulator temperature is reached) will decrease with increasing ambient temperature and increase with increasing wind speed. This is shown schematically for the ambient temperature in Figure 4.27. The setting of the overload protection, which determines in the end how much power can be transported, is the limit for a design temperature. As the overload protection is nonflexible and should protect the component during any weather, the design temperature is taken on the high side, for example, the once in 10 years highest temperature. This is the right choice from a protection point of view, but it does result in a limitation of the amount of power that can be transported also during periods of colder weather, which means most of the time.

Using dynamic loadability, that is, changing the settings of the overload protection based on the ambient temperature and wind speed, would allow higher power transfer and could allow more renewable energy to be transported. The latter is possible only when dynamic loadability is combined with another method. Using the basic approach for the hosting capacity (maximum consumption plus minimum consumption) will still result in the worst case, that is, the lowest hosting capacity. However, using curtailment of generation could be very well combined with dynamic overload limits.

CHAPTER 5

VOLTAGE MAGNITUDE VARIATIONS

This chapter will address the impact of distributed generation on another important parameter in the design of distribution systems: the voltage along the feeder. Here, the impact of distributed generation is more direct and the hosting capacity might even be zero in some cases. As we will see in this chapter, the hosting capacity depends strongly on the choice of the performance index and on the model used to estimate the impact of distributed generation.

The structure of this chapter is similar to that of the previous chapter: Section 5.1 provides a general overview of the way in which distributed generation impacts voltage magnitude variations; Section 5.8 provides an overview of the various methods to allow more distributed generation to be connected to the system. Next, the phenomena are discussed in various levels of detail in Sections 5.2–5.7. A rather simple but generally applicable method for estimating the hosting capacity is presented in Section 5.2. This is a deterministic method; its stochastic equivalent is presented in Section 5.7. The design of distribution feeders without and with distributed generation is discussed in Section 5.3. The use of a model of the feeder allows us to study the impact of different parameters, such as the location of the generator, on the hosting capacity. An alternative method for describing the design of a distribution feeder is presented in Section 5.4. This method is less suitable to study real feeders, but it allows a better insight into the impact of different parameters. The method shows to be especially useful for calculating the hosting capacity for small generation spread over many customers. Tap changers with line-drop compensation are discussed in detail in Section 5.5 and probabilistic methods for the design of distribution feeders in Section 5.6.

Most of the examples in this chapter are based on "European-style" distribution networks with large distribution networks and up to a few kilometers of low-voltage distribution. North American distribution networks contain smaller distribution transformers and much smaller low-voltage networks (often called "secondary distribution"). The methods discussed here do however also apply to those networks.

5.1 IMPACT OF DISTRIBUTED GENERATION

In the design of distribution networks, the prevention of excessive voltage drops is a serious issue. It sets a limit to the maximum feeder length and different methods are

Integration of Distributed Generation in the Power System, First Edition. Math Bollen and Fainan Hassan.

in place to prevent or mitigate voltage drops. Overvoltages have traditionally been less of a concern. If the voltage at the start of a medium-voltage feeder is toward the upper limit of the permissible voltage range, the risk of overvoltages at the terminals of end-user equipment would be small.

The introduction of distributed generation has changed this and overvoltages have become a more serious concern. Especially for locations some distance away from a main substation, the voltage rise is the main limiting factor for the connection of distributed generation [123, 125, 233, 337, 432]. According to Ref. 233, the maximum capacity for generation is 2–3 MW out on the feeder and 8 MVA on the main 11 kV bus.

The voltage rise due to the injected active power is proportional to the resistive part of the source impedance at the point of connection. For remote locations, the voltage rise can be significant. The amount of active power needed for a 1% voltage rise is calculated for different feeder cross sections and lengths in Section 5.2. At 10 kV, 1.7% voltage rise is obtained by connecting a 1 MW unit at 5 km from the main bus along a 50 mm^2 feeder. For a 240 mm^2 feeder, the rise is only 0.4%. At low voltage, rather small units give a significant voltage rise. Connecting a 10 kW unit to a 50 mm^2 feeder at 500 m from the distribution transformer results in a voltage rise of 3%. These unit sizes are much smaller than the maximum load of the feeder; the voltage rise will limit the amount of generation that can be connected. Only for short feeders is the hosting capacity set by the maximum load. The biggest voltage rise is obtained for generators operating at unity power factor. Generators consuming reactive power, similar to induction generators, result in less voltage rise.

Calculations of the voltage rise due to a 500 kW wind turbine operating at 0.95 power factor are presented in Ref. 399. The voltage rise depends on the fault level and on the phase angle of the network impedance. The results are summarized in Table 5.1. The biggest rise in voltage (positive in the table) occurs for a small phase angle (low X/R ratio) and a low fault level (generator far away along a feeder). For high phase angle (high X/R ratio), the reactive power consumption by the wind turbine will even result in a voltage drop.

The hosting capacity strongly depends on the method used to prevent undervoltages. Using off-load tap changers for the distribution transformer gives high voltage magnitude during low load for locations far out along the feeder. This seriously restricts the amount of generation that can be connected. The hosting capacity becomes smaller quickly when the generation is connected further along the feeder. For an example feeder (18 km, 10 kV, 95 mm^2 Al), with commonly used design values, the

TABLE 5.1 Impact of Fault Level and Phase Angle of the Source Impedance on the Voltage Rise Owing to a 500 kW Wind Turbine [399]

Angle	2 MVA	5 MVA	10 MVA	50 MVA
85°	−12%	−4%	−2%	−1%
70°	−3%	0	0	0
50°	+6%	+3%	+2%	0
30°	+13%	+6%	+3%	+1%

hosting capacity is, in Section 5.3, shown to be as low as 250 kW for a generator connected toward the end of the feeder. However, the hosting capacity strongly depends on a number of design parameters, where the choice of overvoltage limit has the biggest influence. Using a 110% overvoltage limit instead of 108% results in a hosting capacity equal to almost 1 MW toward the end of the feeder.

As shown in Section 5.3.5, the use of shunt or series capacitors for voltage control typically results in a higher hosting capacity. This, however, also depends on the control algorithm used. When the capacitor banks are switched on based on the reactive power flow, this may further deteriorate the overvoltages [432]. Reference 432 also mentioned the problem with step voltage regulators equipped for bidirectional operation that allow the voltage to be controlled when the feeder is used for backup operation. These regulators will detect the wrong direction when reverse power flow occurs. This will result in severe overvoltages [432].

Tap changers equipped with line-drop compensation are less prone to overvoltages owing to distributed generation. However, the presence of a generator on one feeder will result in a voltage drop on the other feeders. The risk of undervoltages might limit the amount of generation that can be connected. This is discussed in detail in Section 5.5.

Simulation results for a 230 V low-voltage feeder are shown in Ref. 424. The total feeder length is 800 m and the furthest point is 500 m away from the distribution transformer. The maximum load is 2.5 kW and the feeder impedance is $0.27 + j0.27\ \Omega$. The simulation results show that a 90 kW photovoltaic installation can be connected, 350 m from the distribution transformer, before the overvoltage limit of 106% is reached. A stochastic approach is used in the study; a deterministic approach would have resulted in a hosting capacity equal to 65 kW.

Reference 31 refers to two photovoltaic plants in Austria, with 37 and 101 kW rated power. The calculated voltage rise, at maximum production, is 2.2% and 2.9%, respectively. Measurements on the 110 kW site, however, show that the increase in maximum voltage is only 1.5%. This is because minimum load occurs at different times of the day than maximum production. It shows that the daily variations of both load and production should be included, otherwise the hosting capacity can be strongly underestimated, by a factor of 2 in this example.

The potential voltage rise owing to large numbers of photovoltaic panels connected to low voltage has been calculated for the city center of Leicester, UK [408]. Stochastic models are used for the load in every house. A clear sunny day is used as this gave the worst-case voltage rise, although it is not typical for this part of England. When 50% of the homes are equipped with solar power of about 2 kW rating, the maximum voltage magnitude rises from 250 to 254 V. With a 110% (253 V) overvoltage limit, this would be an unacceptable overvoltage, although this occurs only for a small number of days per year and only for a few customers. An overvoltage limit at 110% would set the hosting capacity to 30% of the homes, which is still a significant penetration. The same network is used in Ref. 221 to study the impact of micro-CHP in combination with solar power. Only the impact on the average voltage is shown in the study: 107.0% without generation, 107.6% when 50% of homes is equipped with solar panels, 107.5% when 100% is homes is equipped with micro-CHP, and 108.1% when 50% of homes is equipped with solar panels and 100% with micro-CHP.

Measurements are shown in Ref. 24 of the voltage variations on a low-voltage feeder in Manchester, UK, supplying 69 homes, each equipped with a 1 kW micro-CHP unit. The feeder was specially designed to cope with large amounts of distributed generation and a 330 mm^2 cable was used as a precautionary measure, instead of the 95 mm^2 that would have been used normally. The measurements showed that the load currents are strongly unbalanced and that the unbalance increases toward the end of the feeder. Strongly unbalanced load increases the risk of overvoltages due to distributed generation; it also increases the risk of undervoltages due to high load. The measurements showed a small increase in voltage along the feeder, with the maximum voltage being close to 251 V at the start of the feeder, about 252 V near the middle of the feeder, and about 253 V at the end of the feeder. The latter could be a concern when an overvoltage limit of 110% (253 V) is used. It should also be noted that a 95 mm^2 cable would have given a voltage rise about three times as high, thus resulting in maximum voltage around 258 V or 112%. Measurements for the same network are also shown in Ref. 455: the increase in the highest 10 min voltage magnitude was only 1 V and the increase in the highest short-duration voltage was 3.6 V. If a 95 mm^2 cable is used, the latter would be 11.4 V or almost 5%. Unacceptable overvoltages would likely have resulted from that.

A study of the voltage rise along a 230 V low-voltage network in the suburbs of Dublin, Ireland, is presented in Ref. 365. The network consists of a number a feeders with a total of 3.1 km three-phase and 2.3 km single-phase copper cable with cross sections 120, 70, and 50 mm^2. The network supplies 307 low-voltage customers. A minimum load of 0.1 kW per customer was assumed. Micro-CHP was added to the network, where a maximum production of 1.2 kW at 0.98 power factor was assumed. For 100% of houses equipped with microgeneration, the maximum voltage during normal operation was 110.8% of nominal. This is at the borderline of acceptability. Setting the overvoltage limit at 110% would result in a hosting capacity between 62% and 100% per feeder. During backup operation, the feeder length is longer, resulting in a higher voltage rise. Voltage magnitudes up to 115.9% are possible during backup operation. To keep the maximum voltage below 110% even in backup operation, the hosting capacity should be as low as 23% of the houses on some of the feeders.

5.2 VOLTAGE MARGIN AND HOSTING CAPACITY

5.2.1 Voltage Control in Distribution Systems

The commonly used basic principle of voltage control in a distribution network is shown in Figure 5.1. The voltage on secondary side of an HV/MV transformer (e.g., 70/20 kV) is kept within a certain deadband by means of the automatic tap changer of the HV/MV transformer.

The active and reactive power consumption will result in a voltage drop along the medium-voltage feeder, which is lowest at minimum consumption and highest at maximum consumption. The further away the location from the main substation, the lower the voltage on the MV feeder will be. To compensate for this, MV/LV transformers with different turns ratios are used (these are sometimes referred to as

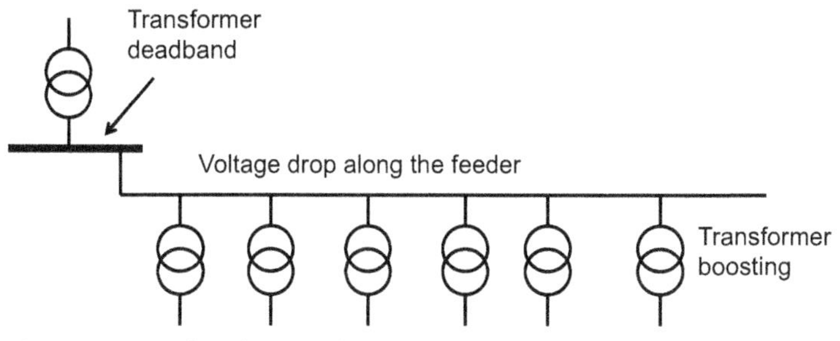

Figure 5.1 Basic principle of voltage control in a distribution network.

"off-load tap changers"). Toward the end of the feeder, the turns ratio is smaller (e.g., 9.5 kV/400 V) than toward the start of the feeder (e.g., 10.5 kV/400 V). The smaller the turns ratio, the higher the voltage on secondary side. In this way, the voltage in the distribution network can be boosted by up to 5%.

The voltage control should be such that the voltage magnitude remains within a band set by an "undervoltage limit" and an "overvoltage limit." There is a lot of discussion going on at the moment about what are appropriate overvoltage and undervoltage limits. Traditionally, many distribution network operators (they were called "utilities" in those days) used a range of 5–8% around the nominal voltage in their design of distribution feeders. Currently, most European network operators refer to the European voltage characteristics standard EN 50160 [134], where limits of 90% and 110% of nominal voltage are given. No matter what the limits are, the basic coordination principles remain the same. These principles are shown schematically in Figure 5.2.

The deadband is normally chosen somewhat above the nominal voltage to allow voltage drop along the medium-voltage and low-voltage feeders. As the next criterion, the voltage for the most remote customer during maximum load should be above the undervoltage limit. The voltage for remote customers can be boosted up to 5% by using a distribution transformer with a different turns ratio. At the same time, the voltage during low load should not exceed the overvoltage limit for any of the customers. The critical case for overvoltages is not the most remote customer but the one nearest to the main substation with voltage boosting. The latter case is shown in Figure 5.3.

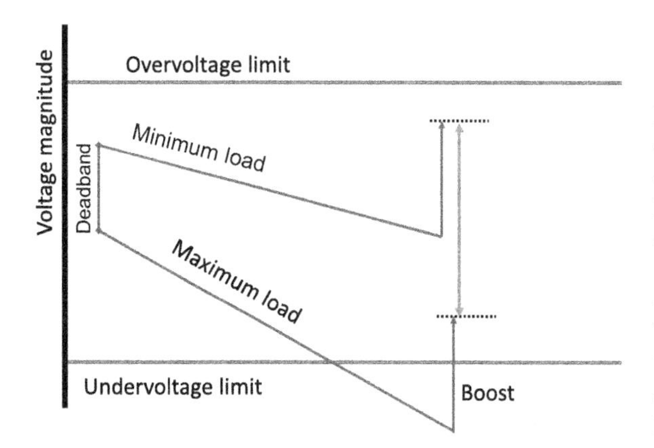

Figure 5.2 Voltage coordination along a distribution feeder. The horizontal axis gives the distance from the main MV substation on secondary side of the HV/MV transformer. The vertical axis gives the voltage on secondary side of an MV/LV transformer at this location.

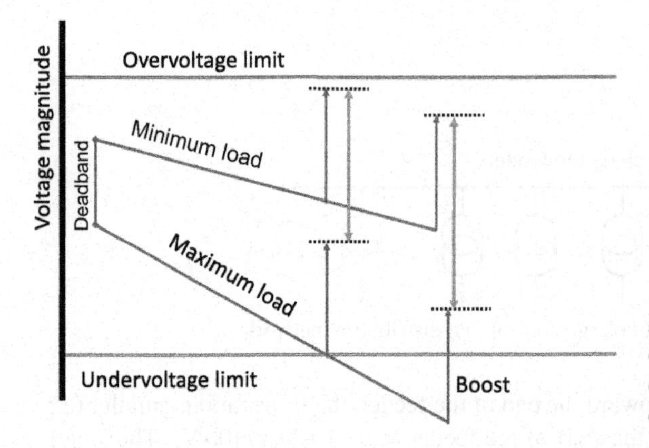

Figure 5.3 Voltage coordination along a distribution feeder: worst case for overvoltage.

The difference between the highest voltage magnitude, for a given customer, and the overvoltage limit is referred to as the "overvoltage margin."

The introduction of distributed generation will result in less voltage drop whenever the generator is producing active power. The result is that the overvoltage margin becomes smaller and the maximum voltage may even exceed the overvoltage limit. In that case, the power system no longer fulfills its primary aim to maintain the voltage magnitude within limits. In other words, the hosting capacity is exceeded.

5.2.2 Voltage Rise Owing to Distributed Generation

The connection of a generator to the distribution network will result in a voltage rise at the terminals of the generator. The relative voltage rise is approximately equal to

$$\frac{\Delta U}{U} = \frac{R \times P_{\text{gen}}}{U^2} \tag{5.1}$$

where R is the source resistance at the terminals of the generator, P_{gen} the injected active power, and U the nominal voltage. This approximation holds for all practical cases at distribution level. For more accurate expressions, see, for example, Chapter 2 of Ref. 45. We will assume here that the generator injects only active power at unity power factor. The case for nonunity power factor will be discussed in Section 5.2.4. The same relative voltage rise as according to (5.1) is experienced by all customers connected downstream of the generator, as shown in Figure 5.4, where the generator is connected to the medium-voltage feeder.

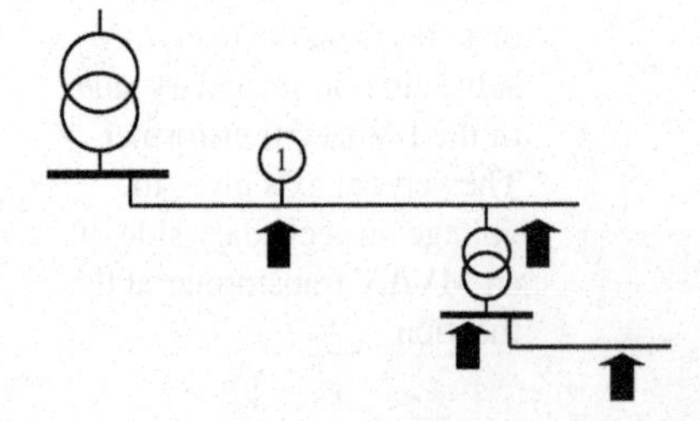

Figure 5.4 The relative voltage rise due to the generator is the same for all indicated positions.

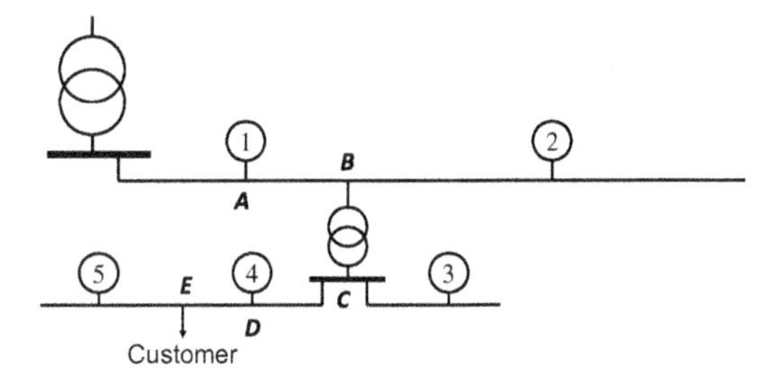

Figure 5.5 Points-of-common coupling (A–E) between a customer and generators at five different locations (1–5).

When, for example, the voltage rise is 2.5% at the point where the generator feeds into the medium-voltage feeder, all customers at the indicated positions in Figure 5.4 will also experience a voltage rise equal to 2.5% due to the power injected by the generator.

What matters for the voltage rise experienced by a given customer owing to a generator at a given location is the source resistance at the point-of-common coupling between the generator and the customer. The points-of-common coupling are shown in Figure 5.5 for a fixed customer location (somewhere along a low-voltage feeder) and five different generator locations (1–5). The point-of-common coupling between the generators 1–5 and the indicated customer location is at location A–E, respectively.

The point-of-common coupling is an important point because it is the resistive part of the source impedance at the point-of-common coupling that determines the voltage rise experienced by a given customer due to a given generator. It is this resistance that is used in (5.1) to calculate the voltage rise.

When the HV/MV transformer is equipped with automatic tap changers, as is typically the case, the voltage at the main MV bus (i.e., on MV side of the HV/MV transformer) is kept constant. In that case, the resistance R in (5.1) is the resistance between the point-of-common coupling and the main MV bus.

5.2.3 Hosting Capacity

The maximum permissible voltage rise, with connection of a distributed generator, is the one that brings the maximum voltage magnitude exactly at the overvoltage limit. The "overvoltage margin" introduced in Section 5.2.1 is defined as the difference between the maximum voltage magnitude for a given customer and the overvoltage limit. The "hosting capacity" introduced in Chapter 3 is the maximum amount of generation that can be connected without resulting in an unacceptable quality or reliability for other customers. From this it follows that when considering overvoltages, the hosting capacity is the amount of generation that gives a voltage rise equal to the overvoltage margin. Each customer has a different overvoltage margin. The connection of a generator to a distribution feeder gives the same relative voltage rise for every location downstream of the generator. What thus matters is the lowest value of the overvoltage margin downstream of the location at which the generator is connected. Here, we will consider this overvoltage margin and hosting capacity. The hosting capacity can easily be calculated from (5.1) as

$$P_{\max} = \frac{U^2}{R} \times \delta_{\max} \tag{5.2}$$

where $\delta_{max} = \Delta_{max}/U$ the relative voltage margin (in percent) and Δ_{max} the absolute voltage margin (in volt). To get an impression of what impacts the hosting capacity, we further use the relation between the resistance R of a wire, its cross section A, and its length ℓ:

$$R = \rho \times \frac{\ell}{A} \tag{5.3}$$

From (5.2) in combination with (5.3), we find the following relations:

- The hosting capacity is proportional to the square of the voltage level. At 20 kV, a generator four times larger than that connected at 10 kV can be connected.
- The hosting capacity is linear with the size of the line of cable: a 100 mm^2 cable can host a generator twice as large as a 50 mm^2 cable.
- The hosting capacity is linear with the overvoltage margin: 1% overvoltage margin allows twice as much generation to be connected as 0.5% overvoltage margin. It also follows that zero overvoltage margin gives zero hosting capacity.
- The hosting capacity is inversely proportional to the distance between the generator and the transformer. The further away from the transformer, the smaller the hosting capacity.

It should be noted here that, in practice, it is not possible to change only one of these parameters without impacting the others. For example, using a cable with a larger cross section will also impact the overvoltage margin.

The hosting capacity has been calculated for a number of 10 kV feeders of different lengths and with different wire cross sections, assuming a 1% overvoltage margin. The results are shown in Table 5.2. The specific conductivity of 1.68×10^{-8} Ωm (for copper) is used.

We can see from the table that the hosting capacity depends strongly on the feeder size and length. For short feeders, several megawatts of generation can be connected before the voltage rise becomes more than 1%. For those feeders, the thermal capacity will probably be exceeded before the voltage rise becomes excessive. For cables longer than several kilometers, the voltage rise is more likely to be the limiting factor, especially for long, small cross-sectional cables. For comparison, the loadability of a three-phase XLPE cable is shown in the last column [1]. For short cables, it is the maximum load that will limit the amount of power that can be connected, whereas for longer cables, it is more likely to be the voltage rise that sets the hosting capacity.

TABLE 5.2 Hosting Capacity for 10 kV Feeders with 1% Overvoltage Margin

	Hosting capacity for feeder length				
Cross section (mm^2)	200 m	1 km	5 km	20 km	Loadability
25	7.4 MW	1.5 MW	300 kW	75 kW	2.5 MW
50	15 MW	3.0 MW	600 kW	150 kW	3.6 MW
120	36 MW	7.1 MW	1.4 MW	360 kW	6.0 MW
240	71 MW	14 MW	2.9 MW	710 kW	8.9 MW

TABLE 5.3 Hosting Capacity per Phase for 400 V Feeders with 1% Overvoltage Margin

Cross section (mm^2)	50 m	200 m	500 m	2 km	loadability
25	16 kW	3.9 kW	1.6 kW	390 W	33 kW
50	32 kW	7.9 kW	3.2 kW	790 W	47 kW
120	76 kW	19 kW	7.6 kW	1.9 kW	80 kW
240	150 kW	38 kW	15 kW	3.8 kW	120 kW

The calculations have been repeated for low-voltage feeders with 400 V nominal voltage. The hosting capacity per phase is shown in Table 5.3. The hosting capacity ranges from more than 100 kW down to less than 1 kW. In the latter case, voltages exceeding the overvoltage limit can easily occur for small amounts of distributed generation.

Here, it should be noted that the feeder resistance is the same for an overhead line as for an underground cable, for the same length and cross section. However, lines are typically used to connect more remote customers and the maximum length of a line is more typically determined by the maximum voltage drop, whereas cables are typically used to supply customers closer to a main substation and the length is mostly determined by the maximum loading. As a result, overvoltage problem appear more often with generation connected to overhead lines than with generation connected to underground cable feeders.

5.2.4 Induction Generators

Wind power is often produced by induction generators. An important property of induction generators is that they consume reactive power. The reactive power consumption consists of the magnetizing current, proportional to the square of the voltage, and the reactive power losses in the leakage reactance, proportional to the square of the current. The reactive power taken by the magnetizing impedance is typically compensated by a fixed capacitor. What remains is an amount of reactive power proportional to the square of the current, and thus proportional to the square of the active power.

Assume that maximum production gives the highest voltage rise. The reactive power at maximum production is a certain fraction of the active power:

$$Q_{\text{max}} = \alpha \times P_{\text{max}} \tag{5.4}$$

The voltage rise at maximum production is equal to

$$\begin{aligned} \Delta U_{\text{gen,max}} &= RP_{\text{max}} - XQ_{\text{max}} \\ &= R\left(1 - \alpha\frac{X}{R}\right) P_{\text{max}} \end{aligned} \tag{5.5}$$

The second factor in (5.5) is independent of the size of the generator. We can therefore introduce an "equivalent generator size" equal to $1 - \alpha X/R$ times the actual generator size. The hosting capacity, in terms of actual generator size, for induction generators, using (5.2), is

$$P_{max} = \frac{1}{1 - \alpha \frac{X}{R}} \times \frac{U^2}{R} \times \delta_{max} \tag{5.6}$$

For distribution feeders, the X/R ratio varies between 0.2 and 1.5; the reactive power for an induction generator is between 0.15 and 0.30 times the active power, where the higher values hold for units connected through a transformer. The factor $1 - \alpha X/R$ thus varies between 0.97 and 0.55. For small X/R ratio (small cross-sectional wires) and small α (units connected without transformer), the impact of the reactive power is small. However, for large X/R ratio and large α, the hosting capacity may become twice the value for generators without any reactive power consumption.

Exercise 5.1 Repeat the calculations resulting in Tables 5.2 and 5.3 for distributed generation with induction machine interface. Use $\alpha = 0.17$ at low voltage and $\alpha = 0.25$ at medium voltage and obtain data on the feeder inductance from data books. Note that the inductance is different for cables than for lines, so the hosting capacity will no longer be the same.

The same reasoning as given here for induction generators applies for generators operated at a constant nonunity power factor, for example, 0.95 lagging. A lagging power factor will result in less voltage rise than unity power factor. However, a leading power factor will increase the overvoltage problem, that is, reduce the hosting capacity. By choosing a lagging power factor such that $\alpha = R/X$, the voltage rise can be reduced to zero. When the generator is equipped with a power electronics interface, a constant lagging power factor can be chosen such that the voltage rise is zero for any amount of power produced. The disadvantage is that the risk of overload and the losses somewhat increase. However, when the voltage rise is the limiting factor for the hosting capacity, that does not reduce the amount of distributed generation that can be connected. Another disadvantage is that the converter rating should be somewhat higher to accommodate the consumption of reactive power even at maximum production of active power.

Exercise 5.2 Calculate the optimal power factor for distributed generation connected to the feeders in Tables 5.2 and 5.3. Obtain the feeder inductance from data books, as in Exercise 5.1. Estimate the increase in losses and the need to overrate the generator inverter due to this.

5.2.5 Measurements to Determine the Hosting Capacity

To be able to estimate the hosting capacity, the overvoltage margin should be known. A long-term measurement of the voltage magnitude variations for a large number of customers would be the most appropriate method. This could be a time-consuming and expensive method unless advanced metering equipment is available. Such metering equipment allows the registration of the voltage and current magnitude at regular intervals (e.g., once every minute). Such data are extremely valuable, not only for determining the hosting capacity, but also for voltage coordination of distribution networks in general. Network operators and customers in several countries have started installing energy meters that allow the recording of voltage, current, active power, and reactive power with a considerable time resolution even for domestic customers. Within a few years, large amounts of data may be available that allow accurate estimation of the hosting capacity. Processing of these data is, however, still a challenge that is not unsurpassable but will have to addressed. For Sweden, with 5.2 million customers, 10-minute values will result in 275 GB of data per year. Processing this is, in 2010, still a challenge but not impossible.

At this moment, the above-mentioned data are not available. However, it is possible to perform a limited measurement campaign, that is, measurements at a limited number of locations during a limited period of time. In that case, the selection of the measurement locations is very important. The aim of the measurement should be kept in mind during the selection process. If the aim is to determine the hosting capacity of the feeder, the interest is primarily in the locations with the smallest overvoltage margin. The measurement campaign could, of course, be part of a larger campaign to map the voltage magnitude variations along the feeder. In that case, other locations would be more appropriate. Here, we, however, assume that estimating the hosting capacity is the only aim.

From (5.2), we see that the hosting capacity is smallest for small overvoltage margin δ_{max} and for large (electrical) distance to the main substation (large R). Both parameters vary along the feeder and it is not possible to give general rules for which location determines the hosting capacity. When the generation is connected to the medium-voltage feeder, the hosting capacity is most likely set by the overvoltage margin on secondary side of a distribution transformer with a small turns ratio (i.e., one that boosts the voltage magnitude). The transformer with the lowest turns ratio, closest to the main MV bus, is likely to have the smallest overvoltage margin. Measurements on secondary side of this transformer are strongly recommended.

Further along the feeder, the overvoltage margin will increase, depending on the voltage drop during minimum load, but at the same time the source resistance will increase, so the hosting capacity might still become smaller. Therefore, it is recommended to measure also at some remote locations, especially those with small minimum load.

Measurements should preferably take place during one whole year to cover seasonal variations. When this is difficult, measurement should take place at least a few weeks during the period of known minimum load. This could be during summer, winter, or somewhere in between. When industrial loads are supplied from the feeder, it is best to measure when this load is lowest as this will give the highest voltage magnitude.

When planning to use simple measures to increase the hosting capacity, such as changing the reference value of the tap-changer controller or requiring a nonunity power factor for generators, it is recommended to perform measurements at more locations. Those locations for which the lowest voltage magnitudes are expected should be measured at.

5.2.6 Estimating the Hosting Capacity Without Measurements

When no measurements are available, the overvoltage margin should be estimated. Performing measurements is very strongly recommended. However, it is not always

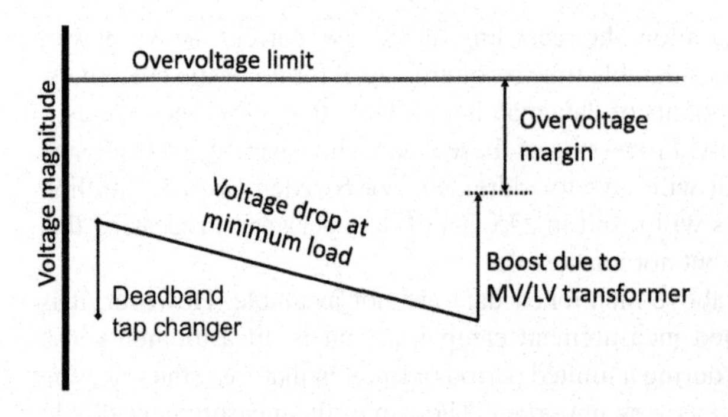

Figure 5.6 Estimating the overvoltage margin when no measurements are available (see text for explanation).

possible to perform measurements owing to time or budget constraints. Where time is the main constraint, it is recommended to perform the measurements even after a decision about the hosting capacity has been made. A method for estimating the overvoltage margin without measurements is illustrated in Figure 5.6.

This method proceeds as follows:

- The upper limit of the deadband is used as a starting point. This is the highest voltage at the main medium-voltage bus.
- The voltage drop along the medium-voltage feeder during lowest load is estimated. This gives the highest voltage on medium-voltage side of the distribution transformers.
- Where needed, the voltage drop over the distribution transformer and along the distribution feeder during low load is estimated as well.
- The boost due to the distribution transformers (up to 5%) is added. This gives the highest voltage on low-voltage side of the distribution transformers.
- The overvoltage margin is the difference between this voltage and the overvoltage limit.

Example 5.1 Consider the following numerical example to calculate the overvoltage margin:

- The deadband is between 102% and 104%.
- The voltage drop at low load is 3%.
- The boosting due to the distribution transformer is 5%.
- This gives a maximum voltage in the low-voltage network equal to $104 - 3 + 5 = 106\%$.
- The overvoltage limit is 110%.
- The resulting overvoltage margin is $110 - 106 = 4\%$.

The hosting capacity in this case is four times the amount given in Tables 5.2 and 5.3.

Example 5.2 Consider an alternative numerical example for the same feeder as in Example 5.1, but using more different estimates:

- The deadband is between 102% and 104%.
- The voltage drop at low load is 1%.

- The boosting due to the distribution transformer is 5%.
- This gives a maximum voltage in the low-voltage network equal to $104 - 1 + 5 = 108\%$.
- The overvoltage limit is 108%.
- The resulting overvoltage margin is $108 - 108 = 0\%$.

The hosting capacity is zero and no distributed generation at all can be connected to the feeder.

These two examples illustrate two different ways of looking at overvoltages. In Example 5.1, the engineer uses what he considers a reasonable value for the voltage drop during low load. He also uses the voltage characteristic (110% of nominal voltage) as the overvoltage limit. The result, as we can see above, is 4% overvoltage margin. In Example 5.2, another engineer uses a low value of the voltage drop to be sure that he is not overestimating the voltage drop. Overestimating the voltage drop could result in overvoltages with end customers that are not anticipated by the network operator. This second engineer also uses the planning level (in that case at 108% of nominal voltage) as the overvoltage limit, so as to have an additional safety margin. The approach in Example 5.2 gives the best protection against overvoltages, whereas the approach in Example 5.1 allows more distributed generation to be connected to the feeder. From the examples, we can see that these two approaches result in rather different hosting capacities.

There is unfortunately no absolutely correct way of estimating the hosting capacity without detailed measurements. Many distribution network operators prefer the safe approach and use a lower estimate of the voltage drop during low load. The result is a lower hosting capacity than that when a "middle estimate" would be used. But even when the voltage drop is accurately known (from measurements or from accurate calculations), the choice of overvoltage limit remains a point of discussion. The planning level (for harmonics, flicker and unbalance) is often used when evaluating the connection of industrial installations to the power system. Using the planning level for the connection of distributed generation would simply be a continuation of this approach. It would, however, for some distribution networks, result in a barrier to the introduction of renewable sources of energy.

5.2.7 Choosing the Overvoltage Limit

The choice of the overvoltage limit very strongly determines the hosting capacity of a feeder for distributed generation. The lower the overvoltage limit, the smaller the overvoltage margin. As the hosting capacity is directly proportional to the overvoltage margin, the choice of the overvoltage limit plays an important role. The overvoltage margin is rarely more than a few percent of the nominal voltage. A 1% reduction in overvoltage limit can easily reduce the hosting capacity by a factor of 2.

When looking at the overvoltage limit, it is important to distinguish between, on one hand, voltage characteristics and other requirements placed on the network operator, and, on the other hand, planning levels and other internal limits used by the network operator in the design of the distribution network.

The European voltage characteristics standard EN 51060 [134] states that for the supply to low-voltage or medium-voltage customers, "Under normal operating conditions excluding the periods with interruptions, supply voltage variations should not exceed ±10% of the nominal voltage." This general requirement is further specified by giving the following measurement-based requirements:

- At least 99% of the 10 min mean rms values of the supply voltage shall be within the limits
- None of the 10 min mean rms values of the supply voltage shall be outside the limits ±15% of the nominal voltage

In a number of European countries, the regulatory authority responsible for the electricity network sets more strict requirements on the voltage supplied to low-voltage and medium-voltage customers. The following four countries have more strict requirements than that according to EN 50160 [82]:

- In France, the voltage should be within 10% of nominal voltage for low-voltage and medium-voltage customers. Medium-voltage customers can sign a contract that gives them a voltage within 5% of a declared voltage. The declared voltage in turn should be within 5% of the nominal voltage.
- In Hungary, for low-voltage customers, 95% of the 10 min rms voltages should be within 7.5% of the nominal voltage, 100% of the 10 min rms voltages should be within 10%, and 100% of the 1 min rms voltages should be within 15%.
- In Norway, 100% of the 1 min rms voltages should be within 10% of the nominal voltage.
- In Spain, 95% of the 10 min rms voltages should be within 7% of the nominal voltage.

The above limits, ranging between 107% and 115% for the overvoltage limit, indicate the requirements set on the network operator. On the basis of these limits, the network operator will set internal limits to be used in the design of the network. This is similar to what is referred to as "planning levels" when discussing harmonics, flicker, and unbalance in IEC 61000-3-6 [211], IEC 61000-3-7 [212], and IEC 61000-3-13 [213], respectively. In these technical reports, indicative values for the planning levels are given that are below the voltage characteristics. This gives the network operator a margin to prevent the voltage characteristics from being exceeded, for example, when the emission or the network properties change or are different from that used in the calculations.

It is important to notice that it is the planning level that is used when evaluating the connection of a disturbing installation to the power system. When deciding about the connection of distributed generation (i.e., when determining the hosting capacity), the planning levels (in this case of the overvoltage limit) are also considered. There are a number of methods for choosing the planning level:

- The planning level is chosen significantly lower than the requirement; the margin between the requirement and the planning level limits the risk for the network operator. The calculation of the hosting capacity is based on reasonable estimates of the after-diversity maximum production and the after-diversity minimum load.

- The planning level equal to the requirement is chosen. The risk is minimized by considering a worst case in the calculation of the hosting capacity, for example, minimum demand equal to zero and maximum production equal to the rated power.
- The planning level equal to or only slightly lower than the requirement is chosen. The calculation is based on reasonable estimates. Next to this, the regulatory framework ensures that the risk for overvoltages is not carried by the network operator.

The first and second approaches are most common today and can be expected to be used more often when regulatory pressure increases to maintain the voltage magnitude within strict limits. The third approach may be a future solution to prevent network operators from unintentionally setting up barriers to the introduction of distributed generation.

Some examples of planning levels for different countries are given in the following list:

- In Japan, the allowed voltage range at low voltage is 101 ± 6 V. The low nominal voltage used implies that low-voltage feeders are very short in Japan. The restrictions on the medium-voltage network are obviously even more strict.
- The voltage limits at the customer bus, as used by Toronto Hydro in Canada, are 94% and 106% of nominal [457].
- For medium-voltage feeders in the United States, the limits are 95–105% [337]. According to Ref. 70, most network operators limit the difference between the highest and the lowest voltage to 8%, some even to 6%. In Ref. 454, the planning levels are given for 10 U.S. network operators. The overvoltage limit varies from 125 to 127 V (104–106%) and the undervoltage limit from 112 to 123 V (93–102%). Especially the large range in undervoltage limit is interesting. This difference mainly translates into different maximum lengths for the low-voltage feeders.
- For medium-voltage feeders in the United Kingdom, the limits are 94–106% [272, 337].
- In Brazil, the acceptable range of voltage magnitudes in low-voltage networks is between 189 and 233 V (86–106% on a 220 V base); the preferred range is between 201 and 231 V (91–105%). For 13.8 kV medium voltage, the acceptable range is 90–105% and the preferred range is 93–105% [21].
- According to Ref. 57, design limits for low-voltage networks used by EDF in France are 90% and 106%.
- Recommended planning levels in Sweden are 95–105% of nominal voltage up to 24 kV and 90–110% above 24 up to 145 kV [398]. These values should, under the recommendation, not be exceeded during at least 95% of the time.

5.2.8 Sharing the Hosting Capacity

The discussion above and in Section 5.3 focuses on one generator connected to a medium-voltage feeder. When more than one generator is connected, sharing of the hosting capacity is needed. If both units are connected to the feeder at the same location, the rule for sharing is straightforward: the total maximum production should not exceed the hosting capacity at that location. For two or more units at different locations, the situation becomes more complicated.

From (5.2), it follows that the hosting capacity is inversely proportional to the location at which the generator is connected. This is further worked out in Section 5.3 using a more detailed feeder model, resulting among others in Figure 5.10–5.13. The conclusion from these figures is also that the hosting capacity is roughly inversely proportional to the generator location. Using this information, the following sharing rule is proposed:

$$\sum_i \lambda_{\text{gen},i} P_i \leq P_{\text{HC}} \tag{5.7}$$

where P_{HC} is the hosting capacity for a generator connected at the end of the feeder. For N generators, the proposed allocation of each generator is as follows:

$$P_i \leq \frac{1}{\lambda_i} \frac{P_{\text{HC}}}{N} \tag{5.8}$$

The closer to the main bus ($\lambda = 0$), the larger a generator allowed to be connected.

Example 5.3 Assume a feeder with a minimum hosting capacity equal to 750 kW. Two generators will be connected, one at 30% from the start of the feeder and the other at 65% from the start of the feeder. The maximum production of each generator is found from (5.8) to be

$$P_1 = \frac{1}{0.3} \frac{750\,\text{kW}}{2} = 1250\,\text{kW}$$

$$P_2 = \frac{1}{0.65} \frac{750\,\text{kW}}{2} = 575\,\text{kW}$$

The sharing rule according to (5.8) assumes that each generator takes up its allocation and that the maximum production of both units is reached at the same time. If this is not the case, more production can be allowed. Separate calculations are needed in those cases. The above sharing rule can, however, still be used as a guidance.

5.3 DESIGN OF DISTRIBUTION FEEDERS

5.3.1 Basic Design Rules

The basic rules in the design of distribution feeders are as follows:

- The current during maximum load should not exceed the ampacity of the feeder.
- The maximum voltage, for any customer, should not be above the overvoltage limit.

- The minimum voltage, for any customer, should not be below the undervoltage limit.

The first rule is mainly a concern for urban networks with high load density. In this chapter, we will assume that this is not a limitation. In the existing design, the second rule is fulfilled through the setting of the deadband of the tap changer. When maximum 5% voltage boosting is used and the upper limit of the deadband is set at 4–5% below the overvoltage limit, voltages above 108–110% are unlikely. The third rule is the one that sets a limit to the feeder length.

When adding generation to the feeder, the "minimum load" may become negative, so the choice of the deadband is no longer sufficient to prevent overvoltages. This may set a limit to the amount of distributed generation that can be connected (the "hosting capacity") or it may set an additional limit to the feeder length. In either case, the design of the distribution feeder becomes more complicated.

5.3.2 Terminology

We will illustrate the design of distribution feeders with some general calculations and a number of numerical examples. To allow an easy comparison between the voltages at different voltage levels, all voltages will be expressed in percent of the nominal voltage. The following notations will be used:

- u_{min}: the minimum voltage experienced by a customer at low voltage
- u_{max}: the maximum voltage experienced by a customer at low voltage
- $u_{\text{min,limit}}$: undervoltage limit
- $u_{\text{max,limit}}$: overvoltage limit
- $u_{\text{db,max}}$: upper limit of the tap-changer deadband
- $u_{\text{db,min}}$: lower limit of the tap-changer deadband
- Δu_{max}: the maximum voltage drop due to consumption, not including any change in voltage due to generation
- Δu_{min}: the minimum voltage drop due to consumption, not including any change in voltage due to generation
- Δu_{boost}: rise in voltage due to boosting, for example, by a distribution transformer
- Δu_{gen}: rise in voltage due to distributed generation, this is the maximum value, unless otherwise indicated

Using this notation, the second and third rules for the design of distribution feeders read as follows:

$$u_{\text{min}} = u_{\text{db,min}} - \Delta u_{\text{max}} + \Delta u_{\text{boost}} > u_{\text{min,limit}} \tag{5.9}$$

and

$$u_{\text{max}} = u_{\text{db,max}} - \Delta u_{\text{min}} + \Delta u_{\text{boost}} + \Delta u_{\text{gen}} < u_{\text{max,limit}} \tag{5.10}$$

Here, it is assumed that the minimum rise in voltage due to distributed generation is zero; that is, sometimes during peak consumption no active power is produced. It is further assumed that maximum production may occur during minimum consumption. When the production has a strong daily variation, this approach may underestimate the hosting capacity. This is especially the case for solar power; in that case, a more detailed study is needed. Alternatively, a stochastic approach, such as in Section 5.6, could be used. In the following sections, we will apply the above design rules to a number of example feeders. This will be used to quantify the impact of different parameters on the hosting capacity.

5.3.3 An Individual Generator Along a Medium-Voltage Feeder

Consider a linear medium-voltage feeder with a total resistance R and a total reactance X, loaded with active power P and reactive power Q. It is assumed that the load is uniformly distributed along the feeder and that the resistance and reactance per unit length remain the same along the feeder. A generator producing active power P_{gen} at unity power factor is connected to the feeder at a location $\lambda = \lambda_{gen}$, where $\lambda = 0$ corresponds to the beginning of the feeder and $\lambda = 1$ corresponds to the end of the feeder. The voltage drop due to the load is

$$\Delta u = \frac{\Delta U}{U_{nom}} = \frac{RP + XQ}{U_{nom}^2}\left(\lambda - \frac{1}{2}\lambda^2\right) \tag{5.11}$$

The voltage rise due to the generator is linear with the distance along the feeder, up to the generator location, and constant beyond the generator location:

$$\Delta u_{gen} = \begin{cases} \lambda \dfrac{RP_{gen}}{U_{nom}^2} & \lambda \leq \lambda_{gen} \\ \\ \lambda_{gen} \dfrac{RP_{gen}}{U_{nom}^2} & \lambda > \lambda_{gen} \end{cases} \tag{5.12}$$

The voltage profile along a overhead 10 kV feeder without generation is shown in Figure 5.7 for maximum and minimum load. The feeder length is 18 km; the load density is 200 kW plus 100 kvar/km during maximum load and one-fourth of that during minimum load. The feeder impedance is $0.309 + 0.329j\ \Omega$/km (corresponding to 95 mm^2 aluminum overhead line). The transformer deadband is set at 102–104% of nominal voltage; 2.5% and 5% boosts are used to prevent the voltage from dropping below the undervoltage limit of 94%. The undervoltage limit is taken at 94% to allow a few percent additional voltage drop in the low-voltage network. The dotted lines indicate the undervoltage limit and the overvoltage limit (at 108%). We see from the figure that both lowest and highest voltages are reached somewhere along the middle of the feeder, not at the end. However, the range between maximum and minimum voltages increases continuously along the feeder.

The voltage profile is calculated again after adding a 1 MW generator at 45% of the feeder length (i.e., 8.1 km from the start of the feeder). The maximum voltage is now reached for minimum load and maximum generation: the upper solid curve in Figure 5.8. Owing to the connection of the generator, the voltage rises above the

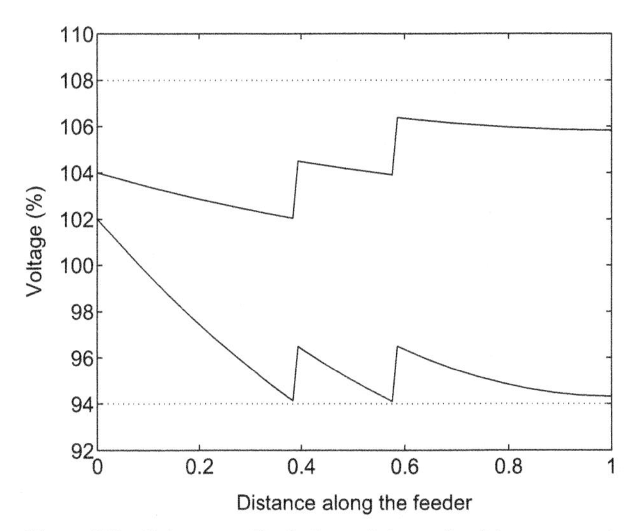

Figure 5.7 Voltage profile during minimum load (upper curve) and during maximum load (lower curve) for a medium-voltage feeder without generation.

overvoltage limit for some of the customers. Note that the maximum voltage is reached neither for customers close to the generator nor for customers at the end of the feeder. Instead, the highest voltage is reached for the location with the highest voltage boost, closest to the main substation.

The curves in Figure 5.9 give the voltage profile for a generator located at 65% of the feeder length. The maximum voltage curves correspond to (from bottom to top)

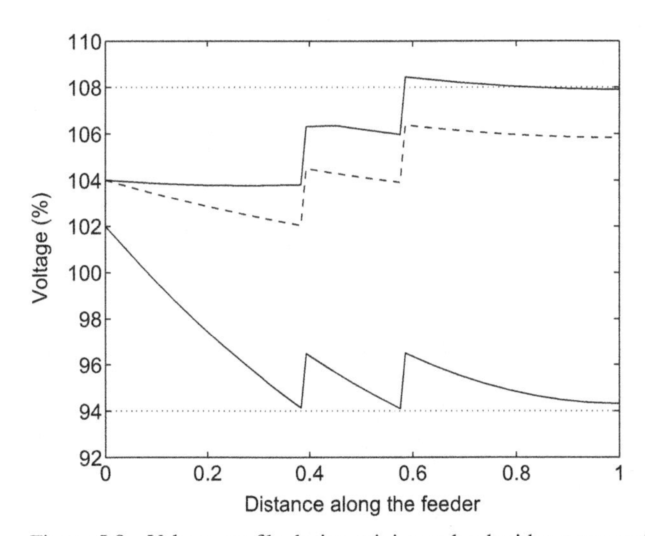

Figure 5.8 Voltage profile during minimum load without generation (upper dashed curve), during minimum load with generation (upper solid curve), and during maximum load without generation (lower curve) for a medium-voltage feeder with generation.

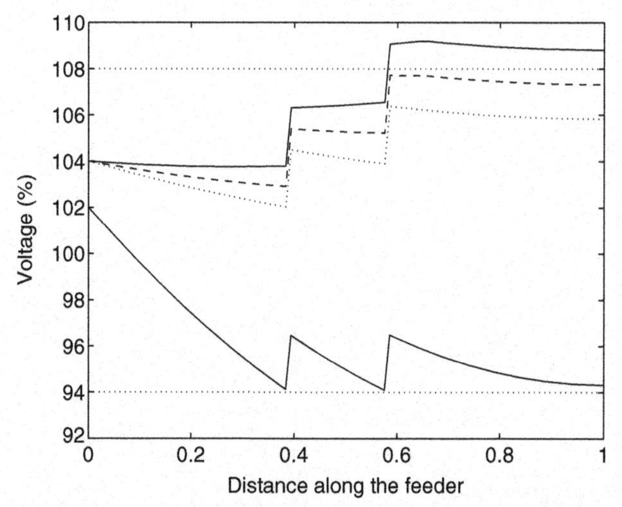

Figure 5.9 Voltage profile during minimum load without generation (upper dotted curve), during minimum load with 500 kW generation (upper dashed curve), during minimum load with 1 MW generation (upper solid curve), and during maximum load without generation (lower curve) for a medium-voltage feeder with generation.

zero, 500 kW and 1 MW generation. For 500 kW generation, the maximum voltage is reached at 60% of the feeder length, and at 1000 kW generation the maximum is reached for 65% of the feeder length. The only conclusion that can be drawn from this observation is that there are no simple rules to determine where the maximum voltage will be reached. It depends on the overvoltage margin, on the location of the boosting points, and on the location of the generator. This presents a complication in the calculation of the hosting capacity. The two locations that often result in the highest voltage are the generator location and the highest voltage boost closest to the main bus.

The hosting capacity has been calculated as a function of the generator location, using (5.2) for both the generator location and the location with the highest voltage before the connection of the generator. The lowest of the two values was chosen. It was verified during the simulations that this indeed results in the hosting capacity for the feeder; that is, none of the other locations experiences a voltage higher than the overvoltage limit as long as the generator size remains below this value. The resulting hosting capacity is shown in Figure 5.10 for four different values of the overvoltage limit. We see that the hosting capacity in general decreases for generation connected further away along the feeder. The decrease becomes less for generation connected beyond the location that had the highest voltage before connection of the generation. The latter holds especially for small hosting capacity. It is also clearly visible in the figure that the choice of overvoltage limit has a significant influence on the hosting capacity. For a 107% limit, the hosting capacity is only a few hundred kilowatts, whereas it is above 1 MW for most locations when a 110% limit is used.

The calculations resulting in Figure 5.10 confirm that the choice of the overvoltage limit has an important influence on the amount of generation that can be

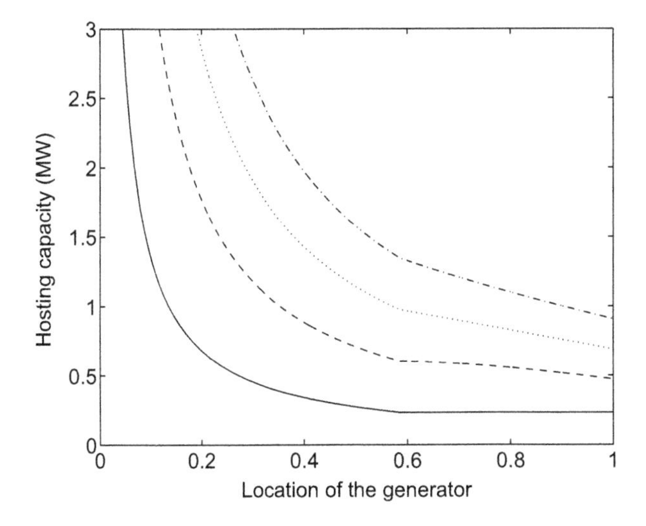

Figure 5.10 Hosting capacity for a single generator connected to the feeder, as a function of the location of the generator: overvoltage limit equal to 107% (solid curve), 108% (dashed), 109% (dotted), and 110% (dash–dot).

connected. A small increase in overvoltage limit may result in a significant increase in the hosting capacity. The impact of feeder length on the hosting capacity is rather complex, as illustrated in Figure 5.11 where the hosting capacity is shown for feeder lengths of 18, 17, 16.5, and 16 km. All other parameters have been kept the same, including the load density, so that the total feeder load decreases proportionally with the decreasing feeder length. The overvoltage limit was chosen at 108%.

For feeder lengths of 18, 17, and 16.5 km, the hosting capacity remains about the same. The small changes are related to the shift of the location at which the maximum voltage is reached before connection of the generator (i.e., the location with the highest voltage boosting closest to the main bus). For a feeder length of 16 km, there is no longer need for 5% voltage boosting, 2.5% is sufficient. The result is a reduction in the maximum voltage by 2.5% and a large increase in hosting capacity.

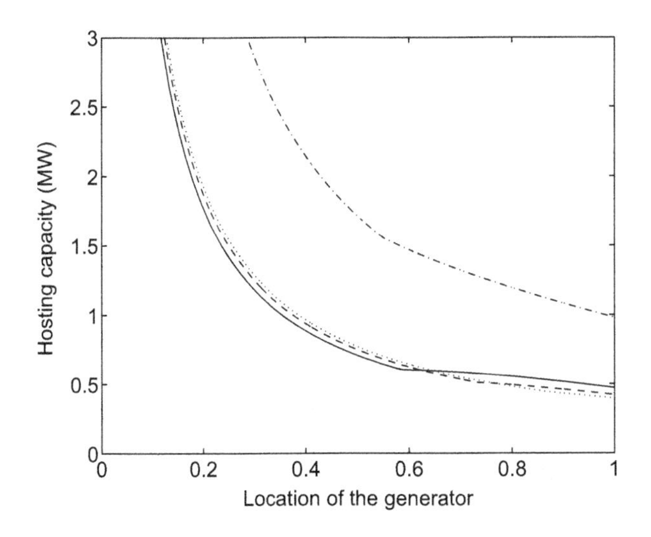

Figure 5.11 Hosting capacity for a single generator connected to the feeder, as a function of feeder length: 18 km (solid curve), 17 km (dashed), 16.5 km (dotted), and 16 km (dash–dot).

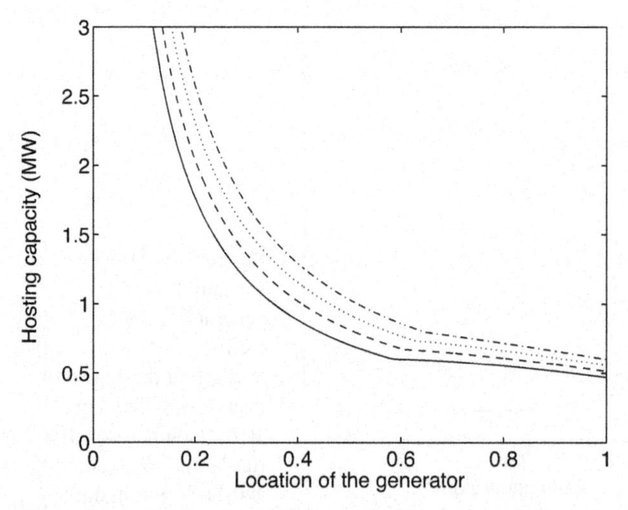

Figure 5.12 Hosting capacity for a single generator connected to the feeder, as a function of the tap-changer deadband: 102–104% (solid line), 102.2–103.8% (dashed), 102.4–103.6 (dotted), and 102.6-103.4 (dash–dot).

Figure 5.12 shows what happens to the hosting capacity when the deadband of the tap changer is made smaller. The smaller the deadband (centered around 103% in all four cases), the larger the hosting capacity. The hosting capacity can be increased even somewhat more by keeping the same lower limit of the deadband (at 102%) and reducing the upper limit. The reduction in the upper limit of the deadband immediately increases the overvoltage margin and thereby the hosting capacity. The disadvantage of reducing the width of the deadband is that normal load variations will cause more regular tap-changer operations. This increases the need for maintenance and will also likely result in a higher failure rate for the transformer.

As discussed previously, the hosting capacity strongly depends on the estimated voltage drop during minimum load. Earlier the minimum voltage drop was directly estimated. In this section, the voltage drop is calculated from the estimated minimum load. The impact of this estimation on the hosting capacity is shown in Figure 5.13. Changing the estimated minimum load from 20% to 30% of maximum load doubles the hosting capacity. This confirms that it is important to have an accurate estimate of the minimum voltage drop so as not to put unnecessary restriction to distributed generation. Measurements again provide the most accurate basis for calculations. The wide-scale introduction of advanced metering will make this possible in the near future.

The impact of the feeder cross section on the hosting capacity is shown in Figure 5.14. A larger cross section gives a larger hosting capacity for the same overvoltage margin; however, it also reduces the overvoltage margin because the voltage drop (during both low and high loads) becomes less. As shown in the figure, the difference in hosting capacity for a 95 mm^2 and a 120 mm^2 wire is small especially toward the end of the feeder. The fact that the hosting capacity for these two cross

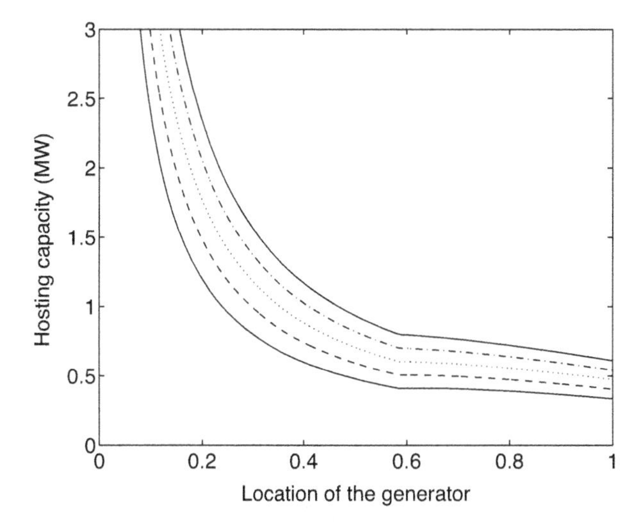

Figure 5.13 Hosting capacity for a single generator connected to the feeder, as a function of the minimum load: 20% (lower solid curve), 22.5% (dashed), 25% (dotted), 27.5% (dash–dot), and 30% (upper solid curve) of maximum load.

sections is almost the same toward the end of the feeder has to do with the shift of the second boosting point further toward the end of the feeder.

The big increase in hosting capacity when using a 150 or 180 mm^2 feeder is because there is no longer a need for the second voltage boosting after the voltage drop during maximum load is reduced significantly.

5.3.4 Low-Voltage Feeders

Calculating the voltage profile along a low-voltage feeder proceeds in a similar way as for a medium-voltage feeder. Starting from the voltage range (maximum and minimum voltage magnitudes) at the connection with the medium voltage, the maximum and minimum voltage drops are calculated. This is illustrated in Figure 5.15, with the

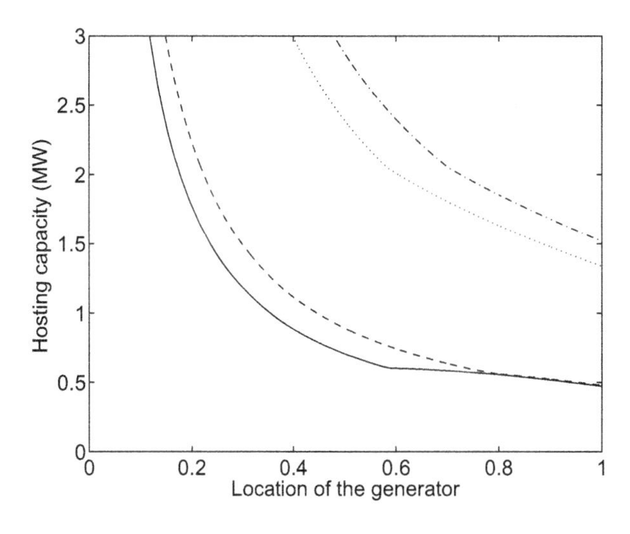

Figure 5.14 Hosting capacity for a single generator connected to the feeder, as a function of the feeder size: 95 mm^2, (solid curve), 120 mm^2 (dashed), 150 mm^2 (dotted), and 180 mm^2 (dash–dot).

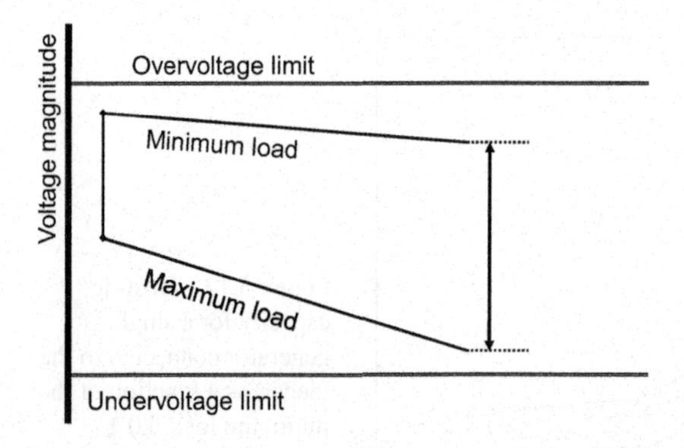

Figure 5.15 Voltage profile along a low-voltage feeder.

vertical band on the left-hand side the voltage range in the medium-voltage network on primary side of the distribution transformer. This is the voltage range that has been calculated before, for example, in Figure 5.7.

As before, the maximum voltage is reached for minimum load in combination with maximum voltage in the medium-voltage network: the minimum voltage is reached for maximum load in combination with minimum voltage in the medium-voltage network. To determine the hosting capacity of the low-voltage feeder, it is the maximum voltage that is of interest. The diversity in the low-voltage network is higher than that in the medium-voltage network due to the lower number of customers. The difference between maximum and minimum loads will therefore also be bigger. For calculations in low-voltage networks, it makes sense to assume zero voltage drop at minimum load. The maximum voltage in the low-voltage network is in that case equal to the maximum voltage in the medium-voltage network. When more accurate data are available, for example, from automatic meter reading, this should be used in the calculations.

The calculation of the hosting capacity proceeds in the same way as before, by using (5.2), which reads as

$$P_{\max} = \frac{U_{\text{nom}}^2}{R} \times \delta_{\max} \tag{5.13}$$

where U_{nom} is the nominal voltage of the low-voltage network, R the source resistance at the point of connection of the generator unit, and $\delta_{\max}$ the overvoltage margin, that is, the difference between the overvoltage limit and the maximum voltage magnitude before connection of the generator. In the calculations, one may further neglect the resistance of the distribution transformer; it is the cable or line resistance that dominates once the generator connection is slightly away from the transformer.

Example 5.4 Consider a generator to be connected to a 50 mm² overhead line feeder, at 500 m from the distribution transformer. The maximum voltage in the medium-voltage network is 107% of nominal. The overvoltage limit is 109%.

The resistance of the wire is

$$\rho \times \frac{A}{\ell} = 1.678 \times 10^{-8}\ \Omega\text{m} \times \frac{500\ \text{m}}{50 \times 10^{-6}\ \text{m}^2} = 0.1678\ \Omega$$

Using (5.13) with $\delta_{max} = 1.09 - 1.07 = 0.02$ gives for the hosting capacity:

$$P_{max} = \frac{400\,\text{V}^2}{0.1678\,\Omega} \times 0.02 = 19\,\text{kW}$$

This holds for a three-phase generator unit. For a single-phase unit, the impedance of the neutral wire (that becomes the return path for a single-phase generator) should be included and 230 V (the line-to-neutral voltage) should be taken as the nominal voltage. Assuming that the resistance of the neutral wire is the same as that of the phase wire, the hosting capacity for single-phase units is

$$P_{max} = \frac{230\,\text{V}^2}{2 \times 0.1678\,\Omega} \times 0.02 = 3.2\,\text{kW}$$

The hosting capacity for a single-phase unit is only one-sixth of the hosting capacity for a three-phase unit.

The hosting capacity for a three-phase generator of the low-voltage feeder has been calculated in this way for all locations along the medium-voltage distribution feeder used for Figure 5.7. It has been assumed in all cases that the three-phase generator is located at 500 m along a 50 mm^2 wire as in Example 5.4. The results are shown in Figure 5.16, where it is further assumed that no other generation is located along the medium-voltage feeder.

We see that the hosting capacity, for the same low-voltage feeder, varies significantly (between 15 and 57 kW for 108% overvoltage limit) depending on where along the medium-voltage network the distribution transformer is located. We also see, as before for generation connection to the medium-voltage feeder, that the lowest values of the hosting capacity strongly depend on the choice of overvoltage limit.

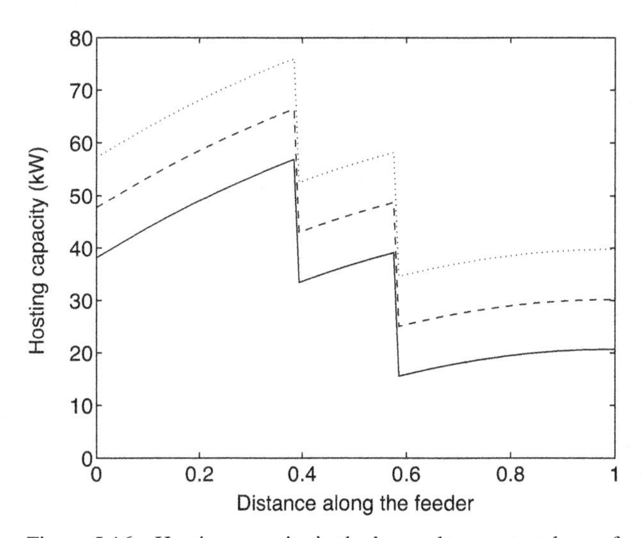

Figure 5.16 Hosting capacity in the low-voltage network as a function of its location along a medium-voltage feeder: overvoltage limit 108% (solid line), 109% (dashed), and 110% (dotted).

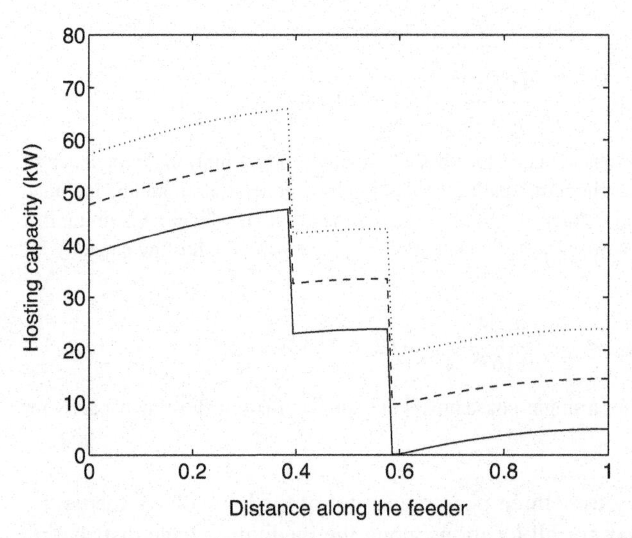

Figure 5.17 Hosting capacity in the low-voltage network as a function of its location along a medium-voltage feeder, with 600 kW generation connected at 60% of the length of the medium-voltage feeder: overvoltage limit 108% (solid line), 109% (dashed), and 110% (dotted).

The calculations have been repeated for the case that a generator is already connected to the medium-voltage feeder. A 600 kW unit at 60% of the feeder length is assumed. As shown in Figure 5.10, this is close to the hosting capacity for 108% overvoltage limit. The remaining hosting capacity in the low-voltage network is shown in Figure 5.17. As expected, the hosting capacity at low voltage is reduced. At around 60% of the medium-voltage feeder length (where the existing generator is located), the hosting capacity becomes zero when 108% is used as an overvoltage limit.

5.3.5 Series and Shunt Compensation

A strong disadvantage of using off-load tap changers is that it easily results in overvoltages. This in turn greatly reduces the hosting capacity of the feeder for distributed generation. This boosting method raises the voltage with the same amount during low load as during high load. To get a sufficient voltage rise during high load, the risk of overvoltages during low load is a consequence. Another disadvantage is that the method is effective only during normal operation of the feeder. Loads that are connected near the beginning of a feeder could become connected near the end of the feeder during backup operation. As voltage boosting is not applied for those loads, very low voltage magnitudes could result.

A boosting method that depends on the loading, that is, the higher the loading, the more the voltage rise, would not have these disadvantages. Two such methods will be briefly described here: series compensation and switchable shunt compensation.

With series compensation, a capacitor is placed in series with the feeder. This capacitor compensates part or all of the voltage drop due to reactive load downstream of the capacitor. The resulting voltage drop due to the downstream load is

$$\Delta u = RP - (X - X_c)Q \tag{5.14}$$

where X_c is the (absolute value of the) reactance of the series capacitor. The location of the series compensation should be carefully chosen. A series capacitor at the start of the feeder will not only compensate all reactive power load, but also result in the highest voltages immediately downstream of the capacitor. Compensation toward the end of the feeder will have less effect because only the reactive power downstream of the capacitor is compensated. A distributed approach, with a number of series capacitors along the feeder, gives the best results from a voltage control viewpoint. When the series capacitor is chosen such that the series reactance is fully compensated, only a voltage drop due to the active power results. This is of course only worth the investment when the voltage drop due to the reactive power is a substantial part of the total voltage drop, or in other words, for feeders with a larger X/R ratio. If we assume an X/R ratio equal to one and 0.9 power factor for the load (so that the reactive power demand is about half the active power demand), about one-third of the voltage drop can be compensated for. We saw before that the maximum allowed voltage drop, using the normal method of voltage boosting, is about 13% (see, for example, Figure 5.7). If one-third can be compensated for, this results in a remaining voltage drop of about 8.5%.

Placing all the compensation at the start of the feeder would result in a voltage boost of about 4.5% at maximum load. Using the same deadband as before (102–104%) would give a maximum voltage of about 108.5%. Note that the maximum voltage would occur at the start of the feeder during maximum load. The lowest voltage would still be reached during maximum load at the end of the feeder. Starting from the lower limit of the deadband (102%) and 4.5% boost by the series capacitor, the 13% voltage drop would result in 93.5% for the minimum voltage. This is just below the undervoltage limit.

At the other extreme, fully distributed series compensation would not result in any voltage rise, instead the voltage would show a continuous drop for both low load and high load. The total voltage drop would be 8.5% during high load and about 2% (one-fourth) during low load. To remain above the undervoltage limit, the deadband could be shifted upward, for example, 104–106%. The lowest voltage would be 95.5% and the highest voltage 106%. Using the same 108% overvoltage limit, the overvoltage margin would be 2% at the start of the feeder and 4% at the end of the feeder.

Accepting the same minimum voltage as before (94%), the lowest deadband is 102.5–104.5%. The resulting overvoltage margin becomes 3.5% at the start of the feeder and 5.5% at the end of the feeder. Even when considering the worst case of zero voltage drop at minimum load (i.e., zero minimum load), 3.5% overvoltage margin still results at the end of the feeder. As we saw previously, this worst-case assumption often results in zero or very small overvoltage margin with the standard method for voltage boosting.

Shunt capacitances result in a constant amount of voltage rise, independent of the load current:

$$\Delta u_c = X_S Q_c \tag{5.15}$$

where X_S is the reactive part of the impedance between the tap-changer location and the location of the capacitor. Any change in voltage over the source reactance at the main MV bus is compensated for by the tap changer of the HV/MV transformer. The voltage rise decays linearly with the distance up to the capacitor location and remains constant after that. For shunt compensation, the highest voltage rise is obtained for a capacitor at the end of the feeder.

When a voltage rise of more than about 3% has to be achieved, it is recommended to check for harmonic resonances. When high levels of fifth harmonics are expected, the capacitor should be part of a harmonic filter or be combined with detuning reactors.

As mentioned previously, the voltage rise is independent of the load current. To prevent overvoltages, the capacitor bank can be disconnected when the voltage exceeds a certain level. A possible control strategy would be to switch on the capacitor when the voltage drops below 95% and to switch it off when the voltage exceeds 105%. The large hysteresis limits the number of switching actions.

5.4 A NUMERICAL APPROACH TO VOLTAGE VARIATIONS

5.4.1 Example for Two-stage Boosting

As was shown in Section 5.3, the design of distribution feeders follows a number of rules. It was also shown that the main limitation to the hosting capacity is due to the boosting of the voltage magnitude to prevent undervoltages. In this section, we will describe the design method in a somewhat different way, with the emphasis more directly on the overvoltage and undervoltage margins. This description is not useful for the design of distribution feeders, but it gives some additional insights into the overvoltage problem. It also provides an easy method for estimating the hosting capacity in case of uniformly distributed generation, as will be shown in Section 5.4.4.

Consider, as an example, the following design parameters for a distribution feeder (all values are in percentage of the nominal voltage):

- The deadband of the tap changer is between 102% and 104%.
- Boosting of the voltage takes place in two steps of 2.5% each.
- The lowest voltage magnitude is 94%, and this is where boosting takes place; this also sets the maximum feeder length.
- The voltage drop during minimum load is one quarter of the voltage drop during maximum load.

The resulting voltage profile for maximum and minimum loads is shown in Figure 5.18.

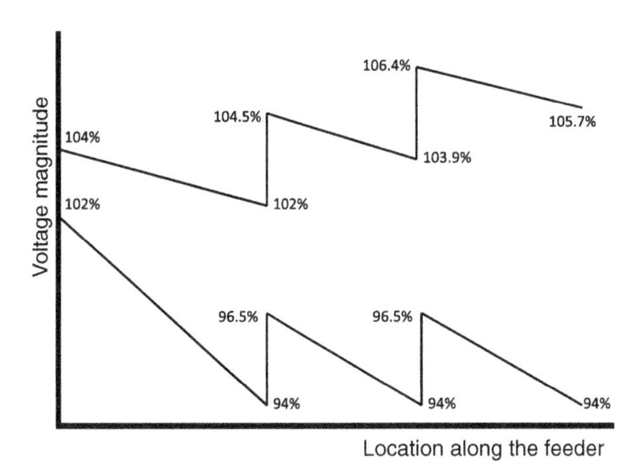

Figure 5.18 Voltage profile along a distribution feeder with voltage boosting in two steps: minimum load (upper curve) and maximum load (lower curve).

The voltage magnitude values in the figure are obtained by using the following reasoning:

- The first boosting position occurs when the minimum voltage is 94%. With a lower limit of the deadband equal to 102%, this corresponds to 8% voltage drop during maximum load.
- The voltage drop during minimum load is one-fourth of this, and thus it is 2%. The maximum voltage magnitude just upstream of the boosting point is thus 104% −2% = 102%. The first value is the upper limit of the deadband.
- Immediately downstream of the first boosting point, the voltage is 2.5% higher. This holds for both the minimum and the maximum voltage, resulting in a minimum voltage of 94% + 2.5% = 96.5% and a maximum voltage of 102% + 2.5% = 104.5%.
- The second boosting takes place when the minimum voltage reaches 94% again, thus for an additional 2.5% voltage drop at maximum load. At minimum load, the voltage drop is one-fourth of this: 0.625%, resulting in a maximum voltage magnitude just before the second boosting point equal to 104.5% −0.625% = 103.875% (103.9% in the figure).
- Immediately downstream of the second boosting point, the voltages are 2.5% higher again, resulting in a minimum voltage of 96.5% and a maximum voltage of 106.375%.
- The voltage drops another 2.5% at maximum load toward the end of the feeder (assuming that the maximum feeder length is used) and another 0.625% at minimum load. The resulting minimum voltage is 94% and the resulting maximum voltage is 105.65%.

The overvoltage margin is smallest immediately downstream of the second boosting, 1.6% if we assume an overvoltage limit equal to 108%. The hosting capacity however may be determined by the overvoltage margin at another location because the source resistance that is, the distance along the feeder, also matters for the hosting capacity. The three most critical points are the end of the feeder and immediately downstream of the two boosting points. The hosting capacity is proportional to the overvoltage margin divided by the source resistance. The latter may increase faster than the distance along the feeder when thinner wires are used toward the end of the feeder where the loading is lower.

5.4.2 General Expressions for Two-Stage Boosting

The calculations for the feeder with two-stage boosting have been repeated using variables instead of fixed values. The following variables have been used:

- Lower limit of the deadband: $u_{\text{db,min}}$
- Upper limit of the deadband: $u_{\text{db,max}}$
- Undervoltage limit: $u_{\text{min,limit}}$
- Voltage boost per stage: Δu_{boost}
- Voltage drop during minimum load: α times the drop during maximum load

The maximum voltage immediately after the first boosting is equal to

$$u_{\max}(\lambda_1) = u_{\text{db,max}} - \alpha(u_{\text{db,min}} - u_{\text{min,limit}}) + \Delta u_{\text{boost}} \tag{5.16}$$

After the second boosting, the maximum voltage is

$$u_{\max}(\lambda_2) = u_{\text{db,max}} - \alpha(u_{\text{db,min}} - u_{\text{min,limit}}) + (2 - \alpha)\Delta u_{\text{boost}} \tag{5.17}$$

At the end of the feeder, assuming the maximum feeder length is used, the maximum voltage is given by the following expression:

$$u_{\max}(\lambda_3) = u_{\text{db,max}} - \alpha(u_{\text{db,min}} - u_{\text{min,limit}}) + (2 - 2\alpha)\Delta u_{\text{boost}} \tag{5.18}$$

These expressions show that the value of α (i.e., the voltage drop during low load) has a strong influence on the maximum voltage. Any reduction in minimum load will result in a further rise in maximum voltage and thus a reduction in hosting capacity. It should further be noted here that during the design process, the minimum and maximum voltage drops over a several-year period are considered. The maximum load is often anticipated only toward the end of the period, whereas the minimum load could already occur during the first year of operation.

Any energy saving campaign, if it is successful, could further reduce the hosting capacity if it will impact the energy consumption during low load. The somewhat remarkable effect could occur that an energy saving campaign becomes a barrier to the introduction of renewable sources of energy. There are also other reasons to address during energy saving campaigns, especially the types of consumption that are high during periods of high load.

The choice of the deadband also impacts the overvoltages. Increasing the upper limit will reduce the overvoltage margin. Every volt increase in the upper limit of the deadband will reduce the overvoltage margin by 1 V. The impact of the lower limit is less. Increasing the lower limit will increase the overvoltage margin (i.e., reduce the maximum voltage), but increasing the lower limit by 1 V will only increase the

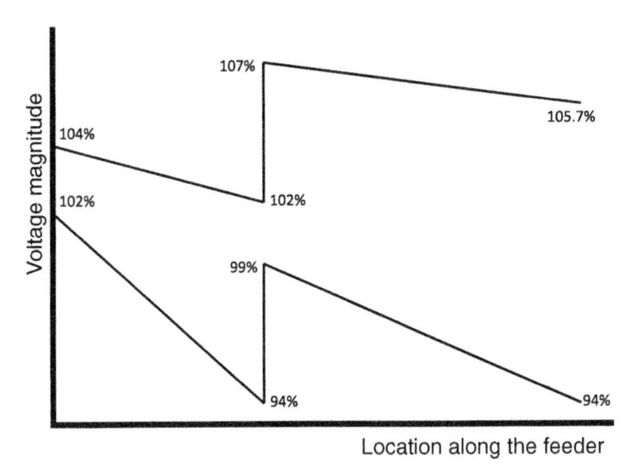

Figure 5.19 Voltage profile along a distribution feeder with voltage boosting in one step: minimum load (upper curve) and maximum load (lower curve).

overvoltage margin by α times 1 V. To increase the overvoltage margin, the deadband as a whole has to be moved down.

Allowing a lower undervoltage limit will also result in more overvoltage margin; however, here the impact is only by a factor of α. It should also be noted that increasing the lower limit of the deadband and allowing a lower undervoltage limit will allow longer feeders and shift the boosting points further toward the end of the feeder.

5.4.3 Single-Stage Boosting

The calculations have been repeated for a feeder with 5% voltage boosting in one step. The results are shown in Figure 5.19. The maximum voltage is higher in this case (107% versus 106.4%), but it is reached closer toward the start of the feeder, where the source resistance is lower.

Using the general expressions, the maximum voltage immediately after the boosting point is equal to

$$u_{\max}(\lambda_1) = u_{\text{db,max}} - \alpha(u_{\text{db,min}} - u_{\text{min,limit}}) + 2\Delta u_{\text{boost}} \tag{5.19}$$

where the boost in voltage is given as $2\Delta u_{\text{boost}}$ in this case to allow an easier comparison with the two-stage boosting case. The maximum voltage at the end of the feeder is the same as that in (5.18).

We see from (5.19) that the maximum voltage is impacted in the same way and by the same parameters as before. The main impact is again by $u_{\text{db,max}}$ and α.

5.4.4 Microgeneration

The numerical approach in the previous sections can be used to estimate the amount of distributed generation that can be connected. It is assumed that the generation is distributed in the same way along the feeder as the demand, so that the voltage rise profile due to the generation is the same as the voltage drop profile due to the demand. This assumption obviously does not hold for large units located at a few places, but it is more realistic for domestic microgeneration such as rooftop solar panels or micro-CHP, where both the production and the consumption are roughly similar per customer.

First, consider the two-stage boosting as shown in Figure 5.18. We assume that 108% is the overvoltage limit. This gives 1.6% overvoltage margin immediately after the second boosting. The maximum demand (both active and reactive power) resulted in a total voltage drop equal to 10.5% at this location. The maximum permissible voltage rise due to microgeneration is 1.6%. This voltage rise at this location is reached by a total production for the feeder $P_{\text{gen},1}$ according to the following expression:

$$RP_{\text{gen},1} = \frac{1.6}{10.5} \times (RP + XQ) \tag{5.20}$$

resulting in

$$P_{\text{gen},1} = 0.152 \times \left(P + \frac{X}{R}Q\right) \tag{5.21}$$

An amount of generation equal to this value, distributed along the feeder in the same way as the demand, will cause the maximum voltage at the second boosting point to reach 108%.

At the end of the feeder, the overvoltage margin is 2.2%. The total demand causes a voltage drop equal to 13% at this location. This sets the following limit to the amount of microgeneration:

$$P_{\text{gen},2} = 0.169 \times \left(P + \frac{X}{R}Q\right) \tag{5.22}$$

Comparing these two values shows that the first boosting point sets the limit to the hosting capacity. We also see that the amount of microgeneration that can be connected is at least 15% of the maximum active power demand. Additional generation can be connected when considering the voltage drop due to the reactive power. For reactive power being 48% of the active power (power factor 0.9) and X/R equal to 0.5, the hosting capacity of the medium-voltage feeder for microgeneration becomes about 20% of maximum demand.

The calculations can be repeated for the single-stage boosting shown in Figure 5.19. For the boosting point, the overvoltage margin is 1%, whereas the voltage drop due to maximum load is 8%. This sets the following limit:

$$P_{\text{gen},1} = 0.125 \times \left(P + \frac{X}{R}Q\right) \tag{5.23}$$

At the end of the feeder, the limit is the same as for two-stage boosting, as in (5.22). The hosting capacity is again set by the boosting point, being at least 12.5% of the maximum active power demand. It is also shown from these calculations that two-stage boosting gives a higher hosting capacity than single-stage boosting (0.152 versus 0.125 times the maximum demand). This is understandable because the overall voltage magnitude is lower when more boosting points are used. In the limit case of an infinite number of boosting points, the minimum voltage would be kept at 94% for a large part of the feeder. The maximum voltage would rise uniformly from 102% to 105.7%. The hosting capacity would be determined by the overvoltage margin at the end of the feeder: 0.169 times the maximum demand according to (5.22).

Keeping the voltage for maximum demand at its minimum acceptable value of 94% for a large part of the feeder will increase the risk of undervoltage with the end customers. It will also not be economically practical to install a large number of small boosters. The theoretical gain in hosting capacity, from 0.152 to 0.169 times the maximum demand, is rather limited.

Continuing along the line of the general expressions, (5.16)–(5.19), expressions can be derived showing the impact of different design parameters on the hosting capacity for microgeneration. For the two-stage boosting case, the overvoltage margin after the second boosting is

$$\delta_{\max} = u_{\max,\text{limit}} - u_{\text{db},\max} + \alpha(u_{\text{db},\min} - u_{\min,\text{limit}}) - (2-\alpha)\Delta u_{\text{boost}} \tag{5.24}$$

The voltage drop at this location for maximum demand is equal to

$$\Delta u = u_{\text{db},\min} - u_{\min,\text{limit}} + \Delta u_{\text{boost}} \tag{5.25}$$

The amount of generation that will raise the voltage to the overvoltage limit at this location is given by the following expression:

$$P_{\text{gen},1} = \frac{u_{\max,\text{limit}} - u_{\text{db},\max} + \alpha(u_{\text{db},\min} - u_{\min,\text{limit}}) - (2-\alpha)\Delta u_{\text{boost}}}{u_{\text{db},\min} - u_{\min,\text{limit}} + \Delta u_{\text{boost}}} \times \left(P + \frac{X}{R}Q\right) \tag{5.26}$$

In the same way, we can find an expression for the amount of generation that will raise the voltage at the end of the feeder to the overvoltage limit.

$$P_{\text{gen},2} = \frac{u_{\max,\text{limit}} - u_{\text{db},\max} + \alpha(u_{\text{db},\min} - u_{\min,\text{limit}}) - (2-2\alpha)\Delta u_{\text{boost}}}{u_{\text{db},\min} - u_{\min,\text{limit}} + 2\Delta u_{\text{boost}}} \times \left(P + \frac{X}{R}Q\right) \tag{5.27}$$

The hosting capacity of the feeder for microgeneration is the lower of the two values $P_{\text{gen},1}$ and $P_{\text{gen},2}$.

Looking at these expressions, we see again that the parameters with the main influence on the hosting capacity are the overvoltage limit $u_{\max,\text{limit}}$, the upper limit of the deadband $u_{\text{db},\max}$, and the ratio of the voltage drop for low and high load α. Using unrealistic values of the overvoltage limit and the voltage drop ratio can put an unnecessary limit to distributed generation.

5.5 TAP CHANGERS WITH LINE-DROP COMPENSATION

To improve the voltage regulation along a distribution feeder, the tap-changer controller may be equipped with the so-called "line-drop compensation" [178, 258]. The line-drop compensation results in a high voltage at the transformer terminals during high load and a low voltage during low load. This is obtained by compensating for the voltage drop along a fictitious impedance. The voltage magnitude at the terminals of the transformer, that is, at the start of the distribution feeder, is kept at approximately the following value:

$$U(0) = U_{\mathrm{ref}} + \frac{r_{\mathrm{s}}P + x_{\mathrm{s}}Q}{U_{\mathrm{nom}}} \tag{5.28}$$

where P and Q are total power through the transformer, U_{ref} the voltage reference of the controller, and r_{s} and x_{s} the impedance settings of the controller. The result of the controller is that the voltage is not kept constant at the terminals of the transformer, but instead the voltage is kept constant for a virtual point at an electrical distance $(r_{\mathrm{s}} + jx_{\mathrm{s}})$ along the feeder.

5.5.1 Transformer with One Single Feeder

Consider a transformer supplying a single feeder with uniformly distributed loads P and Q. The voltage magnitude along the feeder is found from (5.11), resulting in

$$U(\lambda) = U(0) - \frac{1}{2}\frac{RP + XQ}{U_{\mathrm{nom}}}\left(2\lambda - \lambda^2\right) \tag{5.29}$$

Choosing $r_{\mathrm{s}} = 1/4R$ and $x_{\mathrm{s}} = 1/4X$ results in a voltage profile where the maximum voltage (at the start of the feeder) is equally above the set point as the lowest voltage (at the end of the feeder) is below the set point. The resulting voltage profile along the feeder is

$$U(\lambda) = U_{\mathrm{ref}} + \frac{1}{2}\frac{RP + XQ}{U_{\mathrm{nom}}}\left(\lambda^2 - 2\lambda + \frac{1}{2}\right) \tag{5.30}$$

The resulting voltage profile along the feeder is shown in Figure 5.20, where the different curves correspond to different loading situations. It has been assumed in the calculations that active and reactive powers vary together, so that the ratio P/Q remains the same. The voltage reference used is $U_{\mathrm{set}} = 101\%$. The different curves have been obtained by varying the value of $(RP + XQ)/U_{\mathrm{nom}}$.

This kind of voltage control is used for medium-voltage distribution feeders ("primary feeders"). The voltage in the low-voltage network may be further controlled by using (MV/LV) distribution transformers with off-load tap changers. Transformers near the start of the feeder could have a turns ratio of 10.25/0.4 kV, resulting in a 2.5% reduction in voltage, and thus a reduaction in the risk of overvoltage. For transformers near the end of the feeder, a turns ratio of 9.75/0.4 kV could be used, resulting in a 2.5% increase in voltage, and thus a reduction in the risk of undervoltage. The voltages on low-voltage side of the distribution transformers are shown in Figure 5.21. The

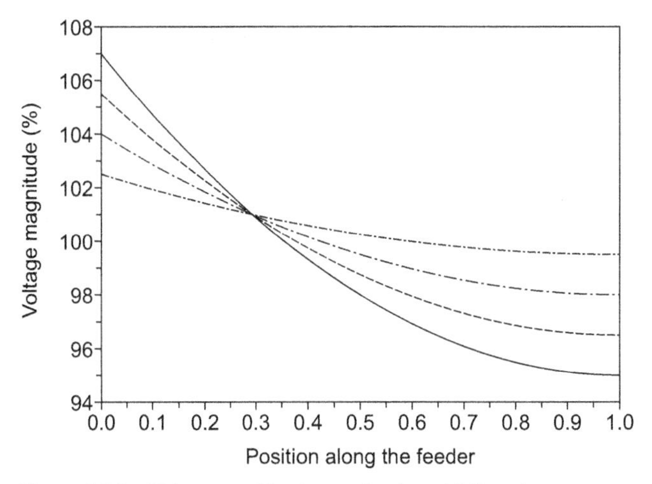

Figure 5.20 Voltage profile along a feeder with line-drop compensation for different loading situations. The solid curve corresponds to the highest loading.

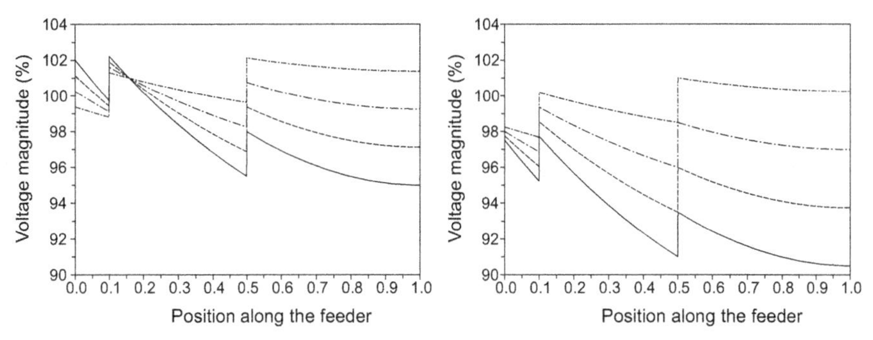

Figure 5.21 Voltage profile on the low-voltage side of distribution transformers with off-load tap changers (left) and on the remote points of the low-voltage network (right), along a medium-voltage feeder with line-drop compensation. The different curves refer to different loading situations, where the solid curve corresponds to the highest loading.

voltage drop over the distribution transformer is taken as 2.5% for full load. Figure 5.21 also shows the voltage drop at remote points in the low-voltage distribution network. The voltage drop over the low-voltage feeder is taken as 4.5% for full load, next to the 2.5% drop over the distribution transformer.

5.5.2 Adding a Generator

Adding a generator with active power P_{gen} to the feeder reduces the active power flow through the transformer, resulting in a lower voltage at the start of the feeder, by the action of the line-drop compensator. The voltage at the start of the feeder, using (5.28) with the same impedance settings as before, becomes

$$U(0) = U_{\text{ref}} + \frac{1}{4}\frac{RP + XQ}{U_{\text{nom}}} - \frac{1}{4}\frac{RP_{\text{gen}}}{U_{\text{nom}}} \tag{5.31}$$

The resulting voltage profile along the feeder, with the generator at location $\lambda = \lambda_{gen}$, is

$$U(\lambda) = U_{ref} + \frac{1}{2}\frac{RP + XQ}{U_{nom}}\left(\lambda^2 - 2\lambda + \frac{1}{2}\right) + \frac{(\lambda - 1/4)\,RP_{gen}}{U_{nom}} \qquad \lambda < \lambda_{gen}$$
$$U(\lambda) = U_{ref} + \frac{1}{2}\frac{RP + XQ}{U_{nom}}\left(\lambda^2 - 2\lambda + \frac{1}{2}\right) + \frac{\left(\lambda_{gen} - 1/4\right)RP_{gen}}{U_{nom}} \qquad \lambda > \lambda_{gen} \tag{5.32}$$

The first two terms in these expressions represent the voltage profile for the feeder without generation. The impact of the generation on the voltage profile can thus be quantified through the following simple expression:

$$\Delta U_{gen}(\lambda) = \frac{(\lambda - 1/4)\,RP_{gen}}{U_{nom}} \qquad \lambda < \lambda_{gen}$$
$$\Delta U_{gen}(\lambda) = \frac{\left(\lambda_{gen} - 1/4\right)RP_{gen}}{U_{nom}} \qquad \lambda > \lambda_{gen} \tag{5.33}$$

The impact of the injected power on the voltage profile is shown in Figure 5.22. The result is a reduction in the voltage magnitude for the first quarter of the feeder and an increase for the rest of the feeder.

Assume that all distribution transformers have the same turns ratio and that no other voltage boosting is used. The voltage profile in Figure 5.20 thus also holds (after correcting for drops in the low-voltage network) for the low-voltage customers. Comparing Figures 5.22 and 5.20 shows that the connection of a generator results in a

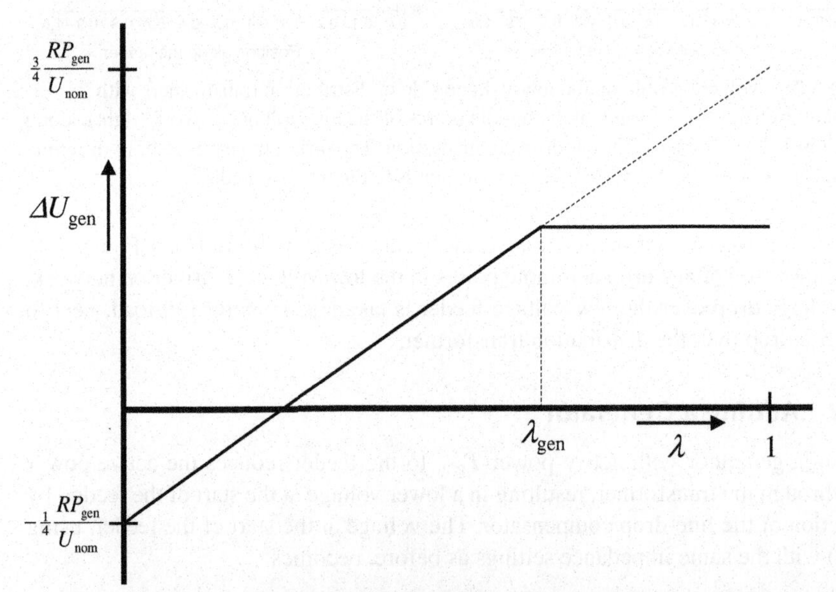

Figure 5.22 Change in voltage magnitude along a feeder with line-drop compensation due to the injection of active power P_{gen} at a location λ_{gen}.

voltage drop where the voltage magnitude is high and a voltage rise where the voltage magnitude is low. The generator thus creates a more flat voltage profile along the whole feeder. From Figure 5.22 and expression (5.33), it also follows that the voltage rise due to the connection of a generator unit is at most (i.e., for a unit at the end of the feeder) 75% of what the voltage rise would be when no line-drop compensation is used. Thus, even for the same overvoltage margin, the hosting capacity is 33% higher than when no line-drop compensation is used. Note however that the value of 33% is determined by our choice of impedance setting.

However, the main impact of line-drop compensation on the hosting capacity is that it is possible to raise the voltage magnitude during high load more than during low load. This results in a larger overvoltage margin that in turns gives a larger hosting capacity.

Comparing Figures 5.22 and 5.20 further identifies the two risks with large generator units connected to the medium-voltage feeder. For units connected near the end of the feeder, there is an overvoltage risk close to the point of connection of the generator (but smaller than that without line-drop compensation). There is further an undervoltage risk near the start of the feeder, independent of where the unit is connected.

5.5.3 Calculating the Hosting Capacity

Consider the situation with a generator connected at the end of the feeder: $\lambda_{gen} = 1$. The voltage magnitude at the point of connection of the generator, that is, at the end of the feeder, is in that case, using (5.32),

$$U(1) = U_{ref} - \frac{1}{2}\Delta U + \frac{3}{4}\frac{RP_{gen}}{U_{nom}} \tag{5.34}$$

with the "voltage drop" ΔU due to the active and reactive power demand defined as

$$\Delta U = \frac{1}{2}\frac{RP + XQ}{U_{nom}} \tag{5.35}$$

The maximum voltage at the end of the feeder is

$$U_{max} = U_{db,max} - \frac{1}{2}\Delta U_{min} + \frac{3}{4}\frac{RP_{gen,max}}{U_{nom}} \tag{5.36}$$

where ΔU_{min} is the minimum value of ΔU according to (5.35) and $U_{db,max}$ the upper limit of the deadband of the voltage controller. The maximum value should not exceed the overvoltage limit $U_{max,limit}$. This results in the following expression for the hosting capacity:

$$P_{gen,max} < \frac{U_{nom}}{R}\left\{\frac{4}{3}\left(U_{max,limit} - U_{db,max}\right) + \frac{2}{3}\Delta U_{min}\right\} \tag{5.37}$$

Compared to the expression for the hosting capacity for a feeder without line-drop compensation, the margin between the maximum permissible voltage and the upper limit of the deadband is 33% larger. However, the impact of the minimum voltage drop is 33% less.

The voltage drop at the start of the feeder, due to generation connected somewhere to the feeder, could result in a voltage magnitude below the undervoltage limit. In the same way as for overvoltages, a hosting capacity can be calculated for this.

The voltage at the start of the feeder is independent of the location of the generator unit:

$$U(0) = U_{\text{set}} + \frac{1}{2}\Delta U - \frac{1}{4}\frac{RP_{\text{gen}}}{U_{\text{nom}}} \tag{5.38}$$

The minimum voltage magnitude is

$$U_{\min} = U_{\text{db,min}} + \frac{1}{2}\Delta U_{\min} - \frac{1}{4}\frac{RP_{\text{gen,max}}}{U_{\text{nom}}} \tag{5.39}$$

where $U_{\text{db,min}}$ is the lower limit of the deadband of the voltage controller. The minimum voltage should be higher than the lowest permissible voltage magnitude $U_{\text{min,limit}}$, resulting in the following expression for the hosting capacity:

$$P_{\text{gen,max}} < 4\frac{U_{\text{nom}}}{R}\left(U_{\text{db,min}} - U_{\text{min,limit}} + \frac{1}{2}\Delta U_{\min}\right) \tag{5.40}$$

Assuming that the deadband is located in the middle of the permissible voltage range, $U_{\text{db,min}} - U_{\text{min,limit}} = U_{\text{max,limit}} - U_{\text{db,max}}$, the hosting capacity according to (5.40) is three times the hosting capacity according to (5.37). Note however that undervoltages at the start of the feeder occur for a generator connected anywhere along the feeder, whereas overvoltages at the end of the feeder occur only for generation close to the end of the feeder. For a generator connected close to the start of the feeder, the hosting capacity according to (5.40) may set the limit to the amount of production.

Like without line-drop compensation, the use of distribution transformers with off-load tap changers reduces the hosting capacity. Consider the hosting capacity for a generator connected to the end of the feeder, in case a voltage boost equal to ΔU_{boost} is used. The maximum voltage at the end of the feeder, compare to (5.36), is

$$U_{\max} = U_{\text{db,max}} + \Delta U_{\text{boost}} - \frac{1}{2}\Delta U_{\min} + \frac{3}{4}\frac{RP_{\text{gen,max}}}{U_{\text{nom}}} \tag{5.41}$$

The hosting capacity for this case is

$$P_{\text{gen,max}} < \frac{U_{\text{nom}}}{R}\left\{\frac{4}{3}\left(U_{\text{max,limit}} - U_{\text{db,max}} - \Delta U_{\text{boost}}\right) + \frac{2}{3}\Delta U_{\min}\right\} \tag{5.42}$$

5.5.4 Multiple Feeders from the Same Transformer

The situation gets more complicated when multiple feeders originate from the same transformer. Rules for setting the line-drop compensator are discussed in detail by Gönen [178]. Those rules are beyond the scope of this book and only a simple case will be discussed here to illustrate the general impact of distributed generation on the performance of the line-drop compensator.

Consider a medium-voltage distribution network with n identical feeders originating from the main substation on secondary side of a transmission transformer with on-load tap changers including line drop compensation. Each feeder has an impedance $R + jX$ and an identical loading P, Q. The voltage reference of the controller is U_{ref}; the impedance settings are $r_s = R/4n$ and $x_s = X/4n$. The voltage magnitude at the start of the feeders is:

$$U(0) = U_{\mathrm{ref}} + \frac{r_s n P + x_s n Q}{U_{\mathrm{nom}}} \tag{5.43}$$

When no generation is present, the voltage along each of the feeders is the same:

$$U(\lambda) = U_{\mathrm{ref}} + \frac{1}{2}\frac{RP + XQ}{U_{\mathrm{nom}}}\left(\lambda^2 - 2\lambda + \frac{1}{2}\right) \tag{5.44}$$

which is exactly the same voltage profile as in the case with only one feeder.

For the line-drop compensation to work properly, it is important that the different feeders have a similar load variation with time. The presence of generation on one of the feeders will intervene with this principle. When a generator injecting active power P_{gen} is added to one of the feeders, the current through the transformer is reduced. To compensate for this, the controller reduces the voltage magnitude at the start of the feeders:

$$U(0) = U_{\mathrm{ref}} + \frac{1}{4}\frac{RP + XQ}{U_{\mathrm{nom}}} - \frac{1}{4n}\frac{RP_{\mathrm{gen}}}{U_{\mathrm{nom}}} \tag{5.45}$$

The reduction due to the presence of the generator is

$$\Delta U_{\mathrm{gen}}(0) = -\frac{1}{4n}\frac{RP_{\mathrm{gen}}}{U_{\mathrm{nom}}} \tag{5.46}$$

This reduction in voltage magnitude is an improvement for the feeder with the generator, but not for the other feeders. Especially during high load, this may result in undervoltages for the feeders without generation. We will come back to this soon, but we will first discuss the voltage profile along the feeder with generation. This voltage profile, for a generator connected at the end of the feeder is

$$U(\lambda) = U_{\mathrm{ref}} + \frac{1}{2}\frac{RP + XQ}{U_{\mathrm{nom}}}\left(\lambda^2 - 2\lambda + \frac{1}{2}\right) + \left(\lambda - \frac{1}{4n}\right)\frac{RP_{\mathrm{gen}}}{U_{\mathrm{nom}}} \tag{5.47}$$

The first two terms are the same as for the feeder without generation, so the voltage change due to the generator reads as

$$\Delta U_{\mathrm{gen}}(\lambda) = \left(\lambda - \frac{1}{4n}\right)\frac{RP_{\mathrm{gen}}}{U_{\mathrm{nom}}} \tag{5.48}$$

With increasing number of feeders, the voltage drop at the start of the feeder quickly drops to zero, whereas the voltage rise at the point of connection of the generator quickly becomes equal to the rise for a feeder without line-drop compensation.

The voltage profile along a feeder without generation, while there is generation present on another feeder, is found by combining (5.44) and (5.46), resulting in the following expression:

$$U(\lambda) = U_{\text{ref}} + \frac{1}{2}\frac{RP + XQ}{U_{\text{nom}}}\left(\lambda^2 - 2\lambda + \frac{1}{2}\right) - \frac{1}{4n}\frac{RP_{\text{gen}}}{U_{\text{nom}}} \tag{5.49}$$

The voltage magnitude at the end of the feeder, where the voltage is lowest, is

$$U(1) = U_{\text{ref}} - \frac{1}{2}\Delta U - \frac{1}{4n}\frac{RP_{\text{gen}}}{U_{\text{nom}}} \tag{5.50}$$

where $\Delta U = (1/2)(RP + XQ)/U_{\text{nom}}$ is the voltage drop along the feeder due to the load. The presence of generation on other feeders may cause undervoltages on this feeder, so the minimum voltage should be compared with the minimum permissible voltage. The minimum voltage occurs for maximum load and maximum generation:

$$U_{\text{min}} = U_{\text{db,min}} - \frac{1}{2}\Delta U_{\text{max}} - \frac{1}{4n}\frac{RP_{\text{gen,max}}}{U_{\text{nom}}} \tag{5.51}$$

This voltage should be higher than the minimum permissible voltage $U_{\text{min,limit}}$. This results in the following value for the hosting capacity:

$$P_{\text{gen,max}} < 4n\frac{U_{\text{nom}}}{R}\left(U_{\text{db,min}} - U_{\text{min,limit}} - \frac{1}{2}\Delta U_{\text{max}}\right) \tag{5.52}$$

Many feeders are voltage drop limited: the minimum voltage is close to the undervoltage limit. Therefore, the above expression is rewritten in terms of the undervoltage margin:

$$P_{\text{gen,max}} < 4n \times \frac{U_{\text{nom}}^2}{R} \times \delta_{\text{min}} \tag{5.53}$$

where δ_{min} is the relative margin between the lowest voltage and the undervoltage limit (the "undervoltage margin"):

$$\delta_{\text{min}} = \frac{U_{\text{min}} - U_{\text{min,limit}}}{U_{\text{nom}}} \tag{5.54}$$

For voltage drop limited feeders, this margin is at most a few percent and in some case close to zero. The hosting capacity depends only on this margin, the nominal voltage, and the feeder resistance. The feeder resistance depends in turn on the feeder size and on its length. Longer, small cross section feeders will have a smaller hosting capacity than shorter high cross section feeders. Note that the location of the generator does not have any influence on this hosting capacity.

The above calculations were based on the assumption that the load varies in the same way for the different feeders. In reality, there will always be some difference between the feeders. This will further reduce the hosting capacity.

In many cases, especially for rural networks, the undervoltage will be close to zero, or the network operator will like to reserve the existing margin for future load growth. In those cases, the hosting capacity is effectively zero and other measures are needed to be able to connect distributed generation. Note that the hosting capacity according to (5.54) refers to the total generation connected to all feeders coming from the same transformer. When one feeder has reached its undervoltage limit, no distributed generation can be connected to any of the other feeders.

5.6 PROBABILISTIC METHODS FOR DESIGN OF DISTRIBUTION FEEDERS

5.6.1 Need for Probabilistic Methods

In the previous sections, the hosting capacity was calculated using so-called "deterministic methods." With those methods, the highest consumption is compared with the lowest production and the lowest consumption with the highest production. The probability that, for example, the overvoltage limit is exceeded is not calculated and typically assumed to be zero. The amount of distributed generation that can be connected is determined by the lowest amount of consumption and the highest amount of production. The deterministic approach for calculating the hosting capacity is hence also referred to as a "worst-case approach."

To be fair, it should be noted that the design of distribution networks is based on stochastic considerations. The loading of a distribution feeder is obtained by multiplying the so-called "after-diversity maximum demand" per customer with the number of customers. The after-diversity maximum demand is the demand per customer that, with high probability, will not be exceeded. Although stochastic methods were used to determine this demand per customer, it is in most studies used as a deterministic value. Introducing distributed generation implies that the original assumptions no longer hold. Instead of after-diversity maximum demand, a number of other parameters could be developed, for example, the "after-diversity maximum production" and the "after-diversity minimum consumption." In case of a strong correlation between production and consumption (e.g., because both show a strong daily variation), the "after-diversity minimum net demand" could be a suitable design parameter.

Developing such methods and simple guidelines for estimating their values will require a large number of measurements and simulations to better understand the stochastic properties of consumption together with different types of distributed generation. We will not attempt to develop these methods here because we suffer from the same lack of suitable measurement data as everybody else. Instead, we will just illustrate the stochastic design method for a simple system.

5.6.2 The System Studied

To introduce the probabilistic approach to describing voltage variations in distribution networks, we consider a simple system supplying one load, as shown in Figure 5.23. Instead of a concentrated load, the loads P and Q can also be considered as an equivalent load, resulting in the same voltage drops. For example, when the load is equally distributed along the feeder, the equivalent concentrated load is half the total distributed load, both in active and in reactive power.

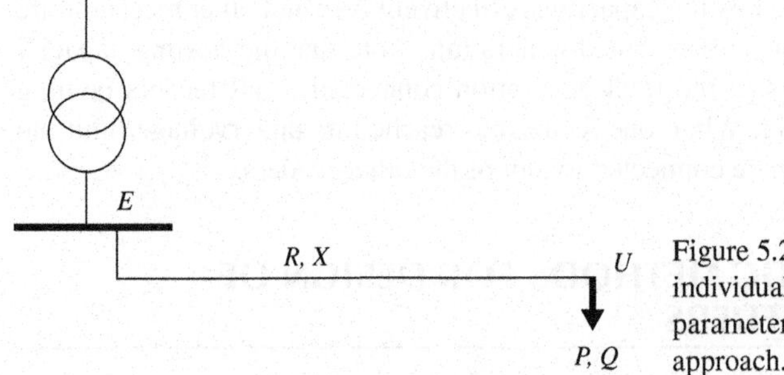

Figure 5.23 Supply to one individual load: introduction of parameters for the stochastic approach.

The voltage at the location of the load, in this case at the end of the feeder, is found by using the following expression:

$$U = E - RP - XQ \tag{5.55}$$

In the probabilistic approach, E, P, and Q are random variables, described by means of a probability density function or a probability distribution function. Furthermore, any correlations between E, P, and Q should be considered. The resistance R and reactance X of the feeder can be considered as deterministic parameters. They do show random variations, for example, the change in resistance with temperature, but those do not impact any of the results. Also, the impact of these variations on the voltage drop is likely to be less than the approximations made in obtaining (5.55). We will consider feeder resistance and reactance as deterministic parameters.

5.6.3 Probability Density and Distribution Functions

For a random variable X, the probability distribution function F_X gives the probability that the random variable does not exceed a certain value:

$$F_X(x) = \Pr\{X \leq x\} \tag{5.56}$$

The probability density function $f_X(x)$ is the derivative of the probability distribution function:

$$f_X(x) = \frac{\mathrm{d}}{\mathrm{d}x} F_X(x) \tag{5.57}$$

The probability density function gives the probability that the random variable is close to a given value. In mathematical terms, this can be presented as

$$f_X(x)\Delta x = \Pr\{x < X \leq x + \Delta x\} \tag{5.58}$$

5.6.4 Distributions of Functions of Random Variables

Consider two random variables X and Y, with joined probability density function $f(x, y)$ and joined probability distribution function $F(x, y)$. A third random variable,

Z, is obtained as a function of the other two variables:

$$Z = g(X, Y) \tag{5.59}$$

The probability distribution function of Z is obtained from the following expression [253]:

$$F_Z(z) = \iint\limits_{g(x,y)<z} f(x, y)\mathrm{d}x\,\mathrm{d}y \tag{5.60}$$

When X and Y are independent variables, so that $f(x, y) = f_X(x) \times f_Y(y)$, we can rewrite (5.60) as

$$F_Z(z) = \iint\limits_{g(x,y)<z} f_X(x) f_Y(y)\mathrm{d}x\,\mathrm{d}y \tag{5.61}$$

For most functions, solving the integral can be done only numerically. However, for some simple functions, analytical simplifications are possible. For the sum of two random variables, $Z = X + Y$, we obtain the following expression:

$$\begin{aligned} F_Z(z) &= \int_{-\infty}^{\infty} f_X(x)\mathrm{d}x \int_{-\infty}^{z-x} f_y(y)\mathrm{d}y \\ &= \int_{-\infty}^{\infty} f_X(x) F_Y(z - x)\mathrm{d}x \end{aligned} \tag{5.62}$$

As a second example, for a linear combination of two random variables $Z = aX + bY$, the probability distribution function is

$$F_Z(z) = \int_{-\infty}^{\infty} f_X(x) F_Y\left(\frac{z - ax}{b}\right) \mathrm{d}x \tag{5.63}$$

5.6.5 Mean and Standard Deviation

Consider two stochastic variables X and Y, with expected values μ_X and μ_Y and, standard deviations σ_X and σ_Y. A new stochastic variable Z is a linear combination of X and Y:

$$Z = aX + bY \tag{5.64}$$

The expected value of the new variable is obtained as the linear combination of the expected values of the original variables:

$$\mu_Z = a\mu_X + b\mu_Y \tag{5.65}$$

When X and Y are stochastically independent, the standard deviation of the new variable is obtained from

$$\sigma_Z = \sqrt{a^2\sigma_X^2 + b^2\sigma_Y^2} \tag{5.66}$$

When the original variables are not stochastically independent, their dependency can be quantified through the so-called covariance:

$$C_{XY} = E(XY) - E(X)E(Y) \tag{5.67}$$

The standard deviation of $Z = aX + bY$ is obtained from the following expression [253, Section 23.8]:

$$\sigma_Z = \sqrt{a^2\sigma_X^2 + b^2\sigma_Y^2 + 2abC_{XY}} \tag{5.68}$$

These expressions hold for any distribution; however, they are most useful when we can assume that the random variables are normally distributed. In the next section, we will make this assumption, but the resulting expressions for expected value and standard deviation hold generally. The calculation, for example, of the 95-percentile, however, holds only for normally distributed variables.

5.6.6 Normal Distributions

With most probability distributions, it is not possible to obtain analytical expression for the distribution function of the voltage magnitude. As shown previously, even for a simple uniform distribution, and even without considering the voltage at the start of the feeder as a random variable, the calculations and the resulting expressions become rather complicated. The main exception is the normal distribution where mean and standard deviation can be obtained easily.

Let the active power consumption P be normally distributed, with mean μ_P and standard deviation σ_P. The reactive power Q is also normally distributed, with mean μ_Q and standard deviation σ_Q. The voltage drop $RP + XQ$ is then also normally distributed, with mean $R\mu_P + X\mu_Q$ and standard deviation $\sqrt{R^2\sigma_P^2 + X^2\sigma_Q^2}$. Adding additional random variables is rather straightforward now. For example, for the voltage at the end of the feeder,

$$U = E - RP - XQ \tag{5.69}$$

the mean and the standard deviation are

$$\mu_U = \mu_E - R\mu_P - X\mu_Q \tag{5.70}$$

and

$$\sigma_U^2 = \sigma_E^2 + R^2\sigma_P^2 + X^2\sigma_Q^2 \tag{5.71}$$

When we consider the correlation between the random variables, the expression for the standard deviation becomes more complicated:

$$\sigma_U^2 = \sigma_E^2 + R^2\sigma_P^2 + X^2\sigma_Q^2 - 2RC_{EP} - 2XC_{EQ} + 2RXC_{PQ} \tag{5.72}$$

When adding distributed generation to the end of the feeder, the voltage at the end of the feeder becomes

$$U = E - RP + RG - XQ \tag{5.73}$$

where G is the active power production by the generator. The reactive power production by the generator is assumed to be zero. Without correlation between the variables, the following expressions hold for the expected value and the standard deviation:

$$\mu_U = \mu_E - R\mu_P + R\mu_G - X\mu_Q \tag{5.74}$$

$$\sigma_U^2 = \sigma_E^2 + R^2\sigma_P^2 + R^2\sigma_G^2 + X^2\sigma_Q^2 \tag{5.75}$$

Considering the covariances, the latter expression reads as

$$\begin{aligned}\sigma_U^2 &= \sigma_E^2 + R^2\sigma_P^2 + R^2\sigma_G^2 + X^2\sigma_Q^2 \\ &\quad -2RC_{EP} + 2RC_{EG} - 2XC_{EQ} - 2R^2C_{PG} + 2RXC_{PQ} - 2RXC_{GQ}\end{aligned} \tag{5.76}$$

For medium voltage feeders, we can assume that the voltage at the start of the feeder E is independent of the current through the feeder. For wind power, we may further assume that the production is independent of the consumption. The expression for the standard deviation simplifies in that case to

$$\sigma_U^2 = \sigma_E^2 + R^2\sigma_P^2 + R^2\sigma_G^2 + X^2\sigma_Q^2 + 2RXC_{PQ}$$

5.6.7 Stochastic Calculations Using Measurements

Strict stochastic calculations would use mathematical expressions for the probability distributions of the random variables. Examples of such distributions are the normal distribution and the uniform distribution, but other distributions could be used when they are deemed more appropriate. In Chapter 2, we discussed stochastic models for wind power and solar power. No such widely used models were found for the consumption. One of the reasons for this is the wide difference in consumption patterns at different locations. The lack of measurement data further contributes to this.

As mentioned previously, the introduction of automatic meter reading will make it possible to obtain sufficient data to develop stochastic models. Such models could be used in future design studies and to more accurately determine the hosting capacity. The presence of large amounts of measurement data will also make it possible to use actually measured voltage and power as inputs to the calculations. In this section, we will use a limited set of measurements just to illustrate how a stochastic model for the

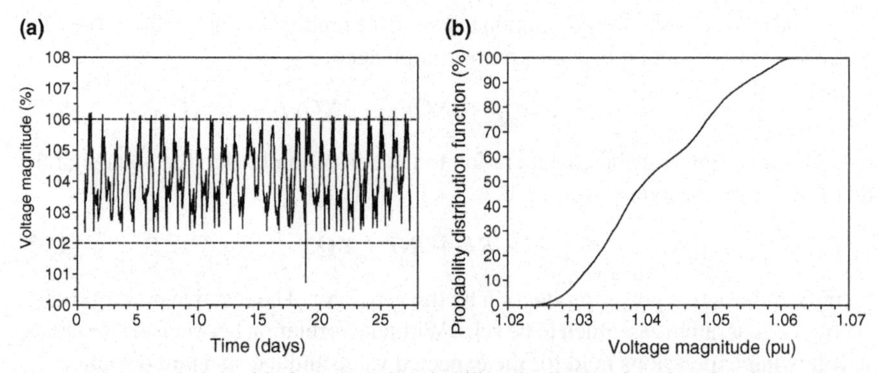

Figure 5.24 Variation with time (a) and probability distribution function (b) of the voltage at the start of the feeder.

production can be used in combination with measurements of the existing situation. Here, it should be noted that for a realistic study, a whole year of data would be more appropriate, unless there is a clear indication that the measurement period is typical for the rest of the year as well.

Figure 5.24 shows the voltage magnitude on the secondary side of a 110/10 kV transformer. The measurements were performed during an almost 4-week period. The voltage magnitude is obtained as the average of the magnitudes for the three phase-to-ground voltages. Its value is given as a percentage of the nominal voltage of 10 kV. The transformer is equipped with on-load tap changers aimed at keeping the voltage between 102.5% and 106% of the nominal voltage. Both the voltage magnitude as a function of time and the probability distribution function for the voltage magnitude are shown in the figure.

The distribution of the voltages reasonably corresponds with a uniform distribution. A uniform distribution would give a straight line. The mean μ_E and standard deviation σ_E of the voltage magnitude are

$$\mu_E = 104.18\%$$
$$\sigma_E = 0.913\%$$

The time variation and the probability distribution function of the active power and the reactive power are shown in Figures 5.25 and 5.26. The active power is obtained as the sum of the active powers per phase. Also, the reactive power is obtained as the sum of the values per phase. The distribution for the active power shows a concentration around two values: about 9 and 15 MW. Note that a high derivative of the probability distribution function corresponds to a high concentration of values.

The mean μ_P and standard deviation σ_P of the active power are

$$\mu_P = 13.105\,\text{MW}$$
$$\sigma_P = 2.743\,\text{MW}$$

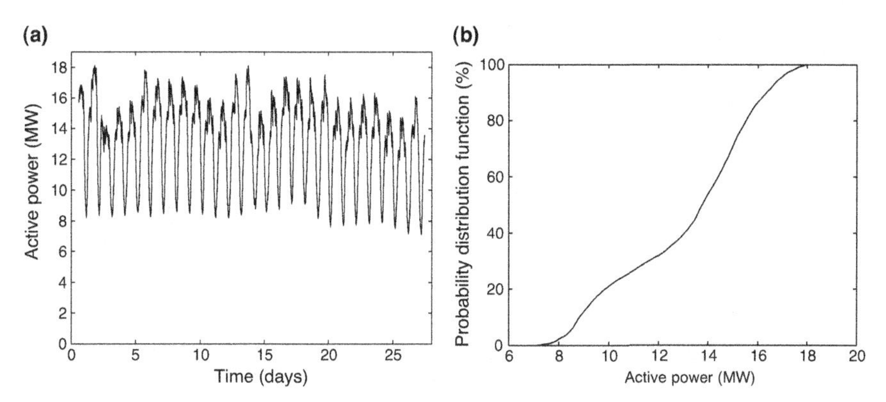

Figure 5.25 Time variation (a) and probability distribution function (b) of the active power through the feeder.

The mean μ_Q and standard deviation σ_Q of the reactive power are

$$\mu_Q = 3.804\,\text{Mvar}$$
$$\sigma_Q = 0.461\,\text{Mvar}$$

From the measurements, it is also possible to calculate the covariance of the three joint probabilities:

$$C_{PQ} = 9.346 \times 10^5$$
$$C_{EP} = -1.007 \times 10^5$$
$$C_{EQ} = -1.361 \times 10^4$$

Although the covariance can be used immediately in the expressions given in the previous section, the level of correlation between the random variables is best interpreted from the correlation coefficient. For two random variables X and Y, the

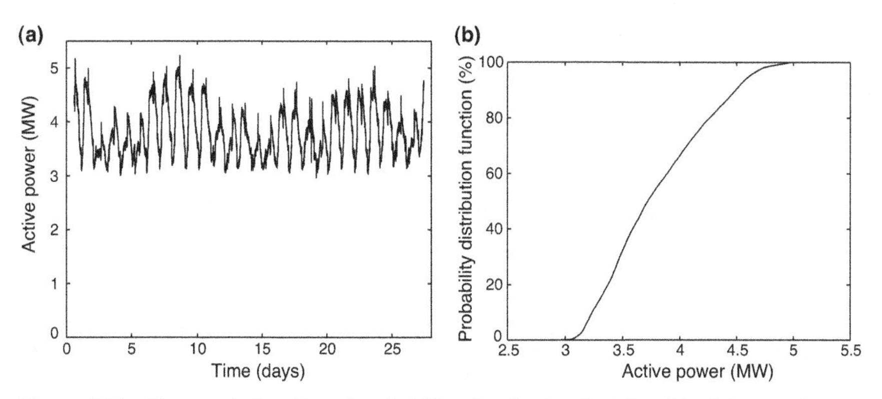

Figure 5.26 Time variation (a) and probability distribution function (b) of the reactive power through the feeder.

correlation coefficient is defined as

$$R_{XY} = \frac{C_{XY}}{\sigma_X \sigma_Y} \tag{5.77}$$

The value of the correlation coefficient is always in the range from -1 to $+1$, where the higher the absolute value, the stronger the correlation. For the three variables in our measurement, the correlation coefficients are as follows:

$$R_{PQ} = 0.7398$$
$$R_{EP} = -0.6960$$
$$R_{EQ} = -0.5605$$

We see that active and reactive powers are positively correlated, whereas the voltage at the start of the feeder is negatively correlated with both active and reactive powers.

From the voltage at the start of the feeder and the active and reactive power flow through the feeder, the voltage at the end of the feeder can be calculated using (5.69). As the time series of all three parameters (E, P, and Q) are known, the time series of the voltage V can be calculated. From this time series, its probability distribution function can be calculated, resulting in Figure 5.27. The resistance and the reactance of the feeder are assumed equal to 0.08 and 0.35 Ω/km, respectively. Expression (5.69) holds only when all quantities are expressed in per unit; to obtain the voltage magnitude in the figure, we used a base voltage of 10 kV and a base power of 10 MVA.

The mean and standard deviation can be obtained from the time series as well for the 3 km long feeder resulting in values of 0.9704 and 0.0182 pu, respectively. Alternatively, the mean and standard deviation can be calculated using (5.70) and (5.71), where the correlations between the parameters have been neglected. A comparison between the two methods of calculating mean and standard deviation is made

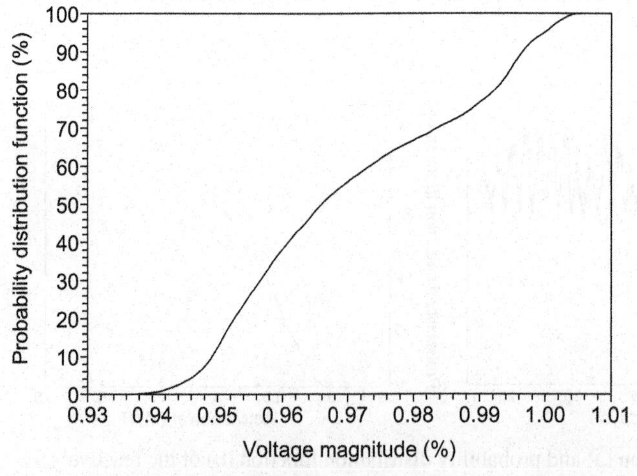

Figure 5.27 Probability distribution function for the voltage at the end of a 3 km long feeder.

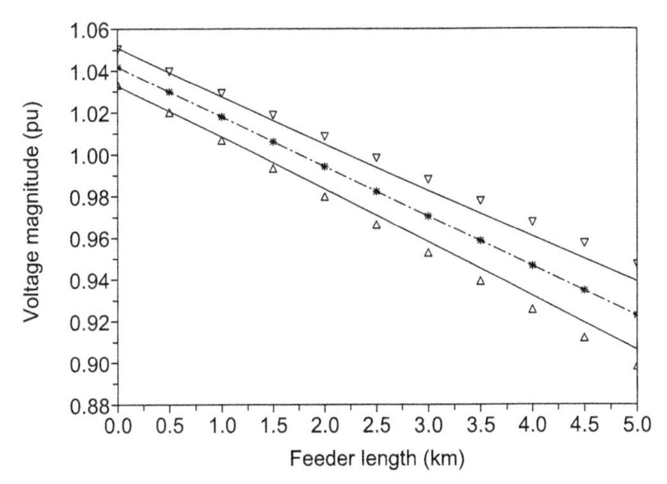

Figure 5.28 Comparison between two methods for estimating the mean and standard deviation of the voltage at the end of a feeder: solid and dashed lines are values obtained by calculation from mean and standard deviation and stars and triangles are values obtained from the calculated time series.

in Figure 5.28. The dashed line in the middle of the plot gives the mean value of the voltage magnitude calculated from (5.70). The stars indicate the values obtained from the calculated time series. The values are identical because (5.70) holds for any stochastic variable. The upper and lower solid lines give the values of $\mu_V + \sigma_V$ and $\mu_v - \sigma_V$, calculated from (5.71), whereas the triangles represent the values obtained from the calculated time series. Assuming active and reactive powers to be independent underestimates the standard deviation.

Of importance for design and operation of the distribution network is not so much the standard deviation of the voltage magnitude, but the probability that it will exceed certain limits. For a normal distribution, these probabilities can be calculated directly from the mean and standard deviation. However, the voltage magnitude is not distributed according to a normal distribution. A comparison between the two methods, for these probabilities, is made in Figure 5.29. The different percentiles can be obtained directly from the calculated time series, where the 5-percentile is the value that is not exceeded by 5% of the voltage magnitude values. In other words, the voltage magnitude is less than the 5-percentile during 5% of the time. The 1-percentile, 5-percentile, 95-percentile, and 99-percentile obtained from the time series are shown as triangles in Figure 5.29.

The different percentiles can also be calculated from the mean and the standard deviation, assuming a normal distribution.

- 1-percentile: $\mu_V - 2.33\sigma_V$
- 5-percentile: $\mu_V - 1.64\sigma_V$
- 95-percentile: $\mu_V + 1.64\sigma_V$
- 99-percentile: $\mu_V + 2.33\sigma_V$

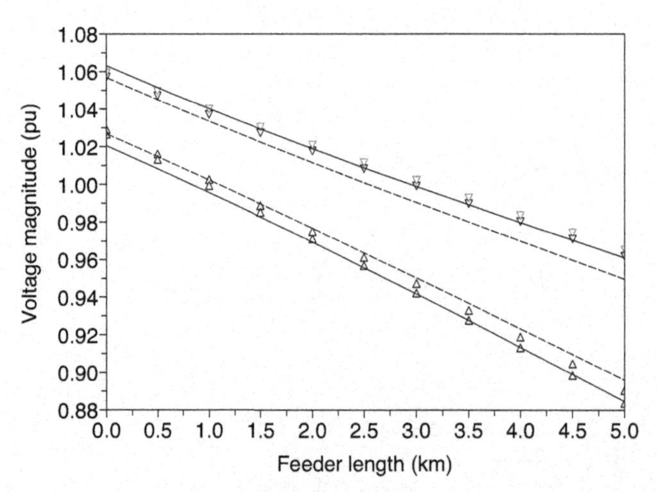

Figure 5.29 1-percentile, 5-percentile, 95-percentile, and 99-percentile of the voltage magnitude at the end of a feeder, as a function of the feeder length. The triangles refer to values calculated from the time series; the continuous lines were calculated assuming normal distributions.

The values calculated in this way, using mean and standard deviation from (5.70) and (5.71), that is, neglecting the correlation, are shown as solid and dashed lines in Figure 5.29. The differences between the values are small: assuming a normal distribution for all variables gives a good indication of the range in voltage magnitudes. This assumption may thus be used in the design process.

5.6.8 Generation with Constant Production

When adding a generator unit to the feeder, the voltage at the end of the feeder will increase. For large amounts of generation, the voltage magnitude may exceed the overvoltage limit. Most generator units inject only active power, so the increase in voltage will take place only over the resistance. For a generator with active power production P_{gen}, the voltage at the end of the feeder is

$$U = E - (P - P_{\text{gen}}) \times R - Q \times X \tag{5.78}$$

This voltage magnitude has been calculated for the same times series of active and reactive power as used before and for a feeder length equal to 5 km. The 1, 5, 95, and 99 percentiles of the voltage magnitude are shown in Figure 5.30 as a function of the size of the generator unit. It has been assumed in the calculations that the voltage is boosted by 5% at the end of the feeder.

For small amounts of distributed generation, the voltage quality actually improves. However, when the size of the generation unit increases more, unacceptable overvoltages occur. The 110% limit is exceeded when more than about 20 MW of generation is connected. The 106% limit is already exceeded for slightly more than 10 MW of generation. This maximum permissible amount of distributed generation should be compared with the active power consumption, which varies between 7

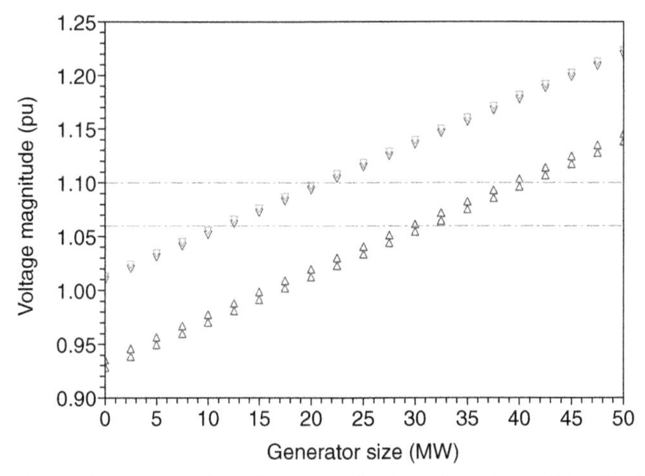

Figure 5.30 Range in voltage magnitude at the end of a 5 km feeder, with 5% boost in voltage, as a function of the amount of active power injected.

and 18 MW (see Figure 5.25). The hosting capacity is thus 55% of peak load when the overvoltage limit is set at 106% of nominal voltage and 110% of peak load for an overvoltage limit at 110%. For comparison, the first hosting capacity with respect to overload is the sum of minimum and maximum loads (see Chapter 4), which would be 25 MW or 125% of peak load in this case.

5.6.9 Adding Wind Power

The probability distribution functions for the active power production by wind power, as obtained in Chapter 2, will be used to calculate the probability distribution function of the voltage as experienced by the customer. As mentioned before, the interest of the customer is in the voltage magnitude, not in the power flows in the network.

The same network situation and consumption pattern as in the previous section has been used, with a 5 km feeder, and time series for the voltage at the start of the feeder and for active and reactive power consumption. The voltage at the end of the feeder, where the customers are connected, is given by the following expression, with all parameters given in per unit:

$$U = E - R \times P - X \times Q + R \times P_{\text{gen}} \tag{5.79}$$

For the distribution of the wind speed, a Weibull distribution has been assumed with shape factor 2.1 and average wind speed equal to 7 m/s. The speed–power curve for a 600 kW wind turbine has been used to obtain the probability distribution function of the generated power P_{gen}. As seen in Chapter 2, the power curves are similar for turbines of different sizes, so the hosting capacity becomes rather independent of the size of the turbine used. The resulting parameters for the voltage distribution are presented in Figure 5.31 as a function of the total wind power capacity (i.e., the number of wind turbines times 600 kW).

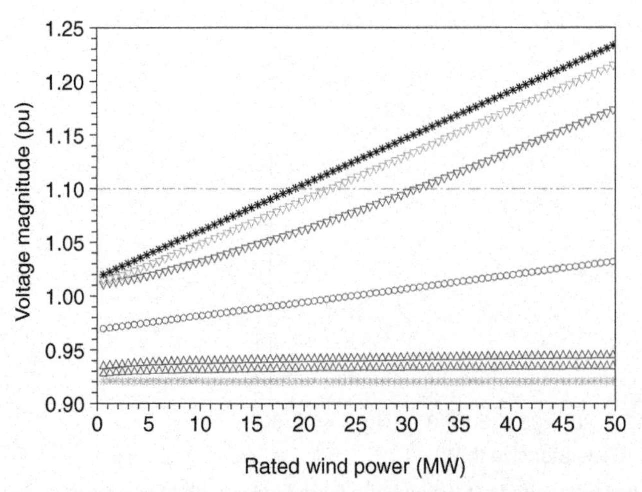

Figure 5.31 Range of voltage at the end of a 5 km feeder, with 5% boost in voltage, as a function of the amount of wind power installed: (top to bottom) maximum, 99%, 95%, 50%, 5%, 1%, and minimum values.

As in the previous example, a boosting of the voltage magnitude with 5% has been assumed. The high-voltage magnitude show the same increase as for the generator with a constant production (Figure 5.30). The low-voltage magnitudes, however, do not show any increase at all. Connecting wind power to a feeder does not solve any undervoltage problems; it does however introduce overvoltage problems. Using the 100% value as a limit, the hosting capacity for wind power is the same as for generation with constant production. Using a 95% or 99% value, the hosting capacity for wind power is about 2 and 10 MW higher. For wind power, the hosting capacity is increased significantly when a stochastic performance criterion is used. We will look further into this in the next section.

5.7 STATISTICAL APPROACH TO HOSTING CAPACITY

The deterministic approach to calculating the hosting capacity as described in Section 5.2 is very much a worst-case approach. The maximum voltage magnitude after connection of the wind power installation is equal to the maximum prewind voltage magnitude and the maximum voltage rise due to the wind power. This approach is the best way to ensure that no equipment damage occurs due to excessive voltage magnitudes. However, the deterministic approach might present a serious barrier to the introduction of renewable energy. Especially for sources such as wind and solar power, the amount of produced power is often less, and even much less than the maximum value. Accepting a small probability that the voltage magnitude exceeds the threshold will increase the hosting capacity a lot, as we will see in this section.

Measurements of the voltage magnitude at a number of locations in several countries have been used to illustrate the statistical approach. In all cases, the 1 or 3 s rms voltages have been used. The method can however be applied equally well for

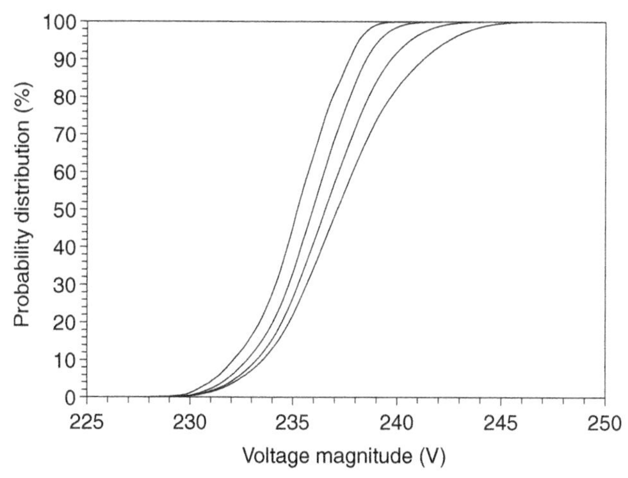

Figure 5.32 Probability distribution function of the voltage magnitude: (left to right) no wind power, 1 MW wind power, 2 MW wind power, and 3 MW wind power.

1 or 10 min rms voltages. The first example is shown in Figure 5.32. The left-hand curve gives the probability distribution function of the prewind voltage, that is, the measured voltage. Three different sizes of wind turbines have been assumed (1, 2, and 3 MW nominal power) with a point-of-common coupling at 10 kV and a source resistance at the point-of-common coupling equal to 1 Ω. The impact of the wind power is a spread of the distribution function over a wider range of voltage values. Of interest for us here is the higher probability of higher voltage magnitudes.

The probability distribution functions shown in Figure 5.32 have been calculated by means of a Monte Carlo simulation. For each distribution, 100, 000 samples were taken. Each sample of the voltage after connection of the wind turbine was obtained as the sum of a sample from the prewind voltage and a sample from the voltage rise. The former was obtained by taking a random value of time, within the measurement time window, and using the voltage magnitude at this random instant. The sample of the voltage rise was calculated from the wind power production, which was in turn calculated from the wind speed using the power curve for a 600 kW turbine scaled to the size of the wind power (1, 2, or 3 MW in this case). For the wind speed, a Weibull distribution with an average value of 7 m/s and shape factor of 2 is used.

Next, four different statistical overvoltage indicators have been calculated as a function of the amount of installed wind power. The results are shown in Figure 5.33. The measured voltage magnitude has been scaled in such a way that the maximum prefault voltage without wind power is 99% of the overvoltage limit. The overvoltage limit corresponds to a value of 100% (the horizontal dashed line) in the figure.

The hosting capacity is the amount of wind power for which the overvoltage indicator exceeds the 100% line. As can be seen from the figure, this value varies strongly between the indicators.

- The maximum voltage exceeds the limit for 1 MW wind power.
- The 99.99% voltage exceeds the limit for 1.4 MW wind power.

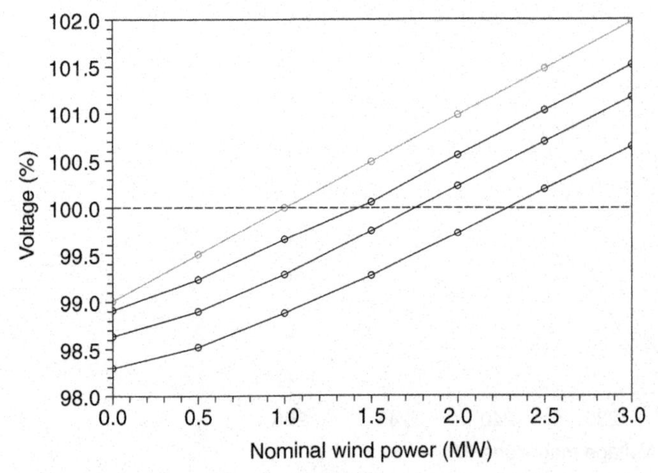

Figure 5.33 Statistical overvoltage indicators as a function of the amount of wind power: (top to bottom) 100%, 99.99%, 99.9%, and 99%.

- The 99.9% voltage exceeds the limit for 1.8 MW wind power.
- The 99% voltage exceeds the limit for 2.3 MW wind power.

The hosting capacity thus varies, for this specific case, between 1 and 2.3 MW depending on which overvoltage indicator is used. It should be clearly noted however that a higher amount of wind power always increases the risk that customer equipment damages due to overvoltages. A risk-based approach accepts that there is a risk and aims at finding the amount of wind power for which the risk is no longer acceptable.

The probability of equipment damage is however lower than the probability of an overvoltage for two reasons. First, overvoltages appear especially during low loads when the amount of equipment connected to the supply is less. Second, most equipment can tolerate voltage magnitudes above the overvoltage limit. How much lower the probability of equipment damage is than the probability of overvoltage is not possible to say without a detailed study of the immunity of equipment against overvoltages. Recent tests of off-the-shelf low-voltage equipment indicate that most equipment can tolerate voltage magnitudes well above 110% of the nominal voltage [378].

Next, the hosting capacity has been calculated for different configurations starting from the base case in Figure 5.33. The results for five different turbine types (all of 600 kW rated power) are shown in Figure 5.34. The base case is turbine type I. It is shown that the impact of the turbine type on the hosting capacity is small. Note that the hosting capacity for maximum voltage is equal to 1 MW in all cases; as this is a deterministic worst-case value, it is not dependent on the probability distribution function of the voltage.

To speed up the calculation of the hosting capacity, the values of the three indicators (99, 99.9, and 99.99%) have been calculated for 1 and 3 MW wind power connected to the feeder. The hosting capacity is next determined by linear interpolation or extrapolation based on these two values. Detailed simulations have been performed

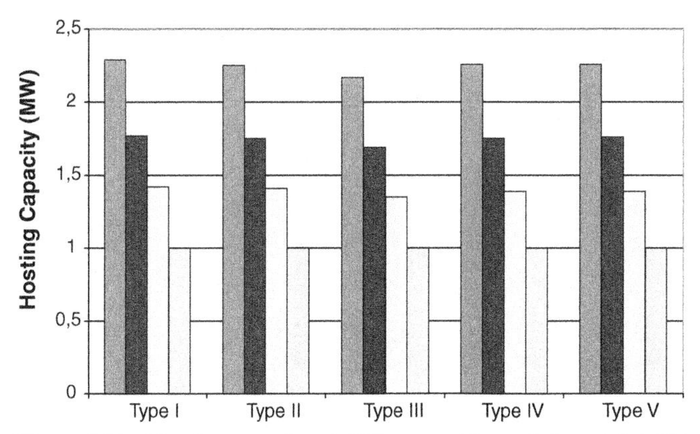

Figure 5.34 Hosting capacity of different turbines: (left to right) 99%, 99.9%, 99.99%, and maximum voltage.

to verify this method and it was shown that this linear approximation gives very accurate results. The hosting capacity for the deterministic case (100%) has been calculated using the analytical expression.

The impact of the wind speed distribution is shown in Figures 5.35 and 5.36. Figure 5.35 shows how the hosting capacity becomes smaller with increasing average wind speed. With higher average wind speed, high production values occur more often and thus high values of the voltage rise. Figure 5.36 shows how the hosting capacity increases for larger shape factor. A larger shape factor corresponds to a smaller spread in wind speed, so larger values occur less often.

The impact of the location, that is, of the prewind voltage distribution, is shown in Figure 5.37. Locations 1A, 1B, and 1C refer to the three phases at the same measurement location. The range of hosting capacity between different locations is much bigger than the impact of any of the other parameters. For example, for the 99% value,

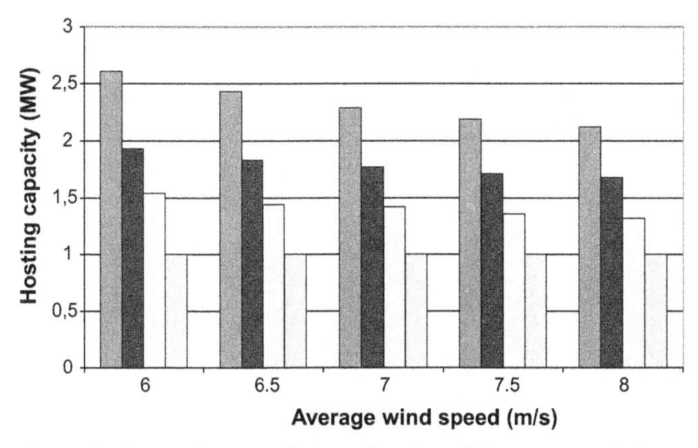

Figure 5.35 Hosting capacity as a function of the average wind speed: (left to right) 99%, 99.9%, 99.99%, and maximum voltage.

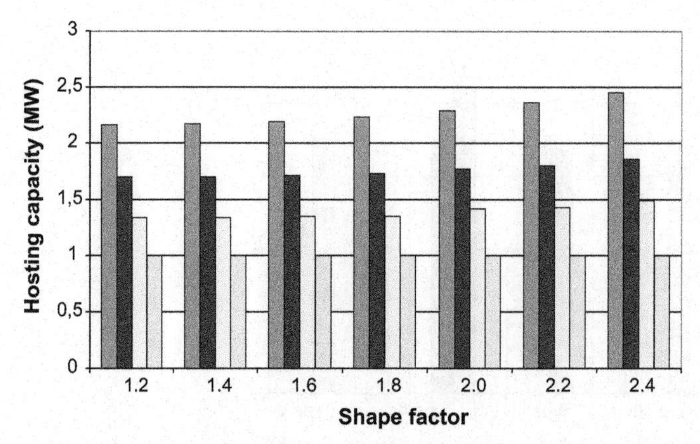

Figure 5.36 Hosting capacity as a function of the shape factor of the wind speed distribution: (left to right) 99%, 99.9%, 99.99%, and maximum voltage.

the hosting capacity varies between 1.56 MW (location 11) and 4.16 MW (location 10 and location 1B). From this figure, the conclusion can be drawn that a measurement is needed at each specific location where high overvoltages are expected.

The hosting capacity has next been calculated for different values of the overvoltage margin. The results for two of the locations are shown in Figures 5.38 and 5.39. The former one is for a location with a low hosting capacity and the latter one for a location with a high hosting capacity.

From both figures, it is evident that the use of statistical indicators gives the same gain in hosting capacity, in absolute terms, independent of the overvoltage margin. For example, when 1% margin gives 1.7 MW for the statistical approach versus 1 MW for the deterministic approach, a 3% margin will result in about 3.7 MW for the statistical approach and 3 MW for the deterministic approach. The relative gain of using a statistical approach thus becomes less for larger overvoltage margin. The

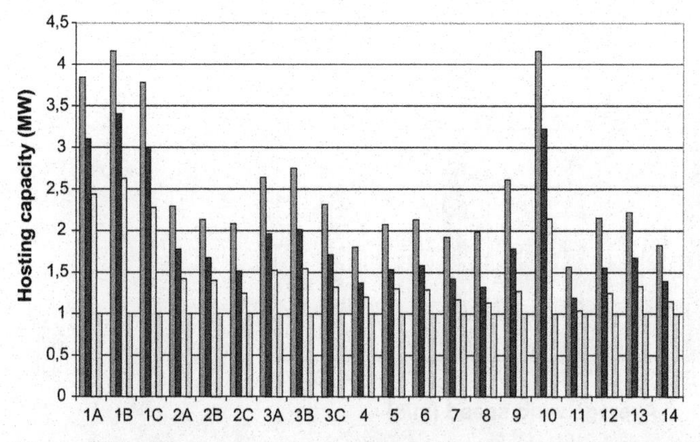

Figure 5.37 Hosting capacity at different locations: (left to right) 99%, 99.9%, 99.99%, and maximum voltage.

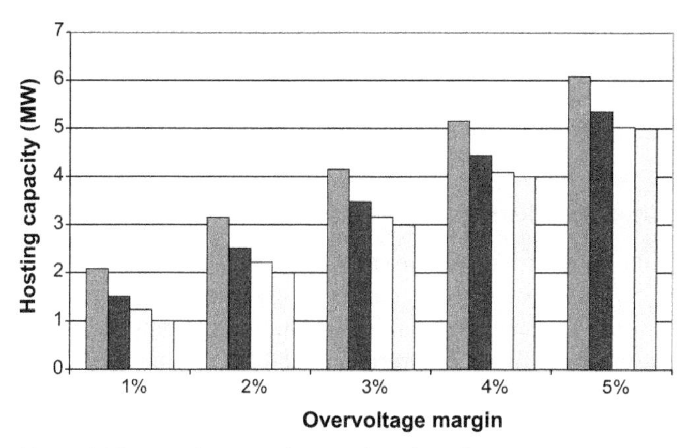

Figure 5.38 Hosting capacity as a function of the overvoltage margin, location 3C: (left to right) 99%, 99.9%, 99.99%, and maximum voltage.

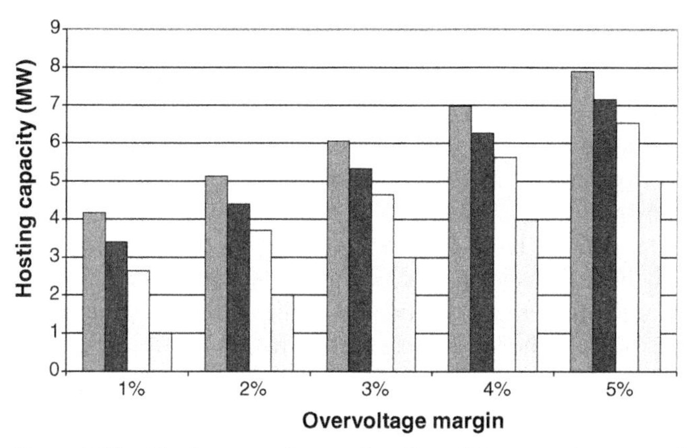

Figure 5.39 Hosting capacity as a function of the overvoltage margin; location 1B: (left to right) 99%, 99.9%, 99.99%, and maximum voltage.

main barrier to the introduction of distributed generation occurs for small overvoltage margins, and this is where the statistical approach gives the biggest gain.

5.8 INCREASING THE HOSTING CAPACITY

In this chapter, the hosting capacity has been calculated for different feeders. It was shown that the hosting capacity varies from zero up to several megawatt for medium-voltage feeders. In the latter case, the hosting capacity is set by the loadability of the feeder, not by the overvoltage limit. Although most feeders can host significant amounts of distributed generation, some tens of percent of the maximum load, in some cases the hosting capacity is very limited. Also, with future expected increases in the amount of distributed generation, a hosting capacity of some tens of percent may not be enough. In this section, a number of methods are discussed for improving the hosting capacity. We start with rather "classical" methods and move toward more advanced and future methods. A lot of research and development on this subject is currently being conducted, and reference to some of which is made in this chapter.

5.8.1 New or Stronger Feeders

Adding more copper, in the form of more or stronger lines, will in almost all cases result in a higher hosting capacity. Dedicated feeders in case of large units connected to weak parts of the grid are a commonly used solution. Building a completely new feeder, in many cases in the form of an underground cable, could be easier than strengthening an existing feeder. Overvoltages might still occur, but because it is a dedicated feeder, only the generator unit is exposed to the overvoltage. However, this can be fully considered during the design of the feeder and the generator unit.

An additional advantage of a dedicated feeder is that it can be designed to prevent any overloading. The solution is, however, practical only for larger units. For small units, the costs of the new feeder will be too high compared to the total costs of the unit. Also, the costs of the new feeder will be highest when the generator is located further away from the main medium-voltage substation (the one equipped with voltage control). These are exactly the locations where the risk of overvoltage is highest (or, in our terminology, the locations with the lowest hosting capacity).

Rewiring the feeder or parts of the feeder with a thicker wire will be cheaper than building a completely new feeder. It will also be easier to obtain permission for rewiring an existing feeder than for building a new overhead feeder. Rewiring may however cause rather long planned interruptions for existing customers. As shown in Figure 5.14, the impact of the feeder cross section on the hosting capacity is not as big as would at first be expected. The lower cross section results in a lower voltage drop during low load. This partly compensates the increase in hosting capacity due to the lower resistance. However, when using sufficiently thick wires, there is no need for a second voltage boosting. This has the main impact on the hosting capacity.

Network operators in several countries are replacing overhead lines with cables. From a voltage control viewpoint, a cable can be seen as a line that is partially series compensated: that is, the series reactance of an underground cable is less than that of an overhead line. The voltage drop along an underground cable is therefore less than that along an overhead line of the same cross section. This reduces the need for voltage boosting, which gives an increase in the hosting capacity.

Replacing lines by cables with the same cross section, without any further measures, might however reduce the hosting capacity. The voltage drop during minimum load, which is partly due to reactive power demand, becomes less, whereas the voltage rise due to distributed generation remains the same. There is thus less generation needed to reach the overvoltage limit.

Building new rural distribution networks completely based on cables might even result in a significant reduction in the hosting capacity. The lesser voltage drop per unit length allows longer feeders, with the boosting points located further away from the main substation. As discussed in Section 5.4, the minimum overvoltage margin is independent of the feeder impedance and loading. With cables, this point will be further out along the feeder, where the source resistance is higher and thus the hosting capacity smaller.

The hosting capacity is also increased by using shorter feeders. The main cost associated with this is that more transformers and substations are needed.

5.8.2 Alternative Methods for Voltage Control

The various calculations in this chapter have shown that one of the main limitations to the hosting capacity is the fact that the voltage boosting used is independent of the load. A load-dependent boosting method would be able to increase the voltage during high load, while not impacting the voltage during low load. The line-drop compensation scheme discussed in detail in Section 5.5 is the first step in the right direction. However, the disadvantage of this scheme is that it assumes all feeders from the same HV/MV transformer to have similar load patterns. When adding larger generator units to the feeder, this no longer holds. Line-drop compensation is used for voltage control with a large wind park connected to 230 kV, where the point of connection is 75 km from the wind park substation [466]. An advanced line-drop compensation relay is presented in Ref. 205. The relay uses as input not only the total current through the transformer, but also the current from the main bus into the feeder with distributed generation. From the difference between the feeders and knowledge on historical load profiles, the amount of generation is estimated. This is used to calculate a correction term for the line-drop compensator. A successful field-trial of the relay in an 11 kV network with 13 feeders and a 5 MW wind park connected to one of the feeders is presented in Refs. 160 and 368.

In Section 5.3.5, both shunt and series compensation have been discussed as alternative methods for voltage control. Concentrated series compensation is most effective when located near the start of the feeder as it compensates only voltage drops due to downstream loads. But this results in high-voltage magnitude at the start of the feeder during high load. As long as these are below the overvoltage limit, the hosting capacity for distributed generation can be high. Doing the compensation in a number of steps, or even distributed, will result in a smoother voltage profile and a higher hosting capacity. Series compensation compensates only for the voltage drop due to the reactive power. When this drop is a substantial part of the total voltage drop, series compensation is a suitable option. In feeders with small cross section, the voltage drop is mainly due to the active power and series compensation is not effective.

Shunt compensation is most effective when connected at the end of the feeder. A switching scheme is needed to prevent overvoltages. To prevent excessive switching, a large hysteresis is preferred. An SVC scheme or a fixed capacitor with a regulated shunt reactor could be an alternative, resulting in small voltage magnitude variations along the feeder.

Another possible solution is the installation of a reactor on secondary side of the distribution transformer. When the voltage magnitude exceeds a threshold, the reactor is switched on and its reactive power consumption results in a voltage drop. As all reactive power-based schemes, this one is most effective when the voltage drop due to reactive power is a substantial part of the total voltage drop.

Automatic tap changers for distribution transformers have been proposed as well, but this would likely be a too expensive solution. Such a scheme requires additional investment costs and there is also an increased need for maintenance and the risk of interruptions or unacceptable voltage quality due to failures of the tap changers. Such tap changers, when used on all distribution transformers, would however remove all limits due to overvoltages for distributed generation connection to the medium-voltage network. The hosting capacity for distributed generation connected to the low-voltage network would be mainly determined by the upper limit of the deadband of the tap-changer controller. Overvoltage margins of 5% could easily be achieved, resulting in a hosting capacity of five times the values shown in Table 5.3, that is, 2 kW and higher. Also, here the hosting capacity would most likely be determined by the thermal loading and no longer by the overvoltages.

Reducing the deadband of the tap changer on the HV/MV transformer gives more space for overvoltages or undervoltages. Reducing the upper limit of the deadband by 0.5% will directly result in 0.5% more overvoltage margin. The big advantage of this solution is that its investment costs are zero. In some cases, the lower limit can also be reduced without resulting in unacceptable undervoltages. However, this is not always the case and the network operator may not be willing to take the risk. Reducing the upper limit without changing the lower limit will result in an increase in the number of tap-changer operations. This will give a higher mechanical loading of the tap changer, with risk of failure and higher maintenance requirements.

Future generations of transformer tap-changers, solid-state switches that allow a more narrow deadband could be used. When the deadband can be reduced from 2% to 0.5%, while the lower limit is kept the same, this gives an additional 1.5% overvoltage limit. This also solves the problem of excessive number of operations due to intermittent generation. The gain in hosting capacity is however most likely not sufficient to justify the investment in a transformer with solid-state tap changing. Installing wires with a higher cross section is likely to be a cheaper solution. Price developments are however not always easy to predict and this solution could still be considered as a possible option.

5.8.3 Accurate Measurement of the Voltage Magnitude Variations

Distribution network operators in many cases do not have detailed information about the voltage magnitudes experienced by their customers. During the design, an estimation is made of the so-called "after-diversity maximum demand." On the basis of this, the minimum feeder cross section (to prevent overload) and the maximum feeder length (to prevent undervoltages) are determined. This approach has shown to result in acceptable voltages for the vast majority of customers.

Adding distributed generation will, in the long term, require new design methods, for example, based on estimates of the "after-diversity minimum consumption" and the "after-diversity maximum production." In the short term, distributed generation will be added to existing feeders, designed in the classical way. The hosting capacity of such feeders is determined by the overvoltage margin experienced by individual customers. This value is however often not known to the network operator and will have to be estimated. To prevent excessive overvoltages, and possible complaints and penalties, the overvoltage margin is underestimated resulting in a low or even zero hosting capacity.

An accurate measurement of the overvoltage margin for individual customers will make it possible to determine the hosting capacity more accurately. This strongly reduces the risk of unexpected overvoltages; it also in most cases will allow more distributed generation to be connected to the feeder. The automatic metering equipment that is being introduced in many countries is often able to measure the voltage magnitude with a time resolution of, for example, 10 min. Although not perfect (voltage magnitudes show time variations at timescales down to less than 1 s), it will give sufficient information to provide an accurate estimate of the hosting capacity.

By obtaining more accurate information on the actual voltages experienced by the customers, the network operator is able to allow more distributed generation to be connected without the risk of unacceptable overvoltages for other customers. Those same measurements at the same time give the network operator information about the available undervoltage margin for future load growth. This is important information for planning future investments in the distribution network.

5.8.4 Allowing Higher Overvoltages

The design of distribution feeders is based on an overvoltage limit and an undervoltage limit. The traditional trade-off has been between the range of voltage magnitude and the maximum length of the feeder. With higher overvoltage and the lower undervoltages, longer feeder lengths would be possible. The saving achieved by allowing longer feeders is especially in the reduction in the number of transformer stations. The longer the low-voltage feeders are, the less distribution transformers are needed; longer medium-voltage feeders require less HV/MV transformers. The relation is more or less square: the number of transformer stations is inversely proportional to the square of the feeder length. A 25% reduction in feeder length would require 78% more transformer stations.

With distributed generation, higher overvoltage and lower undervoltage limits could be used to increase the hosting capacity. The impact of higher overvoltage limits is a direct increase in the overvoltage margin. The impact of lower undervoltage limits is more indirect. It allows the deadband of the tap-changer controller to be shifted to lower values. Shifting the upper limit down directly increases the overvoltage margin. A lower undervoltage limit also reduces the need for voltage boosting. As discussed previously, this greatly increases the hosting capacity.

It should also be noted that higher permissible overvoltages and undervoltages could be used by network operators to build longer feeders. This would most likely reduce the hosting capacity because the location with the lowest overvoltage margin would be further out along the feeder. Allowing higher overvoltages and lower undervoltages would thus not automatically result in more distributed generation being able to connect to a feeder.

The choice of overvoltage and undervoltage limits remains a sensitive issue as many recent discussions between network operators, equipment manufacturers, and regulators have shown. Well-defined and strict voltage characteristics are important for the customers and equipment manufacturers. However, strict voltage characteristics most likely result in even more restrictive planning levels that in turn could form a barrier to the introduction of large amounts of distributed generation. Fortunately, other solutions exist to increase the hosting capacity, but none of them is as easy (and economical) as changing the overvoltage and undervoltage limits when it turns out that these are unnecessarily strict. Serious discussions are needed between different stakeholders to decide about reasonable overvoltage and undervoltage limits.

5.8.5 Risk-Based Approach to Overvoltages

In Section 5.7, it is shown that the hosting capacity can be increased significantly by allowing the overvoltage limit to be exceeded for a small percentage of time. Such an approach is especially effective for sources that have a strongly intermittent character, such as solar power and wind power. For more constant sources, like CHP, the risk-based approach will be less effective.

In the examples given in Section 5.7, the percentage during which the overvoltage limit can be exceeded was taken equal to 0.01%, 0.1%, and 1%. The discussion on what is an acceptable percentage is probably even more difficult that the discussion on what is a suitable overvoltage limit. Both discussions come back to one of the primary aims of the power system: "to supply the equipment with a suitable voltage magnitude."

The risk-based approach as discussed previously is in fact not fully complete. What should be considered is not the probability that an overvoltage limit is exceeded, but the probability that equipment is damaged due to overvoltages. This probability is most likely to be smaller, especially as one takes into consideration that high-voltage magnitudes occur during low load when less equipment is connected to the network. Also, these high voltages occur only with a limited number of customers. What needs to be considered is the probability that those customers experiencing high voltages have sensitive equipment connected when the high voltages occur.

A discussion on the use of a risk-based approach should again involve all stakeholders, where an important point of the discussion should be, who actually carries the risk? If the network operator gets a fine when the overvoltage limit is exceeded, a worst-case approach using a safety margin will likely be used. When a fund is created to pay for damaged equipment due to overvoltages caused by distributed generation, this will encourage network operators to allow more distributed generation to be connected.

The great unknown in the discussion of permissible overvoltage is the immunity of equipment for overvoltages. Although the value 110% of nominal is often mentioned, there are no studies confirming that equipment starts to experience damages when the voltage (slightly) exceeds this limit. On the other hand, customers complaining about damaged equipment often turn out to be experiencing high-voltage magnitudes. The most sensitive piece of equipment appears to be the incandescent lamp. Modern types of lighting are less sensitive to overvoltages, although there are no detailed studies to confirm this. Before a complete risk-based approach can be implemented, equipment manufactures, standard-setting organizations, and regulators have to get involved in the discussion.

A risk-based approach is not simple and may require a significant effort before it can be widely used. However, it has the potential to hugely increase the amount of distributed generation that can be connected without additional investments in the distribution networks. Another barrier to the use of risk-based approaches is the way in which the general public expects companies to ensure a risk-free environment. It is sometimes safer for a company to perform a deterministic study using worst-case limits than to consciously accept a small but nonzero risk. Removing this barrier again requires an industry-wide approach in which risk-based approaches will become accepted as good power engineering practice. The authors are aware that this is no easy task.

5.8.6 Overvoltage Protection

Tripping distributed generation when the terminal voltage comes too close to the overvoltage limit is a very effective way to prevent overvoltages. There are a number of advantages with such an approach. The protection threshold can be set close to the voltage characteristic, that is, the actual maximum voltage allowed on the network. The generator needs only to be tripped when the voltage gets too high for the customers. Instead of the planning level, a value close to the voltage characteristic is used. Even when the unit size exceeds the hosting capacity based on the planning level, the actual instances when tripping is needed may be very limited in number or not even occur at all.

The unit is tripped only when the voltage is high, that is, during periods of low consumption and at the same time high production. The amount of energy produced by the unit is not influenced by estimations on the minimum voltage drop or maximum production. This is obviously a huge gain compared to setting a maximum production limit (maximum unit size) based on a worst-case scenario.

The method is very easy to implement, with none or very limited extra investment needed. What is needed is a voltage measurement, a simple relay, and a breaker. All are typically already present, with the exception of the actual overvoltage relay. In fact, many of the methods for anti-islanding protection (see Section 7.7.4) remove the generator when the voltage becomes too high for too long.

However, there are a number of disadvantages to this method that we will briefly discuss here. The voltage used for the overvoltage protection is the voltage at the terminals of the generator or at the point of connection with the network. This is not the same as the voltage experienced by the customer equipment. In most cases, the customer equipment experiences a lower voltage, but sometimes the voltage at the equipment terminals is higher. During low load, the capacitance of cables and the input capacitance of end-user equipment can result in a small voltage rise. This effect is perceived to be small although no estimation of its magnitude is known to the authors. Of more concern is the fact that the commonly used method for voltage boosting (MV/LV transformers with off-load tap changers) implies that the voltage in the low-voltage network can be up to 5% higher than the voltage in the medium-voltage network. For a generator connected to the medium-voltage network, the overvoltage threshold should therefore be at least 5% lower than the actual overvoltage limit in the low-voltage network. For a unit connected to the low-voltage network, the actual overvoltage limit or a value slightly below this can be used.

Procedures are needed for the reconnection of the generation after an overvoltage trip. The reconnection should not result in the voltage exceeding the threshold again. When the voltage rise is completely or mainly due to one large generation unit, the reconnection voltage can be calculated easily. If δ_{max} is the overvoltage threshold at which the unit is tripped, reconnection can take place whenever the voltage drops below

$$u_{reconnect} = \delta_{max} - \Delta u_{gen} - \delta_{safety} \tag{5.80}$$

where Δu_{gen} is the estimated voltage rise due to the production of the generator and δ_{safety} a safety margin (hysteresis) to prevent reconnection on short-duration voltage below the reconnection threshold. For the latter, a short delay of, for example, a few minutes may be used as well. When multiple units are responsible for the voltage rise, it is no longer possible to calculate a suitable reconnection voltage, unless a very large margin is used (e.g., corresponding to maximum production of all units). Instead, reconnection could be attempted at a fixed interval after disconnection, for example, after 1 h. Together with this, a small hysteresis may be used of, for example, 2% of the nominal voltage. A significant delay is needed to prevent multiple reconnections during a longer period with surplus of generation. However, a long delay could result in significant loss of production (the term "spilled wind" is sometimes used in relation to wind power) when the overvoltage occurs regularly for shorter periods.

Some generation units may not be easy to start and stop. For wind power and solar power, starting and stopping is not a serious concern, but for CHP units, this could be inconvenient. Stopping the electricity production may also require stopping the heat production. When this heat production is used in an industrial process, the economic consequences could be severe.

The disconnection and reconnection can result in significant voltage steps when large amounts of generation are involved. A spread in voltage threshold and a spread in reconnection time could be used to make the transition more smooth.

With large penetration of, for example, solar power in distribution networks, tripping due to overvoltages could become a common phenomenon. With solar power, tripping could occur around noon for every sunny day. This would result in a serious reduction in the amount of energy generated by, in this example, solar power. Even though the generation may only be disconnected during a small percentage of time, this would be exactly when the energy production is highest. The curtailed energy would therefore be a much higher percentage than the percentage of time.

5.8.7 Overvoltage Curtailment

A curtailment scheme in which the production is reduced more when the voltage gets closer to the limit is a more efficient scheme. It will result in less loss of production than overvoltage protection, but some stability concerns remain.

The impact of overvoltage tripping on the voltage and on the energy production is illustrated in Figure 5.40. The difference between the solid curve (the voltage as it

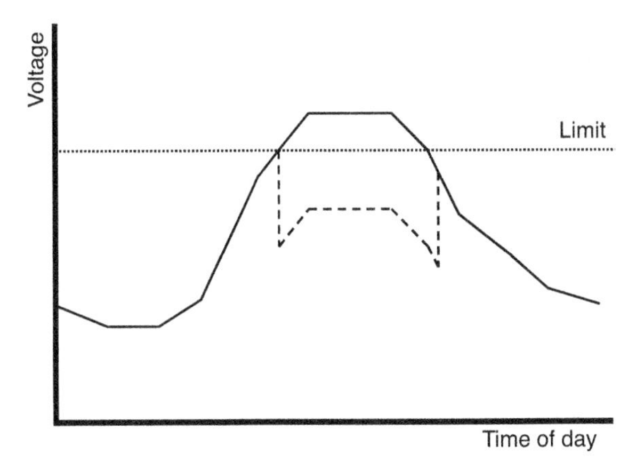

Figure 5.40 Impact of overvoltage tripping on the voltage variations: voltage magnitude versus time without tripping (solid curve) and with tripping (dashed curve). The dotted line indicates the overvoltage margin.

would have been without tripping the generation) and the dashed curve is proportional to the amount of power that is tripped. This implies that the area between the two curves is proportional to the energy not produced, and in case of wind power, this would be the "spilled wind."

The maximum possible amount of production at any moment is the one that would result in a voltage exactly equal to the overvoltage limit, that is, the dotted line in the figure. The area between the dotted line and the solid curve is thus the "necessary lost energy," whereas the area between the dotted line and the dashed curve is the "unnecessary lost energy." The "perfect curtailment algorithm" would keep the voltage exactly at the dotted line and make use of the power system to transport renewable energy as much as possible. Such an algorithm does not exist yet, but a number of algorithms are being discussed in the literature and no doubt more are being developed. In this section, we will discuss some properties of curtailment algorithms and make references to the methods being presented in the literature.

A possible method for power curtailment is shown in Figure 5.41. This method uses only locally measured parameters; no communication between generation units or a central controller is needed. When the voltage at the point of connection exceeds a value U_{ref}, the power production is limited. A second voltage threshold, $U_{\max}$, is where the generated power should be reduced to zero.

In mathematical terms, the curtailment algorithms

$$P_{\text{gen}} = \begin{cases} P_{\max} & U < U_{\text{ref}} \\ P_{\max}\left\{1 - \beta(U - U_{\text{ref}})\right\} & U_{\text{ref}} \leq U \leq U_{\max} \\ 0 & U > U_{\max} \end{cases} \qquad (5.81)$$

where $\beta = 1/(U_{\max} - U_{\text{ref}})$. The principle of voltage control is shown in Figure 5.42, where the dashed curves indicate the voltage rise curves:

$$U = U_0 + RP_{\text{gen}} \qquad (5.82)$$

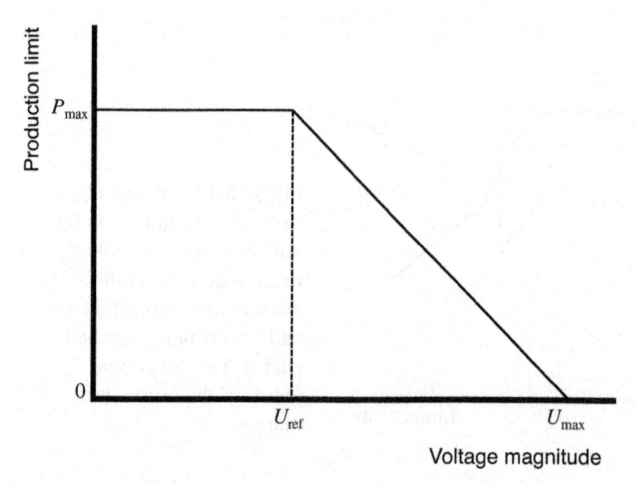

Figure 5.41 Relation between maximum produced power and voltage for a power curtailment method to prevent overvoltages.

for different levels of consumption. Here, it is again assumed that there is no exchange of reactive power between the generator and the grid. The level of consumption is represented by the voltage U_0 for the situation without generation. In the figure, U_0 is the point of intersection with the horizontal axis. The intersection with the horizontal line $P_{gen} = P_{max}$ is the voltage for maximum production: $U = U_0 + RP_{max}$.

The curve indicated by the number "1" represents the case when no curtailment is needed. For curve "2," the voltage with maximum production is equal to the reference voltage: $U_0 + RP_{max} = U_{ref}$, and this is where the curtailment starts. For curve "3," the production is partially curtailed; for curve "4," the production has to be cut completely. The latter occurs when the voltage without generation becomes equal to the upper voltage limit: $U_0 = U_{max}$. This is not a likely situation; it would imply that even without any generation the overvoltage limit is already reached.

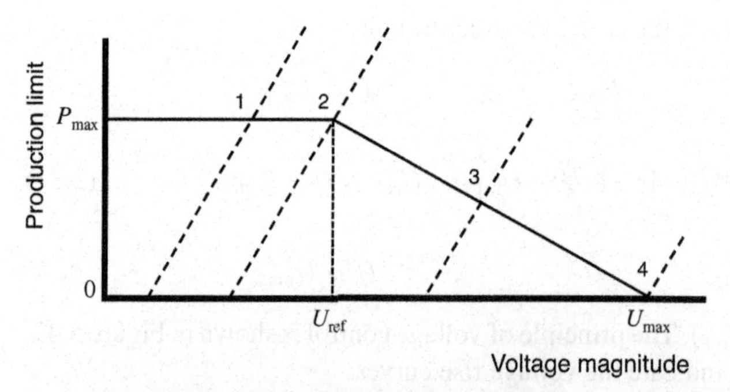

Figure 5.42 Principle of voltage control by curtailment of the production: curtailment curve (solid) and voltage rise curves (dashed).

The operating point is found by combining (5.81) and (5.82), resulting in the following voltage:

$$U - U_{\text{ref}} = \frac{(U_0 - U_{\text{ref}}) + RP_{\text{max}}}{1 + \beta RP_{\text{max}}} \tag{5.83}$$

The denominator gives the voltage rise above the reference voltage as it would have been without curtailment. The curtailment reduces this voltage rise by a factor $1 + \beta RP_{\text{max}}$. The production limit is equal to

$$P_{\text{gen,limit}} = P_{\text{max}} \frac{U_{\text{max}} - U_0}{U_{\text{max}} - U_{\text{ref}} + RP_{\text{max}}} \tag{5.84}$$

When, for example, during the course of 1 day, the no-generation voltage U_0 moves up and down, the voltage with DG is curtailed whenever it rises above U_{ref}. The maximum voltage U_{max} will not be exceeded as long as it is not exceeded by the no-generation voltage. The resulting voltage variations are shown in Figure 5.43. Comparison with Figure 5.40 shows that curtailment instead of tripping will significantly increase the amount of energy produced. The maximum voltage is never reached in this case because the no-generation voltage remains below the overvoltage limit.

The above expressions for voltage and production limit have been derived for one large unit connected to the feeder. However, it can be shown easily that the expressions also hold for more units connected at the same location. In that case, P_{gen} and P_{max} are the sum of the values for the individual generators. The curtailment of the individual generators is by ratio of their maximum production. For example, if (5.84) would result in a reduction by 25% of the total production, each unit would be reduced by 25% compared to its maximum production. This assumes that each generator produces exactly the amount to which it is curtailed. Some units may not be able to curtail production and will instead trip completely. The remaining units will reduce their production by a somewhat smaller amount.

When multiple units are connected at different locations, the curtailment is still shared but no longer in proportion to the unit size. Instead, the local no-generation

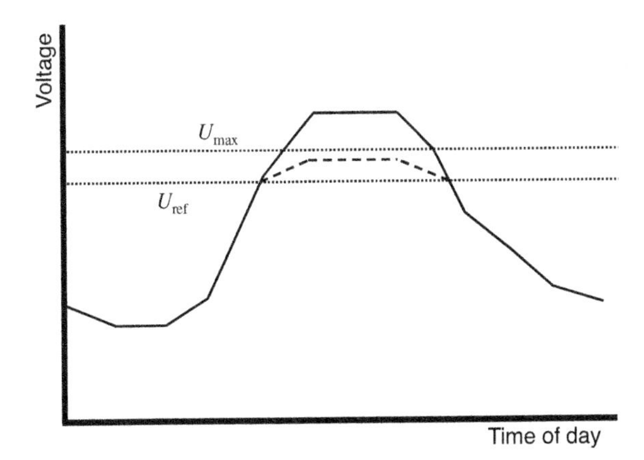

Figure 5.43 Impact of overvoltage curtailment on the voltage variations: voltage magnitude versus time without (solid curve) and with curtailment (dashed curve).

voltage and the local source resistance impact the sharing. It is still possible to obtain closed expressions for the resulting voltages and production limits, but those are complicated expressions and do not give much additional information. Instead, two examples are given. Both examples consider two generator units, with maximum production $P_{max,1}$ and $P_{max,2}$, connected at two different locations with source resistance R_1 and R_2, and no-generation voltages U_{10} and U_{20}. For the generator at location 1 (which is closest to the main substation), the control equation is

$$P_{gen,1} = P_{max,1}\{1 - \beta(U_1 - U_{ref})\} \tag{5.85}$$

The voltage at the point of connection is however also determined by the production of the second generator:

$$U_1 = U_{10} + (P_{gen,1} + P_{gen,2})R_1 \tag{5.86}$$

For generator 2, the control equation is

$$P_{gen,2} = P_{max,2}\{1 - \beta(U_2 - U_{ref})\} \tag{5.87}$$

with the voltage obtained as follows:

$$U_2 = U_{20} + P_{gen,1}R_1 + P_{gen,2}R_2 \tag{5.88}$$

Example 5.5 Consider as an example two generators with maximum production $P_{max,1} = 1$ and $P_{max,2} = 2$, the source resistances $R_1 = 1$ and $R_2 = 2$, and the no-generation voltages $U_{10} = U_{ref} + 1$ and $U_{20} = U_{ref}$. The slope of the curtailment is set equal to $\beta = 1/5$. This results in the following expressions:

$$\begin{aligned} P_{gen,1} &= 1 \times \{1 - 0.2 \times (U_1 - U_{ref})\} \\ P_{gen,2} &= 2 \times \{1 - 0.2 \times (U_2 - U_{ref})\} \\ U_1 &= U_{ref} + 1 + 1 \times (P_{gen,1} + P_{gen,2}) \\ U_2 &= U_{ref} + P_{gen,1} \times 1 + P_{gen,2} \times 2 \end{aligned}$$

Solving this set of equations results in

$$\begin{aligned} U_1 &= U_{ref} + 2.5 \\ U_2 &= U_{ref} + 2.5 \\ P_{gen,1} &= 0.5 \\ P_{gen,2} &= 1 \end{aligned}$$

The voltages at the two locations become the same, and both units are curtailed to half of their maximum production.

Example 5.6 The same network is assumed as in Example 5.5, but the smaller unit is now connected furthest along the feeder: $P_{max,1} = 2$ and $P_{max,2} = 1$. The resulting expressions are now

$$
\begin{aligned}
P_{gen,1} &= 2 \times \{1 - 0.2 \times (U_1 - U_{ref})\} \\
P_{gen,2} &= 1 \times \{1 - 0.2 \times (U_2 - U_{ref})\} \\
U_1 &= U_{ref} + 1 + 1 \times (P_{gen,1} + P_{gen,2}) \\
U_2 &= U_{ref} + P_{gen,1} \times 1 + P_{gen,2} \times 2
\end{aligned}
$$

with solution as

$$
\begin{aligned}
U_1 &= U_{ref} + 2.553 \\
U_2 &= U_{ref} + 2.128 \\
P_{gen,1} &= 0.979 \\
P_{gen,2} &= 0.575
\end{aligned}
$$

In this case, the resulting voltage is somewhat larger with the first (larger) unit, which is also curtailed slightly more (down to 49% of its maximum, versus 57% for the second unit.

5.8.8 Dynamic Voltage Control

The existing method of voltage control in distribution networks is based on fixed settings. The tap-changer deadband of the HV/MV transformer is set once and is kept the same for many years. An occasional change can be made, for example, as part of a large overhaul of the network, but those are exceptional cases. If the reference value (the center of the deadband) could be changed, this would allow a more optimal control of the voltage. Based on information obtained from measurements at a limited number of locations in the distribution network, the voltage reference could be changed. This would allow more distributed generation to be connected to the distribution network. The drawback of this type of control is that it requires communication, which has traditionally been seen as something that should be avoided because of its limited reliability and high costs. However, communication becomes more common (e.g., for remote meter reading) and has become much more reliable through the years. Also, the costs of communication equipment have dropped significantly.

With dynamic voltage control, the tap-changer controller remains an autonomous controller, using only local measurements. But the reference value might be changed in case the voltage at one of the measurement locations is outside the acceptable range. This amount of communication needed is very limited. Several methods for this have been published in the literature and are under different levels of development.

References 204, 381 and 409 present a method for dynamically adjusting the setting of the tap changer on the HV/MV transformer. The so-called "segment controller" collects information from the load of individual feeders and from voltage and power flows at selected locations on the feeders. This information is used to estimate the voltage profile along the feeders and based on this the optimal voltage at the main bus is determined. Field experience for an 11 kV network in the United Kingdom with five feeders is presented in Ref. 409. The voltage and power with a 2.25 MW wind park are used as additional input to the controller. Both overvoltage and undervoltage margin are calculated assuming limits of 94% and 106%. If one of the margins becomes negative (i.e., if the limit is exceeded), the reference voltage of the tap changer is changed by 1.5% steps. Measurements during the field trial show that the method works well and that an additional 2–3 MW can be added to the wind park without the voltage limits being exceeded.

Another method for setting the reference value of the tap changer is proposed in Ref. 272. Voltage measurements at a number of "essential locations" along the feeder are used as input. The essential locations should include the points where the maximum voltage or the minimum voltage is likely to occur. When one of the voltages is outside its limit, the reference value of the tap-changer controller is adjusted.

For any form of dynamic voltage control, it is important to be able to estimate the voltage magnitude throughout the network. However, the number of measurement locations available is limited, so the voltage magnitude at other locations should be estimated. A number of algorithms for this "distribution state estimation" are being discussed in the literature. The algorithm described in Ref. 28 combines load profiles, uncertainty in load profiles, and power produced by the generators. The latter requires communication and to limit the demands on communication, the power production is transferred only to the network operator when it deviates more than a predefined amount from the last transferred value. Simulations in an 80-node network with seven generators show that an accuracy of about 1% in voltage magnitude can be achieved. It is also shown in Ref. 89 that with a limited number of observation points (sometimes just one point) an accuracy of 1% can be achieved. However, it should be noted that both Ref. 28 and 89 base their conclusions on simulations only. No comparison with measurements has been presented yet.

5.8.9 Compensating the Generator's Voltage Variations

A generator producing a certain amount of active and reactive power impacts the voltage magnitude for each location along the feeder. As shown in, for example, (5.12), the voltage rise due to the injection of active power increases linearly from the source up to the generator location and remains constant beyond the generator location. When the generator also injects or consumes reactive power, the voltage may rise or drop. In that case, the change in voltage along the feeder is equal to

$$\Delta u_{\text{gen}} = \begin{cases} \lambda \dfrac{RP_{\text{gen}} + XQ_{\text{gen}}}{U_{\text{nom}}^2} & \lambda \le \lambda_{\text{gen}} \\ \\ \lambda_{\text{gen}} \dfrac{RP_{\text{gen}} + XQ_{\text{gen}}}{U_{\text{nom}}^2} & \lambda > \lambda_{\text{gen}} \end{cases} \tag{5.89}$$

By keeping $RP_{\text{gen}} + XQ_{\text{gen}}$ equal to zero, the impact of the generator on the voltage variations is minimized. Only a small impact remains due to the active and reactive power losses associated with the transport of the power between the main grid and the generator.

To strongly limit the impact of distributed generation on the voltage variations, each unit should consume an amount of reactive power equal to

$$Q_{\text{gen}} = -\frac{R}{X} P_{\text{gen}} \tag{5.90}$$

where the minus sign indicates consumption of reactive power and R and X are the real and imaginary parts of the source impedance at the point of connection of the generator unit. With multiple units, each unit can compensate its own contribution to the voltage magnitude variations and in that way limit the total voltage variations. The risk of overvoltages will no longer be a limitations for the amount of distributed generation that can be connected.

The disadvantage of this approach is that the unit should be able to consume reactive power. This requires a power electronics converter or a synchronous machine. Especially for weak feeders (large resistance), the factor R/X could reach large values, up to five or higher. This would require a converter with a rating five times as large as for injecting active power only.

To limit the reactive power requirements of the converter, the reactive power consumed by the downstream load could be used to compensate part of the voltage rise. With P and Q the active and reactive power consumption downstream of the generator location, the voltage rise due to generator plus downstream consumption is equal to

$$\Delta U = R(P_{\text{gen}} - P) + X(Q_{\text{gen}} - Q) \tag{5.91}$$

To compensate for this, the reactive power exchange between the generator and the feeder should be

$$Q_{\text{gen}} = Q - \frac{R}{X}(P_{\text{gen}} - P) \tag{5.92}$$

In this way, the generator even contributes to an improvement in the voltage profile along the feeder.

Another option is to combine this approach with the overvoltage curtailment discussed in Section 5.8.7. In that case, the curtailment curve in Figure 5.41 would not refer to the active power production but to a combination of active and reactive powers, $P_{\text{gen}} + (R/X)Q_{\text{gen}}$. Instead of reducing active power production, a generator unit can consume reactive power to an amount resulting in the same voltage drop.

5.8.10 Distributed Generation with Voltage Control

Distributed generators with synchronous machine, DFIG, or power electronics interface have the ability to control the amount of reactive power consumed or injected into the grid. This allows their use to control the voltage in the medium- or low-voltage network. Most existing wind turbines, even when they have the ability for reactive power control, are disconnected when the wind speed drops below the cut-in speed. As a result, the reactive power capability is not available during 20–40% of time. A decision to have reactive power support available continuously should be made early in the discussions between the network operator and the owner of the wind turbine. In this case, the extra costs can be kept low.

The ability of different generator types to provide reactive power is presented in Ref. 63. It is shown that a small overrating of the inverter will result in a significant reactive power capacity. This is in fact a common practice with conventional power stations, where the rating of the synchronous machine, generator transformer, and so on is somewhat larger than the maximum of active power that can be produced. In this case, the generator can contribute to reactive power and voltage control under all circumstances. It is also shown in Ref. 63 that even without overrating, inverters for solar and wind power have a significant reactive power capacity most of the time. For example, a 110 kW photovoltaic installation, with a 100 kVA inverter, has 80 kvar or more reactive power available during 93% of time.

Reference 57 discusses possible methods for letting distributed generation actively contribute to the control of the voltage in distribution networks. The methods were selected for further development by the main French distribution network operator:

- Using distributed generation to control the voltage at a limited number of locations in the network
- A central regulator calculating the required voltage or reactive power references for the individual generators
- Separate active and reactive power control for the generators

The latter method is developed further in Refs. 412 and 413: the control method depends on the voltage at the point of connection. When this voltage is within a "desired range," the generator does not contribute to the voltage control. When the voltage is outside the desired range, but still within the limits, the reactive power control is used with the border of the desired range as a voltage reference. When the generator reaches its reactive power limit and the voltage is outside the limits, the active power is reduced or increased. In most cases, the latter will not be possible, but the option is available in the controller. A similar algorithm, but with more restrictive bands, is discussed in Ref. 352. There is no need for very restrictive control bands because equipment connected to the grid is able to operate for a wide range of voltage magnitude variations.

The impacts have been studied of a large penetration of microgeneration (CHP and PV) on 1262 properties in central Leicester, UK [408]. A simple integral controller that increases the reactive power consumed per property by 50 var per minute per volt error was implemented. This simple local controller without communication is shown to give good results. The maximum voltage is reduced from 256 to 253 V; also the minimum voltages are improved. The drawback is that the increased reactive power flows result in an increase in losses. In the network without generation, the losses were 3%. They were reduced to 2% by the introduction on noncontrolled generation. Increasing the reactive power consumption to prevent overvoltages resulted in an increase in losses to 4%. However, as the energy produced by the microgeneration is 60% of the consumption during winter and 69% during summer, the increase is losses is not a concern from an environmental point of view.

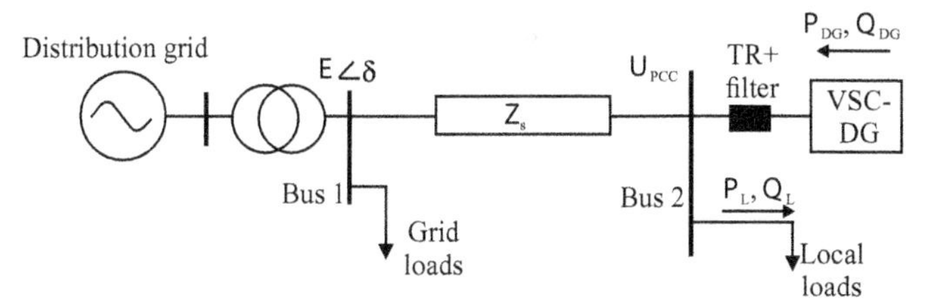

Figure 5.44 Grid impedance effect on voltage control possibility using a power electronics-interfaced DG.

The increase in hosting capacity for a 38 kV feeder in Ireland by using alternative methods for voltage control has been studied in Ref. 327. Operating the wind turbines at inductive instead of unity power factor increases the hosting capacity from 12.1 to 32.1 MW. Coordinated voltage control further increases this value to 43.0 MW. Allowing 2% and 5% curtailment of the production increases the hosting capacity to 51.1 and 55.7 MW, respectively.

A power electronics-interfaced DG unit provides independent control over the injected active and reactive powers, which promotes it as a good candidate to accomplish local voltage control. However, there are some limitations that should be considered to maintain the stability of the system and the connectivity of the DG unit. Four basic limitations are important to pinpoint: short-circuit ratio at the connection point, size of the interfacing converters, DC link voltage amplitude (assuming the use of voltage source converter (VSC) interface), and control system response time.

To better understand such limitations, a Thévenin equivalent circuit is considered as shown in Figure 5.44. In the figure, Z_S represents the short-circuit impedance as seen from the PCC. For simplicity, this impedance is assumed purely reactive (i.e., $Z_S = X_S$).

The power flow through the feeder (P_F and Q_F) represents the power mismatch between the DG injected power (P_{DG} and Q_{DG}) and the consumed local load power (P_L and Q_L). According to the directions indicated in the figure, the power mismatches are

$$P_F = P_{DG} - P_L \tag{5.93}$$

$$Q_F = Q_{DG} - Q_L \tag{5.94}$$

Also, the power mismatch can be read as

$$P_F = \frac{U_{PCC} E}{X_S} \sin \delta \tag{5.95}$$

and

$$Q_F = \frac{U_{PCC}^2}{X_S} - \frac{U_{PCC} E}{X_S} \cos \delta \tag{5.96}$$

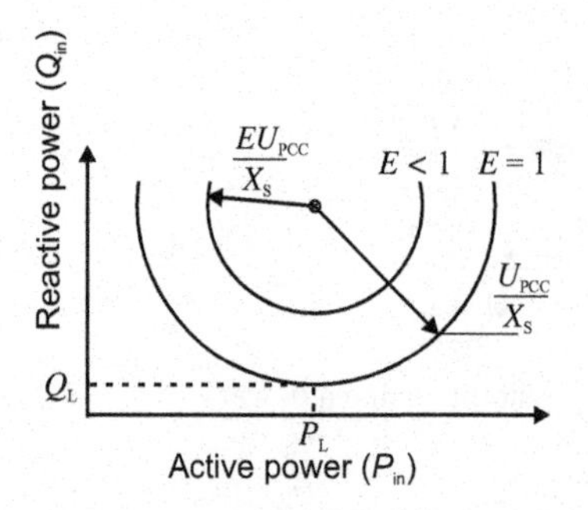

Figure 5.45 Injected DG power for different remote bus voltages.

The grid voltage angle, δ, can be eliminated by squaring and adding (5.95) and (5.96). Using the trigonometric identity, the following expression is obtained:

$$P_F^2 + \left(Q_F - \frac{U_{PCC}^2}{X_S}\right)^2 = \left(\frac{U_{PCC}E}{X_S}\right)^2 \tag{5.97}$$

Equation (5.97) represents the equation of a circle with a radius of $(U_{PCC}E/X_S)$ and centered at $\left(0, U_{PCC}^2/X_S\right)$. Different circles, referred to as power circles, can be obtained for different parameters. The power circles are used here to define the operational limits for the DG connected at the PCC regarding the short-circuit ratio [195].

Two main cases can be considered. In the first case, the change in the remote bus voltage (e.g., due to voltage dips or switching grid loads) is considered, while the PCC voltage is regulated. The power circles represented by the injected DG power with constant local loads (i.e., constant P_L and Q_L) are shown in Figure 5.45. In this case, the injected active power of DG should be kept within the following limits (assuming $U_{PCC} = 1$ pu), in order to have a real solution for the reactive power:

$$P_L - \frac{E}{X_S} < P_{DG} < P_L + \frac{E}{X_S} \tag{5.98}$$

Using either limit of (5.98) and substituting for P_{DG} in (5.93), the reactive power flow limit to compensate for the voltage variations in E is

$$Q_{F,\text{limit}} = \frac{1}{X_S} \tag{5.99}$$

This case is impractical, however, since compensating for a remote bus voltage drop needs remote measurement signals and it may also cause local overvoltages.

In the second case, the remote bus voltage is assumed constant (e.g., by adjusting the tap changer of the transformer), while the PCC voltage is not regulated and is directly affected by the change of the local loads. To simplify the analysis in this case, the local loads are assumed to be initially not connected. Then at the connection of these loads, the PCC voltage will instantly drop. Two cases are further assumed to visualize the impact of the short-circuit impedance at the PCC (X_S) on the control capability of the converter-interfaced DG.

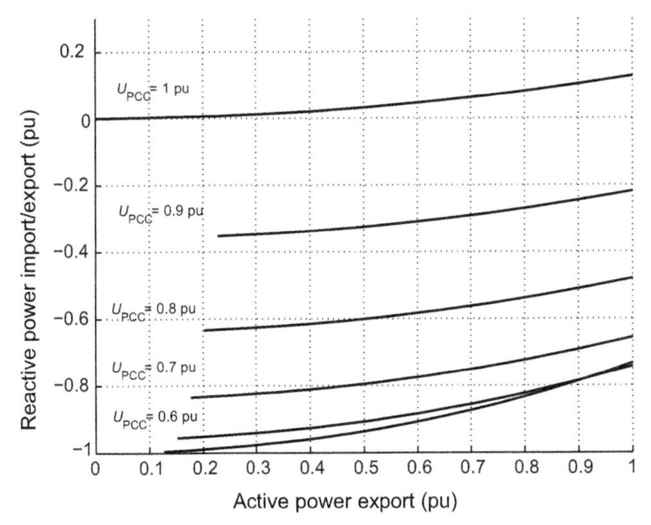

Figure 5.46 Operational region of the DG regarding various voltage amplitudes at the PCC for SCR $= 4P_{\mathrm{DG}}$.

Typically, at the transmission level, the short-circuit ratio is higher than four times the DC link transmitted power [297, Chapter 6]. Assuming the same limit at distribution results in the power circles shown in Figure 5.46 ($X_{\mathrm{s}} \approx 0.25$ pu). In the figure, the only parameter that changes the radius and center of the power circles is U_{PCC}. Only the possible operational area is depicted in the figure, where the active power from the DG is only injected and the injected reactive power is not to exceed 1 pu in order to respect the physical limitations of the DG.

Equation (5.97) can be rewritten as (P_{L} and Q_{L} are both equal to zero)

$$Q_{\mathrm{DG}} = \frac{U_{\mathrm{PCC}}^2}{X_{\mathrm{S}}} - \sqrt{\left(\frac{U_{\mathrm{PCC}}E}{X_{\mathrm{S}}}\right)^2 - P_{\mathrm{DG}}^2} \tag{5.100}$$

This expression can be interpreted as follows: for a given grid voltage, E, short-circuit impedance, X_{s}, and active power input, P_{DG}, the reactive power input, Q_{DG}, can be varied to obtain a regulated voltage at the PCC.

Now, back to Figure 5.46, assuming that the DG is injecting 0.5 pu active power at the PCC and the PCC voltage drops to 0.9 pu (e.g., due to switching on the local load), then approximately 0.35 pu reactive power needs to be injected to bring the voltage back to 1 pu. This is the vertical distance between the circles for $U_{\mathrm{PCC}} = 0.9$ pu and $U_{\mathrm{PCC}} = 1$ pu at $P_{\mathrm{DG}} = 0.5$ pu. By increasing the injected active power, less reactive power is available due to the size limitation of the interface converter. For instance, if the DG is injecting 0.8 pu active power and the voltage at PCC drops to 0.7 pu, a voltage regulator would fail to boost the voltage up to 1 pu. This is basically because the required reactive power is about 0.8 pu, which results in apparent power exceeding 1 pu. This may trip the overcurrent protection of the converter and electrically disconnect the DG from the grid. However, if the voltage regulator is adaptive in the sense that the

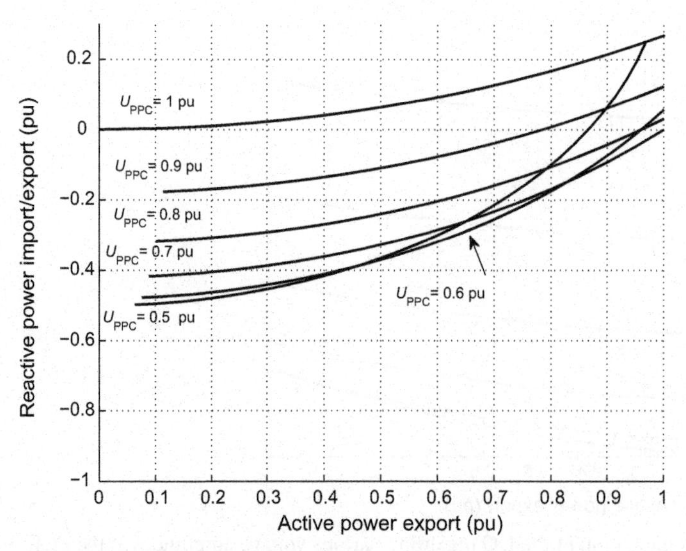

Figure 5.47 Operational region of the VSC regarding various voltage amplitudes at the PCC for short-circuit ratio (SCR) $= 2P_{DG}$.

reference voltage value may change according to the DG operating point, the system may survive through boosting the voltage to a level lower than 1 pu. A different situation may result for a higher short-circuit impedance. The power circles for short-circuit ratio of $2P_{DG}$ are shown in Figure 5.47. It is obvious that the reactive power needed for voltage compensation is lower than the case with that of short-circuit ratio of $4P_{DG}$. Regarding the same example as above, if the DG is injecting 0.8 pu active power and the voltage at the PCC drops to 0.7 pu, about 0.4 pu reactive power is needed to boost the voltage level to 1 pu. And this is a possible operation, as opposed to the case shown in Figure 5.46.

The operational limitations mentioned above are due to the network parameters and they give the theoretical boundaries for local voltage regulation at the PCC for an ideal generation plant. There are also operational limitations due to the physical configuration of the DG itself and its control arrangements.

Regarding a VSC-interfaced DG unit, there are basically two limitations that affect the transfer of reactive power through the feeder. First, the current amplitude is limited in order to protect the semiconductors of the VSC through overcurrent protection. This limitation will highly impact the compensation of the PCC voltage by using the injected reactive power of the DG. Since the maximum available active power from the DG is usually injected into the grid, a smaller amount of reactive power is available without violating this current limit. This eventually means that certain voltage quality phenomena at the PCC may not be compensated because of the lack of reactive power. The second limitation is the converter DC link voltage amplitude limitation that basically prevents the saturation of the VSC modulator in order to reduce the injected harmonics into the grid. In other words, the amplitude of the AC converter terminal voltage will not increase much above the DC link voltage. Usually, a ratio of 1.12 can be applied. This also limits the reactive power transfer,

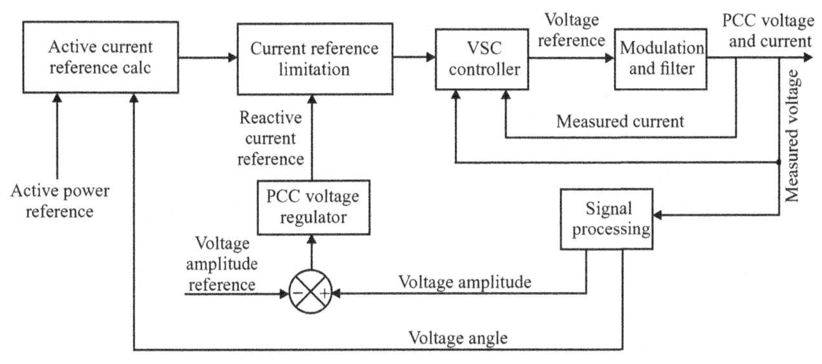

Figure 5.48 Simplified schematic of VSC-DG overall control system.

which is basically controlled through changing the converter voltage amplitude as was discussed in Chapter 2.

The first limitation is discussed further next, while the second limitation is more elaborated in Section 8.12. Since the controls of active and reactive powers can be performed independently, the active and reactive currents can also be treated independently. A simple schematic diagram of the overall VSC-DG control system is shown in Figure 5.48. The current reference limitation block is further described in Figure 5.49. It will be assumed that initially the converter is injecting the current vector "a." This means that only active power is injected. The quarter circle in the figure represents the maximum allowed converter current. If the voltage at the PCC is reduced, with voltage regulation using the DG, an injection of the reactive power is needed. Assuming that the current command for the converter moves to "b" implies the increase in the current amplitude above the allowed value. Since such an operation is not possible, the current limitation protection may change the command current vector to "c." This vector provides less reactive current than required for the voltage regulation, and hence the voltage is not expected to be fully regulated.

Other current limitation algorithms are also possible if reducing the active current (or injected power) of the DG unit is acceptable. Assuming that the energy

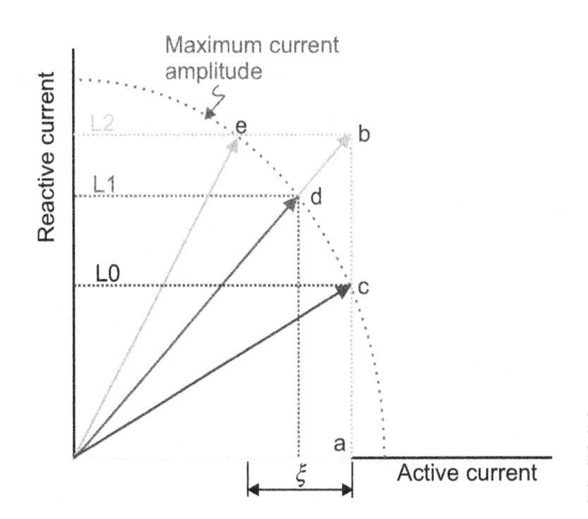

Figure 5.49 Reference current limitation: ξ represents possible active power curtailment.

production of the DG unit is allowed to reduce by a value of ξ, two current limitations algorithms referred to as L1 and L2 in Figure 5.49 are possible.

With L1, the reactive current reference is limited in such a way to reduce the injected current amplitude and maintain the current vector angle. Using L2, both active and reactive currents are reduced in such a way to inject as much reactive current as possible into the grid. The limitation over the maximum allowed reduction in the active current (ξ) will affect, in this case, the amount of the injected reactive current. Referring to Figure 5.49, if the current vector originally lies at "a" and in case of a change in the voltage amplitude moves to "b," using L1 it will be limited to "d" while using L2 it will be limited to "e." Generally, any limited current vector that lies between "c" and "e" is possible.

5.8.11 Coordinated Voltage Control

As discussed in Section 5.8.7, the overvoltage may be curtailed by reducing the power production of the generation. It has been assumed that the generation neither produces nor consumes reactive power. When many units are connected at close proximity, it has been shown that the curtailment of the individual generators can be expressed as a ratio of their maximum production. When different types of generation units are included, there might appear a condition that some units cannot curtail their share, and hence other units would have to increase their share of curtailment. Moreover, some units may have the capability to produce or consume reactive power, which in turn will impact the voltage profile. If two identical units are connected close to each other and both are equipped with voltage control that takes the same terminal voltage as input, they would produce exactly the same correction signal at the same time, which might lead to adverse effects. For instance, if the two units are injecting reactive power to compensate for terminal voltage decrease, double the amount needed for correction will be injected, and hence a voltage rise may result at the connection point or even a control instability that in turn will lead to unit tripping. This is of course if a unit does not know that the other one exists. Again, sharing the required amount of the correction signal between the two units will do the job. This is basically referred to as coordination.

Voltage control coordination is essential in transmission and distribution grids in order to maintain the voltage within its acceptable limits and at the same time prevent counteracting operation of individual active devices and maintain stability [262]. Generally, having parallel devices with the same assigned function is also beneficial from the operational point of view regarding the power sharing and reliability while maintaining the $(N-1)$ criterion [201,301]. The droop scheme is the classic way of coordinating different active devices without the need for any communication signals between them [199,235]. It is derived from the power flow equations. Assuming a inductive feeder with reactance X and a power angle where the power flows from A to B,

$$P_{\mathrm{A}} = \frac{U_{\mathrm{A}} U_{\mathrm{B}}}{X} \sin\delta \tag{5.101}$$

$$Q_{\mathrm{A}} = \frac{U_{\mathrm{A}}^2 - U_{\mathrm{A}}U_{\mathrm{B}}}{X} \tag{5.102}$$

By further assuming that δ and $(U_{\mathrm{A}} - U_{\mathrm{B}})$ are small, it is predominant that the power angle δ is more sensitive to the changes in the active power, while the voltage amplitude is more sensitive to the change in the reactive power. Hence, by adjusting P and Q independently, the frequency and amplitude of the grid voltage are determined, respectively. This is the basis for the well-known frequency and voltage droop control through active and reactive powers, respectively, which is expressed as

$$f - f_0 = -k_{\mathrm{p}}P - P_0 \tag{5.103}$$

$$U - U_0 = -k_{\mathrm{q}}Q - Q_0 \tag{5.104}$$

where f_0, U_0, P_0, and Q_0 are nominal operating points. In Figure 5.50, P_0 and Q_0 are assumed equal to zero. The constants k_{p} and k_{q} determine the amount of active and reactive powers to be injected or consumed at the connection point. They actually represent a virtual impedance between the active unit and the grid, implying that increased values (more steep slope) will decrease the effect of the active unit on the grid voltage. Assigning different droop curves for different active units in close proximity means different amounts of the corrective signal will be injected. Moreover, each unit does not aim at full compensation of the voltage due to the droop. Hence, even if all the units have the same droop curve and operate at the same time, the voltage amplitude will probably not exceed its limit.

However, such a situation can be avoided by applying a master–slave criterion in combination with the droop scheme. This is basically implemented through introducing a time delay between the operation of the master unit (or units), which could be a big generator or any other active device such as a transformer tap changer, and the other active units. The master unit first reacts to a voltage change and then sends control signals to the other units. Control interconnections are essential in this case, though it is rather in a simple form. It could be only between the master units and its slaves, without the need for the slave units to communicate with each other.

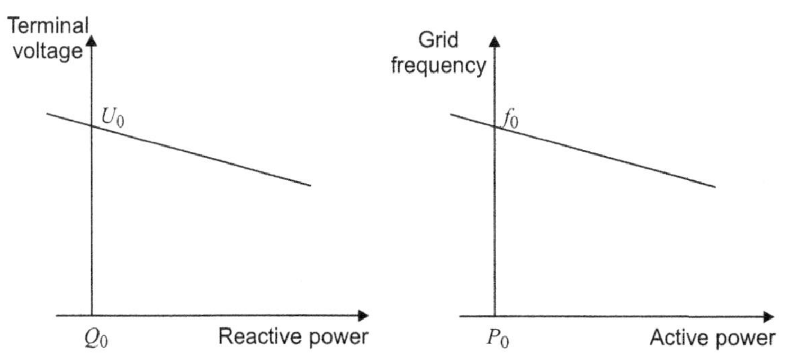

Figure 5.50 Principle of droop in voltage amplitude and frequency using active and reactive powers.

A communication-based controller to maintain the voltage magnitude in a low-voltage network within limits is presented in Ref. 438. A central controller is located with the 20 kV/400 V transformer and remote measurement and control units with individual loads, generators and storage units. The target function of the controller is to maintain the 10 min rms voltage between 90% and 110% of the nominal voltage.

Operator sends a reference value for the power factor to the supervisory controller of a large wind park. The wind park is connected through a 138/34.5 kV transformer to a radial overhead line. The voltage control equipment consists of a STATCOM and four switched capacitor banks. The supervisory controller of the wind park calculates appropriate settings for the StatCom and for the position of the capacitor banks. The reference value of the power factor at the point of connection is updated by the network operator when needed. The default value is unity power factor.

An algorithm for coordinated voltage control is presented in Ref. 411. Reference values are calculated for the tap changer and for the individual generator units. Input to the algorithm are the active and reactive power flows measured at a number of locations in the distribution networks. A tolerance band of ±2% in voltage is used for each inverter. The tap-changer deadband is ±1.25% around the reference value. The reference values are updated at regular intervals, 50 s in the case study.

Reference 206 combines autonomous control by individual photovoltaic units, with a central controller that calculates the reference values for each unit. The method is applied to a model of a 5.5 km, 6.6 kV feeder with 40% generation connected along the feeder. Each unit maintains the voltage magnitude within a deadband and sends information on the reactive power production or consumption to the central controller. The central controller next recalculates the reference values.

The so-called "contribution matrix approach" is proposed in Ref. 425 to determine the impact of individual generator units on the voltage at different locations in the distribution network. The voltages at a number of critical nodes are measured. If one of them exceeds the limit, the contribution matrix is used to calculate new voltage references. In the coordinated control algorithm, the tap changer has priority, followed by reactive power control by the generator. If neither of that is sufficient to bring the voltages within the limits, combined control of active and reactive powers is used.

A coordinated voltage control method is also proposed in Ref. 423. The coordination takes place between the on-load tap changers, switched capacitor banks in the main substation, and switched capacitor banks along the feeder. For every 10 min interval, the settings of capacitor banks and tap changer are chosen such that the network losses are minimized. For time coordination, the capacitor banks along the feeder are switched first, followed by the capacitor bank in the main substation, with the tap changer coming last. Distributed generation with voltage control options is coordinated by making it faster than the switching of the capacitor banks along the feeder. The optimization algorithm is next adjusted to the presence of the generation. Although the authors of Ref. 423 suggest not to use any communication, the set points have to be communicated to the capacitor banks. Also, is it important to know the generation and consumption patterns.

The algorithm proposed in Ref. 379 goes a step further by also considering step voltage regulators and shunt reactors in the optimization. Assuming widespread availability of communication, the optimum settings are obtained from a genetic algorithm. Here, the objective of the controller is the minimization of the losses with as much constraints that the node voltage should remain within preset limits.

There is, however, a reliability issue because the communication between the master and slave units might fail. Moreover, with unpredictable generation and consumption models, the units may not be able to provide their share, which would result in degraded operation. Hence, a high-level control system seems inevitable, where a reliable communication infrastructure plays a vital role.

With the recent advances in the information and communication technologies (ICT), more advanced coordination schemes are possible. ICT enables universal connectivity between a large variety of grid devices, including power production resources, network nodes, and local loads [152]. This provides new and better technical foundations for distant control of highly distributed networks on an increasingly large scale. Universal connectivity is a key enabler for the proper management of any future energy network. There are different views on how this should work in the future grid. The European vision SmartGrid [153] and its American counterpart vision GridWise [87] are both attempts to provide a platform for the future grids in order to incorporate the latest technologies to provide more flexible, accessible, reliable, and economic solutions.

The future grids in both visions are able to provide certain features, such as plug-and-play, self-healing, and self-organizing in order to facilitate the integration of distributed generation and increase the hosting capacity, while providing improved performance and reliability. Regarding voltage control, a new distributed generation in the future grid should be connected through communication ports to the other components of the grid where automatic identification and control upgrades are instantly performed to allow the new unit to contribute to the existing control function. The same thing happens with generation tripping, where this condition is also communicated and the control system is upgraded to cope with the new condition without any interruption in its function. Here, the power system components talk to each other through intensive data processing and real-time state monitoring.

5.8.12 Increasing the Minimum Load

Increasing electricity consumption might seem a strange method to allow more renewable energy to be connected. There are however cases where this makes sense. The energy from some types of distributed generation will go unused when they are not producing. Wind, sun, wave, and flow of river are obvious examples. In other cases, it is not possible or practical to turn of the unit. CHP is such an example where switching off the generator would mean that the heat demand is not fulfilled.

Some types of electricity consumption can be shifted in time without having a huge impact on the process that is powered by the electricity. The charging of electrical cars is often mentioned in discussions, but this obviously holds for all types of charging. Also, with heating and cooling processes, the use of the electricity can be somewhat shifted in time. Dedicated battery storage has been proposed as a way of preventing overvoltages.

There are a number of methods for linking the excess production to an increase in load. Direct communication between the production and the consumption is possible. This has been proposed as a way of using solar power to charge cars. The charger would normally operate below its capacity, but when the voltage at the generator terminals gets too high, the charging process is accelerated to reduce the voltage. When this is still not enough, the production is reduced.

Another option is to use the voltage directly to control the load, in a similar way as, the power-frequency control with conventional generators. A high voltage points to a surplus of power, which is compensated by an increase in consumption. Also, undervoltages could be mitigated in this way. With the electric car example, the charging would go faster when the voltage would be high and slower when the voltage would be low. Similar control algorithms could be implemented in other chargers and in heating and cooling load. As a result, the indoor temperature would depend on the voltage of the grid.

A third option is to use a market-based scheme where different loads and generators put in bids for consumption and production. A surplus of production would result in a low price, possibly even a negative price. This in turn would increase the consumption and reduce production. This scheme offers more freedom to individual users, for example, charging a car (or a mobile phone) quickly would require to put in a higher bid for electricity. When the voltage is low, it will be more expensive to charge the car quickly than to charge it slowly. The disadvantage of such a scheme is however the extensive need for communication.

Local battery storage in combination with solar power has been studied in Japan [189]. In one of the experiments, a total of 4 MW of photovoltaic generation is combined with a 1.5 MW battery system for a 5-year test. The installation is connected to the 33 kV network. One of the aims of the study is to develop control algorithms for optimal use of the battery storage. Conflicting requirements have to be fulfilled, including the participation in the electricity market and the prevention of overloads and overvoltages. The optimal scheduling of the battery storage depends on the load and voltage patterns on the feeder.

In Ref. 337, a combination of solar power and battery storage is described. The following control algorithm is used for voltage control:

- Voltage below 119 V: discharge battery
- Voltage between 119 and 121 V: deadband
- Voltage between 121 and 124 V: charge battery
- Voltage above 124 V: disconnect generator

A STATCOM is combined with energy storage in Ref. 20 to compensate fast voltage fluctuations and maintain more constant production over longer timescales. Also, Ref. 112 mentions that local energy storage can be used to keep the voltage profile within its limits. Calculations referred to in the study have shown that a few percent change in voltage profile along a medium-voltage feeder would require between a few hundred kW and a few MW during 2–10 h.

The use of a large battery storage for voltage control on an 11 kV feeder with a 2.25 MW wind farm is described in Ref. 427. The feeder peak load is 2.3 MW.

CHAPTER 6

POWER QUALITY DISTURBANCES

Power quality was introduced in Chapter 3 as covering all deviations from the ideal voltage and current. Such a deviation is referred to as a "power quality disturbance." The terms "voltage quality disturbance" and "current quality disturbance" are also used, when addressing either the voltage or the current. Some phenomena that may be classified as power quality disturbances have already been discussed in earlier chapters: slow variations in voltage magnitude in Chapter 5; variations in current magnitude leading to overload or additional losses in Chapter 4. In this chapter, we will discuss the main remaining power quality disturbances. Short and long interruptions may also be considered as part of power quality. They are not discussed in this book.

6.1 IMPACT OF DISTRIBUTED GENERATION

The introduction of generation to the distribution network will impact the power quality in a number of ways. Connection of small amounts of generation will mainly have local effects, whereas a massive introduction of distributed generation will also have global effects (i.e., at subtransmission and transmission levels).

The following developments associated with the introduction of distributed generation will have an impact (positive or negative) on the power quality.

- The emission of disturbances by the generator units may result in increased emission levels. This concerns especially harmonics, voltage fluctuations, and unbalance, as well as disturbances due to switching of the generators. The emission by distributed generation will be discussed in all the forthcoming sections. With some exceptions, the emission of power quality disturbances due to distributed generation is not a concern. Flicker due to fast voltage fluctuations and harmonics up to about 1 kHz are mentioned in the literature mostly as serious emission sources; they will be discussed in detail in Sections 6.2 and 6.4, respectively. We will see, however, that the emission of these harmonics and of flicker is limited in most cases. Harmonics at higher frequencies are not so commonly addressed in the literature, probably because of the lack of standards and other information. But several types of interface for distributed generation result in emission at these higher frequencies. Also do several types of distributed generation show more of a broadband spectrum than the existing equipment. The lack of knowledge about the consequences of this indicates

Integration of Distributed Generation in the Power System, First Edition. Math Bollen and Fainan Hassan.

that further studies are essential. Emission of high-frequency distortion (above about 1 kHz) will be treated in detail in Section 6.5.

- Large single-phase generators or many small single-phase generators will result in an increase in the voltage unbalance. This is discussed in Section 6.3.
- The increased strength of the distribution network will limit the spread of disturbances. The impact is different for different types of interface and for different types of disturbances. Synchronous machines are in general more advantageous than other types of interface. The impact can become complex for harmonics because of the frequency dependence of the source impedance. A reduction in source impedance at one frequency may go together with an increase in source impedance at another frequency. We will discuss this further in Section 6.4.
- The shift of generation from transmission to distribution will reduce the strength of the transmission network. This will result in a wider spread of disturbances that originate at the transmission level or that reach the transmission system from a lower level. This could concern voltage fluctuations, unbalance, or harmonics due to large industrial installations, and also voltage dips due to faults at transmission level. The impact of a weakening transmission system on the number of voltage dips is discussed in Ref. 353 in association with the shift of generation due to the deregulation of the electricity market. In Refs. 46, 52, and 391, this is further discussed in association with the shift from large conventional generation to other types of generation. The flicker level due to a large steel installation in Northern Europe increased significantly when a local power station no longer was in operation whenever the steel plant was producing. This is a consequence of deregulation; however, massive penetration of distributed generation will have similar impacts.
- Several types of distributed generation are associated with capacitors or capacitor banks. Also, long cables at transmission and subtransmission levels will result in additional capacitance connected to the grid. This will increase the risk of resonances at harmonic frequencies and shift existing resonances to lower frequencies. At distribution level, the impact appears to be limited to the introduction of resonances at new locations and a shift of resonances to lower frequencies. Resonances around 1 kHz have been reported with solar power installations in domestic environments. Calculations indicate that resonances can be expected around the seventh harmonic due to induction generators equipped with capacitors. The main concern are new resonances, possibly as low as 100 Hz, due to the introduction of long cables at subtransmission and transmission levels. Resonances may also occur in the collection grid of a wind park. Harmonic resonances are discussed in detail in Section 6.4.5.

All the mentioned impacts are unintended impacts; even if some of the impacts are positive, they remain unintended. It is, however, possible to use distributed generation intentionally to improve the power quality. The presence of distributed generation may be used by the network operator to defer investments. Some possible examples are as follows:

- The presence of distributed generation has a positive impact on certain disturbance levels. This holds among others for voltage dips, harmonics, and voltage fluctuations. The presence of a voltage source makes the distribution system stronger and introduces damping for these disturbances. This will be discussed in more detail in the forthcoming sections. The location of a generator can be chosen in such a way that the improvement in voltage quality is sufficient to defer investment in the distribution network. This obviously requires cooperation between the network operator and the owner of the generator unit.
- The voltage at the terminals of the generator unit can be controlled by actively controlling the reactive power flow between the generator and the grid. This is possible when synchronous machines or generators with power electronics interface are used. The latter provides the most flexible and fastest control possibilities. By using active control, voltage fluctuations and voltage dips can be mitigated. Algorithms have even been discussed to reduce the voltage unbalance, for example, in Ref. 355. Also, the voltage magnitude can be controlled as is discussed in Chapter 5.
- Harmonic distortion can be mitigated by means of advanced control options that are offered by modern power electronics converters. The converters might simply compensate for the downstream harmonic emission using open-load control. The converters may also provide damping at harmonic resonances and in that way limit the voltage distortion at the resonance frequency. A wide range of publications address possible algorithms for stand alone active filters and for filters integrated in the converter of a distributed generator; for example, Refs. 349, 358, 361, and 373.

6.2 FAST VOLTAGE FLUCTUATIONS

Fast changes in voltage magnitude are referred to as voltage fluctuations or sometimes "voltage flicker." Their main concern is that they result in a phenomenon called "light flicker" with the frequency range between about 1 and 10 Hz. Flicker is a sensation of unsteadiness in the light intensity where the observer notices that the light is not of constant intensity over a longer period, but cannot observe the individual changes. Most people experience light flicker as uncomfortable even if they not always notice the flicker. Studies have shown that even nonnoticeable flicker will result in increased activity in certain parts of the brain. Long-term exposure to flicker will result, for example, in headache and tiredness. The amount of flicker a voltage fluctuation causes for a standard incandescent lamp is quantified by the "flicker severity," P_{st}, where $P_{st} = 1$ corresponds to a level that is experienced as uncomfortable by 95% of persons. The relations between the voltage fluctuations and the flicker and the calculation of flicker indices are discussed in detail in Section 2.4 in Ref. 45.

Light flicker should not be confused with occasional changes or dips in light intensity due to voltage steps or voltage dips. These are observable for voltage steps as low as a few percent of the nominal voltage [377], but are not continuous and do not have the same impact as continuous flicker. The terms "voltage fluctuations" and "rapid voltage changes" are sometimes used with the following meaning:

- *Voltage fluctuations* are continuous changes in voltage magnitude at timescales up to several minutes.
- *Rapid voltage changes* are fast and stepwise changes in voltage magnitude.

Gönen [178] distinguishes in this context between "abrupt flicker" and "sinusoidal flicker."

There are no clear definitions or limits for these phenomena. The only limits that exist concern voltage fluctuations at timescales up to a few seconds and rapid voltage changes between two voltage magnitude levels. The former limits are based on observable light flicker with incandescent lamps due to the voltage fluctuations. The latter limits are related and are often found in the same standard documents. The replacement of incandescent lamps by other types of lighting that has started recently in the European Union and in some other countries will require a new look at the limits for changes in voltage magnitude. The discussion on this subject has however not really started yet, but there are clear indications that nonincandescent lamps [45, Section 2.4.8; 74, 188] are less susceptible to voltage fluctuations than the incandescent lamps.

6.2.1 Fast Fluctuations in Wind Power

Fast variations in generated power may lead to voltage fluctuations. These are a concern for those sources for which the available power strongly varies with time, notably wind power and solar power. Wind turbines produce a continuously varying output. In Ref. 232, three timescales are distinguished:

- Variations with a frequency of several hertz due to the turbine dynamics, the tower resonance, and the gearbox.
- Periodic power pulsations at the frequency at which the blades pass the tower, typically around 1 Hz for a large turbine. These are referred to as 3p oscillations for three-blade turbines. Detailed measurements presented in Refs. 405 and 406 show a range of frequencies associated with the rotation of the blades with respect to the tower—from 1p through 18p.
- Slower variations due to changes in wind speed.

It is, however, pointed out in Ref. 232 that the increased emission will at least be partly compensated by the increase in fault level due to the installation of the generators.

There is some indication that the turbines in a wind park may reach a state of "synchronized operation," thus amplifying the power pulsations due to the tower. The cause of this synchronous operation is not fully clear, but it is thought to be due to interactions between the turbines through the network [232]. Synchronous operation can only be expected for sites with a rather constant wind speed not affected by turbulence due to the terrain. This was often mentioned as a serious problem in the past, but more recently the risk of synchronous operation is perceived as small.

A study of the different fluctuations in the power generated by a fixed speed wind turbine is presented in Ref. 406. Measurements were performed after the frequency components in the power fluctuations, between 0.1 and 10 Hz. The following components were found, next to a continuous spectrum:

- A 1.1 Hz fluctuation corresponding to the tower resonance.
- A 2.5 Hz fluctuation corresponding to the rotation speed of the gearbox.
- Four different components related to the rotation of the blades: 1p, 2p, 3p, and 6p. The 1p fluctuations are due to unbalance in the rotor and small differences between the blades. The 3p oscillations are due to the passing of the blades in front of the tower. The 2p and 6p components are probably harmonics of the 1p and 3p fluctuations, respectively.

For low wind turbulence (wind from sea in this case), these discrete components dominate the spectrum. For high wind turbulence, the fluctuations form a continuous spectrum. In Ref. 405, additional components are found at 4p/3, 4p, 14p/3, 5p, 9p, 12p, and 18p.

Measured or simulated flicker due to wind turbines is presented in a number of publications [259, 303, 335, 372, 397, 406, 407, 426]. An overview of the results from these studies is shown in Figure 6.1. The flicker values are shown as a function of the short-circuit ratio (SCR), that is, the ratio between the fault level and the size of the turbine. Most of the values are below 0.2. The two exceptions occur both for a short-circuit ratio equal to 20 and for simulations. The value of 20 for the short-circuit ratio is often used as a typical value in simulations. From the figure, one could draw the conclusion that the contribution of wind power to the flicker level is about 0.2, rather independent of the size of the installation in relation to the fault level.

In case there are several turbines connected close to each other, the flicker level will be higher than with one turbine. According to IEC 61400-21, the short-term

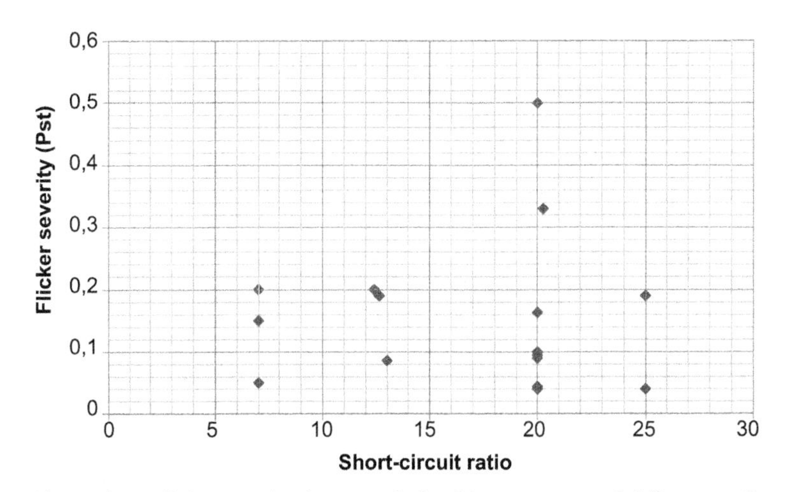

Figure 6.1 Flicker severity due to a wind turbine: summary of different studies.

flicker severity should be added by using the following expression:

$$P_{st} = \sqrt{\sum_{i=1}^{N} P_{st,i}^2} \tag{6.1}$$

where $P_{st,i}$ is the contribution from each individual turbine. With identical turbines, the contribution from N units is $\sqrt{N}$ times the contribution from one unit. If the emission due to one unit is a short-term flicker severity equal to 0.2, a level equal to 1.0 is obtained by 25 units. A large wind park can easily contain more units than this. As a consequence, the flicker emission due to one unit should be kept well below 0.2 for large parks. Detailed information on the impact of large wind parks on the flicker level is rare, but the general impression is that the increase in flicker level is small in most cases.

6.2.2 Fast Fluctuations in Solar Power

Variations in the production by a solar power installation occur at a range of time scales, from seasonal variations, due to the tilt of the earth axis together with its orbit around the sun, down to variations at a timescale of seconds due to passing clouds. The latter are the ones that cause fast fluctuations in voltage magnitude, possibly resulting in light flicker. The measurements and simulations given in Chapter 2 indicate that for a single solar panel, the power production can change by 50% of rated power in 5–10 s. For a number of panels spread over a distance of a few hundred meters (i.e., a low-voltage feeder), such changes take place in 30–60 s. The former are at the limit of the flicker range, the latter do not have any impact on the flicker severity.

The relation between flicker severity (P_{st}) and power fluctuations due to large PV installations was measured in Ref. 31. Measurements were performed for a 101 kW plant, consisting of 55 inverters of 1.8 kW each, and for a 37 kW installation consisting of four converters of 1.5 kW and four of 4.6 kW. For one of the installations, no correlation was found between flicker severity and power fluctuations, which indicates that the background level dominates. The other installation showed a clear increase in flicker severity with injected power. The cause of this was traced back to the control system used for the inverters. A network impedance measurement is performed every second after which the current reference value of the inverter is adjusted. This results in current fluctuations with a frequency of exactly 1 Hz, within the flicker range.

6.2.3 Rapid Voltage Changes

The intentional or accidental switching of wind turbines will result in fast changes in voltage magnitude, referred to as "voltage steps" or "rapid voltage changes." Rapid voltage changes exceeding a few percent of the nominal voltage cause visible changes in the light intensity of incandescent lamps. According to a study, with about 100 test persons of different age [377] are rectangular changes visible to the majority of observers when they exceed 1.5% of the nominal voltage. Changes exceeding 4% are visible to 95% of the observers. For slower changes with a ramp rate less than 5 V/s, the visibility becomes less. Voltage dips due to motor starts become more visible when the start-up time is longer. There are no very strict requirements on rapid voltage changes in power quality standards. For example, values up to 6% are allowed at medium voltage under the Norwegian power quality regulations. But as shown by the examples above, the resulting change in light intensity becomes visible already when the change in voltage magnitude exceeds a few percent. A large part of the complaints on "bad power quality" are in fact the light intensity changes due to rapid voltage changes. This is in most cases incorrectly classified as "flicker."

There are no clear limits on frequency and magnitude of rapid voltage changes; however, we still give some general guidance:

- If repeated changes in voltage magnitude result in a short-duration flicker value exceeding 1.0 over a longer period (1 h or longer), the majority of persons will consider this as uncomfortable when incandescent lamps are used. Most other types of lighting are less sensitive, but a detailed investigation is still missing.
- Rapid voltage changes of 2% or higher give noticeable changes in light intensity of an incandescent lamp [377]. The impact of rapid voltage changes on other types of lighting, to the authors' knowledge, has not been studied.
- Different standards limit the number of rapid voltage changes exceeding 2% to at most several times a day.
- Rapid voltage changes exceeding 5% should only occur rarely, at most a few times per year.

A wind turbine can take a high current when energized. The concern is most with induction generators that take a significant amount of reactive power. Larger machines are equipped with soft starters to limit the inrush current and the associated voltage change. A wind turbine starts to produce power when the wind speed exceeds a value around 3–4 m/s. If the wind speed is below this value, the turbine is not connected to the network. Repeated switching actions, energizing and de-energizing, can occur when the wind speed remains around 3–4 m/s for a longer period. The local conditions will determine how likely it is that such a situation occurs. The system to control the starting of a wind turbine will typically delay a new start to prevent multiple starts within a short period. However, a too long delay could result in loss of energy production.

Disconnection of wind or solar power at high production is not likely to occur. A failure may occur during high production prompting the disconnection of the generator. Failures are, however, relatively rare (at most a few times per year) and the production is high only during a limited part of the time (at most 10–20%) so that a disconnection during high production is not expected to occur more often than once every few years. Wind power installations also need to be disconnected when the wind speed gets too high. Typical limits are in the 20–25 m/s range. Such high wind speed occur at most a few times a year, but at most locations only once every few years.

From the above reasoning we can conclude that a reasonable limit for the size of a wind power or a solar power installation would be such that its disconnection at full power will not result in a rapid voltage change of more than 5%.

Small generation based on combined heat-and-power (the term "micro-CHP" is often used to refer to units with size up to a few kilowatt connected to the

TABLE 6.1 Size of a Single-Phase Micro-CHP Unit That Results in a Rapid Voltage Change of 3%, as a Function of the Distance to the Distribution Transformer and the Cross section of the Wire

	50 m	200 m	500 m	2000 m
25 mm^2	45 kW	11.4 kW	4.5 kW	1.1 kW
50 mm^2	90 kW	23 kW	9.0 kW	2.3 kW
120 mm^2	210 kW	54 kW	22 kW	5.4 kW
240 mm^2	420 kW	110 kW	44 kW	10.8 kW

low-voltage network) may switch on and off with full power regularly. The control of these units is rather simple: the unit is producing at full power for a certain period and not producing anything at all for the next period. The heat demand is controlled by varying the duration of the on-period. A simple thermostat (as used for many years in domestic heating systems) is all that is needed for the control. The consequence for the power system is a load that causes a high number of rapid voltage changes. A reasonable limit for these would be 2–3% of the nominal voltage. When higher values are allowed, a significant increase in the number of complaints can be expected.

Stepwise changes in production (active power) result in voltage changes over the resistive part of the source impedance. The resistance mainly depends on the distance to the distribution transformer and the areal cross section of the cable or line. The values in Table 6.1 indicate the maximum size of a micro-CHP unit (the hosting capacity) when rapid voltage changes have to be limited to 3%. The values hold for single-phase generators. For three-phase generators, the hosting capacity is three times as high.

As shown in Table 6.1, the resulting rapid voltage change will be significantly less than 3% for most locations. Only when many units are connected close together in weak parts of the network can flicker problems occur due to too many such changes within short periods of time.

The discussion on rapid voltage changes has picked up again recently [64, 187]. Among the issues that are being discussed are methods for quantifying rapid voltage changes and suitable limits for rapid voltage changes, besides the overall need for requirements on rapid voltage changes especially after the replacement of the majority of incandescent lamps by other types of lighting. Even this issue remains far from being completely solved; however, there are clear indications that at least the majority of nonincandescent lamps are less sensitive to flicker, so less strict requirements on rapid voltage changes could be considered [73, 74, 188]. Before new limits on continuous voltage fluctuations and rapid voltage changes are introduced, the impact of these disturbances on other equipment should be further studied as well. Especially rotating machines, including wind turbines, may be adversely impacted by repeated rapid voltage changes.

6.2.4 Very Short Variations

In the previous sections, we discussed the fastest variations in voltage magnitude. Next to these fast fluctuations, the power production by distributed generation varies over a range of timescales, from seconds through years. Of most concern at distribution level are variations in solar and wind power production at timescales between seconds and minutes. Voltage variations at timescales of minutes and longer have already been discussed in Chapter 5.

There are at this moment no limits on voltage magnitude variations at the timescale between several seconds (the slowest variations resulting in flicker) and 1–10 min (the standard averaging time for voltage magnitude variations). The impact of such variations is unknown, but high levels of variations are expected to especially impact motor loads.

The measurements presented in Chapter 2 show that severe variations in current magnitude occur at a timescale between 3 s and 10 min. This timescale is normally not considered in power quality discussions. In Refs. 44 and 48, the "very short variation" (VSV) is introduced as a way to quantify variations at this timescale. This index has recently been taken over by an international working group studying rapid voltage changes [187] and extended to also cover the time range shorter than 3 s.

Reference 44 shows measurements of the very short variations at a limited number of locations in different countries. The measurements show two components in the very short variations:

- A continuous level due to very small changes in voltage magnitude caused by load variations and large numbers of switching actions of small loads (corresponding to "continuous voltage fluctuations" discussed before).
- An occasionally higher level due to fast and larger changes in voltage magnitude (corresponding to "rapid voltage changes").

The following performance criteria were concluded from these measurements [40]:

- The 50-percentile of the 10 min very short variations during 1 week should not exceed 1.0 V.
- The 95-percentile of the 10 min very short variations during 1 week should not exceed 2.5 V.

It is important to note here that these performance criteria are based on a very limited number of measurements at a very limited number of locations. The above levels are, however, certainly safe limits, so they can be used to estimate a lower limit to the hosting capacity with respect to very short variations.

In Ref. 32, measurements are shown of the very short variations in voltage and current at the terminals of a 100 kW solar power installation connected to a low-voltage network. The fault level at the point-of-common coupling with other customers is 3 MVA with an X/R ratio about 0.9. On a variably clouded day, when the variations in irradiance are at their biggest, the measured very short variations in current were up to about 20 A or about 25% of the current magnitude. However, the very short variations in voltage were similar at daytime as at nighttime, with highest values up to about 1.5 V.

To calculate the very short variations due to distributed generation, for example, wind power, the variations in active power, VSV_P, and in reactive power, VSV_Q, have

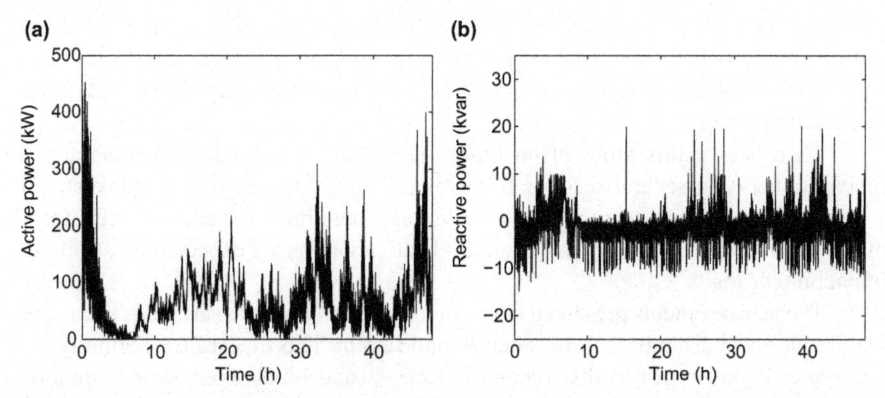

Figure 6.2 Active power (a) and reactive power (b) produced by a 650 kW turbine during a 48 h period.

to be calculated separately. The variations in voltage, VSV_V, are next obtained from

$$\mathrm{VSV}_V = R \times \mathrm{VSV}_P + X \times \mathrm{VSV}_Q \tag{6.2}$$

Measurements with a group of three 650 kW wind turbines, with active power converter, have been used to illustrate the fluctuations in wind power at a timescale less than a few minutes. The active and reactive power as a function of time for one turbine are shown in Figure 6.2. The active and reactive power have been obtained by the measurement equipment every second during the measurement period, which was in this case slightly longer than 2 days. The active power shows variations over a wide range, both slow and fast variations. The reactive power is, however, on average close to zero, but with fast fluctuations of 10–20 kvar both positive and negative.

To quantify the fluctuations in active and reactive power at a timescale less than a few minutes, so-called "very short variations" of the active power production have been calculated. The very short variations have been introduced in Ref. 44 to quantify variations in voltage magnitude. The integration of wind power and other renewable sources of energy was mentioned as an important reason for introducing this concept [44, 48]. The 1 s very short variation is in this section calculated as the difference between the 1 s average power and the 10 min average power. The latter is determined over a window centered around the 1 s value. The results for a single turbine are shown in Figure 6.3. Note the large difference in vertical scale between the figures: the very short variations in active power are much bigger than in reactive power. As feeder reactive and resistance are similar for medium-voltage distribution systems, we can conclude from (6.2) that the very short variations in voltage will be determined mainly by the variations in active power.

The 1 s very short variation is next used to calculate the 10 min very short variation. The 10 min very short variation is the root mean square of the 1 s very short variation over a 10 min period. The result of this is shown in Figure 6.4. The curves show the same variation with time as the 1 s variations, but more smooth. Again, the very short variations in active power are much larger than in reactive power.

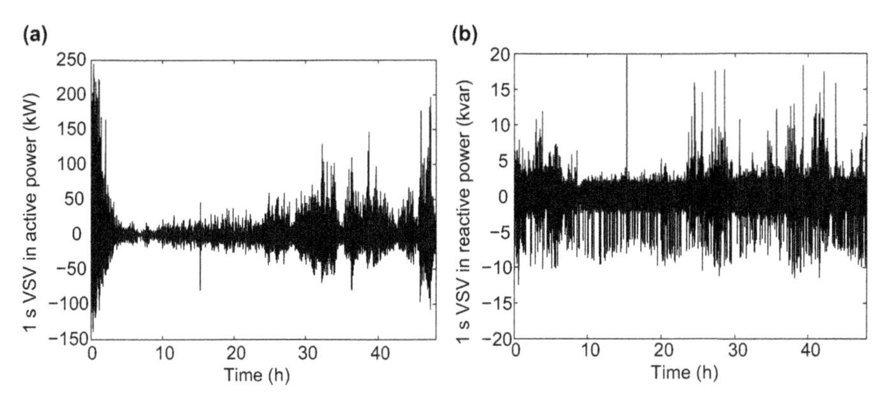

Figure 6.3 The 1 s very short variations in active power (a) and reactive power (b) produced by a 650 kW turbine during a 48 h period.

The active and reactive power production by the three turbines together has been measured at the same time as the production by one turbine. The 10 min very short variations have also been calculated for the three turbines together. The variations in active power for one and three turbines are compared in Figure 6.5. The variations for three turbines are a bit less than twice the variations in one turbine. The ratio of the maximum values is 1.92, while the ratio of the average values is 1.99.

The comparison has also been made for the reactive power in Figure 6.6. The correlation between one and three units is less than that for the reactive power, and the highest values for three units are only slightly higher than that for a single unit. The ratio is 1.33 for the maximum value and 1.79 for the average value.

6.2.5 Spread of Voltage Fluctuations

Voltage fluctuations are caused by the fluctuating active and reactive power at the terminals of the generator. The largest voltage fluctuations may, however, occur

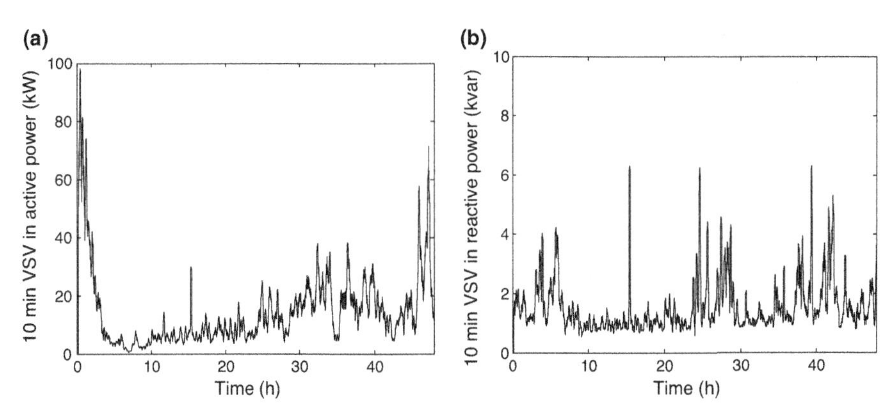

Figure 6.4 The 10 min very short variations in active power (a) and reactive power (b) produced by a 650 kW turbine during a 48 h period.

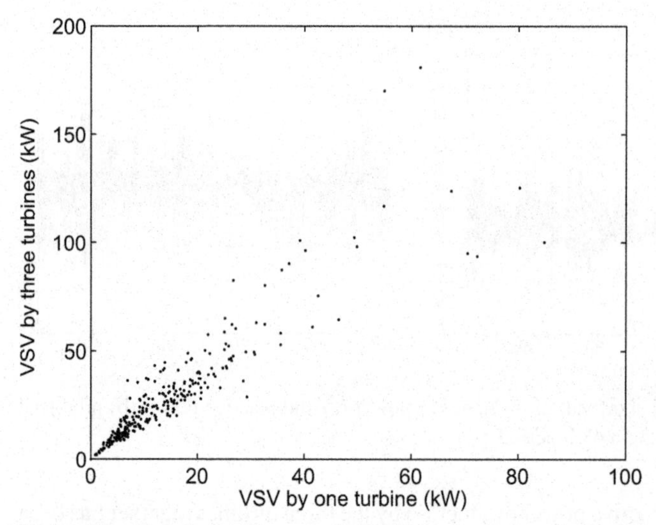

Figure 6.5 Correlation between the 10 min very short variation in active power for one turbine and three turbines.

somewhere else. To illustrate this, consider a generator that supplies active power P to the grid and that at the same time consumes reactive power Q. The source impedance is $R + jX$ and the source voltage is E. At an "electrical distance" α from the generator, the source impedance is $(1 - \alpha)R + j(1 - \alpha)X$; $\alpha = 1$ corresponds to the voltage source; and $\alpha = 0$ corresponds to the generator. The active and reactive

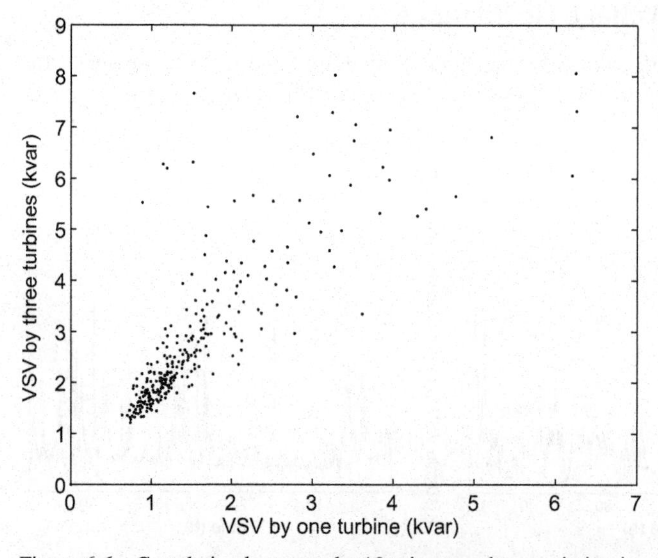

Figure 6.6 Correlation between the 10 min very short variation in reactive power for one turbine and three turbines.

power flow at this location consist of the power flow at the generator terminals and the losses:

$$P(\alpha) = P - \alpha R S^2 \tag{6.3}$$

$$Q(\alpha) = Q + \alpha X S^2 \tag{6.4}$$

where $S^2 = P^2 + Q^2$ is the square of the apparent power. All variables are expressed in per unit with the nominal voltage as voltage base. The voltage at this location is approximately

$$V(\alpha) = E + (1-\alpha)(RP - XQ) - \alpha(1-\alpha)Z^2 S^2 \tag{6.5}$$

with $Z^2 = (R^2 + X^2)$ being the square of the magnitude of the impedance. The highest or lowest voltage magnitude occurs for $\mathrm{d}V/\mathrm{d}\alpha = 0$, which gives

$$\alpha_{\mathrm{ex}} = \frac{1}{2} + \frac{1}{2}\frac{RP - XQ}{Z^2 S^2} \tag{6.6}$$

The extreme value of the voltage magnitude is obtained by substituting (6.6) into (6.5), resulting in the following expression:

$$V_{\mathrm{ex}} = E + \frac{1}{2}(RP - XQ) - \frac{1}{4}Z^2 S^2 - \frac{1}{4}\frac{(RP - XQ)^2}{Z^2 S^2} \tag{6.7}$$

When active and reactive power at the generator terminals are such that the voltage at the generator terminals is not impacted, $RP - XQ = 0$, the lowest voltage magnitude is reached electrically halfway between the generator and the source, $\alpha = 1/2$. This lowest value is obtained by substituting $RP - XQ = 0$ into (6.7):

$$V_{\min} = E - \frac{1}{4}\frac{Z^4}{X^2}P^2 \tag{6.8}$$

This is an important case, as it corresponds to voltage control by the generator or by additional control equipment such as an SVC, close to the terminals of the generator. We see that keeping the voltage at the generator terminals constant does not prevent all voltage variations. The reactive power losses result in a small voltage drop along the feeder, with its minimum electrically halfway between the generator and the infinite source.

For $RP - XQ > 0$, the voltage at the generator terminals rises and the minimum is obtained for $\alpha > 1/2$, that is, it moves toward the source. The minimum voltage is obtained at the source, that is, the voltage rises everywhere along the network, when

$$RP - XQ > Z^2 S^2 \tag{6.9}$$

In the same way, for $RP - XQ < 0$, the voltage at the generator terminals drops and the minimum voltage is obtained for $\alpha < 1/2$, that is, it moves toward the generator. The minimum voltage is obtained at the generator terminals, that is, the

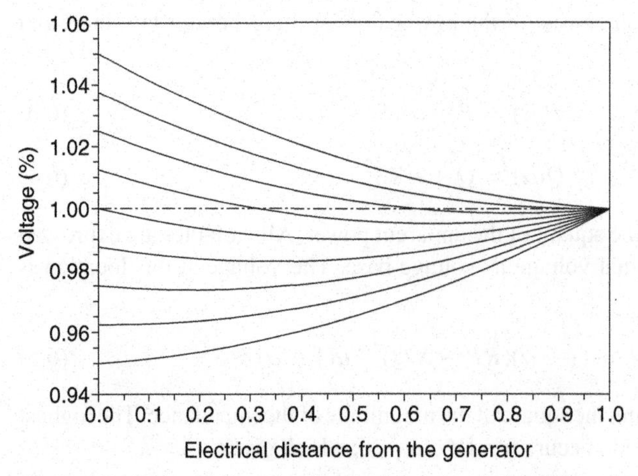

Figure 6.7 Voltage along the distribution network for different values of the voltage rise at the generator terminals. The generator is on the left and the source on the right.

voltage drops everywhere along the network, when

$$RP - XQ < -Z^2S^2 \tag{6.10}$$

The voltage magnitude along the feeder is shown in Figure 6.7. For a large voltage rise or voltage drop at the generator terminals, the extreme value is obtained at the generator terminals. However, for a small rise or drop, the minimum voltage is reached further into the distribution network.

The calculations resulting in Figure 6.7 were obtained for a source impedance of 0.2 per unit on a base equal to the injected apparent power, that is, for a short-circuit ratio of 5. The larger the source impedance, that is, the weaker the grid, the more the voltage drop along the feeder. The impact of the generator on the voltage can be minimized by allowing a small voltage rise at the generator terminals. The effect is, however, small and even in a weak grid, as in Figure 6.7, not more than 1% of the nominal voltage.

The above calculations can be repeated considering the change in voltage due to changes in active and reactive power. The voltage changes along the distribution feeder are obtained from (6.5):

$$\Delta V(\alpha) = (1-\alpha)(R\Delta P - X\Delta Q) - \alpha(1-\alpha)Z^2(2P\Delta P + 2Q\Delta Q) \tag{6.11}$$

The maximum occurs for

$$\alpha = \frac{1}{2} + \frac{1}{2}\frac{R\Delta P - X\Delta Q}{Z^2(2P\Delta P + 2Q\Delta Q)} \tag{6.12}$$

When the voltage at the generator is kept constant by a control system, $RP - XQ$ is also kept constant. Therefore, the following relation holds between the changes in active and reactive power:

$$R\Delta P = X\Delta Q \tag{6.13}$$

From (6.11), we get the following expression for the voltage changes due to the changes in active power:

$$\Delta V(\alpha) = \alpha(1-\alpha)\frac{Z^4}{X^2}2P\Delta P \tag{6.14}$$

We see that the change in voltage increases with decreasing feeder reactance. The reason for this is that with decreasing feeder reactance, increasing amounts of reactive power are needed to compensate for the voltage rise due to the injection of active power. The transport of this reactive power over the feeder results in changes in voltage magnitude.

6.3 VOLTAGE UNBALANCE

The connection of distributed generation can impact the voltage and current unbalance in a number of ways. Large single-phase units or many randomly distributed small units will result in an increase in unbalance. Single-phase generator units can only be expected in low-voltage networks, most likely with domestic and small commercial customers. The connection of three-phase generators will limit the voltage unbalance either by creating a low-impedance path for the negative-sequence impedance (induction and synchronous generators) or by creating a balanced voltage source (power electronics converters). At transmission and subtransmission levels, the shift from conventional generation to distributed generation could result in an increase in the unbalance. All this will be discussed in more detail in the following sections.

6.3.1 Weaker Transmission System

When more distributed generation is producing power, less large conventional generator units will be operating. The result is that the short-circuit capacity in the transmission and subtransmission systems becomes less. The same unbalanced (negative-sequence) current will give more unbalanced (negative-sequence) voltage. The two main sources of unbalanced voltage at transmission level are as follows:

- Unbalanced (nontransposed or not well transposed) transmission lines generate a negative-sequence voltage even with only a positive-sequence current flowing through the line. For long transmission lines, especially when not transposed, a negative-sequence voltage up to 2% may occur.
- Large single-phase loads such as certain types of arc furnaces and railway traction take a heavily unbalanced current. High-speed railways are reported to be the main source of unbalance in some countries.

The former contribution is less affected by the amount of conventional generation connected to the system. The latter contribution is inversely proportional to the fault level. When the fault level halves, the voltage unbalance doubles. A significant increase in unbalance is expected at transmission level when conventional power stations close to large single-phase loads are in operation only part of the time or not at all. It is difficult to predict beforehand for which amount of distributed generation this will take place. It should also be noted that the consumption also shows variations so that the amount of generation connected to the transmission grid will change with time anyway. Next to that does the opening up of the electricity markets result in power transport over longer distances. All this will result in temporary or local weakening of the transmission grid. The shift from large power stations to distributed generation will further contribute to this, but it is not the only cause.

The actual impact will strongly depend on the way in which the connection of large installations with unbalance current is assessed. In the IEC technical report for the connection of unbalanced installations [213], it is recommended to use the lowest fault level that could occur during normal operation. With a large penetration of generation not contributing to the fault level (i.e., distributed generation, but also large wind parks), the lowest fault level to be considered corresponds to minimum consumption coinciding with maximum production. This could locally result in a significant reduction in fault level. Maintaining the same planning level for unbalance could make the connection of the installation much more expensive.

6.3.2 Stronger Distribution System

The presence of three-phase generators connected to the distribution system gives a decrease in negative-sequence voltage. The negative-sequence impedance of induction machines and synchronous machines is less than 0.2 per unit. When a generator transformer is used, its impedance should be added to the negative-sequence impedance. The reduction in unbalance becomes less in that case.

Example 6.1 Consider a 400 V low-voltage bus supplied through a 200 kVA, 5% transformer from a 70 MVA medium-voltage grid. The fault level at the low-voltage bus is equal to 3.78 MVA. This same value will hold for the positive-sequence as for the negative-sequence.

A 100 kVA induction generator is connected to the same bus, with a negative-sequence impedance equal to 0.17 per unit. The negative-sequence source impedance after connection of this generator corresponds to a fault level of 4.37 MVA. The 16% increase in fault level will give about 14% decrease in negative-sequence voltage. In the extreme case of a 250 kVA machine connected to the bus, the fault level will increase to 5.26 MVA; a total increase by 39% reducing the negative-sequence voltage by 28%.

We see from Example 6.1 that even for a large amount of distributed generation, the impact on the fault level remains small. This holds for a generator connected to the main low-voltage bus. The situation becomes different when the unit is connected along a feeder at some distance away from the main bus.

Example 6.2 Assume that the generator is connected to an overhead line, at 1 km from the main low-voltage bus. The impedance of the line is 350 mΩ/km. At the point of connection of the generator, the fault level for the negative sequence increases from 408 to 996 kVA by the connection of the 100 kVA generator. The negative-sequence voltage due to emission downstream of the generator will reduce to only 40% of its original value.

From Example 6.2, we can conclude that the impact of generation on the negative-sequence voltage is significant when the unit is connected at some distance from the main low-voltage bus.

Exercise 6.1 Calculate the improvement in fault level for the negative sequence when the same generator is connected at 2 km from the main low-voltage bus.

Distributed generation is not the only form of rotating machines that are connected to the distribution network. A significant part of the load also consists of induction motors. The amount of three-phase induction motor load, as a fraction of the (positive-sequence) fault level will vary strongly between different locations. These induction motors also result in a reduction of the negative-sequence source impedance and thus in an improvement of the voltage unbalance.

Example 6.3 Consider a medium-voltage bus supplied through a 10 MVA, 15% transformer from a 2000 MVA subtransmission grid. The fault level at the low-voltage bus is equal to 64.5 MVA. A 4 MVA induction generator is connected to the same bus through a 4 MVA, 7% transformer. The negative-sequence impedance of the induction generator is 0.18 per unit. The (negative-sequence) fault level after connection of this generator becomes 80.5 MVA, which is an increase of about 25%.

The increase in fault level reduces the voltage unbalance due to unbalanced current originating downstream of the generator connection. The same low negative-sequence impedance of the rotating machine reduces the unbalance that originates in the transmission system.

In Ref. 58, the fault level is calculated for a 20 kV medium-voltage network with 35 MVA load and 17 MW distributed generation. The distributed generation consists of three wind parks with six turbines each and one hydroelectric plant with three turbines. Without distributed generation, the fault level is 239 MVA; the distributed generation increases the fault level by 25% to 300 MVA. As a result, the voltage unbalance due to unbalanced downstream load will reduce by about 25%.

Distributed generators connected through a power electronics interface will also improve the unbalance locally. Seen from the grid, the generator can be represented as a (positive-sequence) voltage source behind a reactance. For the negative-sequence, it is only this reactance that matters. The smaller the impedance, the more the generator unit reduces the negative-sequence voltage. This comes, however, at the expense of large negative-sequence currents flowing through the converter. Typically, values of the reactance used in power electronics converters are between 0.1 and 0.2 per unit, that is, in the similar range as the negative-sequence impedance of induction and synchronous machines. Again, the use of a generator transformer will increase the negative-sequence impedance and thus reduce the ability of the distributed generator to reduce the voltage unbalance.

As mentioned above, the presence of voltage unbalance will result in negative-sequence currents flowing through the converter of the stator of the rotating machine. This will require derating of the converter or machine in terms of active power production. The nameplate rating often includes up to 1% unbalance in the terminal voltage. For higher voltage unbalance, derating of the generator is needed.

6.3.3 Large Single-Phase Generators

The connection of a large single-phase generator will result in the injection of positive-, negative-, and zero-sequence currents to the same amount. Production of active power P_{gen} and zero reactive power will give a current equal to

$$I = \frac{P_{gen}}{U} \tag{6.15}$$

where U is the phase-to-neutral voltage at the point of connection. It has been assumed here that the generator is connected phase to neutral. The negative-sequence current injected by the single-phase generator is equal to

$$I_2 = \frac{1}{3} I = \frac{P_{gen}}{3U} \tag{6.16}$$

This negative-sequence current will result in a negative-sequence voltage, the ratio between voltage and current being determined by the negative-sequence impedance Z_2 at the point of connection.

$$U_2 = I_2 \times Z_2 = \frac{P_{gen} \times Z_2}{3U} \tag{6.17}$$

The negative-sequence impedance Z_2 is the parallel connection of the series impedance of the grid and the shunt impedance due to local load. This is shown schematically in Figure 6.8.

The series impedance Z_{S2} is mainly the impedance of the distribution transformer and the wires between the distribution transformer and the point of connection. The shunt impedance Z_{L2} is mainly determined by the negative-sequence impedance of the load in the neighborhood of the point of connection. Three-phase motors with their low negative-sequence impedance will dominate this impedance.

Example 6.4 Consider a single-phase generator in a suburban network producing 10 kW, with the point of connection connected by 1 km of underground cable to a 400 kVA, 5% distribution transformer. The impedance of the distribution transformer is equal to 20 mΩ. The impedance of 1 km of underground cable is assumed to be 150 mΩ.

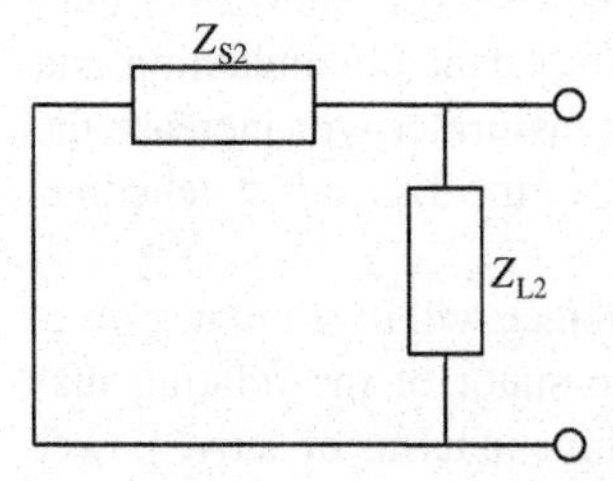

Figure 6.8 Negative-sequence impedance seen by a single-phase load.

In the worst case, with the voltage magnitude being equal to 90% of its rated value (i.e., 207 V), the negative-sequence voltage is

$$U_2 = 0.17\,\Omega \times \frac{10\,\text{kW}}{207\,\text{V} \times 3} = 2.74\,\text{V}$$

This is 1.3% of the voltage, using the actual voltage (207 V) as a reference.

Exercise 6.2 A 3 kW single-phase generator is connected to a rural low-voltage network. The point of connection is supplied by 4 km of overhead line (300 mΩ/km) to a 50 kVA, 5% distribution transformer. Calculate the negative-sequence voltage at the point of connection for 230 V.

From (6.17), the hosting capacity can be calculated, that is, the largest single-phase generator that does not result in an unacceptable value for the negative-sequence voltage. Assume that the maximum permissible value of the negative-sequence voltage is equal to $u_2 \times (U_{\text{nom}}/\sqrt{3})$, with U_{nom} the nominal phase to phase voltage. The lowest voltage at the point of connection for which this criterion should hold is equal to $u \times (U_{\text{nom}}/\sqrt{3})$. The hosting capacity is given by the following expression:

$$P_{\text{gen}} = uu_2 \times \frac{U_{\text{nom}}^2}{Z_2} \tag{6.18}$$

Most of the parameters in this expression will not vary much between different locations. The voltage unbalance limit u_2 will be between 1% and 2%, depending on the background level and the amount of risk the network operator is willing, or allowed, to take. The minimum voltage u will be between 0.9 and 1.0, having only a small influence on the hosting capacity. The nominal voltage is 400 V throughout Europe and large parts of the rest of the world. The only parameter that varies a lot is the negative-sequence impedance at the point of connection.

The negative-sequence voltage due to single-phase equipment matters in networks that have mixture of single-phase and three-phase equipment. In single-phase low-voltage networks (as are common in North America) or in three-phase networks with only single-phase customers (feeders with only domestic load in several European countries), voltage unbalance is not an issue.

Example 6.5 Consider the same supply as in Example 6.4. Assume that 1.5% negative-sequence voltage is allowed due to the generator and that this should hold for a voltage as low as 90%. That gives for the hosting capacity:

$$P_{\text{gen}} < 0.015 \times 0.9 \times \frac{400^2}{0.17} = 12.6\,\text{kW} \tag{6.19}$$

Exercise 6.3 Calculate the hosting capacity for the point of connection in Exercise 6.2.

For remote customers, the negative-sequence impedance of the grid is almost exclusively determined by the length of the overhead line. For lines of several kilometers, the hosting capacity may be only a few kilowatt and thus even a small unit will

TABLE 6.2 Worst-Case Hosting Capacity for Large Single-Phase Generators: 1% Maximum Voltage Unbalance at Nominal Voltage

	400 kVA underground	200 kVA underground	100 kVA overhead	50 kVA overhead
0.5 km	16 kW	13 kW	6.9 kW	5.2 kW
1 km	9.3 kW	8.4 kW	4.2 kW	3.4 kW
2 km	–	4.7 kW	2.3 kW	2.0 kW
3 km	–	3.2 kW	1.6 kW	1.4 kW
4 km	–	–	1.2 kW	1.1 kW
5 km	–	–	1.0 kW	0.9 kW

exceed the hosting capacity. The hosting capacity has been calculated for a number of low-voltage feeders, with the results shown in Table 6.2. Equation (6.19) has been used, with 1% allowable negative-sequence voltage ($u_2 = 0.01$) at nominal voltage ($u = 1.0$). A transformer impedance of 5% has been assumed as well as a cable impedance of 0.15 Ω/km and a line impedance of 0.3 Ω/km.

Feeders longer than 2 km are very uncommon, but they are included here for completeness. The calculations resulting in Table 6.2 hold independent of whether the equipment is producing or consuming active power. The hosting capacity would thus be the same for single-phase loads. Single-phase consuming equipment of 1 kW or higher is not uncommon, so the large voltage unbalances will be normal in some of the networks represented in the table, even without distributed generation.

The results in Table 6.2 are worst-case values, that is, lower limits for the hosting capacity, as the shunt impedance (Z_{L2} in Figure 6.8) has not been considered. Voltage unbalance is a concern only when three-phase load is present. For example, induction motors cannot cope with large negative-sequence voltage; the same holds for adjustable-speed drives. Such loads also have a small negative-sequence impedance; the negative-sequence impedance of an induction motor is around 0.18–0.20 per unit. A 25 kVA of induction motor load (assuming 0.2 per unit negative-sequence impedance) would result in a shunt impedance as low as 1.28 Ω.

Example 6.6 Assuming 25 kVA motor load and a point of connection at 4 km from a 50 kVA transformer would result in a negative-sequence impedance of 0.66 Ω (1.36 Ω in parallel with 1.28 Ω). The hosting capacity would increase from 1.1 kW given in Table 6.2 to 2.4 kW.

It is difficult to make an accurate estimate of the amount of three-phase load connected to a low-voltage feeder, but for long feeders a small amount will allow significantly larger single-phase generators to be connected.

6.3.4 Many Single-Phase Generators

The presence of large numbers of single-phase generators (i.e., up to a few kilowatt) also results in an increase in unbalance in low-voltage networks. Examples are domestic CHP and rooftop solar panels.

When multiple single-phase generator units are connected to the same feeder, they will most likely be randomly distributed over the three phases. The probability that they are all connected to the same phase will decrease quickly with an increasing number of generators. For N units, the probability that they are all in the same phase is equal to

$$\Pr(n = N) = \frac{1}{3^{N-1}} \tag{6.20}$$

The probability that they are all in one specific phase, for example, in phase A, is one-third of this: $1/3^N$. For three units, the probability that they are all in the same phase is, according to (6.20), equal to 11%; for six units, the probability is only 0.4%; and for nine units, this probability is as small as 0.015%. With large numbers of small single-phase generators, there is thus obviously no need to design the network for the situation that all units are connected to the same phase. The units will, however, in most cases not spread equally over the three phases, so some unbalance remains. We will estimate this unbalance below.

The spread of the generators over the phases is illustrated in Figure 6.9. A Monte Carlo simulation has been done to estimate the probability that a specific phase would contain a certain number of generator units. These calculations have been performed when a total of 6 units are connected to the feeder and when a total of 30 units are connected to the feeder. In both cases, 100, 000 simulations were performed to obtain an estimate of the probabilities. For six generators, the number of units connected to one phase varies with 95% confidence between zero and four. For 30 units, this varies between 5 and 15.

With the increasing number of single-phase generator units, the negative-sequence current increases. Note that the current unbalance (ratio of negative- and positive-sequence current) decreases with increasing number of units. What matters to the grid is the absolute value of the negative-sequence current as this determines the negative-sequence voltage. Using the above-mentioned Monte Carlo simulation, the probability distribution function of the negative-sequence current has been calculated. From this distribution, the average value, the standard deviation, and the 95-percentile (the value not exceeded by 95% of the samples) have been calculated and plotted in

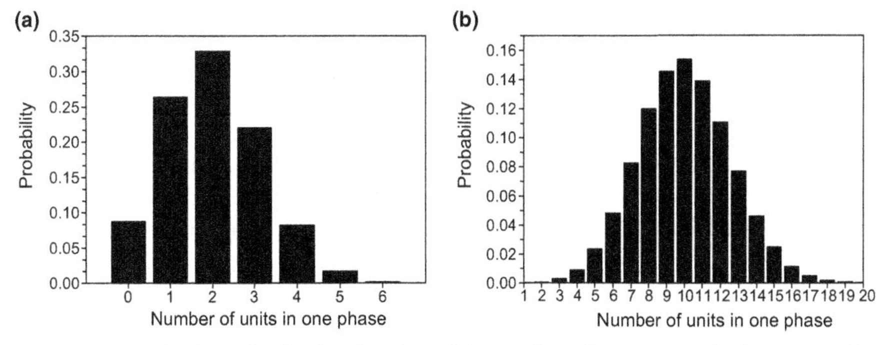

Figure 6.9 Probability distribution function of the number of generator units in one specific phase, for a total of 6 units (a) and for a total of 30 units (b).

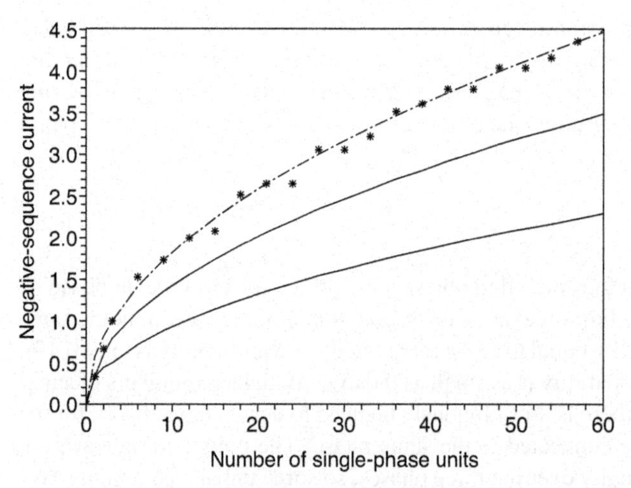

Figure 6.10 Negative-sequence current due to multiple single-phase generator units: average (bottom solid line), average plus standard deviation (upper solid line); 95-percentile (stars); and approximation for the 95-percentile (dash-dotted line).

Figure 6.10. The unit along the vertical scale is the rated current of one generator. The dotted line is an approximation that will be discussed later, resulting in (6.21).

Example 6.7 A total of 30 single-phase units are connected to a three-phase feeder, with each unit consuming 500 W, at 230 V. The current taken by one unit is 2.17 A. From the figure, we find the following values for the negative-sequence current:

- Average: $1.62 \times 2.17 = 3.5$ A
- Average plus standard deviation: $2.47 \times 2.17 = 5.4$ A
- 95-percentile: $3.06 \times 2.17 = 6.6$ A

Exercise 6.4 How many 400 W units can be connected to a three-phase feeder before the probability that the negative-sequence current exceeds 5 A becomes more than 5%?

From a design viewpoint, the 95-percentile value is the most important parameter. When this value is kept below a certain value, the probability that the negative-sequence current exceeds this value is seen as sufficiently small. There is no need to use the worst case (i.e., all units in one phase) in the design. Using the average value on the other hand would be overly optimistic as there is a rather large probability (somewhere around 50%) that the actual unbalance is more than the average value. Using the average value as a performance indicator will, however, allow much more single-phase generators to be connected before investments in the network would be needed. In the forthcoming discussion, we will, however, use the 95-percentile as the value setting the limit.

Example 6.8 Assume that the limit for negative-sequence current is equal to the current taken by one single-phase generator unit. Using the 95-percentile as an indicator would allow

3 units to be connected; using the average value instead would increase this hosting capacity to 12 units.

Exercise 6.5 Redo the calculations in Example 6.8 when the limit is 1.5 and 2.0 times the current taken by one unit.

As the 95-percentile plays an important role in the design, a simple expression for it would be helpful. The following expression gives a good approximation for the 95-percentile of the negative-sequence current.

$$I_{95} = \sqrt{\frac{1}{3}N} \times I \tag{6.21}$$

where N is the number of single-phase generator units and I is the current produced by one unit. This expression can next be used to estimate the hosting capacity. To illustrate the accuracy of this approximation, it is shown as a dotted line in Figure 6.10. In the same way, the following approximation can be used for the average value of the negative-sequence current:

$$I_{\mathrm{mean}} = \sqrt{\frac{1}{12}N} \times I \tag{6.22}$$

and the following one for the standard deviation

$$I_{\mathrm{mean}} = \sqrt{\frac{1}{42}N} \times I \tag{6.23}$$

When discussing the impact of single-phase generator units on the unbalance, it is important to realize these square root relations. The (average or 95-percentile) negative-sequence current is proportional to the square root of the number of units and directly proportional to the size of one unit. This means that 5 units of 1000 W cause the same amount of unbalance as 20 units of 500 W.

The negative-sequence current due to the unequal spread of the single-phase generators over the three phases gives a negative-sequence voltage U_2 over the negative-sequence impedance Z_2. Using the approximation (6.21), we get for the negative-sequence voltage:

$$U_2 = \sqrt{\frac{N}{3}} \times \frac{P_{\mathrm{gen}} \times Z_2}{U} \tag{6.24}$$

From (6.24), the hosting capacity is calculated, that is, the amount of generation $N \times P_{\mathrm{gen}}$ that does not result in an unacceptable value for the negative-sequence voltage. Assume again, like in Section 6.3.3, that the maximum permissible value of the negative-sequence voltage is equal to $u_2 \times (U_{\mathrm{nom}}/\sqrt{3})$, with U_{nom} being the nominal phase-to-phase voltage. The lowest voltage for which this criterion should hold is assumed to be equal to $u \times (U_{\mathrm{nom}}/\sqrt{3})$. This gives the following inequality for number and size of the single-phase generators:

$$\sqrt{N} P_{\mathrm{gen}} < \frac{u u_2}{\sqrt{3}} \times \frac{U_{\mathrm{nom}}^2}{Z_2} \tag{6.25}$$

Example 6.9 Consider a location with a negative-sequence source impedance equal to 0.45 Ω. The negative-sequence voltage should not exceed 1.5% of the nominal positive-sequence voltage (230 V) whenever the voltage magnitude is more than 95% of the nominal voltage. Equation (6.25) gives in this case:

$$\sqrt{N} P_{\text{gen}} < \frac{1}{\sqrt{3}} \times 0.95 \times 0.015 \times \frac{400^2}{0.45} = 2900 \text{ W}$$

This gives the following hosting capacity: 3 units of 1500, or 8 units of 1000 W, or 14 units of 750 W.

Exercise 6.6 What will be the hosting capacity for 750 W units when only 1% negative-sequence voltage is allowed for a voltage magnitude equal to 90% of the nominal voltage?

The calculations resulting in Table 6.2 have been repeated assuming 500 W single-phase units instead of one large single-phase unit. The results are shown in Table 6.3 for a 400 V system. It has been assumed that 1% negative-sequence voltage is allowed at nominal voltage. In weak systems, less than three units can be connected. In this case, the hosting capacity for small units is the same as for large units.

From (6.25), we get the number of units of size P_{gen} that can be connected:

$$N < \frac{u^2 u_2^2}{3 P_{\text{gen}}^2} \times \frac{U_{\text{nom}}^4}{Z_2^2} \tag{6.26}$$

From this equation, the following rules can be derived to calculate the hosting capacity for other parameters than the ones used in Table 6.3:

- The hosting capacity increases with the square of the permissible amount of negative-sequence voltage. When 1.5% negative-sequence voltage is allowed, the hosting capacity will increase by a factor of 2.25.
- The hosting capacity is also proportional to the square of the lowest nominal voltage for which the requirement should hold. Requiring the negative-sequence voltage to be below 1.0% for voltage magnitudes down to 0.9 per unit would reduce the hosting capacity to 81% of the values in the table.

TABLE 6.3 Hosting Capacity for 500 W Single-Phase Generators: 1% Maximum Voltage Unbalance at Nominal Voltage

	400 kVA underground	200 kVA underground	100 kVA overhead	50 kVA overhead
0.5 km	186 kW	126 kW	31 kW	17 kW
1 km	58 kW	46 kW	11.5 kW	7.5 kW
2 km	–	14.5 kW	3.5 kW	2.5 kW
3 km	–	7.0 kW	1.5 kW	<1.5 kW
4 km	–	–	<1.5 kW	<1.5 kW
5 km	–	–	<1.5 kW	<1.5 kW

- The hosting capacity (in kilowatt) is inversely proportional to the square of the generator size. For 700 W units, the hosting capacity is 51% of the values in the table.
- When the hosting capacity becomes less than three times the size of a single unit, the values in Table 6.2 should be used instead.

6.4 LOW-FREQUENCY HARMONICS

Distributed generation does not produce a completely sinusoidal current waveform, just like the majority of other equipment. The harmonics injected into the distribution grid by the generator will result in some increase in the voltage distortion. The emission by distributed generation is, however, smaller than the emission by modern consuming equipment, so the increase in voltage distortion will be small and rarely a problem. This will especially be the case for those frequency components that have traditionally been dominant in power systems: harmonics 5, 7, 11, 13, 17, 19, and so on, as well as 3, 9, and 15 at low voltage. The presence of distributed generation may, however, result in a significant increase in the level of frequency components that have traditionally been small, even harmonics, higher order triplen harmonics, and interharmonics. The permissible levels for these frequency components have traditionally been low, so the hosting capacity could actually turn out to be rather low. Also, these frequencies have traditionally been used for power line communication (PLC), based on the fact that the disturbance levels were low here. Increasing amounts of distributed generation could, therefore, result in interference with power line communication. The interaction between the new equipment and the power line communication is, however, more complicated, depending on the distortion level. See Refs. 370 and 371 for a more discussion on this.

The harmonic emission of distributed generation is low for the frequencies that have traditionally been of concern for harmonics (third, fifth, seventh, up to about 1 kHz). A number of measurement examples and simulation results will be discussed below. The results from the different sources are summarized in Table 6.4.

In Ref. 27, an assessment is made of the hosting capacity when harmonic voltage distortion is the limiting factor. A maximum permissable voltage distortion of 3% (for each harmonic) is assumed, in accordance with IEEE standard 519 [217]. It is further assumed that the emission of a generator unit is equal to the limit in IEEE standard 519 [217] for large customers (short-circuit ratio less than 20). Three typical medium-voltage feeders of different length are considered. A distinction is further made between generation distributed uniformly along the feeder, concentrated at the beginning of the feeder, and concentrated at the end of the feeder. The limits are reached first for triplen harmonics (9, 15, and 33 to be more specific). The results are summarized in Table 6.5, as a percentage of the feeder capacity. High harmonic voltage levels are expected first for triplen harmonics at long feeders when the generation is concentrated toward the end of the feeder.

When interpreting Table 6.5, it should be kept in mind that the actual emission of distributed generators was used nowhere during the analysis. Instead, it was simply

TABLE 6.4 Harmonic Emission from Distributed Generation: Summary of Different Sources

Sources	Description	THD (%)	Individual harmonics
31	37 and 110 kW solar power installation	3.5	
109	30 kW wind turbine with capacitor bank	5–6	
133	800 kW wind turbine	1	up to 0.4% for 3, 5, 7, 11, 13, and 16
149	650 kW full converter		4% at order 29 and 31
222	Inverter typical for microgeneration	3–6	3% second harmonic; 1.5%, third, fifth, and seventh
257	30 kW wind turbine	1.5–4	
277	DFIG		1% slot harmonics; 3% at 220, 250, 320, and 350 Hz
279	660 kW induction machine	6–7	
399	Wind turbine with full converter		1.6% fifth harmonic, 0.6% seventh harmonic

assumed that the emission is exactly equal to the limit set by IEEE standard 519 for large customers. Without further information about the actual emission, it is difficult to know if these limits are conservative or nonconservative. However, if the network operator imposes the IEEE standard 519 limits on the generators, there will rarely be high levels of voltage harmonics due to distributed generation.

6.4.1 Wind Power: Induction Generators

In Ref. 109, measurements are shown of the harmonic emission of a 30 kW wind turbine together with a capacitor bank. The current total harmonic distortion (THD) is between 5% and 6% with harmonics 5, 11, and 13 being the dominant ones. The current distortion is measured between the grid and the capacitor bank. The main harmonics are probably due to the capacitor bank. The harmonic voltage distortion was about 2.5% close to the capacitor bank, with the fifth harmonic being by far the dominant component.

Measurements of the harmonic emission of a 30 kW wind turbine in a laboratory environment [257] show a THD between 1.5% and 4% for the current to the wind

TABLE 6.5 Hosting Capacity According to Ref. 27

Length	Triplen			Non triplen		
(miles)	Uniform (%)	Beginning (%)	End (%)	Uniform (%)	Beginning (%)	End (%)
2 m	> 82	> 100	> 64	> 100	> 100	> 100
5 m	37–93	> 52	28–72	> 99	> 100	> 83
10 m	19–50	28–72	14–38	> 62	> 83	49–91

TABLE 6.6 Typical Harmonic Currents Produced by a Wound Rotor Induction Motor [16]

Frequency (Hz)	Current (% fundamental)	Cause
20	3.0	Pole unbalance
40	2.4	Rotor-phase unbalance
50	100	Fundamental
80	2.3	Pole unbalance
220	2.9	5th and 7th harmonic components
320	3.0	5th and 7th harmonic components
490	0.3	11th and 13th harmonic components
590	0.4	11th and 13th harmonic components

turbine. The harmonic distortion increases from 1.5% for zero wind up to 4% for wind speeds around 12 m/s.

In Ref. 277, measurements are shown of the harmonics in the stator current and rotor current of a double-fed induction generator (DFIG) machine. The spectrum is shown for frequencies up to 2 kHz. According to Ref. 277, the switching frequency of the converter is sufficiently high, so the resulting harmonics fall outside this frequency range. An interesting observation of this paper is the presence of so-called "slot harmonics" in the stator current due to the finite number of slots in the induction machine. The measurements show that for 10% slip, these harmonics appear at 1030 and 1130 Hz; whereas for −10% slip, they appear at 1270 and 1370 Hz. In both cases, the magnitude of these components is about 1% of the fundamental current. At lower frequencies, harmonics appear at 220, 250, 320, and 350 Hz with a magnitude of about 3%.

Measurements of the harmonic distortion due to one and four 660 kW turbines are shown in Ref. 279. All four are identical induction generators. At full power production, the THD of the current is about 4% for one turbine and about 3.5% for the four turbines together. The current of one turbine shows peaks in THD up to about 16%. This occur for low power production. An estimated translation to rated values (based on the graphs in Ref. 279) gives that the THD values can be up to about 7% of the rated current. For the four units together, the highest THD values are about 6% of the fundamental current.

The spectrum of a modern 800 kW wind turbine is given in Ref. 133. The harmonics 3, 5, 7, 11, 13, and 17 are present at 0.3–0.4% of the rated current. The other harmonics are 0.1% or lower.

Typical current components produced by a wound rotor induction motor are given in Table 6.6 [16]. The values hold for a six-pole machine operating at 10% slip. They are judged to be applicable also for induction generators. Current source type interharmonics appear in the frequency range of 750–1000 Hz, due to rotor slots in four- and six-pole induction generators rated tens of kilowatt.

6.4.2 Generators with Power Electronics Interfaces

Simulations of the harmonic emission due to different switching patterns with a voltage source converter are presented in Ref. 149. A 650 kW unit connected to 6 kV has been modeled. The harmonic voltage distortion up to order 40 is shown. For hysteresis control, the highest distortion (about 1.2%) occurs for harmonic 40. A broadband emission is visible from order 20 upward. The spectrum clearly shows an increasing trend toward higher frequencies, but unfortunately harmonics only up to order 40 are shown. As the maximum switching frequency is 5 kHz, emission above order 40 is expected to be even higher. For sinusoidal PWM (with switching frequency 1.5 kHz), the harmonic voltage distortion is highest (around 4%) at order 29 and 31, that is close to the switching frequency.

According to a measurement example shown in Ref. 399, a wind turbine with power electronics converter shows a current distortion of 1.6% at fifth harmonic and 0.6% at seventh harmonic. All other harmonics are below 0.2%.

Measurements of the harmonic emission due to solar power installations of 37 and 101 kW are presented in Ref. 31. The dominant harmonics are of order 3 and 5, with levels equal to about 2% of rated current. The measured total harmonic distortion of the current is about 3.5% of the rated current.

Measurements are shown in Ref. 222 of the harmonic distortion of an inverter that is expected to be typical for future PV or microgeneration installations. The second harmonic, with 3%, is the highest; harmonics 3, 5, and 7 are around 1.5% and harmonic 9 is just below 1%. The high level of second harmonic was traced back to an unnecessarily high distortion in the lookup table used to generate the reference sine wave. Reference 222 further shows that the background voltage distortion has a big influence on the current distortion. The THD of the current was shown to increase from slightly above 3% for a clean supply to around 6% for a supply with a normal voltage distortion. The interaction between inverters was also shown in the paper. Up to six inverters in parallel, the THD remained between 2.5% and 3%; but for seven and eight inverters in parallel, the THD became 4% and 5%, respectively.

Interharmonics due to power electronics converters are mainly due to frequencies that occur on the nongrid side of the converter. The coupling between the two sides of the converter is strongest for cycloconverters, but even two converters connected via a DC link will transfer frequency components. Resonances in the DC link may even amplify certain interharmonic frequencies. Considering six-pulse input as well as output stage (the most typical configuration) results in a spectrum with the following frequencies being present:

$$f = (6m \pm 1) f_{\text{in}} \pm 6n f_{\text{out}} \tag{6.27}$$

where f_{in} is the grid frequency, f_{out} is the nongrid-side frequency, $n = 0, 1, 2, \ldots$, and $m = 1, 2, \ldots$. The input spectrum contains the characteristic harmonics of the power system frequency (5th, 7th, 11th, 13th, etc.) with sidebands due to the characteristic frequencies (6th, 12th, etc.) of the nongrid-side frequency. Examples of small generator units that possibly lead to interharmonics are wind turbines (double-fed or with DC link) and microturbines. There are indications that active power electronics converters (IGBT or GTO based) generate more of a continuous spectrum than the existing power electronics converters. This will lead to more interharmonics and an increased risk of high distortion due to resonances.

Certain types of power electronics converters may amplify interharmonics that are present in the voltage at the converter terminals. A well-known effect of interharmonic voltage components is that they result in a modulation of the voltage amplitude, which in turn may lead to light flicker [86, 121, 240]. An interharmonic voltage component may also lead to phase modulation, which shows up as a slow oscillation in the apparent frequency of the voltage. The phase-locked loop (PLL) may adjust the frequency of the injected current to follow the apparent oscillation in voltage frequency. The result is that the current contains interharmonic components as well. Such a phenomenon could occur for generator units with power electronics interface.

Another concern with interharmonics is the possible interference with power line carrier systems for controlling load. Such a kind of communication via the grid may be a way of starting or stopping generator units in a certain part of the system. High levels of interharmonics may interfere with PLC systems. PLC systems, on the other hand, have the same impact on equipment as interharmonics, including the generation of additional current interharmonics, as already mentioned in the previous paragraph. The impact of waveform distortion on power line communication remains a point of discussion. According to some, this is the main cause for interference between end-user equipment and power line communication. According to others, the main cause is the low-impedance path due to the presence of capacitances at the interface of the end-user equipment.

The spectra of a number of modern wind turbines (with rating up to over 1 MW) are shown in Ref. 403 and 404. Interestingly, the spectra do show rather different spectra. Three machines with full power converter have been investigated. One of them shows a broadband spectrum with an amplitude of about 0.4% of rated current (per 50 Hz band) around 2.5 kHz, whereas the other two show a spectrum dominated by the characteristic harmonics of a six-pulse converter (5, 7, 9, 11, and 17 in both cases) superimposed on a lower broadband spectrum. The amplitude of the discrete harmonics is in both cases around 1%. The spectra for two double-fed induction generators are also shown: one showing mainly a broadband spectrum up to about 2 kHz, while the other showing almost exclusively fifth and seventh harmonic distortion.

6.4.3 Synchronous Generators

When using synchronous generators, their design should be such that they produce little third harmonic distortion. The low-voltage system is almost a short circuit for triple harmonics, which would cause large third harmonic currents with standard synchronous generators (where fifth and seventh harmonic components are minimized). The standard way of designing the winding of synchronous machines causes a relatively large third harmonic voltage: about 10% of the fundamental component, whereas the fifth and seventh harmonic voltages are around 1% [430]. As the third harmonic component has a zero-sequence character, it is not transferred through the Dy-connected generator transformer. However, when connecting a rotating machine directly to the low-voltage network (i.e., without a transformer) or through a star–star-connected transformer (as is the requirement with many U.S. utilities [123]), the third harmonic distortion may spread through the system. A high penetration of such generators may lead to excessive currents through neutral wires and delta windings in low- and medium-voltage networks.

Another source of harmonic distortion related to synchronous machines is mentioned in Ref. 16. In a balanced system, the fifth harmonic component has a negative-sequence character and the seventh harmonic component a positive-sequence character. Both are linked to a sixth harmonic component in the rotor current. Therefore, a fifth harmonic current will induce both a fifth and a seventh harmonic voltage. The same holds for a seventh harmonic current. The situation gets more complicated for unbalanced loading of the machine. For example, a negative-sequence fundamental frequency current induces a negative-sequence fundamental frequency voltage as well as a positive-sequence voltage at the third harmonic. The effect is normally neglected; but in special cases when either the load is asymmetrical or the generator feeds static converter equipment, the machine can be an important source of harmonic generation [16]. Simulations as well as experimental results in Ref. 4 show that a synchronous machine with a linear but unbalanced load can produce a significant amount of harmonic distortion, especially the third and fifth harmonics. The situation could become even more severe due to resonance between the winding inductance and load capacitance.

6.4.4 Measurement Example

Measurements have been performed on a number of wind turbines in northern Sweden [422]. The spectrum has been calculated for the currents taken by a group of three 600 kW turbines with full power converter. The measurements were performed at 10 kV side of the current from the three turbines together. The waveform during a 20 ms period is shown in Figure 6.11. The spectrum obtained by applying a 200 ms rectangular window is shown in Figure 6.12.

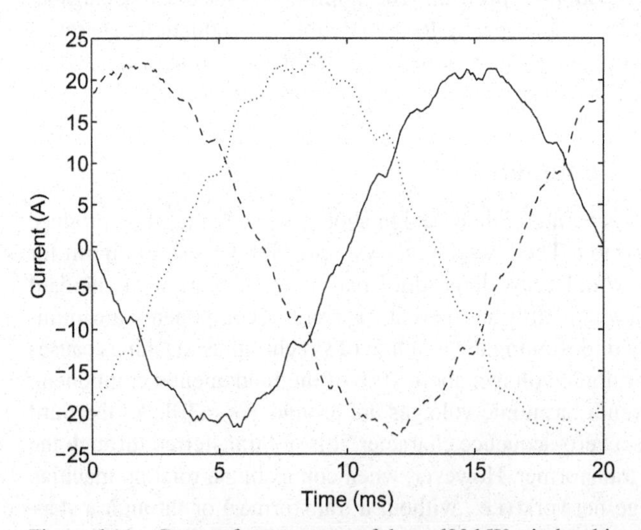

Figure 6.11 Current from a group of three 600 kW wind turbines with full power converter.

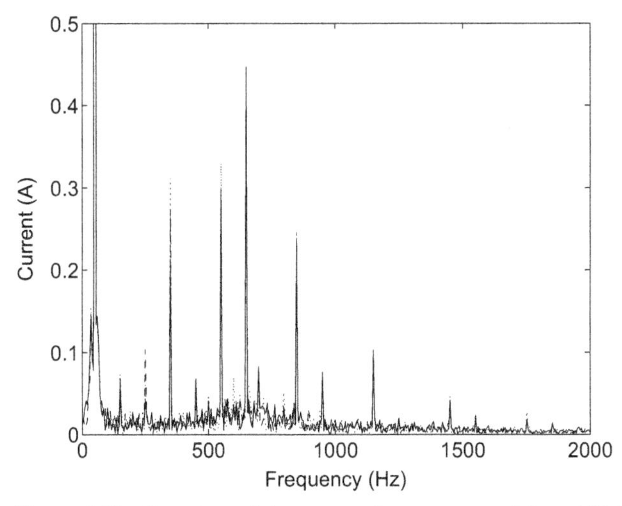

Figure 6.12 Spectrum of the current from a group of three 600 kW wind turbines with full power converter.

The spectrum shows the dominant 50 Hz component (14.9, 15.2, and 15.4 A in phase A, B and C, respectively, far beyond the upper limit of the vertical scale), next to which a limited number of discrete harmonics are visible. The main discrete harmonics are of order 13 (650 Hz), 11 (550 Hz), 7 (350 Hz), and 17 (850 Hz). Some other discrete harmonics are visible, but the most important feature of the spectrum is the broadband emission that covers the whole frequency range with a flat maximum somewhere around 700 Hz. This broadband spectrum continues beyond 2 kHz with a second flat maximum around 4 kHz.

The variation over time of the four dominant harmonics is shown in Figure 6.13 for a 24 h period. The figures show that not only the harmonic distortion varies strongly over time but also the spectrum changes with time. It is not always the same harmonic that dominates. Harmonic 13 is the highest most of the time, but harmonic 11 shows the overall highest values.

The two strongest interharmonic subgroups are shown as a function of time in Figure 6.14. The interharmonic subgroups represent the broadband spectrum visible in Figure 6.12. The two interharmonics are strongly correlated. Also the other interharmonic subgroups vary with time in a similar way.

The harmonic emission shows only a minor variation with variations in power production. This is illustrated in Figure 6.15 where the vertical scale gives the rms value of the harmonic distortion, that is, the root sum square of all the harmonic components. Several publications show that the waveform distortion is very high for low power production. This result is obtained when the THD is used, where the root sum square value is divided by the fundamental component of the current. As the latter is small when the produced power is small, high THD values result. From a network viewpoint, it is however the "emission in ampere" that is relevant.

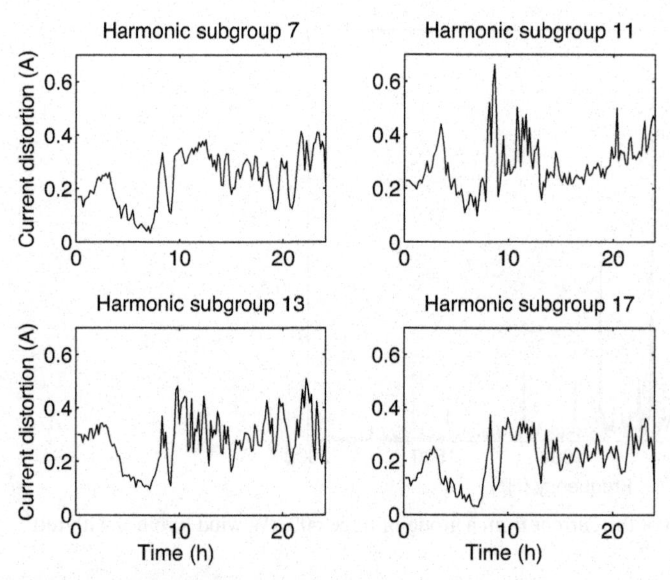

Figure 6.13 Harmonic distortion for a group of three wind turbines as a function of time: harmonic subgroups 7, 11, 13, and 17.

6.4.5 Harmonic Resonances

A potential problem with high penetration of distributed generation is the occurrence of resonances due to the increased amount of capacitance connected to the distribution grid. The interface with distributed generation often contains a capacitor.

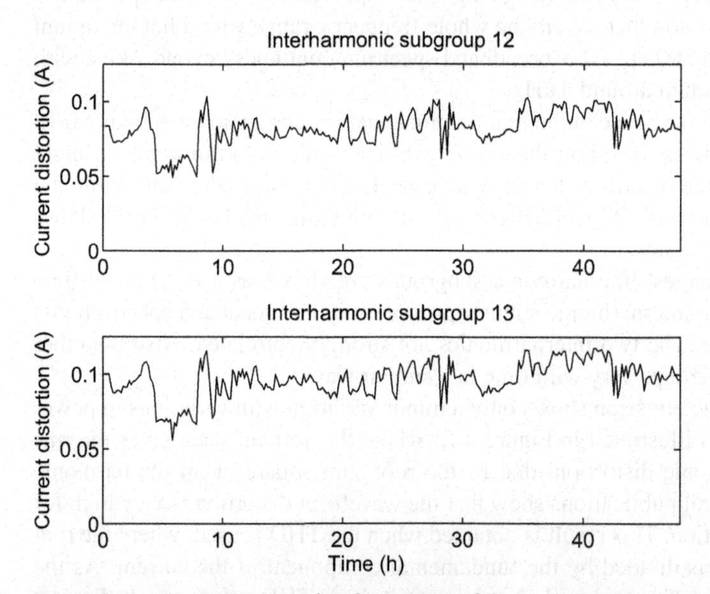

Figure 6.14 Harmonic distortion for a group of three wind turbines as a function of time: interharmonic subgroups 12 and 13.

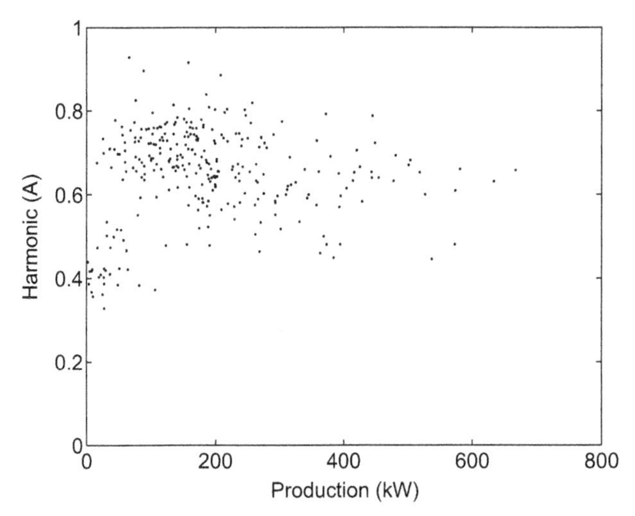

Figure 6.15 Correlation between wind power production and harmonic emission.

This capacitor may be involved in series or parallel resonances that cause amplification of harmonic distortion produced elsewhere [396]. Manufacturers as well as consultants recommend that capacitor banks be installed with small AC reactors and to be tuned to a resonance frequency lower than the lowest harmonics in the actual grid.

The increased capacitance is, however, *not* the source of the harmonic distortion. The harmonic currents may be generated by the units themselves, by other local equipment, or by equipment elsewhere. As the current generated by the units does not contain significant low-frequency harmonics, the source is most likely found in other local equipment or in equipment elsewhere. Despite this, the potential increase of harmonic voltage distortion should be treated as an impact of increasing levels of distributed generation.

6.4.5.1 Parallel and Series Resonances A harmonic *parallel resonance* occurs when a harmonic current source sees the parallel connection of a capacitance and an inductance. Consider the situation shown in Figure 6.16. A generator is connected to a low-voltage feeder, with a capacitor close to the point of connection. The generator could be an induction machine or a generator with power electronics interface. Any harmonic emission originating downstream of the point of connection

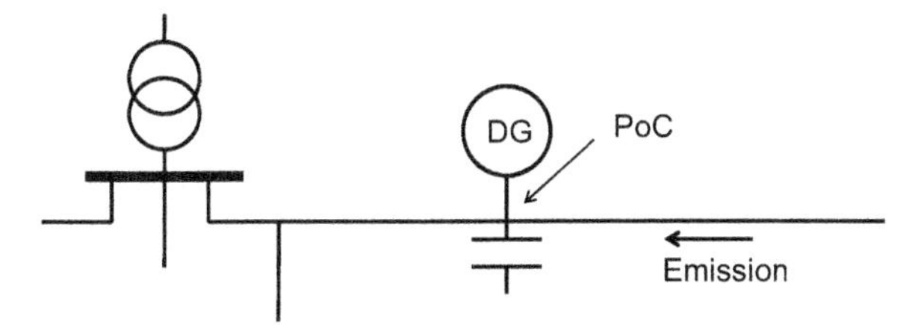

Figure 6.16 Distributed generator connected to low-voltage network, possibly resulting in harmonic resonance.

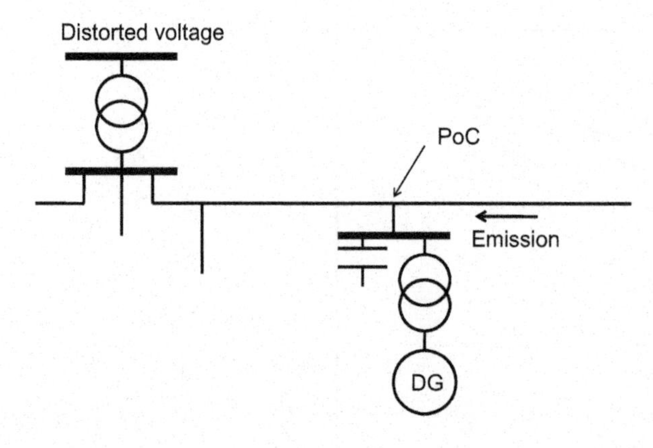

Figure 6.17 Example of series resonance for a distributed generator connected to a medium-voltage network.

will see the parallel connection of the capacitor and the inductive part of the source impedance.

The harmonic voltage at the point of connection is equal to the product of the harmonic current and the impedance at the harmonic frequency. This impedance has a maximum at the resonance frequency. If this resonance frequency corresponds to a frequency that is present in the emission, high harmonic voltage levels can occur.

A *series resonance* occurs when the local capacitance is in resonance with the inductance between the local bus and a remote bus with a high harmonic voltage distortion. As a result of the series resonance, the harmonic voltage distortion at the local bus could get high. A network configuration where high harmonic levels due to series resonance may occur is shown in Figure 6.17. The distributed generator is connected via a generator transformer to a medium-voltage feeder. A capacitor is located close to the point of connection to compensate for the reactive power consumed by the generator.

The harmonic voltage at the remote bus with distorted voltage, indicated in the figure, may be amplified by the series resonance. The capacitance involved in the series resonance is the capacitor at the point of connection. Cables and lines in the neighborhood also add to the capacitance; but at distribution levels, fixed capacitor banks dominate. The inductance involved in the series resonance is the inductance between the point of connection and the bus with the distorted voltage. At medium-voltage level, for locations not too far away from the main medium-voltage bus, the transformer impedance dominates. Let S_{tr} be the transformer rating, ϵ the transformer impedance, and Q the size of the capacitor bank. The resulting resonance frequency is equal to

$$f_{\mathrm{res}} = f_0 \times \sqrt{\frac{S_{\mathrm{tr}}}{\epsilon Q}} \tag{6.28}$$

Example 6.10 Consider a weak medium-voltage network, supplied by a 10 MVA, 10% transformer. A 2 MW induction generator (magnetizing reactance 4.1 per unit; leakage reactance 0.17 per unit) is connected to this medium-voltage network through a 2 MVA, 6% generator transformer. A switched capacitor bank is installed at the medium-voltage feeder close to the point of connection to compensate for the reactive power consumption of the generator.

The reactive power at no load is $10/4.1 = 2.44$ Mvar. The reactive power at full load is $2.44 + (0.17 + 0.06) \times 10 = 4.74$ Mvar. The capacitance connected to the medium-voltage network will thus range between 2.44 and 4.74 Mvar. Using (6.28) gives resonance frequencies between 230 Hz (for 4.74 Mvar) and 320 Hz. There is thus a risk for a fifth harmonic resonance.

Exercise 6.7 For which size of capacitance will resonance occur at the fifth harmonic? What conclusion do you draw from this for the design of the capacitor bank?

The source impedance at medium-voltage level is mainly determined by the transformer impedance. As a result, the series resonance occurs at about the same frequency as the parallel resonance. A high level of harmonic voltage distortion may, therefore, be due to an amplification of emission downstream through parallel resonance or amplification of voltage distortion at a higher voltage level through series resonance.

6.4.5.2 *Practical Examples*

Various harmonic resonance issues are studied in Ref. 146 for a housing estate with a large number of photovoltaic (PV) inverters. The housing estate contains about 200 houses with photovoltaic installation, fed from an underground cable network. Measurements and simulation show that both parallel and series resonances occur, with resonance frequencies between 250 and 2000 Hz. The resulting high voltage distortion at the terminals of the inverters resulted in tripping of the inverter or in high distortion of the inverter current. Reference 146 gives the following values for the capacitance that should be considered for resonance studies: 0.6–6 μF per household; 0.5–10 μF per inverter. As the range is rather large, it will be difficult to determine resonance frequencies without measurements.

A real-world case of harmonic resonance due to the capacitors with induction generators is described in detail in Ref. 19. The measurements show high levels of 11th harmonic voltage in a wind park with 50 induction generators of 500 kW each. A capacitor of 100 kvar is connected to the 690 V side of the turbine transformer. The result is a high level of voltage distortion that interfered with the turbine controller and with protection. The harmonic resonance was found only when searching for the cause of speed instabilities and protection maloperation. The 630 kVA, 5% turbine transformer corresponds to a fault level of 12.6 MVA at the generator terminals, neglecting the source impedance at medium voltage. Together with the 100 kvar capacitor, this gives a resonance frequency of 561 Hz. This is close to the 11th harmonic that was high during the measurements. The high levels of harmonic voltage distortion were observed only for power levels below 25% of rated power. This is most likely a series resonance, where the series connection of an inductance and a capacitance attracts harmonics from a higher voltage level.

Another example of resonance in a wind park is described in Ref. 273. A wind park consists of 100 double-fed induction generators of 2 MW rated power each. To the main medium-voltage bus is connected 72 Mvar capacitance, switchable in steps of 12 MW. The harmonic emission of a generator at the low-voltage bus is very low, as shown in Table 6.7. Despite the low emission of the individual generator units, high

TABLE 6.7 Harmonic Emission of a DFIG Machine [273]

Harmonic	Current (%)
5	0.5
7	0.54
11	0.21
13	0.10
17	0.13
19	0.19
23	0.19
25	0.02

harmonic voltage distortion occurs in the medium-voltage network due to harmonic resonance, as shown in Table 6.8. The resonance frequency and highest voltage harmonic are given for different amount of capacitance connected to the medium-voltage bus. The values in the first three rows have been obtained from Ref. 273. The values for 36 Mvar and higher have been calculated by extrapolating from the value for 24 Mvar, assuming that the resonance frequency is inversely proportional to the square root of the capacitor size. No simulation results for larger capacitance values are given in Ref. 273.

The simulations have been repeated in Ref. 273 when adding a suitable tuning reactor to the capacitor banks. The result is that the harmonic voltage distortion is significantly reduced, with the THD being less than 2%. As is clearly shown in this paper, the risk of harmonic resonance is higher when capacitor banks can be switched in multiple steps. The number of possible resonance frequencies is equal to the number of possible combinations. In this example, six identical capacitors were used, resulting in seven possible combinations. When the capacitor sizes are chosen to obtain maximum control (2.25, 2.25, 4.5, 9, 18, and 36 Mvar), 73 different resonance frequencies result and some kind of filtering is certainly needed.

An example showing the complexity of harmonic resonances is presented in Ref. 374. A prototype 5 MW, 3.3 kV synchronous machine with full power converter

TABLE 6.8 Harmonic Resonance Frequency and Harmonic Voltage Distortion, according to Ref. 273

Capacitance (Mvar)	Resonance order	Highest voltage harmonic
0	13	6.3% (harmonic 13)
12	8.9	6.0% (harmonic 7)
24	7.1	9.3% (harmonic 7)
36	5.8	–
48	5.0	–
60	4.5	–
72	4.1	–

is connected to the 110 kV grid through a 110/20 kV transformer, 2700 m of 20 kV cable, and a 20/3.3 kV transformer. A 7th and 11th harmonic-tuned filter is located on 20 kV side of the generator transformer. Measurements of the current through the cable showed that the sixth harmonic was dominating with a value up to 10% of rated current. This was traced back to the resonance between the seventh harmonic filter and the 110/20 kV transformer. Simulations showed that the current shows an amplification by a factor of 26 at a frequency of 298 Hz. In other words, only 0.4% emission by the converter results in 10% current being injected into the grid. As mentioned before, an active power electronics converter injects a broad spectrum, so that resonances are easily excited.

The resonances due to the cable connecting a large offshore wind park to the 150 kV grid are discussed in Ref. 443. Simulations have been performed of the 220 MW wind park, the 100 km underground cable to connect the wind park to the 150 kV grid, the local 150 kV grid, and the connection to the 400 kV grid. The 150 kV cable is fully compensated by a 75 Mvar shunt reactor. Seen from the wind side of the cable, the parallel resonance frequency of the grid is only 140 Hz.

Example 6.11 The simulations presented in Ref. 443 show a source impedance of 650 Ω for the third harmonic, seen at the wind-site of a 100 km cable connecting an offshore wind park to the 150 kV grid. The rated power of the installation is 220 MW, which corresponds to a rated current of 850 A at 150 kV. A third harmonic current equal to 1% of rated power, for example, due to the magnetizing current for a mild overvoltage would result in $0.01 \times 850\,\text{A} \times 650\,\Omega = 5.5\,\text{kV}$ of third harmonic voltage, which corresponds to 6.5% of the nominal phase-to-neutral voltage.

6.4.5.3 *Induction Generators*

Consider an induction generator with rated power S_{gen} and magnetizing reactance x_{m} (in per unit with the generator rating as base). The reactive power taken by the induction generator at nominal voltage is equal to

$$Q = \frac{S_{\text{gen}}}{x_{\text{m}}} \tag{6.29}$$

A capacitor bank is connected to compensate for reactive power consumption at no load. The size of the capacitor bank is thus also given by (6.29). The induction generator and the capacitor are connected to a grid with short-circuit capacity S_k and nominal voltage U_{nom}. The resonance frequency is obtained from the following simple expression [45, p. 154]:

$$f_{\text{res}} = f_0 \times \sqrt{\frac{S_k}{Q}} \tag{6.30}$$

where f_0 is the power system frequency (50 or 60 Hz). Inserting (6.29) in (6.30) gives the following expression for the resonance frequency:

$$f_{\text{res}} = f_0 \times \sqrt{\frac{S_k}{S_{\text{gen}}} \times x_{\text{m}}} \tag{6.31}$$

which can be further simplified by introducing the short-circuit ratio, $\mathrm{SCR} = S_k/S_{\mathrm{gen}}$:

$$f_{\mathrm{res}} = f_0 \times \sqrt{\mathrm{SCR} \times x_{\mathrm{m}}} \tag{6.32}$$

The resonance frequency decreases with decreasing short-circuit ratio (i.e., with weaker network at the point of connection) and with decreasing magnetizing reactance (i.e., with higher reactive power demand in no load). Note that the lower the resonance frequency, the higher the risk of high harmonic voltages. We first take two examples to get some impression of the order of magnitude of the resonance frequencies.

Example 6.12 A 200 kVA induction generator with magnetizing impedance 3.9 pu is connected to a strong point in a 50 Hz low-voltage network: at the secondary side of an 800 kVA distribution transformer with 5% impedance. Neglecting the impedance of the medium-voltage network, the short-circuit impedance at the point of connection is 16 MVA, resulting in a short-circuit ratio equal to 80. Using (6.31) gives for the resonance frequency,

$$f_{\mathrm{res}} = 50 \times \sqrt{\frac{16\,\mathrm{MVA}}{200\,\mathrm{kVA}}} \times 3.9 = 883\,\mathrm{Hz}$$

Resonance may occur for the 17th harmonic at 850 Hz.

Example 6.13 Assume that the induction generator in Example 6.12 is connected to an overhead feeder at 500 m from the secondary side of the distribution transformer. Due to the feeder impedance, the short-circuit capacity has dropped to 1000 kVA, so the short-circuit ratio equals 5. Using (6.32) results in the following resonance frequency:

$$f_{\mathrm{res}} = 50 \times \sqrt{5 \times 3.9} = 221\,\mathrm{Hz}$$

The fifth harmonic at 250 Hz is prone to resonance in this case.

When the resonance frequency is close to a harmonic frequency, this can result in high levels of harmonic voltage distortion. The curves in Figure 6.18 indicate the combinations of short-circuit ratio and magnetizing reactance that result in resonance frequencies for odd harmonics, that is, for the frequencies that are present in the emission by common low-voltage load.

The main concern are resonance frequencies at 250 and 350 Hz as these are where the main emission takes place for most equipment. These can occur for short-circuit ratios less than 15. When an induction generator is connected to a location with short-circuit ratio less than 15, it is worth to check if any increase in fifth or seventh harmonic voltage can be expected. The presence of a resonance frequency close to these harmonics does not necessarily cause high harmonic voltages. A source of harmonic emission should be present to excite the resonance and the damping should be sufficiently low. In low-voltage networks, damping is present in the form of feeder resistance and ohmic loads.

In medium-voltage networks, induction generators are connected through a transformer. The capacitor to compensate for the reactive power consumption at no load can be found at the low-voltage side (i.e., with the generator) or at the medium-voltage side (i.e., at the point of connection) of this transformer. In both cases, (6.32)

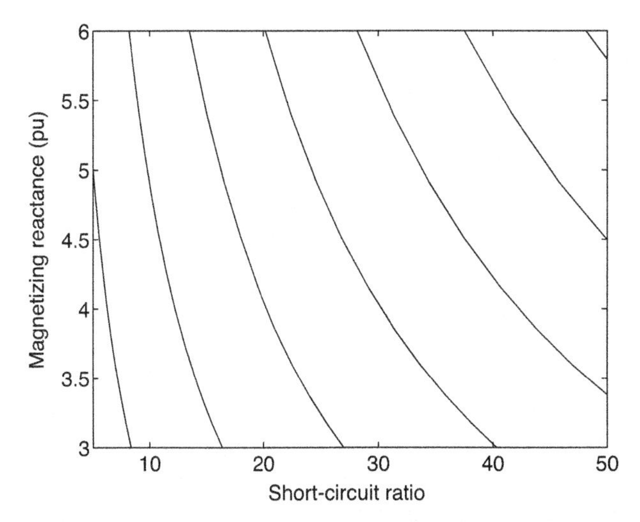

Figure 6.18 Harmonic resonances at (left to right) 250, 350, 450, 550, 650, 750, and 850 Hz, as a function of the short-circuit ratio (SCR) and the magnetizing reactance of an induction generator.

still holds. For a capacitor at low-voltage side, the short-circuit ratio at the generator terminals should be used in the expression. This ratio is lower than that at the point of connection, hence resonances occur at lower frequencies. The resonance can be excited only by the emission from the induction machine, which is normally low (see Table 6.6). We, therefore, do not expect high levels of harmonic voltage distortion due to parallel resonance in this case. Series resonance does however occur at similar frequencies and could result in high harmonic voltage distortion at the machine terminals.

Exercise 6.8 Consider a generator connected to a medium-voltage network with a fault level equal to 100 MVA. The generator transformer has an impedance of 6.5% and a rating equal to the generator rating. The magnetizing reactance is equal to 4.2 pu. At which generator ratings does resonance occur at 250 and 350 Hz. Perform the calculations for a capacitor connected at medium-voltage and low-voltage sides of the generator transformer.

When the capacitor is connected at medium-voltage side, the short-circuit ratio is higher, so resonances occur at higher frequencies. But the emission is also higher: all load downstream of the point of connection contributes. Also the damping is lower than that for connection of the generator to the low-voltage network. Induction generators of a few megawatt are connected to the medium-voltage network. In weak parts of the network, short-circuit ratios lower than 15 could result from such a connection, especially when multiple machines are connected to the same location.

Example 6.14 A 2.5 MVA induction generator, used as a wind turbine, is connected on secondary side of a 10 MVA, 130/15 kV transformer with 10% impedance. The fault level on secondary side, neglecting the impedance of the 130 kV grid, equals 100 MVA. The short-circuit ratio for the connection of this generator is 40. From Figure 6.18, we see that resonance frequencies are around the 11th harmonic. With domestic load, the emission at these frequencies is small. But when an industrial installation is connected to the same distribution grid, a further investigation of the harmonic situation is recommended.

Exercise 6.9 What resonance frequencies are to be expected when a 2 MVA unit is connected at the same location as the unit in Example 6.14? What could be the possible emission sources exiting this resonance?

Example 6.15 The induction generator from Example 6.14 is connected at 15 km from the transformer. The feeder reactance is 275 mΩ/km. The resulting fault level at the point of connection is 35.3 MVA, with a short-circuit ratio of 14. From Figure 6.18, resonances around the seventh harmonic could occur.

We see from the examples that resonances around the 11th or 13th harmonic are likely to occur. These could give high harmonic voltages when industrial installations are connected downstream of the induction generator. Resonances around the fifth or seventh harmonic are likely only when the generator is connected to a part of the network with short-circuit ratio being 15 or less. In such cases, the connection may not be possible for other reasons, as discussed in other chapters. Also, the load present at remote locations, and thus the harmonic emission, will normally be small.

The calculations in this section have all been based on the assumption that only the magnetizing current is compensated by the capacitor bank. In some cases, a switched capacitor bank is also used to keep the reactive power exchange between strict limits for high power production. For high production levels, a large amount of capacitance is connected to the grid, resulting in resonance frequencies lower than the ones calculated here.

6.4.5.4 ***Cables at Transmission Level*** New harmonic resonances may also be introduced by long underground cables connecting a wind park to the transmission grid [95]. To calculate the impedance of a cable as a function of frequency, a complicated set of equations with so-called “hyperbolic functions” is needed; alternatively, a suitable power system analysis package can be used (this package should be able to model the cable correctly even for higher frequencies). The impedance on the grid side of the cable depends, among others, on the length of the cable and on the impedance with which the cable is terminated. When the cable is short (shorter than one-quarter of a wavelength is an often used criterion) and when the cable is lightly loaded (which is typically the case at harmonic frequencies), the impedance seen from the grid side is mainly capacitive and linearly proportional to the cable length. The capacitance per kilometer depends on the geometry of the cable and on the dielectric properties of the insulation material. Values for capacitance range between 100 and 250 nF/km for transmission cables [1].

Consider a cable with capacitance C per kilometer and length ℓ. The cable is connected to the transmission system at a location with nominal voltage U_{nom} and fault level S_k. The inductance of the source is

$$L = \frac{U_{\text{nom}}^2}{\omega S_k} \tag{6.33}$$

TABLE 6.9 Range of Fault Levels for the UK Transmission System

Voltage level (kV)	Lowest (MVA)	5% (MVA)	50% (MVA)	95% (MVA)	Highest (MVA)
66	526	740	1630	2642	2902
132	686	988	2316	4106	5649
275	3434	5930	11270	14937	16076
400	2806	9062	20521	32001	37204

The total capacitance of the cable is equal to $\ell \times C$. The resulting resonance frequency is found from the following expression:

$$f_{\text{res}} = \frac{1}{2\pi U_{\text{nom}}}\sqrt{\frac{\omega S_k}{\ell C}} \tag{6.34}$$

Note that the cable length ℓ represents the total length of all cables connected at the location.

Example 6.16 A 20 km cable, with 180 nF/km capacitance, is connected to a 175 kV, 1200 MVA grid. Using (6.34) we obtain

$$f_{\text{res}} = \frac{1}{2\pi 175\,\text{kV}}\sqrt{\frac{314 \times 1200\,\text{MVA}}{25 \times 180\,\text{nF}}} = 263\,\text{Hz}$$

Exercise 6.10 Calculate the resonance frequency when a 27 km cable, with 150 nF/km capacitance, is connected to a 220 kV grid with a fault level of 4000 MVA.

Exercise 6.11 What will be the resonance frequency when a second cable, of 8 km length, is connected at the same point?

From the above equations and examples, it is clear that the resonance frequency, for a given cable length, strongly depends on the fault level at the connection point of the cable. Typical ranges of fault level at different transmission voltages were obtained from Ref. 320. The results are shown in Table 6.9. Assuming a cable capacitance of 150 nF/km, the resonance frequencies have been calculated for different cable lengths and 5% and 95% fault levels from Table 6.9. The results are shown in Table 6.10.

TABLE 6.10 Resonance Frequency for Cables Connected at Different Transmission Voltages at Connection Points with Different Fault Levels

	66 kV		132 kV		275 kV		400 kV	
	0.75 GVA	2.6 GVA	1 GVA	4.1 GVA	6 GVA	15 GVA	9 GVA	32 GVA
5 km	1350 Hz	2520 Hz	780 Hz	1580 Hz	917 Hz	1450 Hz	772 Hz	1460 Hz
15 km	780 Hz	1450 Hz	450 Hz	912 Hz	529 Hz	837 Hz	446 Hz	840 Hz
50 km	–	–	247 Hz	500 Hz	290 Hz	459 Hz	244 Hz	461 Hz
100 km	–	–	–	–	205 Hz	324 Hz	173 Hz	326 Hz

From the table, we see that resonance frequencies below 1 kHz can easily occur when long cables are connected to transmission or subtransmission systems. According to Ref. 95, resonance frequencies below the 10th harmonic (500 Hz in a 50 Hz system) require further attention. We would like to go a step further and indicate the following resonance frequencies that could give high harmonic voltage levels:

- The frequencies 550, 650, 850, and 950 Hz are characteristic harmonics of six-pulse converters as are used in industrial installations and for HVDC converters. Especially for the higher two frequencies, no filtering may be present near the source. When the resonance frequency comes below 1 kHz and industrial installations or HVDC converters are present in the neighborhood of the cable, a further investigation is recommended.
- The frequencies 250 and 350 Hz are the dominating harmonics emitted by domestic and commercial installations. The current through a transmission transformer is in almost all cases mainly distorted by these harmonic components. Resonance frequencies close to 250 or 350 Hz will likely result in significantly higher harmonic voltage distortion than before the connection of the cable.
- The frequency 150 Hz is due to the magnetizing current of transmission transformers. When the voltage magnitude is higher than nominal, as is often the case with lightly loaded cables, the magnetizing current becomes higher. Together with a resonance frequency close to the third harmonic, this will result in high third harmonic voltage distortion.
- The frequency 100 Hz is normally not present in any significant amount at transmission level. However, when present, it can be amplified by HVDC converters. A positive feedback mechanism exists with HVDC links for the second harmonic [16]. Such low resonance frequencies will not likely occur during normal operation: the capacitance of 300 km of cable would be needed for the 9 GVA, 400 kV connection point. However, during abnormal operation, for example, the loss of one or more lines or generators, low resonance frequencies could occur.

6.4.5.5 *Power Electronics Converters* Consider a situation where an MV/LV transformer supplies a low-voltage bus to which a certain amount of distributed generation with power electronics interface is connected. The rating of the transformer is S_{tr} and the impedance is ϵ. The amount of converter-based distributed generation connected to the low-voltage bus is P_{DG} with capacitance C_1 per unit of power. The transformer inductance is equal to

$$L_{\mathrm{tr}} = \epsilon \frac{U_{\mathrm{nom}}^2}{\omega S_{\mathrm{tr}}} \tag{6.35}$$

The total capacitance due to distributed generation connected to the low-voltage network is

$$C_{\mathrm{LV}} = P_{\mathrm{DG}} C_1 \tag{6.36}$$

This results in the following expression for the resonance frequency (neglecting the already existing capacitance connected to the low-voltage network, as well as the capacitance of the cables to the generator units):

$$f_{\text{res}} = \frac{1}{2\pi\sqrt{L_{\text{tr}}C_{\text{LV}}}} = \frac{1}{2\pi U_{\text{nom}}\sqrt{(\epsilon C_1/\omega)\Pi_{\text{DG}}}} \tag{6.37}$$

where $\Pi_{\text{DG}} = P_{\text{DG}}/S_{\text{tr}}$ is the fraction of distributed generation with respect to the rating of the distribution transformer. High harmonic distortion is mainly present for $f_{\text{res}} = nf_0$, with $n = 5, 7, 11$, or 13, and the power system frequency f_0. This corresponds to the following amount of distributed generation:

$$\Pi_{\text{DG}} = \frac{1}{2\pi f_0 n^2 \epsilon C_1 U_{\text{nom}}^2} \tag{6.38}$$

The capacitance of distributed generator units in the 1–3 kW power range varies between 0.5 and 10 μF per converter, according to Ref. 145. If we assume that the nominal power of the converter is 1 kW, we get $C_1 = 0.2 \cdots 10$ nF/W. An experimental setup of an 80 kVA unit is shown in Ref. 295. A 90 μF grid-side capacitor is used, corresponding to 1.1 nF/W, which is in the middle of this range. The capacitance of a low-voltage cable varies between 0.5 and 2 μF/km (for medium-voltage cables this range is 0.14–1.1 μF/km). By supposing a cable length of maximum 500 m for a 1 kW converter, the increment of C_1 is less than 1 nF/W due to the supply cable. The equivalent capacitance of a home connection at low voltage (household capacitance) can vary in the range of 0.6–6 μF [145].

Consider a high-capacitance situation to estimate a worst-case hosting capacity. Taking the upper range of the indicated capacitance of the converters (10 nF/W) plus additional capacitance for the cables, plus capacitance for other low-voltage equipment, we come to a value of 17 nF/W. Using the values $n = 7$, $U_0 = 230$ V, $\epsilon = 0.05$, and $C_1 = 17$ nF/W gives $\Pi_{\text{DG}} = 1.44$ for the hosting capacity. Thus, the amount of generation that can be connected to the low-voltage bus before a harmonic resonance at the seventh harmonic occurs is 1.44 times the rating of the transformer. The already existing (but not yet considered) capacitance connected to the low-voltage bus, for example, in the form of additional cables, power electronics equipment, and lighting, can further reduce the hosting capacity.

Resonance at the 11th harmonic order occurs when the amount of generation is 0.6 times the rating of the transformer. Note however that the above selection of the capacitance C_1 was pessimistic.

The following practical example from Ref. 145 shows the interactions between the distribution network, household capacitance, and generator units. In this example, there are 12, 000 m^2 roof-mounted photovoltaic arrays installed on 500 homes in a suburb near to Amersfoort, the Netherlands. The PV inverters are connected directly to the 230/400 V network. A part of this network has been simulated (36 homes, 18 inverters) and compared with a simplified analysis using the following data:

- Transformer and cable reactances, $L_{\text{tr}} = 80$ μH
- Household capacitance (3 μF/household), $C_{\text{home}} = 108$ μF
- Inverter capacitance (6 μF/inverter), $C_{\text{PV}} = 108$ μF

The most dominant resonance frequency is found from a simplified analysis as being equal to

$$f_{\text{res}} = \frac{1}{2\pi\sqrt{L_{\text{tr}}\left(C_{\text{home}} + C_{\text{PV}}\right)}} \approx 1200\,\text{Hz} \tag{6.39}$$

The actual results from the network simulation [145] showed resonances around 21st–23rd harmonics, that is, between 1050 and 1150 Hz. The simplified analysis gives a result that is not very far away from this.

6.4.6 Weaker Transmission Grid

The replacement of large conventional power stations connected to the transmission grid by small generators connected at distribution level will result in a weaker transmission grid. At the power system frequency, weakening of the grid implies an increase in the source impedance. At harmonic frequencies, this is not always the case. This is illustrated in Figure 6.19. The source impedance has been calculated for a 132 kV network, with a 30 Mvar capacitor bank connected and damping provided by 100 MW of resistive load. Two values of the fault level have been assumed: 1000 MVA ("weak grid" in the figure) and 3000 MVA ("strong grid").

The reduction in fault level results in a shift in resonance frequency to a lower value (from harmonic 10 to harmonic 6 in this case). For low frequencies, the source impedance increases and the same amount of emission will result in higher voltage distortion. For high frequencies, the situation is just the other way round: the source impedance gets smaller resulting in lower voltage distortion. This of course assumes that the emission remains the same.

The situation is not as severe as it may look. Because of the variation in consumption through the year, the amount of generators connected to the transmission

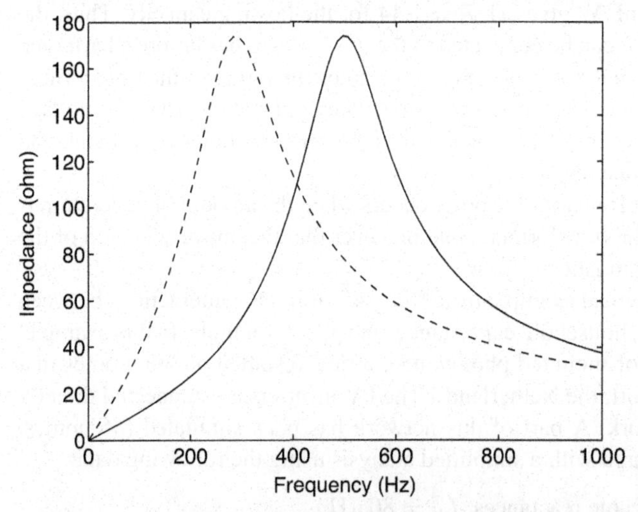

Figure 6.19 Source impedance as a function of frequency for a strong grid (solid line) and for a weak grid (dashed line).

already shows a variation by about a factor of 2 in the existing situation. Every generator will also be out of operation for maintenance or due to a failure, typically around once every year. The resulting local reduction in fault level also results in an increase in harmonic voltage distortion at lower frequencies.

A high penetration of distributed generation could result in situations where the voltage distortion close to large industrial installations becomes too high during a too high part of the year. There are a number of ways to address this:

- Higher levels of harmonic voltage distortion may be accepted during certain parts of the year. The impact of this on the end customer is likely to be limited because only a part of the voltage distortion originates at higher voltage levels. Capacitor banks at transmission level may however age faster. The overall higher voltage distortion could result in some increased aging for motor loads.
- Existing capacitor banks could be equipped with harmonic filters or additional harmonic filters could be installed. This will result in additional costs for the transmission system operator. The costs cannot be allocated to a specific generator unit, so they have to be spread over all customers.
- Emission limits for industrial installations could remain to be set in terms of voltage distortion, as is common practice for several network operators. This would result in additional costs for the industrial installations.

6.4.7 Stronger Distribution Grid

Connecting a generator unit to the distribution grid will make the grid locally stronger. Any generator acts as a voltage source (behind some impedance) at the power system frequency (50 or 60 Hz). Its impedance at harmonic frequency will not be the same as at the fundamental frequency, but the source behind impedance model will in most cases still hold. The smaller the impedance, the more the generator helps in limiting the local voltage distortion. In this section, we will quantify this impact in more detail. A first distinction that has to be made is between the harmonics that are predominantly of a zero-sequence character and the harmonics that are predominantly of positive- or negative-sequence character. The impedance of rotating machines is very high for the zero-sequence component. The impact of distributed generation on zero-sequence harmonics can, therefore, be neglected. The discussion below concerns only the positive- and negative-sequence components. In a system with more or less balanced load, the 3rd, 9th, 15th, and so on, harmonics are predominantly of zero-sequence character. All other odd harmonics are mainly of positive- or negative-sequence character.

At harmonic frequencies, both induction and synchronous machines can be modeled as a constant inductance. For induction machines, this inductance is equal to the leakage inductance; for synchronous machines, it is equal to the subtransient inductance. Neglecting the magnetizing impedance gives the following expression for the impedance of the induction machine at harmonic order h:

$$Z(h) = jhX \tag{6.40}$$

where X is the leakage reactance at power system frequency.

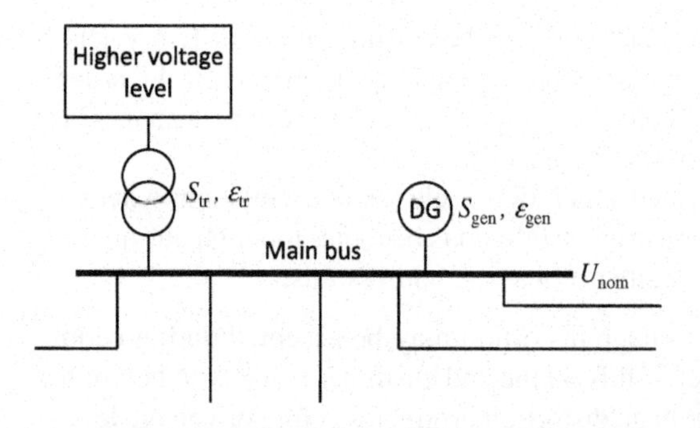

Figure 6.20 Generators connected close to the main low- or medium-voltage bus.

For synchronous generators, the harmonic impedance is close to the negative-sequence impedance, which in turn is close to the subtransient reactance of the machine. The only difference is again the frequency dependence of the impedance. We can, thus, write for synchronous generators:

$$Z(h) = h \times X'' \tag{6.41}$$

Comparing (6.40) and (6.41) shows that the harmonic behavior is the same for both induction and synchronous machines. Even the impedance values are similar. The impedance of a rotating machine increases linearly with frequency up to the frequency at which the input capacitance of the generator (including all cables and other equipment) takes over. Resonances due to other capacitance in the distribution network will already occur at lower frequencies, so there is no real need to consider the capacitance of the generator itself. In the examples below, we will only consider synchronous generators, but the calculations hold equally well for induction generators.

Consider a number of generators connected to a low- or medium-voltage network, close to the main bus, as shown schematically in Figure 6.20. The impedance seen by a harmonic current generated downstream of the main bus is the parallel connection of the transformer impedance and the generator impedance. Here we have neglected the impedance of the grid at the higher voltage level; also, we do not consider any capacitance.

The impedance of the transformer for harmonic order h is equal to

$$Z_{tr}(h) = h \times \epsilon_{tr} \frac{U_{nom}^2}{S_{tr}} \tag{6.42}$$

The impedance of the generator is

$$Z_{gen}(h) = h \times \epsilon_{gen} \frac{U_{nom}^2}{S_{gen}} \tag{6.43}$$

The source impedance, being the parallel connection of (6.42) and (6.43), is given by the following expression:

$$Z(h) = \frac{hZ(1)}{1 + (\epsilon_{tr}/\epsilon_{gen})(S_{gen}/S_{tr})} \tag{6.44}$$

where $Z(1) = \epsilon_{tr}(U_{\mathrm{nom}}^2/S_{\mathrm{tr}})$ is the source impedance at the power system frequency without distributed generation.

For connection without generator transformer, which is possible only at low voltage, ϵ_{gen} is the subtransient impedance of the machine, typically in the range of 0.15–0.20 per unit. When a generator transformer is present, the impedance of this transformer should be added to ϵ_{gen}, resulting in a value between 0.2 and 0.3 per unit.

Example 6.17 For connection to the low-voltage network, ϵ_{tr} is the impedance of the distribution transformer, typically 0.04–0.05 per unit. For a connection without generator transformer, ϵ_{gen} is about four times ϵ_{tr}. This gives for the source impedance:

$$Z(h) = \frac{hZ(1)}{1 + (1/4)K_{\mathrm{DG}}}$$

where $K_{\mathrm{DG}} = S_{\mathrm{gen}}/S_{\mathrm{tr}}$ is the total installed capacity of distributed generation as a fraction of the transformer rating. When the installed capacity is equal to the transformer rating, $K_{\mathrm{DG}} = 1$, the source impedance is reduced by 20%. With the same amount of harmonic (current) emission, the harmonic voltage is reduced by 17%.

Exercise 6.12 Repeat the calculations in Example 6.17 for generation capacity equal to 25% and 50% of the rating of the distribution transformer.

Exercise 6.13 What will be the reduction in harmonic voltage when all units are connected via a generator transformer. Assume that the total impedance (machine plus generator transformer) is 30% and the impedance of the distribution transformer is 5%.

For synchronous machines connected further away along a low-voltage feeder, the impact of the generator on the harmonic source impedance can be much bigger than that for a connection close to the main low-voltage bus. This is good news for the network operator as the main concern with harmonic voltage is further out on the network. It is, however, bad news for the owner of the generator as the harmonic currents emitted by the source will mainly flow through the rotating machine.

Example 6.18 Consider a 50 kVA generator with 18% impedance connected to a 230 V network supplied by a 200 kVA, 5% impedance transformer. The transformer impedance for harmonic order h is $0.013h\ \Omega$; the generator impedance is $0.19h\ \Omega$. Connecting the generator close to the secondary side of the transformer results in an impedance equal to $0.012h\ \Omega$, a reduction by about 8%.

Considering 1 km line with impedance at the power system frequency of $0.28\ \Omega$, the source impedance at the end of that line is $0.293h\ \Omega$ without distributed generation. Connecting the 50 kVA generator to the end the line gives an impedance of $0.115h\ \Omega$, a reduction by 60%.

At medium voltage, the impedance values become different. The impedance of the transmission transformer is in the range of 10–25%; the impedance of the generator (including the generator transformer) is in the range of 20–30%. The ratio between the impedances is less than that at low voltage, so the generator will give a higher reduction in harmonic impedance. The transformer impedance is highest in urban networks with large transformers and smaller for the smaller transformers in urban areas.

Example 6.19 An urban medium-voltage network is supplied by a 40 MVA transformer with 22% impedance. A 10 MVA generator is connected close to the main medium-voltage bus. The impedance of the generator, including the generator transformer, is 30%. Using (6.44) gives a reduction in harmonic source impedance of 14%.

Exercise 6.14 What will be the reduction when a total of 40 MVA generation is connected close to the main medium-voltage bus?

The impact of a generator with active power electronics interface depends strongly on the control algorithm used. Using the right control algorithm, it is possible to filter harmonics as long as their frequency is significantly below the switching frequency of the converter. But even when no filtering algorithm is used, the converter will have a positive impact on harmonics. The control algorithm will generate a voltage waveform at the terminals of the voltage source converter. This converter is connected to the grid through an inductance, which can be part of a filter, a transformer, or both. For harmonic frequencies significantly below the switching frequency (the low-order harmonics up to about order 20 would typically fit in this category), the generator can again be modeled as a voltage source behind reactance. The size of the reactance is around 0.1–0.2 per unit, thus of the same order of magnitude as the leakage reactance for rotating machines.

From this we draw the conclusion that the harmonic impedance of distributed generators is rather independent of the type of interface. The difference is, however, in the emission generated by some types of interface and in the active filtering capabilities of power electronics converters.

6.5 HIGH-FREQUENCY DISTORTION

Voltage source converters are known as a source of high-frequency harmonics. The switching frequency and multiples of the switching frequency (1 kHz and up) are present in the spectrum of the current. A systematic approach to determine the amplitude of high-frequency harmonics is given in Ref. 110, where it is also shown mathematically that pulse width modulation leads to groups of frequency components around the integer multiples of the switching frequency.

A problem reported among others in Ref. 125 is that the switching frequency may be close to a system resonance causing a rather large high-frequency ripple on the voltage. Most switching schemes will result in broadband emission, which has also been confirmed by measurements. Resonances cannot be avoided for these higher frequency ranges, so there is a high probability that a resonance will be exited resulting in high voltage distortion. Fortunately, the damping in general increases with frequency, for example, due to the skin effect in the wires. Despite this damping effect, repetitive failure of cable connections has been reported due to high distortion in the kilohertz range. High voltage distortion was in that case due to resonance in a cable connection close to the switching frequency of a power electronics converter connected at medium voltage.

The German association of network operators (VDN, Verband der Netzbetreiber) gives guidelines for the connection of renewable energy to the grid. This includes the following requirements for the emission per 200 Hz band in the frequency range of 2–10 kHz, depending on the voltage level at which the installation is connected:

- $I_\mu < (16/\mu)$ for connections at 110 kV,
- $I_\mu < (8/\mu)$ for connections at 220 kV, and
- $I_\mu < (4.5/\mu)$ for connections at 380 kV,

where μ is the harmonic order of the center frequency of the 200 Hz band. The current is given in ampere per 1000 MVA.

Example 6.20 Consider a 200 MVA installation connected to a 110 kV grid. At 4 kHz (harmonic order 80), the emission limit is $16/80 = 0.5$ A/GVA. For the 200 MVA installation, this emission limit at 4 kHz is 0.1 A.

An increasing level of distributed generation, with power electronics interfaces, will lead to an increasing level of high-order harmonics. There are two schools of thought here: a positive and a negative. The positive way of looking at the problem is that high-frequency disturbances do not propagate far due to their high damping. In addition, high-frequency harmonics can be easily filtered. Some experts, however, warn that the propagation and effects of high-frequency disturbances remain an unknown territory. They call for caution and further studies before installing large amounts of interfaces producing high-frequency disturbances. Potential victims of these high-frequency disturbances are, among others, the capacitors that are common with generator interfaces. The number of studies on high-frequency distortion due to distributed generation is very limited. Some studies have, however, been conducted for smaller equipment with a power electronics interface. Measurements with smaller end-user equipment are much easier to perform and, in the lack of more specific results, some of the conclusions may be applied to distributed generation as well. An overview of emission in the frequency range of 2–150 kHz is given in Ref. 261. Measurements of the interaction between different devices are presented in Refs. 370 and 371. These measurements show that high-frequency currents mainly flow between neighboring devices. The emission from a group of power electronics devices to the grid is small. Simulations and measurements given in Ref. 284 indicate that there is a risk of resonance between EMC filters at higher frequencies. A real case of such a resonance is shown in the paper. According to a recent CIGRE report [95], the harmonics originating from the PWM switching of a DFIG converter or full-size converter are usually filtered sufficiently. This statement is, however, not further documented in the report, neither is there any information in the report about what is "sufficient" in this case.

6.5.1 Emission by Individual Generators

In Ref. 469, a number of filtering techniques are compared for a 110 kW converter to connect a wind turbine. It is shown that the control algorithm and the filter configuration have a huge influence on the amount of current distortion at the switching frequency. The paper does not give any quantitative results but instead gives a visual comparison of the current waveforms. A rough estimation of the waveforms shown gives a ripple of several percent on the current waveform for the best method of limiting distortion. For the other methods, the ripple is clearly higher. Similar levels of ripple in voltage as well as in current are shown in Ref. 356 for a 400 kW double-fed induction generator. Also, in this case, the control algorithm has been optimized to reduce waveform distortion. The ripple in voltage is about 5% and in current it is somewhat less. The current waveform for a full power converter connected to a 33 kV feeder is shown in Ref. 289. The total harmonic distortion is reported to be about 6.5%. A significant ripple is visible in the waveform shown in the paper.

Reference 203 presents measurements of the harmonic current emission of a 2250 W PV inverter. The harmonics up to order of 50 are presented. The value given for the highest harmonic orders is about 0.4% of the rated current. This would result in a value around 1.5% for the 200 Hz group.

Reference 350 presents measurements for a 10 kVA experimental inverter with isolating transformer. The switching frequency of 8192 Hz results in a distortion band between 8 and 8.5 kHz with amplitude around 0.1% of rated current. The paper does not explicitly mention over which time window the DFT is taken. By assuming a 50 ms window, as the voltage and current waveforms are presented in the same paper, the corresponding value would be 0.32% for the 200 Hz group.

Note that the distortion amplitude is a result of the filter design done by the authors of Ref. 350, where 0.1% was a requirement set by them. A different choice of filter parameters will result in a different emission level.

Measurements at a wind turbine with double-fed induction generator are given in Ref. 302. The switching frequency of the inverter part was 2250 Hz. The spectrum up to 2.5 kHz is shown, consisting of a continuous part with individual harmonic components superimposed. The continuous part of the spectrum of the inverter current is reproduced in Figure 6.21 for low, medium, and high power. The spectrum changes

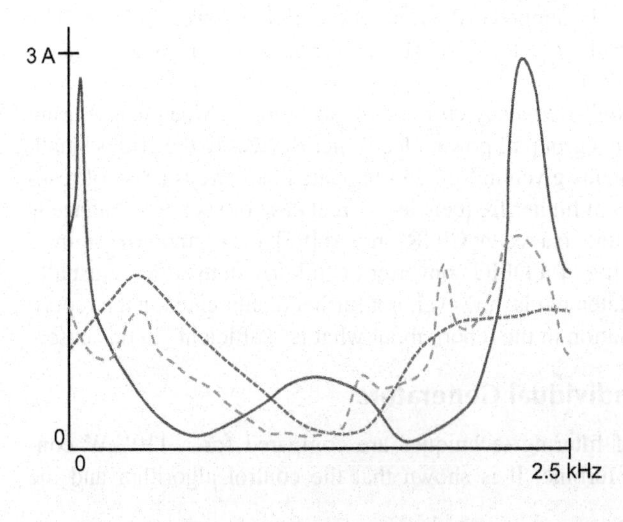

Figure 6.21 Continuous part of the harmonic spectrum of the inverter current with a double-fed induction generator [302]. The three curves correspond to power production of 200 kW (solid curve), 900 kW (dashed curve), and 1500 kW (dotted curve).

somewhat, but the main emission remains around low-order harmonics and around the switching frequency. Estimation from the figures in the original report resulted in a value of about 12% for the worst 200 Hz group.

Individual frequency components appear at different frequencies for different power levels. Some components appear at harmonic frequencies (5, 7, 11, 13) especially for low power, but many components appear at interharmonic frequencies. The amplitude of individual harmonic frequencies ranges from a few amperes up to 10 A. The highest amplitude occurs around the switching frequency.

Reference 228 gives measurements from a microturbine of size 100 kW electric and 167 kW thermal. The spectrum at 5 Hz intervals at 80% loading of the turbine is shown in the original report. The spectrum consists of a line spectrum resembling that of a six-pulse rectifier together with a continuous (noise) spectrum. The continuous spectrum has a peak around 1 kHz (harmonic 20). The highest 200 Hz value around 1 kHz is estimated to be about 3.3%.

Measurements on a 110 W photovoltaic inverter based on hysteresis control are given in Ref. 276. The measured spectrum of the current injected by one individual inverter shows a peak between 4 and 5 kHz. The highest value for the 200 Hz group is estimated to be around 5%. Individual frequency components are present for low-order harmonics (3, 5, 7, 9, 11, and 13), with the fifth and seventh harmonic dominating at about 2.5%.

Reference 250 gives measurements of small PV inverters (around 150 W) connected to weak and strong grids. The harmonic spectrum, over a 20 ms window, is presented up to harmonic order 50. The resonance of the filter capacitance and the grid impedance gives high harmonic current around 1.7 kHz. In the case of a strong grid, the highest amplitude is about 0.1% for individual harmonics; for a weak grid, this increases to over 1%. The highest 200 Hz group is estimated at 2.3% in the latter case.

Measurements of a 100 kW PV system are given in Ref. 330. A typical spectrum is shown in Figure 6.22. The third and fifth harmonics are dominating, but they are much smaller than those for normal low-voltage equipment. The rest of the spectrum has a "noise-like character" with even and odd harmonics of similar magnitude. The 200 Hz group obtained from harmonic orders 27 through 30 is equal to 3.1%.

From the examples above, it can be concluded that the harmonic spectrum of converters applied by distributed generation, such as microturbines of photovoltaic converters, may be filled with sub- and interharmonics besides the characteristic harmonics. This spectrum can vary during the fundamental period and depend on the generated power. Moreover, the harmonic content may also be influenced by the background harmonics and the unbalance of the power network.

The discussed measurement results have been compared to obtain an estimate of the levels of harmonic emission that can be expected from distributed generation with power electronics interface. The measured distortion per 200 Hz group has been summarized in Table 6.11. The two deviating results are a laboratory setup and a DFIG inverter. From the other results, it is concluded that the distortion is between 1% and 5%. It is unclear if the laboratory result is simply a better design or that the lower current distortion is due to the cleaner background voltage. Further research on this is needed.

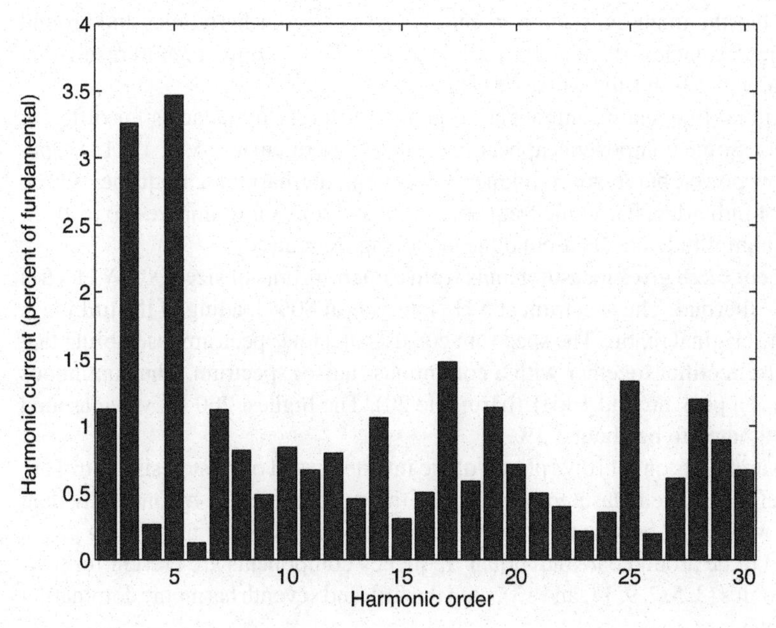

Figure 6.22 Typical spectrum of a photovoltaic system, according to Ref. 330.

6.5.2 Grouping Below and Above 2 kHz

An informative annex with IEC 61000-4-7 discusses a characterization method for disturbances in the frequency range of 2–9 kHz [45, 215]. Applying a discrete Fourier transform (DFT) algorithm to a 200 ms window results in frequency values with 5 Hz separation. The resulting values are grouped into 200 Hz bands, using the following

TABLE 6.11 Overview of Measured Current Distortion: Estimated Level of Highest 200 Hz Band

Sources	200 Hz group current (%)	Comments
203	1.5	
228	3.3	
250	2.4	
276	5	
289	5	Amplitude of ripple
302	12	DFIG inverter
330	3.1	
350	0.3	Laboratory setup
356	6	Amplitude of ripple
469	3	Amplitude of ripple

grouping algorithm:

$$G_b = \sqrt{\sum_{f=b-95\,\text{Hz}}^{b+100\,\text{Hz}} C_f^2} \tag{6.45}$$

where C_f is the rms value of the frequency component at frequency f. This grouping algorithm results in 35 frequency bands centered around 2100, 2300, ..., 8700, and 8900 Hz. An older version of the standard uses a 100 ms time window. However, the size of the frequency window has remained unchanged at 200 Hz. Note that the distortion level per 200 Hz band can be obtained independent of the window length.

The choice of frequency bands in IEC 61000-4-7 is based on CISPR 16-1 [98], where the same 200 Hz bandwidth is used. The characterization method is, however, not compatible with the method used for harmonic and interharmonic distortion below 2 kHz [45]. The grouping algorithms for harmonics and interharmonics, as prescribed in the main body of IEC 61000-4-7, add the signal energy of all frequency components within each group or subgroup [45, 215]. The grouping algorithm used in the frequency band 2–9 kHz also adds the energy over all components within each frequency band but with a different bandwidth. For a single-frequency signal (a line spectrum), the result will be consistent for both methods. Thus, a 1 V signal of one single frequency (e.g., 1999 Hz or 2001 Hz) will result in a 1 V value for the 40th harmonic, as well as a 1 V value for the 2100 Hz band.

The situation is different for broadband signals, the kind of signals that are more common for the higher frequency ranges. To compare different frequency bands, it is important to consider Parseval's theorem. Parseval's theorem relates the energy of the signal in time domain and in frequency domain. For digital (sampled) signals, Parseval's theorem reads as

$$\frac{1}{N}\sum_{i=1}^{N}\{g(t_i)\}^2 = \sum_{k=0}^{N/2}|C_k|^2 \tag{6.46}$$

The left-hand side is the square of the rms value of the signal, which is independent of sampling rate or measurement window. Doubling the measurement window will double the number of frequency components but not the energy over a fixed frequency band. Let the amplitude of a 5 Hz component (obtained from a 200 ms window) be equal to 1. The amplitude of a 200 Hz band is equal to $\sqrt{40} = 6.32$. All this assumes a flat continuous spectrum within the 200 Hz frequency band. Table 6.12 gives the resulting amplitude for the harmonic and interharmonic groups and subgroups in a 50 and 60 Hz system when the 200 Hz band results in a value of 1 V, or 158 mV per 5 Hz component.

6.5.3 Limits Below and Above 2 kHz

As shown above, the distortion levels in the range of 2–9 kHz cannot be immediately compared with the levels for lower frequencies (harmonics 2 through 40). Harmonics are treated as predominantly a line spectrum superimposed on a continuous spectrum.

TABLE 6.12 Relation Between Amplitude of Groups and Subgroups for a Flat Continuous Spectrum When the 200 Hz Band Results in a Value of 1 V

	50 Hz	60 Hz
Harmonic group	0.5 V	0.55 V
Harmonic subgroup	0.27 V	0.27 V
Interharmonic group	0.47 V	0.52 V
Interharmonic subgroup	0.42 V	0.47 V

In IEC 61000-4-7, harmonic and interharmonic groups and subgroups are defined to separate the harmonic spectral lines from the rest of the distortion.

In the frequency range of 2–9 kHz, the spectrum is treated as a continuous spectrum. The energy in every 200 Hz band is used as a performance index. Note that there is no sharp transition between a line spectrum and a continuous spectrum at 2 kHz. The actual change is smoother and may start at 1 kHz and some discrete frequency components will also be present above 2 kHz. However, in international standards, the change in analysis methods is chosen, somewhat arbitrarily at 2 kHz.

To compare the levels and limits below and above 2 kHz, the values in Table 6.12 should be used. For the extension of the indicative planning levels to higher frequency, a constant value of 0.2% has been chosen. When translating this value to a value for the 200 Hz bands, the method used for obtaining the harmonic distortion becomes important. According to IEC 61000-4-30, the harmonic subgroups should be used to characterize harmonic distortion. Using Table 6.12, a level of 0.2% for the harmonic subgroup corresponds to a level of $0.2/0.27 = 0.74\%$ for the 200 Hz band.

Technical document IEC 61000-3-6 further gives indicative planning levels for interharmonics equal to 0.2%. According to IEC 61000-4-30, the interharmonic subgroup shall be used to characterize interharmonic distortion. Using again Table 6.12, the level of 0.2% for the interharmonic subgroup corresponds to a level of $0.2/0.42 = 0.48\%$ for the 200 Hz band.

Based on this extension of the limits in IEC 61000-3-6, planning levels between 0.4% and 0.75% would be reasonable above 2 kHz. A planning level for the frequency range between 2 and 9 kHz is proposed that is equal to 0.5% of the nominal voltage for each 200 Hz band.

In order to easily estimate the maximum allowable harmonic voltage emission level from the harmonic current limits (or vice versa), information about the harmonic network impedance is needed. This impedance is usually calculated from simulations of the harmonic power network or directly accessible from measurements. The harmonic network impedance varies considerably between the nodes of a power network and also during the day. Measuring results are available for the LV network in the frequency range of 2–9 kHz in Ref. 25. According to the measurements, the impedances were higher in overhead systems than in cable systems by a factor of 3–5. The ratio of "phase-to-phase" impedance to "phase-to-neutral" impedance was in the range of

1.5–2. The 90–95% probability curve of the phase-to-neutral impedance is approximated as

$$Z(f) = \frac{f}{1000} + 1\,\Omega \tag{6.47}$$

where f is given in Hertz. For example, the phase-to-neutral impedance in the LV network at 5 kHz is below 6 Ω with 90–95% probability.

When both the harmonic current spectrum of an inverter and the allowed voltage emission level at the PCC are known at a considered frequency, then the hosting capacity of the same type of inverters can be estimated by assuming a suitable summation law. The harmonic voltage level, due to a single inverter with relative harmonic current I_{nrel} at harmonic order n is equal to,

$$V_{\mathrm{n}} = Z_{\mathrm{n}} I_{\mathrm{nrel}} I_{\mathrm{nom}} \tag{6.48}$$

where Z_n is the network impedance at harmonic order n and I_{nom} is the nominal current. For N identical inverters with noncorrelated harmonic currents and harmonic orders $n > 10$ (random signals), the attenuated harmonic voltage at the PCC is

$$V_{nN} = \sqrt{N} \times V_n = \sqrt{N} \times Z_n I_{\mathrm{nrel}} I_{\mathrm{nom}} \tag{6.49}$$

By supposing that the attenuated harmonic voltage V_{nN} is equal to the maximum permissible voltage emission level E_{nrel} at harmonic order n and nominal phase voltage U_{nom}, the hosting capacity can be calculated:

$$N = \left(\frac{E_{\mathrm{nrel}} U_{\mathrm{nom}}}{Z_n I_{\mathrm{nrel}} I_{\mathrm{nom}}} \right)^2 \tag{6.50}$$

For the low-voltage network and the frequency band 2–9 kHz (harmonic orders n from 40 to 180), this equation can be written as

$$N = \left(\frac{E_{\mathrm{nrel}} U_{\mathrm{nom}}}{(0.05n + 1) I_{\mathrm{nrel}} I_{\mathrm{nom}}} \right)^2 \tag{6.51}$$

From the study after performance indices and limits presented above, a limit of $E_{\mathrm{nrel}} = 0.5\%$ was concluded for the voltage distortion per 200 Hz group in the frequency band between 2 and 9 kHz. This value, together with the values for the current distortion per 200 Hz group, can be used to estimate the hosting capacity.

Example 6.21 Consider a voltage distortion limit of 0.5% at 230 V. How many 1000 W single-phase units (4.3 A rated current) can be connected when the current distortion is 3% at 5 kHz ($n = 100$)?

Using (6.51) gives

$$N < \left(\frac{0.005 \times 230}{(0.05 \times 100 + 1) \times 0.03 \times 4.3} \right)^2 = 2.2$$

At most two such units can be connected.

The resulting hosting capacity, as calculated in Example 6.21, is very sensitive to the different parameters and to the summation law used. For example, with 5% emission, one unit will result in the voltage distortion limit that is being exceeded: the hosting capacity here is zero. When the emission is only 1%, the hosting capacity becomes 19 units. For a cubic summation law, giving a third power in (6.51), 4 units with 3% emission, 11 units with 2% emission, or 80 units with 1% emission can be connected. Increasing the permissible voltage distortion from 0.5% to 1.0% will allow four times as many units to be connected.

In the above calculations, no consideration was made for interference due to the harmonic emission. Instead, safe limits were estimated by extrapolating the limits below 2 kHz. Whether these are borderline-safe limits or overly safe units, we do not know. There are voices that claim the latter, sometimes using good arguments, but the lack of knowledge will make it difficult to set any limits at this stage. As the hosting capacity strongly depends on the voltage distortion limit used, this is an important question.

6.6 VOLTAGE DIPS

Distributed generators do not result in any significant direct increase in the number of voltage dips. The only possible impact is the voltage dip due to the connection of the generator to the grid. This is of concern only for generation with rotating machine interface, and even in this case, typically a soft starting is used to limit the voltage drop.

The presence of distributed generation, however, impacts the number of dips in several indirect ways:

- Distributed generation connected to the distribution system will locally strengthen the grid, which will result in a decrease of the number of dips experienced by local customers. We will consider this impact in more detail, looking at synchronous machines in Sections 6.6.1 and 6.6.2. The damping effect of induction generators on unbalanced dips is treated in Section 6.6.3.
- Replacement of large conventional power stations by distributed generation will weaken the transmission grid, which will result in an increase of the number of dips experienced by the customers. This has been studied in Ref. 52 for the UK transmission system. It is estimated that 20% distributed generation (assumed constant through the year) will for no customer result in a doubling or more of the number of dips. As the number of dips will vary much more from year to year and due to the lack of any guidance on what are acceptable levels, this is seen as a moderate increase.
- Large penetration of distributed generation will require enforcement of the power system in the form of new cables or overhead lines. Especially the integration of large wind parks into the subtransmission system is reported to require significant amounts of new lines. These lines will result in more voltage dips for customers connected close to these lines. This may especially

impact large industrial customers that have traditionally been most sensitive to voltage dips. The authors are not aware of any study in which this impact has been quantified.

- The weakening of the transmission system may also result in a longer fault-clearing time and thus longer voltage dips. Distance protection and differential protection are not much impacted by the fault level, over a broad range of fault levels. But overcurrent protection, sometimes used in subtransmission system, may be impacted.
- After fault clearing, rotating machines take a large reactive power. In a weak system, this will pull down the voltage after the fault and result in longer voltage dips. Here, we enter the realm of transient and short-term voltage stability. We will further discuss this in Chapter 8, where one of the conclusions will be that the impact very strongly depends on the system.
- The power electronics converter, with which many distributed generators are equipped, can be used to maintain the voltage for the local load. Two different approaches are being discussed: temporary island operation of the generator with its local load and injection of reactive power during the dip, while remaining connected to the grid. More details about the different algorithms can be found, for example, in Refs. 49, 196, 197, 287, 358–360.

6.6.1 Synchronous Machines: Balanced Dips

Consider the simplified system shown in Figure 6.23: a synchronous generator (SM) is connected to a medium-voltage bus (MV) close to the secondary side of the HV/MV transformer. During a fault, the synchronous machine can be modeled as a voltage source behind impedance. If we further neglect the load influence on the voltage and the influence of the synchronous machine on the voltage at transmission level, we get the model on the right to calculate the voltage as experienced by the end customers.

The residual voltage (i.e., the voltage with the end customers during the fault) is found from this model:

$$U_{\mathrm{MV}} = \frac{Z_{\mathrm{gen}}}{Z_{\mathrm{gen}} + Z_{\mathrm{tr}}} U_{\mathrm{HV}} + \frac{Z_{\mathrm{tr}}}{Z_{\mathrm{gen}} + Z_{\mathrm{tr}}} E_{\mathrm{gen}} \tag{6.52}$$

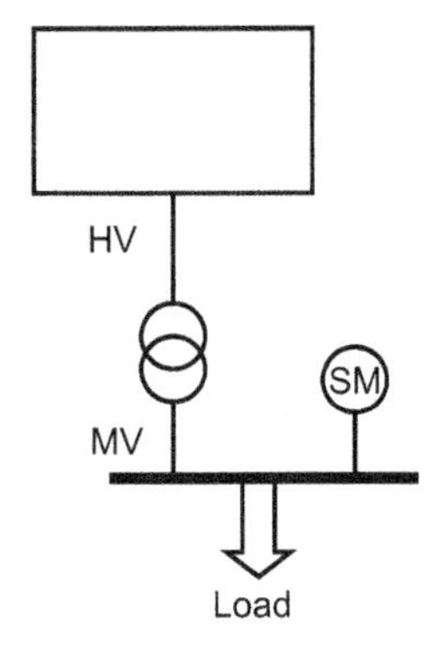

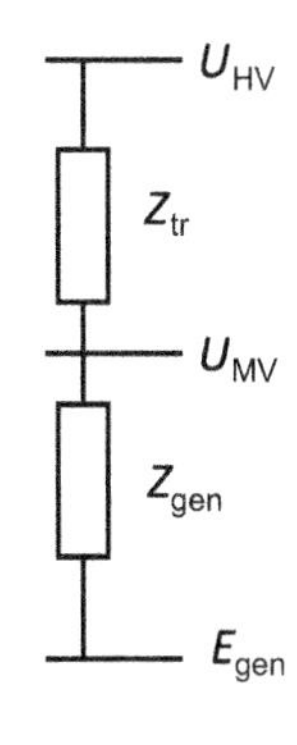

Figure 6.23 Network with synchronous generator at medium voltage, and equivalent circuit during a transmission system dip.

If we further assume that the back-emf E_{gen} of the synchronous machine is equal to 1 pu, we get the following expression for the drop in voltage experienced by the end customers during the fault:

$$(1 - U_{\mathrm{MV}}) = \frac{Z_{\mathrm{gen}}}{Z_{\mathrm{gen}} + Z_{\mathrm{tr}}}(1 - U_{\mathrm{HV}}) \tag{6.53}$$

The voltage experienced by the end customers in case there is no generator connected at medium voltage is equal to U_{HV}, so the reduction in voltage drop is related to the impedance ratio between the transformer and the synchronous generator.

Example 6.22 Consider a 10 MVA synchronous generator connected to the medium-voltage bus. The transient reactance is 25%. The HV/MV transformer is rated at 40 MVA with 20% impedance. This gives for the impedances at a 10 MVA base:

$$Z_{\mathrm{gen}} = 0.25 \text{ pu}$$
$$Z_{\mathrm{tr}} = 0.05 \text{ pu}$$

In this case, the reduction in voltage drop is

$$(1 - U_{\mathrm{MV}}) = 0.83(1 - U_{\mathrm{HV}})$$

The remaining voltage due to one specific fault is not of direct relevance to a customer. The power quality experienced by the customer is instead quantified as the (expected or measured) number of voltage dips over a longer period with a residual voltage below a certain threshold. This gives information to the customer on the number of times per year that a device or a process is expected to trip.

To estimate the impact of distributed generation on the number of dips experienced by the end customers, we use a simplified relation between the number of dips and the residual voltage:

$$N_{\mathrm{dips}} = N_{50}\frac{V}{1 - V} \tag{6.54}$$

where N_{50} depends on the system configuration and the fault frequency. This expression has been derived from theoretical models and has been confirmed by both simulations and measurements [42].

The number of dips below U_{MV} at the medium-voltage bus is equal to the number of dips below U_{HV} at the high-voltage bus, with U_{MV} and U_{HV} related as in

$$(1 - U_{\mathrm{MV}}) = \frac{Z_{\mathrm{gen}}}{Z_{\mathrm{gen}} + Z_{\mathrm{tr}}}(1 - U_{\mathrm{HV}}) \tag{6.55}$$

Inverting this expression leads to the following expression for the number of dips experienced by the end customer:

$$N_{\mathrm{MV}}(V) = N_{50}\frac{V'}{1 - V'} \tag{6.56}$$

where $V' = V - (Z_{tr}/Z_{gen})(1 - V)$. This latter expression is interesting because the ratio (Z_{tr}/Z_{gen}) can be easily linked to a percentage of distributed generation. Let z_{tr} be the transformer impedance in per unit on a base equal to the transformer rating S_{tr} and z_{gen} be the per unit generator impedance on a base equal to the generator rating S_{gen}. This results in

$$\frac{Z_{tr}}{Z_{gen}} = \frac{z_{tr}}{z_{gen}} \times \Pi_{DG} \tag{6.57}$$

where $\Pi_{DG} = \frac{S_{gen}}{S_{tr}}$ is the fraction of distributed generation (often abbreviated as %DG). Combining (6.56) and (6.57) results in the following expression for the number of dips below a threshold V at the medium-voltage bus:

$$N_{MV}(V) = N_{50}\frac{V'}{1 - V'} \tag{6.58}$$

with

$$V' = V - \frac{z_{tr}}{z_{gen}}(1 - V)\Pi_{DG} \tag{6.59}$$

The consequence of the second term in (6.59) is that a dip threshold at medium voltage corresponds to a lower dip threshold at high voltage and thus to a lower dip frequency (SARFI value). The bigger the second term, the more the reduction in dip frequency. An increasing amount of distributed generation with synchronous machine interface leads to an increased reduction of the voltage dip frequency.

Example 6.23 Consider the number of dips with a residual voltage less than 90% (often referred to as SARFI-90). According to (6.54), the number of dips without distributed generation is

$$N_{HV}(90) = 9 \times N_{50}$$

Due to the damping by the synchronous generator in Example 6.22, a dip with residual voltage 90% with the end customer corresponds to a dip with residual voltage 88% at HV. (According to (6.53), a drop of 12% at HV gives a drop of 10% at MV.) This gives for the number of dips:

$$N_{MV}(90) = 7.3 \times N_{50}$$

This is a reduction by 19%. When we consider the number of dips with a residual voltage below 70% (a value more relevant for equipment performance), we get the following values for the dip frequency without and with distributed generation:

$$N_{HV}(70) = 2.33 \times N_{50} \tag{6.60}$$

$$N_{MV}(70) = 1.78 \times N_{50} \tag{6.61}$$

a reduction by 24%. Thus, the presence of the generator reduces the number of severe dips experienced by the end customers by about one-quarter.

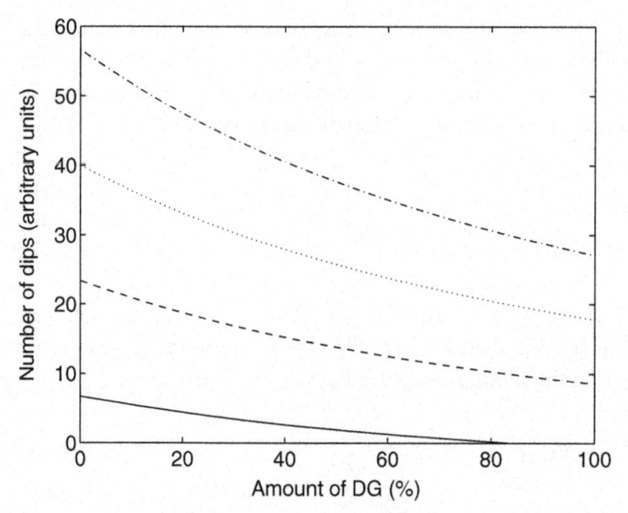

Figure 6.24 Dip frequency as a function of the amount of distributed generation with synchronous machine interface: SARFI-40 (solid), SARFI-70 (dashed), SARFI-80 (dotted), and SARFI-85 (dash-dotted).

The number of dips as a function of the amount of distributed generation is shown in Figure 6.24, where a transformer impedance of 0.20 pu and a generator impedance of 0.25 pu have been used. The presence of a large amount of distributed generation results in a significant reduction in the dip frequency.

The value of, for example, SARFI-70 is the number of dips per year with residual voltage less than 70% of the nominal voltage. For equipment or an installation that can tolerate dips with residual voltage below 70% of the nominal voltage, SARFI-70 gives the number of times per year that the equipment will trip due to voltage dips. A reduction in SARFI-70 is thus a direct improvement in voltage quality for this equipment or installation. The value of SARFI-40 drops to zero for more than 80% of distributed generation. This means that there are no longer any dips with residual voltage below 40% of the nominal voltage.

The operator of an industrial installation often gets accustomed to a certain number of production stoppages due to voltage dips. The relative reduction in the number of dips could be a better indicator of how the perception of improvement is than the absolute reduction. This relative reduction is shown in Figure 6.25. For SARFI-85, SARFI-80, and SARFI-70, the relative reduction is about the same: about 35% for 50% distributed generation and, about 60% for 100% distributed generation. For SARFI-40, the reduction is much more, up to 100%, when the amount of distributed generation becomes more than 80% of the transformer rating.

6.6.2 Synchronous Machines: Unbalanced Dips

A simplified theoretical model to describe the impact of rotating machines on voltage dips is introduced in Refs. 37 and 38. Although the model has been developed for induction motors, it is equally valid for induction and synchronous generators.

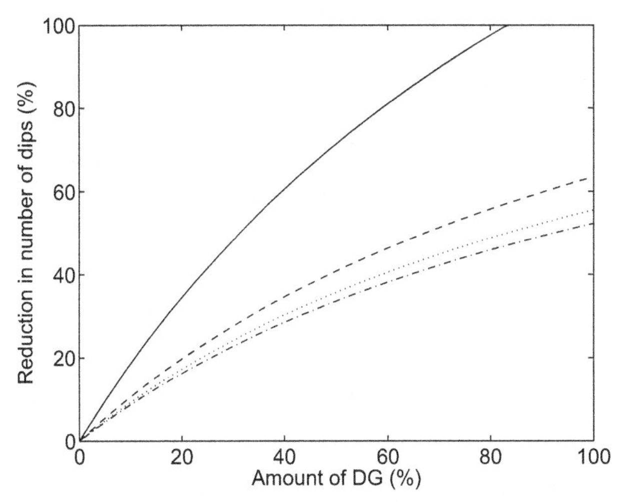

Figure 6.25 Relative reduction in dip frequency due to increasing amount of distributed generation with synchronous machine interface: SARFI-40 (solid), SARFI-70 (dashed), SARFI-80 (dotted), and SARFI-85 (dash-dotted).

The impact on the positive- and negative-sequence impedances is treated separately. Here, we will apply this model to distribution systems with synchronous or induction generators. The basic model is shown in Figure 6.26. The rotating machine (being an induction machine or a synchronous machine) is modeled as a voltage source behind reactance for the positive-sequence voltage and as a constant impedance for the negative-sequence voltage. This model has been confirmed by both measurements and simulations.

For a synchronous generator, the voltage $E_M(t)$ is constant. This voltage depends on the reactive power exchange between the generator and the grid. The more the reactive power produced by the generator, the higher the magnitude of this

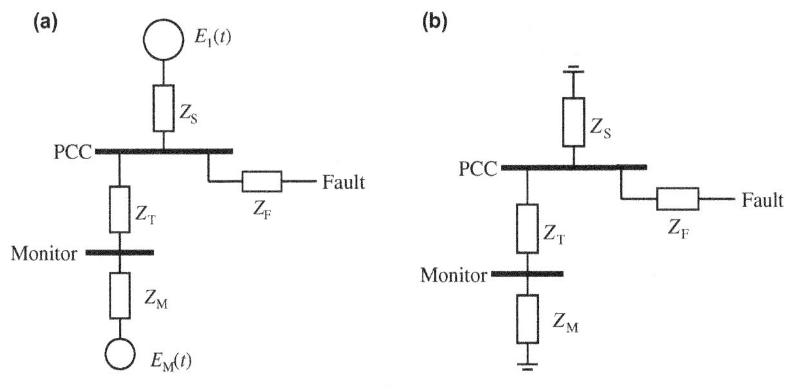

Figure 6.26 Positive-sequence model (a) and negative-sequence model (b) for calculation of voltage dip characteristics when including distributed generation.

voltage. Here, we will assume, as we did for balanced dips, that the voltage behind the reactance is equal to the source voltage of the grid $E_1(t)$.

Consider first the synchronous machine case. A voltage dip occurs at the point-of-common coupling (PCC) with positive-sequence voltage U_{1p} and negative-sequence voltage U_{2p}. The positive-sequence voltage close to the generator ("monitor location" in the figures) is obtained from

$$U_{1m} = \frac{Z_M}{Z_T + Z_M} U_{1p} + \frac{Z_T}{Z_T + Z_M} E_M \tag{6.62}$$

The negative-sequence voltage close to the generator is equal to

$$U_{2m} = \frac{Z_M}{Z_M + Z_T} U_{2p} \tag{6.63}$$

A complete characterization of unbalanced voltage dips is given in Refs. 55 and 468. The method distinguishes between unbalanced dips of two basic types, referred to as type C and type D, and uses two characteristics to quantify the severity of the dip: the characteristic voltage V and the so-called "PN factor" F. The characteristic voltage is a generalization of the residual voltage for single-phase events; the PN factor is a rather complicated concept where the difference between characteristic voltage and PN factor quantifies the amount of unbalance in the voltages. Both characteristic voltage and PN factor are complex numbers. This basic classification has recently been taken over by an international working group [97] where type D is referred to as type I and type C as type II.

A type C (or type II) dip constitutes a drop in magnitude for two of the phase-to-neutral voltages. The following expressions hold for the three phase-to-neutral voltages in case of such a dip:

$$\begin{aligned} U_a &= F \\ U_b &= -\frac{1}{2}F - \frac{1}{2}jV\sqrt{3} \\ U_c &= -\frac{1}{2}F + \frac{1}{2}jV\sqrt{3} \end{aligned} \tag{6.64}$$

A type D (or type I) dip constitutes drop in magnitude for one phase-to-neutral voltage, with the following expressions for the phase-to-neutral voltages:

$$\begin{aligned} U_a &= V \\ U_b &= -\frac{1}{2}V - \frac{1}{2}jF\sqrt{3} \\ U_c &= -\frac{1}{2}V + \frac{1}{2}jF\sqrt{3} \end{aligned} \tag{6.65}$$

At transmission and subtransmission levels, the PN factor is close to the pre-fault positive-sequence voltage for dips due to single-phase and phase-to-phase faults. Closer to the loads, the PN factor becomes smaller especially due to the impact of induction motor load. Here, we assume the PN factor to be equal to the prefault voltage at the PCC and will calculate the impact of distributed generation on both characteristic voltage and PN factor.

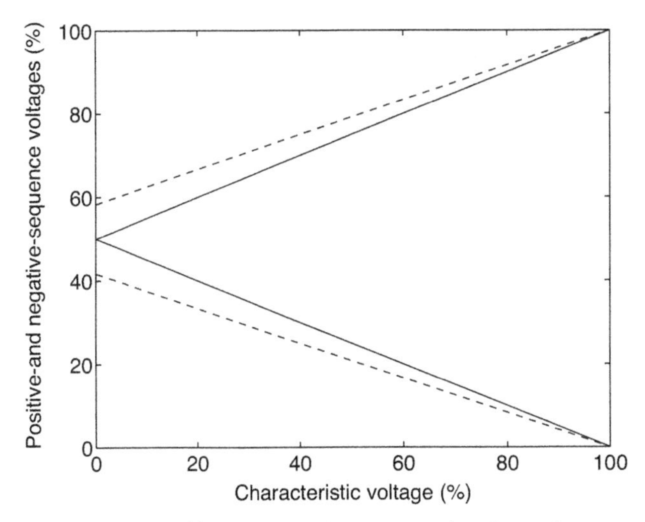

Figure 6.27 Positive-sequence (upper curves) and negative-sequence (lower curves) voltages at the PCC (solid) and as experienced by MV and LV customers (dashed).

To illustrate the impact of synchronous generators on unbalanced voltage dips, we consider the same system as in Example 6.22. A 10 MVA synchronous generator is connected to a medium-voltage network. The reactance of the machine including the generator transformer is 25%. The HV/MV transformer is rated at 40 MVA with 20% impedance. The positive- and negative-sequence voltages in the medium-voltage network, for a voltage dip originating at a higher voltage level, are shown in Figure 6.27.

The voltage dip at the PCC is calculated according to (6.64) with F equal to 1 per unit. Positive- and negative-sequence voltages are calculated using the standard expressions; see for example, Ref. 42. The transfer from HV to MV is calculated by using (6.62) and (6.63). The positive- and negative-sequence voltages are shown in Figure 6.27 as a function of the characteristic voltage at the point-of-common coupling.

Because of the presence of the synchronous machine, the positive-sequence voltage increases in magnitude (upper curves) and the negative-sequence voltage reduces in magnitude (lower curves). The calculated positive- and negative-sequence voltages at the medium-voltage bus have next been used to calculate the phase-to-neutral voltages, which are shown in Figure 6.28 for type C dips. The calculations have been repeated for type D dips, with the results shown in Figure 6.29.

For both type C and type D dips, with the presence of synchronous machines, the drop in magnitude as experienced in the distribution network becomes less. Distributed generation thus improves the voltage quality for the customers.

For type D dips, combining the above expressions give the following relation between the lowest phase-to-neutral voltages at the PCC, U_{ap}, and in the distribution network, U_{am}:

$$U_{\mathrm{am}} = \frac{Z_{\mathrm{T}}}{Z_{\mathrm{M}} + Z_{\mathrm{T}}} + \frac{Z_{\mathrm{M}}}{Z_{\mathrm{M}} + Z_{\mathrm{T}}} U_{\mathrm{ap}} \tag{6.66}$$

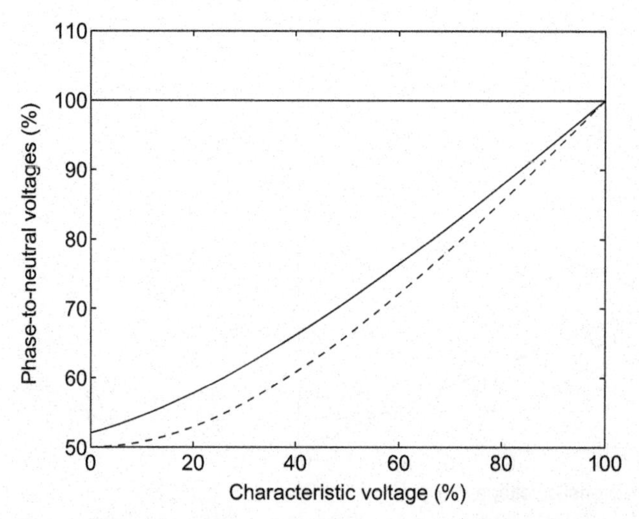

Figure 6.28 Phase-to-neutral voltages at the medium-voltage bus (solid line) and at the PCC (dashed line), as a function of the characteristic voltage at the PCC, for type C dips.

Example 6.24 For the above example, this gives

$$U_{\mathrm{am}} = 0.17 + 0.83U_{\mathrm{am}}$$

The number of dips with residual voltage less than U_{am} in the distribution network corresponds to the number of dips with residual voltage less than U_{ap} at the PCC, where U_{am} and U_{ap} are related according to (6.66). This expression can be inverted to give the voltage at the PCC as a function of the voltage in the distribution

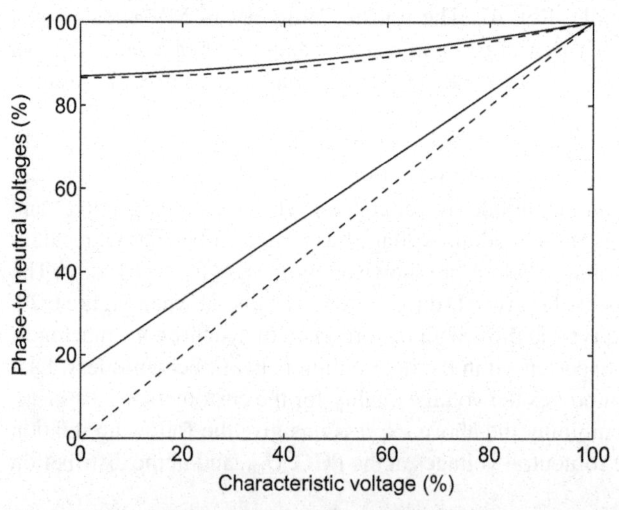

Figure 6.29 Phase-to-neutral voltages at the medium-voltage bus (solid line) and at the PCC (dashed line), as a function of the characteristic voltage at the PCC, for type D dips.

network:

$$U_{ap} = \frac{Z_M + Z_T}{Z_M} U_{am} - \frac{Z_T}{Z_M} \tag{6.67}$$

This is exactly the same expression as the one derived earlier for balanced dips, (6.55): the reduction in the number of equipment trips due to type D dips is the same as was calculated before for balanced dips, and shown in Figure 6.24. It has to be assumed, however, that the tripping of equipment is determined by the lowest of the three phase-to-neutral voltages. When another voltage, or another combination of voltages, determines the equipment behavior, the number of equipment trips may change in another way with increasing amount of distributed generation. However, as none of the three phase-to-neutral voltages gets lower in magnitude, it is unlikely that the number of equipment trips will increase.

For type C dips, there is no simple expression relating the magnitude of the voltages in the distribution network with the magnitude of the voltages at the PCC. Numerical calculations are needed to estimate how much the number of dips is reduced due to the introduction of distributed generation.

Example 6.25 The same configuration as in the previous examples is considered. It has been assumed that the number of dips at the PCC with characteristic voltage less than V is obtained from

$$N_p(V) = \frac{V}{1 - V} \times N_{50}$$

The number of dips with lowest phase-to-neutral voltages below 70%, 80%, and 85% at the PCC is found to be 1.30, 2.59, and 3.85 times N_{50}, respectively. In the distribution network, these values are 0.92, 1.99, and 3.04 times N_{50}, a reduction by 29%, 23%, and 21%, respectively.

The calculations in Example 6.25 have been repeated for different amounts of distributed generation (see also Figure 6.24). The results are shown in Figures 6.30 and 6.31. The reduction in the number of dips below the various thresholds is significant, especially for SARFI-70.

6.6.3 Induction Generators and Unbalanced Dips

When the majority of distributed generation consist of induction machines, the equations become somewhat different. The impact of induction generators on the positive-sequence voltage is small, because they do not contribute to the fault current beyond the first one or two cycles of the fault. Induction machines do however take a large negative-sequence current and thereby reduce the magnitude of the negative-sequence voltage. The expressions for the propagation from the PCC to the distribution network read as follows for induction generators:

$$U_{1m} = U_{1p} \tag{6.68}$$

$$U_{2m} = \frac{Z_M}{Z_M + Z_T} U_{2p} \tag{6.69}$$

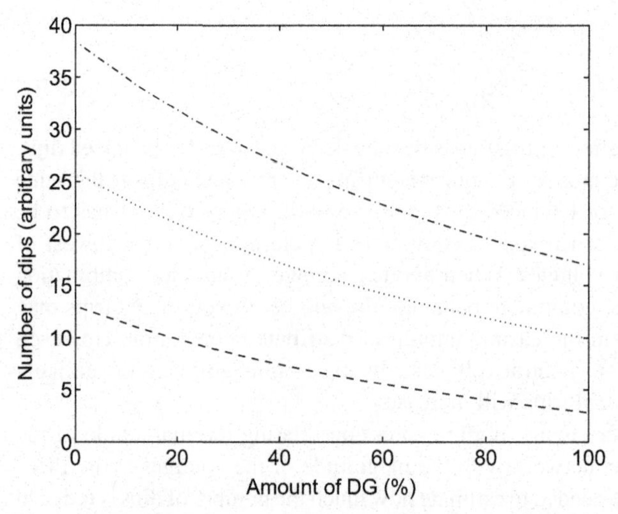

Figure 6.30 Dip frequency, for type C dips only, as a function of the amount of distributed generation with synchronous machine interface: SARFI-70 (dashed), and SARFI-80 (dotted), and SARFI-85 (dash-dotted).

The calculations given in the previous section for synchronous machines can be repeated for induction machines. Some of the results are presented below. The impact on the phase-to-neutral voltages is illustrated in Figures 6.32 and 6.33. Note that the voltage magnitudes actually become smaller for type C dips in some cases.

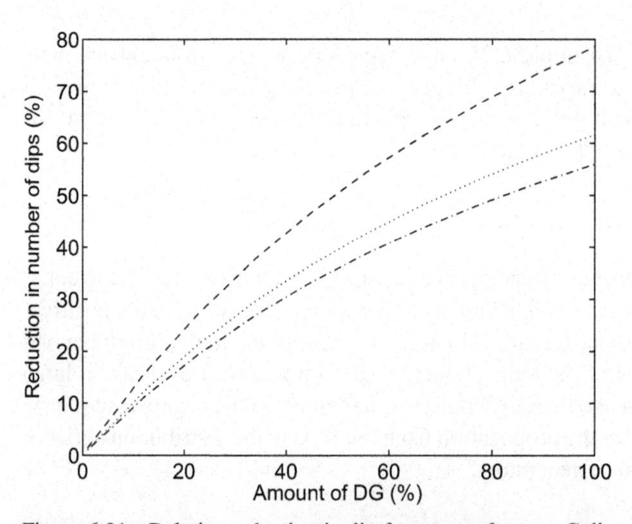

Figure 6.31 Relative reduction in dip frequency, for type C dips only, due to increasing amount of distributed generation with synchronous machine interface: SARFI-70 (dashed), SARFI-80 (dotted), and SARFI-85 (dash-dotted).

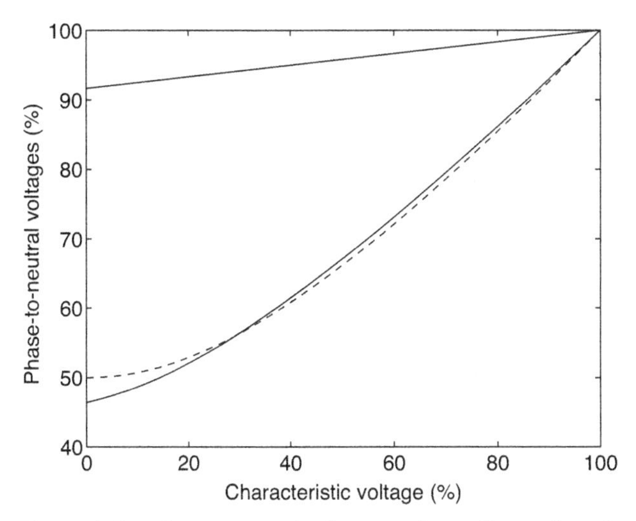

Figure 6.32 Phase-to-neutral voltages at the medium-voltage bus (solid line) and at the PCC (dashed line), as a function of the characteristic voltage at the PCC, for type C dips.

The reduction in negative-sequence voltage impacts the different phases in different ways. For type D dips, the presence of induction generators always results in an improvement in the voltage.

The reduction in the number of dips, with increasing amounts of induction generators, is illustrated in Figures 6.34 and 6.35 for type C dips and in Figures 6.36 and 6.37 for type D dips. The reduction in the number of dips is not as big as

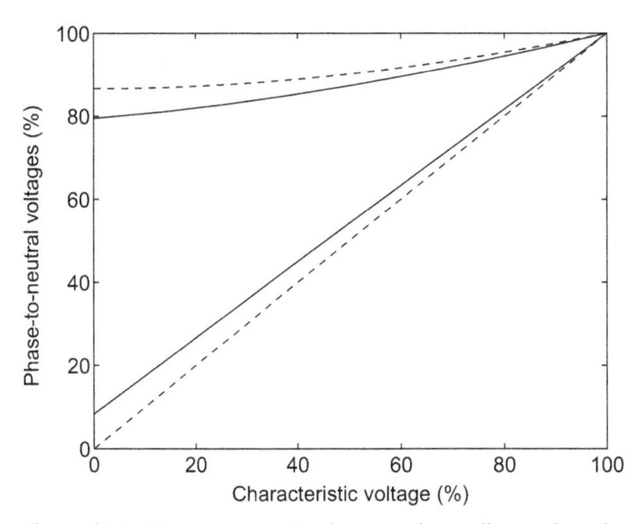

Figure 6.33 Phase-to-neutral voltages at the medium-voltage bus (solid line) and at the PCC (dashed line), as a function of the characteristic voltage at the PCC, for type D dips.

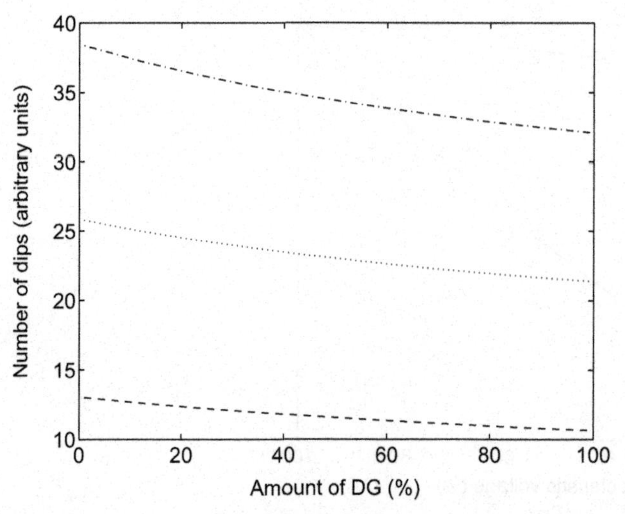

Figure 6.34 Dip frequency, for type C dips only, as a function of the amount of distributed generation with induction machine interface: SARFI-70 (dashed), SARFI-80 (dotted), and SARFI-85 (dash-dotted).

for synchronous generators, due to the fact that induction machines impact only the negative-sequence impedance. It should further be noted that induction motors, which constitute a significant part of the load, impact voltage dips in the same way as do the induction generators.

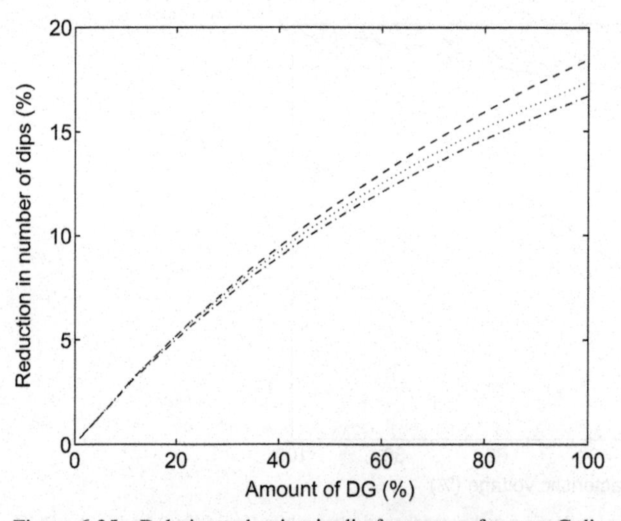

Figure 6.35 Relative reduction in dip frequency, for type C dips only, due to increasing amount of distributed generation with induction machine interface: SARFI-70 (dashed), SARFI-80 (dotted), and SARFI-85 (dash-dotted).

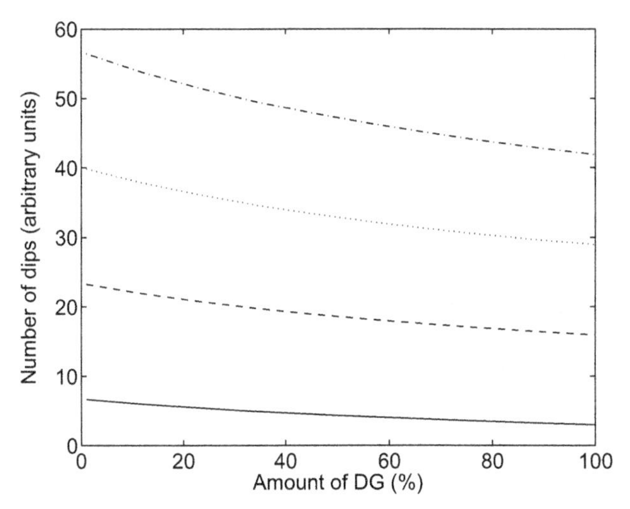

Figure 6.36 Dip frequency, for type D dips only, as a function of the amount of distributed generation with induction machine interface: SARFI-40 (solid), SARFI-70 (dashed), SARFI-80 (dotted), and SARFI-85 (dash-dotted).

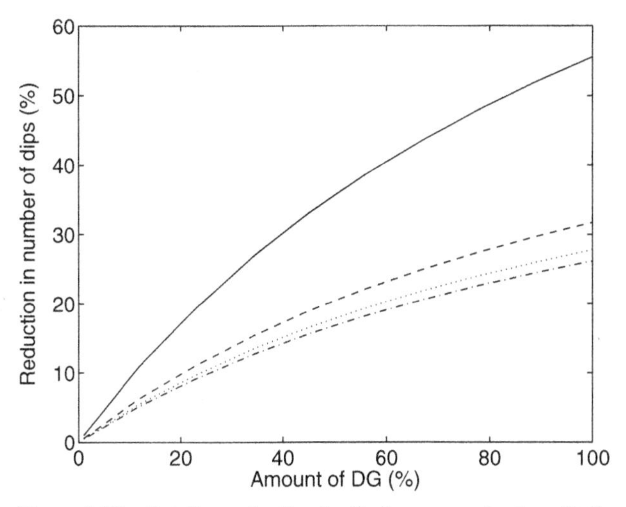

Figure 6.37 Relative reduction in dip frequency, for type D dips only, due to increasing amount of distributed generation with induction machine interface: SARFI-40 (solid), SARFI-70 (dashed), SARFI-80 (dotted), and SARFI-85 (dash-dotted).

6.7 INCREASING THE HOSTING CAPACITY

As was shown in this chapter, in some cases, the level of power quality disturbances sets a limit to the amount of distributed generation that can be connected to the power system. Power quality can be a local issue, like high disturbance due to one or a few generators, or a global issue, like resonances or the weakening of the transmission system. When power quality sets the limit, there are a number of measures that can be taken to allow more distributed generation to be connected, that is, to increase the hosting capacity. A number of these measures will be discussed in this section.

6.7.1 Strengthening the Grid

With power quality disturbances, the voltage disturbance level is often proportional to the source impedance at the location of the source of the disturbance. This is the case, for example, for voltage fluctuations (flicker), rapid voltage changes, and unbalance. For all these disturbances, it is the impedance for the power system frequency that matters. Increasing the fault level will reduce the voltage disturbance level. Also for harmonics, the source impedance impacts the voltage distortion in the same way, but the frequency dependence of the source impedance makes the situation more complex for harmonics. Increasing the fault level may increase or decrease the harmonic distortion, depending on the frequency. For low-order harmonics, strengthening the source will in almost all cases reduce the voltage distortion. For high-order harmonics, no general conclusions can be drawn.

There are various ways of strengthening the grid: adding lines or cables, adding transformers, reducing feeder length, and increasing the feeder cross section. These have traditionally been important tools in the design of distribution and transmission systems. It has become very difficult, however, to build new lines, especially at higher voltage levels. As a result, other solutions are more and more being discussed.

An additional reason why strengthening the grid is not always considered as a first option has to do with the fact that multiple stakeholders are involved in power quality. Network operators have a duty to maintain an acceptable level of power quality. To ensure this, they place limits on the emission by individual customers. Also within standard-setting organizations, like IEC and IEEE, emission limits for individual customers and devices are defined. It is considered that the emission by devices or individual customers is such that in a well-designed grid, the voltage disturbance levels do not exceed acceptable levels. Although there are a number of problems with this approach, such as what is a well-designed grid, this approach is rather general used by network operators as well as during the development of new standards. The result is that strengthening the grid is not seen as an appropriate tool for mitigating power quality disturbances. For power quality events, such as voltage dips, the situation becomes even more complicated. We will come back to this later.

6.7.2 Emission Limits for Generator Units

The phenomena discussed in the previous sections, overloading, losses, and voltage rise, are related to the transport of active power by the new generator units. It is difficult to imagine a generator unit that does not result in a change in active power transport. Setting limits on the active power transport would immediately put a limit on the size of the generator. With several of the power quality disturbances, the situation is different. For example, setting a limit on the amount of harmonics that can be emitted does not directly limit the size of the unit. It might only make the unit more expensive because a different grid interface could be required.

Emission limits for generator units could be set by network operators or in international or local standards. The IEEE harmonics standard, IEEE standard 519, sets requirements on the emission of any generator that are equal to the requirements for the largest customers. Also, the connection agreements for large wind power installations often contain emission limits. The European standard for the connection of microgeneration, EN 50438, refers to the generic emission standard, IEC 61000-6-3, for emission limits [135]. For harmonics and voltage fluctuations, a specific reference is made to IEC 61000-3-2 and IEC 61000-3-3, respectively. This means that the same emission limits hold for microgeneration (rated power less than or equal to 16 A per phase) as for other equipment connected to the low-voltage network.

The setting of emission limits for generator units is rather easy for low-frequency harmonics (up to 2 kHz), for very fast voltage fluctuations (the ones resulting in flicker), for unbalance with three-phase installations, and for rapid voltage changes. For most of these disturbances, relatively easy measures can be taken to limit the emission. Reducing flicker and rapid voltage changes will require some kind of energy storage that could be difficult to achieve for large installations. However, the flicker emission from such installations is in most cases sufficiently small, so no further reduction is needed.

The situation becomes more complicated for unbalance due to large, or large numbers of, single-phase generators, for slower voltage fluctuations (timescale several seconds or longer) and for high-frequency distortion. Reducing unbalance due to single-phase generators would require limits on the size and number of generator units. Reducing slower voltage fluctuations would require some larger energy storage. Also, there are not yet any suitable performance indices for slower voltage fluctuations. Such indices and the associated limits of what are acceptable levels are needed before emission limits can be placed on individual generator units. The same holds for high-frequency distortion.

Setting limits on the emission by individual generator units will not help against the weakening of the transmission system or against harmonic resonances.

6.7.3 Emission Limits for Other Customers

The emission by distributed generation is, in general, small as seen throughout this chapter. However, it does result in an overall increase in the level of voltage disturbances. This could be a problem locally when a large generator unit is connected at a location with a large existing disturbance level. A further increase is to be expected from the weakening of the transmission system and from harmonic resonances.

When the acceptable levels are not changed, this increase simply leaves less space available for new customers. The rules for allocating emission limits to individual customers are discussed in a number of documents. The most widespread ones are the three technical reports issued by IEC [211–213], but several countries and even individual network operators have their own rules. A common approach is to choose a level that is seen as the highest acceptable level after connection of the new customer. This is called the "planning level." Next, the planning level is compared with the existing level, that is, before the connection of the new customer. The difference between the existing level and the planning level is the permitted increase due to the new customer, using some kind of allocation algorithm. This emission limit, in terms of voltage, is next translated into an emission limit in terms of current. A complete approach for this is given in Ref. 36.

A rising disturbance level will increase the "existing level." Leaving the planning level unchanged will reduce the emission allocation to the new customer. This new customer may actually be a distributed generator or a wind park. When the existing level comes close to the planning level or even exceeds the planning level, connecting new installations could become difficult and expensive.

6.7.4 Higher Disturbance Levels

Instead of putting stricter emission limits on individual customers, generating and nongenerating customers, the choice of planning levels and other objectives can be reevaluated. This is a sensitive issue because it increases the probability of electromagnetic interference, that is, equipment damage or malfunctioning due to high levels of voltage disturbances. Before we discuss ways of allowing higher disturbance levels, it is good to have a look at the complete framework shown in Figure 6.38. The ultimate aim is to ensure that equipment connected to the power grid works as intended or as stated in the IEC standards to ensure a high probability that electromagnetic compatibility is achieved.

The reference level in the figure could be the electromagnetic compatibility level according to IEC standards, voltage characteristics according to EN 50160, or the highest permitted levels set by the local regulator. The reference level is chosen such that the immunity of equipment, with high probability, is higher than the reference level. Requirements on equipment immunity (i.e., "immunity limits") are set in international standards and are difficult to change. It is, however, the network operator's responsibility to keep the disturbance levels below the reference level.

To be able to ensure this, the network operator is putting requirements on the emission from individual customers connected to the grid. During the design of the network and when assessing the connection of new customers, the aim is to keep the disturbance level below a planning level. There are indicative values for these in IEC technical reports, but in the end it is up to each network operator to set appropriate planning levels. The planning levels are always chosen below the reference level so

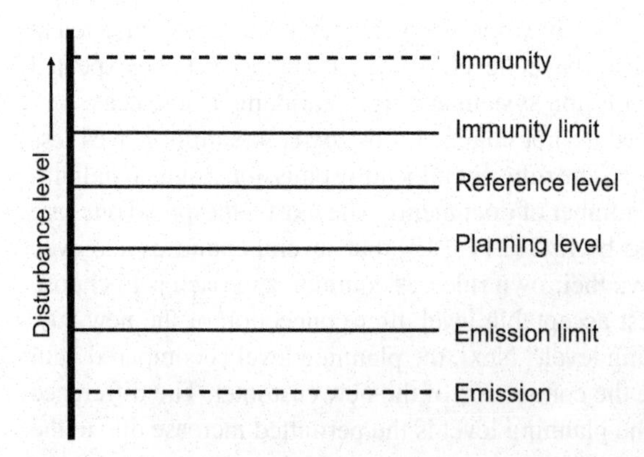

Figure 6.38 Different objectives and disturbance levels related to the compatibility between equipment and the supply.

as to have a safety margin covering, among others, uncertainties in the calculation models, future growth in emission, and emission from small customers to which no emission limits can be allocated.

The amount of emission allowed for a given customer (i.e., the "emission limit") is determined from the planning level using some kind of allocation algorithm. For individual devices, emission limits are set in the IEC product standards and in other standards or regulations.

From Figure 6.38 it becomes clear that raising the planning level or using an allocation method resulting in a higher emission limit will not immediately result in widespread interference. There are sufficient margins between the different levels and interference occurs only when the actual disturbance level exceeds the actual immunity of a device. Raising the planning level, however, immediately allocates more space for emission by individual customers and devices. This would simplify (i.e., make cheaper) the connection of new industrial installations and the connection of distributed generation.

Some network operators use rather strict and low planning levels and may be very reluctant to raise them. The reason for this is to ensure that the disturbance level does not exceed the reference level. This will reduce the risk of interference, but it could result in an excessive burden on the emission side. This in turn could be a barrier to the introduction of renewable sources of generation. An important part of any discussion on raising the planning levels or the emission limit is the question of who carries the risk when the reference level is exceeded? When this risk is carried solely by the network operator, the incentive to raise the planning level will obviously be very small.

The next step to consider is related to increasing the reference level. This is more difficult because it will probably require a change in international standards on electromagnetic compatibility and voltage characteristics. There has been strong objection to this from many sides, including from the authors of this book. Allowing the reference level to be raised when the actual disturbance level becomes high, would remove any incentive for network operators to improve their network. The various power quality and EMC standards are coordinated such that a well-designed grid and well-designed customer equipment would, with high probability, result in electromagnetic compatibility between the equipment and the grid. The setting of compatibility levels and other reference levels is in turn based on a trade-off between the costs for improving immunity and the costs for improving emission. Even with the introduction of distributed generation would these basic rules hold.

The introduction of large amounts of distributed generation might, however, require a reconsideration of the various limits. The first change is that emission takes place at frequencies that were not a concern before: interharmonics and even harmonics, high-frequency harmonics, and slower voltage fluctuations. The reference levels for these are either low (interharmonics and even harmonics) because it was easy to reduce the emission or no limits exist at all (slower voltage fluctuations and high-frequency harmonics) because the disturbance was not seen as a concern. The introduction of large amounts of distributed generation will result in an increase in the levels of interharmonics and even harmonics. As those harmonics were rarely emitted in the past, the reference levels are low and can easily be exceeded. The same will occur if the reference values for distortion above 2 kHz are extrapolated from the ones below 2 kHz, as was shown in Section 6.5. In fact, more and more consumer electronics and lighting equipment also generate interharmonics and high-frequency harmonics. Also for triple harmonics, the disturbance levels now exceed or come close to the reference values in many cases, even in the existing grid without distributed generation. The discussion on the choice of reference levels, therefore, also concerns distributed generation.

The impact of interharmonics, even harmonics, and high-frequency harmonics on equipment remains unknown. Being careful has been to a large extent the guiding principle when setting limits, that is, setting the limits for voltage distortion somewhat below the highest levels measured at locations where no problems were reported. Investigations are needed to decide, for example, if it is reasonable to set more strict limits at 370 Hz than at 350 Hz (in a 50 Hz system).

6.7.5 Passive Harmonic Filters

Passive harmonic filters consist of inductances and capacitances and sometimes resistances. The two most commonly used configurations are the series filter and the high-pass filter. The series filter, consisting of an inductance and a capacitance, creates a low-impedance path for harmonic currents at a single frequency. The filter is typically tuned at or close to harmonics 5 and 7 or 11 and 13. Passive high-pass filters, containing also resistances, create a low-impedance path for all frequencies above a certain cutoff frequency. The cutoff frequency is typically rather high, above the 15th harmonic, to prevent increased losses at the power system frequency.

Passive series filters can be used in all cases where the voltage distortion for an individual frequency would otherwise become too high. This is most likely not due to the emission from distributed generation but more likely due to harmonic resonances or weakening of the transmission grid. Passive high-pass filters would be a possible solution to high-frequency distortion. The disadvantage of passive series filters is that the capacitance of the filter, together with the source inductance, creates a parallel resonance, with a high impedance, at a lower frequency. This may require additional filters for lower harmonics and has shown to be a problem with broadband emission.

6.7.6 Power Electronics Converters

Mitigation equipment based on power electronics converters has a high potential to improve the power quality. Static var compensators (based on thyristor technology) and STATCOMS (based on voltage source converter (VSC) technology) have been successfully used to mitigate voltage fluctuations. Such converters can be installed at central locations in the system, for example, to limit voltage fluctuations in weak parts of the transmission grid, or close to a source of voltage fluctuations like a wind turbine.

Converters based on VSC converters can also be used to mitigate harmonic distortion as long as the switching frequency is sufficiently high. A whole range of algorithms are being discussed to mitigate voltage distortion without the need to know the current emission. Such equipment could be installed at central locations in the system. Commercially available equipment is still based on open-loop control where the injected current compensates the measured current. The use of such equipment is limited to locations close to the source of the emission.

Power electronics converters based on VSC technology are also more and more used as the grid interface for distributed generation. Examples are the inverter with micro-CHP or solar power and the full power or partial power (DFIG) inverter with wind power installations. The same algorithms for mitigating voltage fluctuations and harmonic distortion can also be implemented in these converters. An interesting application that is being discussed in the literature is to implement damping at selected frequencies. At these frequencies, the converter behaves like a resistance providing damping in case of harmonic resonances. Such an approach would be relatively simple to implement because no additional hardware is needed. The requirements on the accuracy of the algorithm, like the frequency tuning, do not have to be high because the damping is provided by many individual converters. The occurrence of a resonance cannot be prevented this way, but the damping will reduce the resulting voltage and current distortion.

6.7.7 Reducing the Number of Dips

The mentioned increase in the number of dips due to weakening of the transmission grid is a separate issue for which most of the mentioned solutions do not hold. It should first be noted that no reference levels exist for voltage dips (with some exceptions, which we will not discuss here, see Ref. 53 instead). There are no limits on what is an acceptable number of voltage dips. The first "solution" to the increased number of dips would be to just refer to the lack of any standards and simply accept it.

This would put the risk completely on those customers that are sensitive to voltage dips, typically large industrial installations connected to the subtransmission or transmission grid. There are, however, a number of ways to mitigate the voltage dip issue, independent of what the cause of the dips is:

- Reducing the number of faults or reducing the percentage of faults involving more than one phase. There are many different ways of achieving this, like shielding wires, animal protection, and replacing lines by cables. Apart from the costs, this is often one of the best solutions.
- Limiting the spread of dips by changing the way in which the grid is operated. Limiting the number of lines connected to a substation, limiting the number of parallel lines, and opening loops are some options. The disadvantage with most of these solutions is that it weakens the grid, with reliability, stability, and power quality problems as possible consequences. The costs of such solutions could, however, be rather low.
- Reducing the fault-clearing time by using faster protection. This is often a good solution as well, but it could increase the risk of protection mal-trip with consequences for reliability and operational security.
- Installing mitigation equipment such as some local energy storage to allow the installation to ride through the dip. Also, microgrids with the possibility to go very quickly into island operation have been mentioned as a solution.

- Improving the immunity of equipment in the installation and of the industrial process against voltage dips.

6.7.8 Broadband and High-Frequency Distortion

It was shown in this chapter that the harmonic emission by distributed generation has a large broadband component next to the "normal" harmonics (5, 7, etc.) present in existing equipment. This broadband component continues up to high frequencies, where it remains unclear what typical distortion looks like for higher frequencies. The existing standards for measuring and evaluating distortion are based on the presence of narrowband signals near the integer harmonics of the power system frequency. Many of the measurements on distributed generation use standard methods and, therefore, often do not notice the presence of the broadband component in the spectrum. Measurements of longer time windows are needed to map the emission due to various types of distributed generation. Further understanding, describing, and modeling the emission by distributed generation over a wide range of frequencies (e.g., from 51 Hz to 150 kHz) in both time and frequency domains remains an important research challenge.

CHAPTER 7

PROTECTION

The protection of the distribution system is already difficult at the best of times. By using only local current measurements, each protection relay should be able to distinguish between a fault and a nonfault situation. The latter includes faults for which the relay is not supposed to generate a tripping signal. Each relay should be able to make this distinction for all fault locations, for all fault types, for all load levels, with the feeder in normal configuration and in any possible backup supply configuration. The relays should further be able to provide backup for all downstream relays. The presence of generation will further complicate the protection. For more details on the protection of distribution systems (mainly without distributed generation), the reader is referred to several good books on this subject [13, 107, 131, 174].

7.1 IMPACT OF DISTRIBUTED GENERATION

Distributed generation impacts the protection of distribution networks in a number of ways. Some impacts are due to an increase in fault current because of the generator. Other impacts occur because the fault current contribution from a generator is too small. One of the widely discussed consequences of this is the risk of noncontrolled island operation. This and the protection against it will be discussed in Section 7.7.4. The impact of distributed generation on protection strongly depends on the fault current contribution and thus on both the size and type of the interface. Synchronous generators deliver a continuous short-circuit current; induction generators contribute during one or two cycles in case of a three-phase fault and longer in case of a nonsymmetrical fault. Units with power electronics interface have none or a very limited contribution to the fault current. An overview of the impact of distributed generation on protection is given in several publications, including Refs. 123, 124, 172, 232, 238, and 239.

Protection of a distribution network, in the vast majority of cases, is based on overcurrent protection. This can be definite-time or inverse-time overcurrent protection. Each has its own advantages and disadvantages. The setting of the protection should be such that each fault is cleared sufficiently fast with minimum impact on the customers, even when a protection device would fail. Obtaining such settings for a distribution feeder is a complicated task because it has to include all possible system states and because the protection relays used in distribution system only have

Integration of Distributed Generation in the Power System, First Edition. Math Bollen and Fainan Hassan.

a limited amount of information available. The input signals to an overcurrent relay are only the currents in the three phases.

Adding a generator to a distribution feeder will change the fault currents and thus increase the risk of a protection failure. The protection can fail in two different ways:

- **Unwanted operation:** The opening of a circuit breaker when there is no fault that warrants operation of that breaker.
- **Failure to operate:** A circuit breaker that should open to clear a fault does not open or opens too late.

Both failures do occur occasionally, even in distribution networks without generation, for example, due to equipment failure or due to incorrect settings. It is very difficult to get information on the failure rate and probability for protection. An overview of the information available in 1993 is given in Ref. 41; for protection equipment, the reported probabilities of unwanted operation and failure to operate range from 0.01% up to 5%. Times to failure range from a few years up to thousands of years. A further subdivision of protection failures is given in Ref. 50:

- **Failure of the concept:** An unwanted operation or failure to operate due to a trade-off that has been made during the design of the system. One of the demands of the protection might be impossible to fulfill because of too high costs.
- **Failure of the model:** A failure that was already present in the design but that was not known by the designer due to the model being incomplete or inaccurate. Also, protection failures due to changes in load or system, without changing the protection settings, should be seen as failures of the model.
- **Failure of the setting:** Failures of the protection due to a setting not being equal to the design setting. This may be due to human error or due to a change in settings over time.
- **Failure of the device:** All those failures inside a protective device (instrument transformer, protective relay, and circuit breaker) that result in unwanted operation of failure to operate.

The introduction of distributed generation will increase the risk for a failure of the model. In the longer term, the occasional protection failure might be accepted to allow more distributed generation to be connected. In that case, the probability of failure of the concept has increased. Note that this distinction is of no interest to the customer, both will increase the number of interruptions in the same way. It is, however, important for the design process and for reliability analysis to understand the difference.

The following protection failures could occur due to distributed generation (see, for example, Refs. 123, 124, 172, 232, 238, and 239):

- The fault current contribution from the generator could result in a total fault current exceeding the rating of some equipment. This is a concern with both induction and synchronous generators. Synchronous generators increase both the make-current and the break-current, whereas induction generators increase only the make-current. Generators with power electronics converters do not contribute much to the rise in fault current. The amount of generation that can be connected (the "hosting capacity") depends on the existing margin between the maximum fault current and the component rating. If this margin is small, the hosting capacity is small. In most distribution networks, there is sufficient margin, but in some cases the hosting capacity is zero or close to zero. The increase of the fault current due to distributed generation is discussed in Section 7.6.

- The presence of a generator along a feeder will reduce the fault current at the start of the feeder, for a fault beyond the generator. When the fault current for a fault at the end of the feeder drops below the overcurrent setting of the relay at the start of the feeder, a failure to operate may occur. This is especially a concern with synchronous machines and to a lesser extent with induction machines. The problem is most severe for long feeders, where the fault current for a remote fault is already low. With high-impedance faults, the situation becomes even more severe. When using definite-time overcurrent protection, certain faults will simply not result in protection operation. When using inverse-time overcurrent protection, the fault-clearing time may become unacceptably long. We will come back to this in Sections 7.2 and 7.3.2. Calculations of the impact of one or more generators on the relay sensitivity at the start of the feeder are given in Ref. 274. With inverse-time protection, this could result in a very long fault-clearing time, especially when used as backup protection [229]. The worst-case fault-clearing time is in most case shorter for definite-time than for inverse-time overcurrent protection. Simulations of this so-called "protection blinding" are also given in Ref. 294. It is shown that the impact is biggest for a generator connected half-way along the feeder.
- The fault current by the generator, for a fault on another feeder, could result in unwanted opening of the breaker at the start of the feeder. This is again especially a concern with synchronous machines and to a lesser extent with induction machines. This is further studied in Sections 7.2 and 7.3.1. With large amounts of such machines connected to the grid, a situation could occur where the maximum current for an upstream fault (for which the protection should not intervene) is higher than the minimum current for a downstream fault (for which the protection should intervene). In that case, it is no longer possible to obtain selective protection in all cases. Detailed calculations to estimate the hosting capacity are presented in Section 7.4. If a generator is connected through a delta-star transformer, with the star connection on the grid side, it will form a low-impedance path for the zero-sequence current during an earth fault. This will greatly increase the probability of maloperation of the feeder breaker, but it will make it much easier to trip the generator selectively during an earth fault. Most network operators, however, do not allow this connection.
- When using fast reclosing for fuse saving, the coordination between the fuse and the recloser is endangered by the presence of distributed generation along the feeder. The generator contributes to the fault current through the fuse but not to the fault current through the recloser. The design of the protection is based on the

assumption that both will carry the same fault current. The consequence could be that the fuse operates faster than the recloser is able to open. The result is a long interruption for the customers on the lateral protected by the fuse. This is discussed, among others, by Refs. 61, 62, 84, 170, and 176. In Ref. 84, the hosting capacity is calculated for an example network and different cases, including improvements. The impact of distributed generation on fuse–recloser coordination is discussed in Section 7.2.3.

- In distribution networks with fuses at different locations, where fuse–fuse coordination is needed, the impact of distributed generation could be felt rather quickly according to Ref. 62. Although a relay has only one tripping curve, two curves should be distinguished for fuses: the "minimum melting time " and the "maximum clearing time." When the latter is exceeded, the fuse will open, correctly or incorrectly. When the minimum melting time is exceeded but not the maximum clearing time, the fuse will not open but merely age faster. The situation could arise somewhere later that a high but acceptable current, for example, due to the starting of a motor, results in the opening of the fuse. A general rule for fuse–fuse coordination is that the maximum clearing time for the downstream fuse should not exceed 75% of the minimum melting time of the upstream fuse [174, Section 6.2].
- The fault current contribution of some generators is not big enough for the overcurrent protection to detect the fault. A failure to operate could result from this. This is especially a concern for small generators and any generator equipped with a power electronics interface. Nondetection of the fault also occurs for a generator with a delta-star-connected generator transformer and an earth fault in the grid. The consequences of nondetection are plural, including one of the most serious concerns with the integration of distributed generation: the risk of uncontrolled island operation. This will be discussed in more detail in Section 7.7.4. As a consequence of the generator not detecting the fault and, therefore, remaining connected, it will interfere with the autoreclosing scheme in use to limit the number of long interruptions. Even the small fault current supplied by the generator could be sufficient to prevent an arcing fault from extinguishing. The result is an unsuccessful reclosure followed by a permanent trip and a long interruption for all customers connected to the feeder. A second consequence is that the reclosing with the generator still connected could result in phase opposition, with damage to network components or to the generator unit. It is recommended in Ref. 124 to delay autoreclosure to about 1 s for feeders containing distributed generation. This assumes that the anti-islanding protection is able to remove all generators from the feeder within 1 s. According to Ref. 432, fast reclosing is typically 12–18 cycles (at 60 Hz, i.e., 200–300 ms), which may not be enough for the generator protection to detect and clear the fault situation. The situation becomes impossible to solve with very fast reclosing, as fast as 3–4 cycles, as part of a fuse saving scheme [432].
- In Ref. 129, the protection of looped feeders with distributed generation is studied. Such looped feeders are common in subtransmission systems and are also sometimes used in primary distribution networks. Coordination of the

protection using a combination of overcurrent and directional overcurrent relays is difficult even without generation connected to the loop. Adding generation even directional overcurrent protection is not able to solve all coordination problems. It is suggested in Ref. 129 to add fault current limiters to reduce the contribution of the generators to the fault current. This could be a suitable solution in primary distribution networks, assuming that it will not result in stability issues and that it will not cause fail-to-trip of the generator protection. For subtransmission systems, a combination of distance protection and differential protection is a better solution.

- The presence of a generator on a distribution feeder could result in the incorrect operation of fault current indicators. These are commonly used in underground cable networks to facilitate the location of the faulted section and to reduce the duration of an interruption of the supply for the customers. The presence of a generator unit downstream of a fault may result in the fault current indicators between the fault and the generator to become activated. This will make the fault location more difficult and result in longer interruption duration. The hosting capacity for this is similar to the hosting capacity for definite-time overcurrent protection.
- Single-phase reclosing or single-phase fuse clearing can result in ferroresonance conditions with distributed generation. Therefore, fuses and single-phase breakers should be avoided between a three-phase generator and the next upstream three-phase breaker [432]. As we will see in Section 7.7.5, ferroresonance can even occur after three-phase tripping of the feeder.

7.2 OVERCURRENT PROTECTION

7.2.1 Upstream and Downstream Faults

Synchronous machines give a sustained contribution to the fault current for any type of fault. This current can interfere with the correct operation of the overcurrent protection against short circuits or earth faults. Consider first a fault in another distribution feeder than the one to which the generator is connected, as shown in Figure 7.1. For a correct operation of the protection, breaker CB2 should open and breaker CB1 should not.

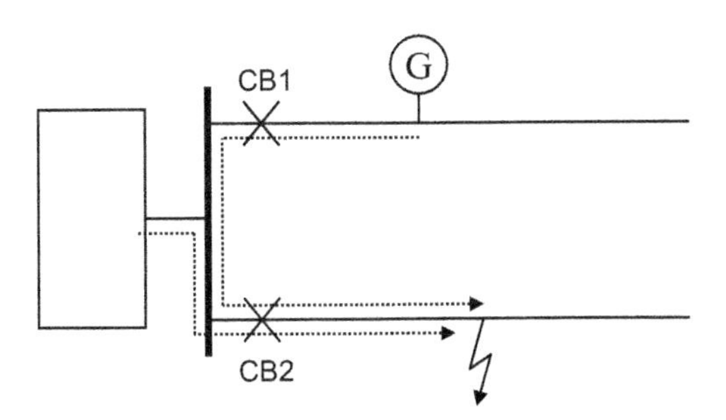

Figure 7.1 Fault current contributions for a fault on a feeder without generation.

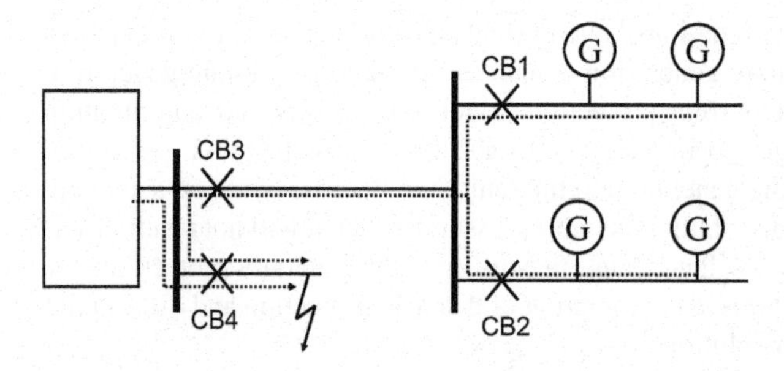

Figure 7.2 Fault current contributions for an upstream fault.

Also, the generator should not trip. The contribution of the generator to the fault may, however, be mistaken for a downstream fault (i.e., a fault on the feeder to be protected) by the protection associated with breaker CB1. The result would be a mal-trip of breaker CB1. Overcurrent protection in distribution systems is rarely equipped with a directional element; the tripping criterion is based on the magnitude of the current only.

The problem may also occur for faults further upstream in the grid as shown in Figure 7.2. Both breakers CB1 and CB3 may experience an erroneous trip. As the fault is electrically further away, the risk of mal-trip for breaker CB1 or CB2 will be less than in the previous example. However, when generators are present at several locations downstream of CB3, the risk of mal-trip could be significant for this breaker. Breaker CB3 may even be at a higher voltage level. In that case, large amounts of generation could result in losing a whole distribution network due to a nearby fault at the transmission level.

For a fault downstream of the generator, as shown in Figure 7.3, the fault current through breaker CB1 becomes less. The result could be that the overcurrent protection no longer detects the fault. Both mal-trip and fail-to-trip of overcurrent protection will be discussed in more detail in Sections 7.3 and 7.4.

7.2.2 Hosting Capacity

When considering the impact of generation on protection in distribution networks, different hosting capacities should be considered, corresponding to different changes to be made in the system.

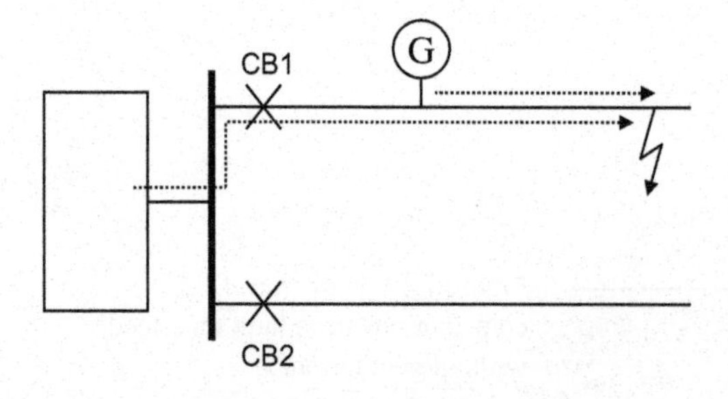

Figure 7.3 Fault current contributions for a downstream fault.

1. ***Changing protection settings*** This may be required for small levels of generation and is limited to a change in the current limits. A new protection coordination is recommended whenever a generator is connected to a distribution grid so as to ensure that no coordination problems occur. However, the costs of the changes are small as no investments in equipment are needed.
2. ***Additional time step in protection coordination*** When a feeder contains a high amount of generation, more than about 25–50% of the transformer rating, according to various studies [114–116, 185], coordination problems occur that require the introduction of a new time step. This is in itself easy and cheap, but it could cause problems related to withstanding fault current for transformers or cables as well as with personnel safety. If such problems occur, significant investments may be needed where a dedicated transformer to the generator could be the easiest solution.
3. ***Additional circuit breaker*** For a long feeder, the difference between the highest load current and the lowest fault current may become too small for a reliable protection. This requires the installation of a circuit breaker somewhere along the feeder, in fact a new substation. For long feeders that are close to their maximum length, the installation of a small amount of generation may require this. It is not clear how common this situation is, but the impression is that there is sufficient margin for most feeders. It should also be noted that for feeders close to their maximum length, a small amount of load growth would also result in the need for an extra circuit breaker.
4. ***New protection concepts*** When two or more feeders from the same transformer have a significant amount of generation, it is no longer possible to obtain selectivity with only overcurrent protection. Other solutions are needed like directional protection and communication between relays.

7.2.3 Fuse–Recloser Coordination

A special case of protection coordination occurs when autoreclosing is used together with a practice called "fuse saving." It is used for overhead medium-voltage feeders in remote areas. With reference to Figure 7.4, the recloser is equipped with an overcurrent relay, which opens the recloser with minimum delay. Any delay needed would be to accommodate for load starting, transformer energizing, and so on. The fuse is chosen such that it will not be impacted by the fault current under the fast stage of the recloser. Next, the recloser closes again and in most cases the fault will have disappeared. About 80% of faults on overhead distribution lines are of a transient nature, so a simple reclosing action is sufficient to remove the fault.

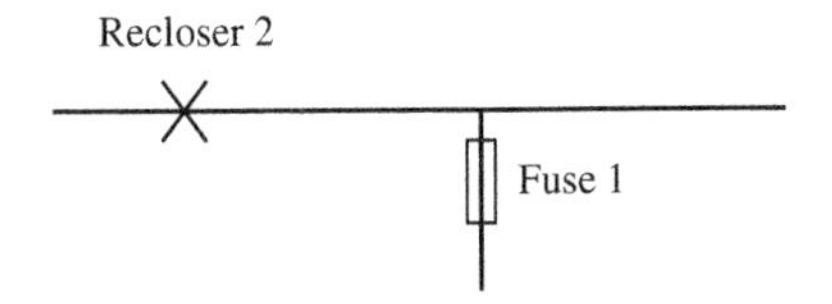

Figure 7.4 Feeder with recloser and fuse.

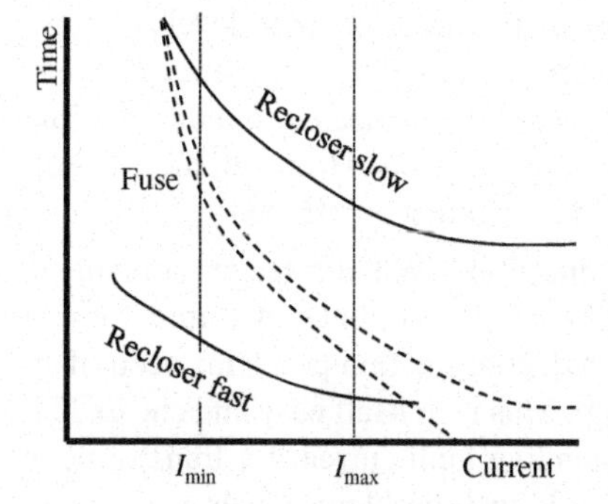

Figure 7.5 Coordination of fuse and recloser in a scheme with autorecloser and fuse saving.

When the fault turns out to be a permanent fault, the recloser will experience again a high current. In some cases, the second fast opening is attempted with a longer time before reclosing, but in other cases, the recloser goes to its slow stage immediately. The time–current curve of the slow stage is such that the fuse will clear the fault before the recloser opens permanently. This scheme minimizes the number of long interruptions experienced by the customers connected to the feeder. See also Refs. 42, Chapter 3; 70, Chapter 4; and 125, Section 3.7, for a discussion on reclosing, short, and long interruptions.

The coordination between the fuse and the recloser is somewhat complicated and is illustrated in Figure 7.5. The two vertical dotted lines indicate the maximum and the minimum fault current for a fault downstream of the fuse. It is for this range of currents that the curves should be coordinated. The two dashed curves are the minimum melting time (bottom curve) and the maximum clearing time (top curve) of the fuse. The following coordination rules apply:

- The minimum melting time of the fuse should be longer than the clearing time of the fast stage of the recloser. A certain minimum distance between the curves should be maintained. In Ref. 174, minimum factors are recommended: 1.25 times for the single reclosing, 1.35 for two fast reclosings with a reclosing time of 1 s or more, 1.8 times for two fast reclosings with a reclosing time of one half second.
- The maximum clearing time of the fuse should be shorter than the clearing time of the slow stage of the recloser.

The presence of a generator on the distribution feeder, as shown in Figure 7.6, will impact the coordination in a number of ways. The main impact is that the current through the fuse is no longer the same as the current through the recloser. The current through the fuse will increase, whereas the current through the recloser will decrease.

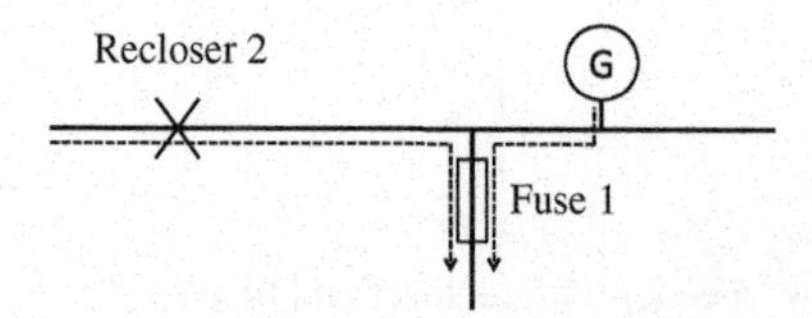

Figure 7.6 Feeder with recloser, fuse, and generator.

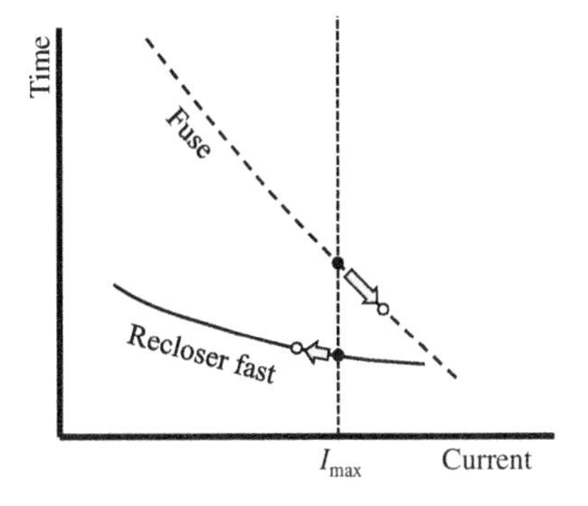

Figure 7.7 Impact of distributed generation on the coordination between fuse and recloser during the fast stage.

The latter occurs because the generator will somewhat maintain the voltage at the start of the lateral, in the same way as discussed in detail in Section 7.3.2. Both changes will reduce the coordination margin.

This is illustrated in Figure 7.7, where only the fast stage curve for the recloser and the minimum melting time of the fuse are shown. Due to the presence of the generator the current through the fuse increases and the current through the recloser decreases. The black dots in the figure show the current and clearing times for the feeder without generator. The setting of the recloser and the selection of the fuse are based on this and the coordination margin should be sufficient in this case.

The open circles show the current and clearing times for the situation when a generator is present. The currents are no longer the same, and especially the minimum melting time of the fuse will be impacted. The difference between the opening time of the recloser and the minimum melting time of the fuse will become less. The fuse could melt or even break the current. Fuse melting without breaking results in aging of the fuse with likely an unwanted operation of the fuse on a future fault or high load. Breaking of the current would immediately result in a long interruption for the customers downstream of the fuse.

The result of the miscoordination would be an increased number of long interruptions for customers downstream of the fuse. The impact is, however, not as severe as one might conclude from the above reasoning, for a number of reasons.

- With protection coordination, a certain safety margin is always included, for example, to accommodate for future load growth. It should, for example, be noted that induction motors connected to the feeder also result in a higher current through the fuse than through the recloser. This is naturally included in the protection coordination.
- The miscoordination occurs only when the fault current is close to the maximum fault current, that is, for faults close to the location of the fuse. For faults further down the fuse, there remains sufficient margin.
- Between 10% and 30% of the faults on an overhead feeder result in long interruptions, even without distributed generation. It is only when this percentage increases significantly that a revision of the protection scheme is needed.

It is not easy to draw any general conclusions about the hosting capacity without performing detailed calculations for a number of distribution feeders. A study for a rather extreme feeder, the IEEE standard 34 node test feeder, known for its length and load unbalance, is presented in Ref. 382. The test network mainly consists of 24.9 kV, but with one 4.16 kV lateral. A total of 12 fuses and 1 recloser were used for the protection of the feeder. It is shown that adding 20% distributed generation does not result in miscoordination. Miscoordination problems started to occur with 30% distributed generation.

The coordination during the slow stage of the reclosing is not adversely impacted by the presence of distributed generation. The fuse becomes faster and the recloser becomes slower, which increases the time margin. Unfortunately, this additional margin during the slow stage cannot be used to mitigate the problems during the fast stage because the generator may not be always present. As was already mentioned at the beginning of this chapter, the protection coordination of distribution feeders should be set such that it covers all situations.

7.2.4 Inverse-Time Overcurrent Protection

The calculation of the hosting capacity in Section 7.4 will be based on definite-time overcurrent protection, where the protection relay generates a tripping signal once the current exceeds the threshold for longer than a certain time. Instead of definite-time overcurrent protection, the so-called "inverse-time overcurrent protection" can be used, where the tripping time is a function of the current. The higher the current, the faster the trip signal generated. The curves for recloser and fuse shown in Section 7.2.3 are examples of inverse-time curves. In fact, one of the advantages of inverse-time relays is that their coordination with fuses becomes easier. Sometimes a distinction is made between "inverse time," "very inverse time," and "extremely inverse time." Here we will use the term inverse-time to refer to all of these.

In this section, we will briefly discuss the impact of distributed generation on the protection coordination when inverse-time overcurrent relays are used. Three cases will be considered: the coordination between upstream and downstream relay, the unwanted tripping during upstream faults as shown in Figure 7.1, and the fail-to-trip during a downstream fault as in Figure 7.3.

When coordinating a fuse or a relay with a downstream fuse or relay, the impact of distributed generation will be the same as for the coordination during the slow stage of a recloser, as discussed in Section 7.2.3. The presence of a generator on the feeder will increase the current through the downstream device, making it faster. At the same time, the current through the upstream device will reduce, making it slower. The time margin between the upstream and downstream devices will increase, thus improving the coordination. Also, it is in general not possible to use here this additional margin to prevent unwanted trip or fail-to-trip elsewhere: the protection should be selective with and without the generator connected. When it can be assumed that the generator will be connected most of the time, it may be decided to accept a certain nonselectivity when the generator is not connected. The impact of this on the reliability experienced by the end customers should be considered in such a decision.

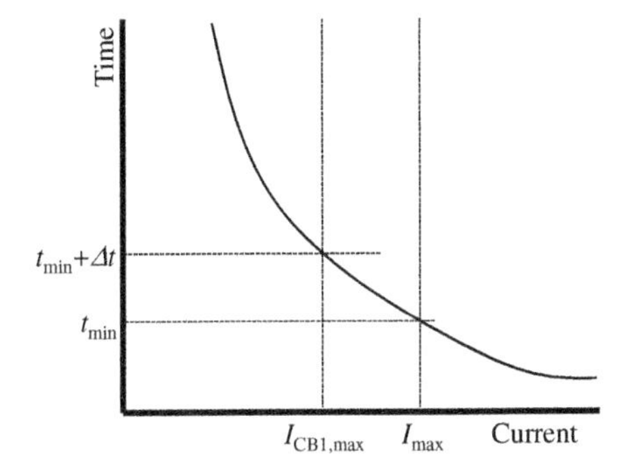

Figure 7.8 Inverse-time overcurrent protection: unwanted operation during upstream faults.

Consider again the situation shown in Figure 7.1, but now it is assumed that both breakers (CB1 and CB2) are equipped with inverse-time overcurrent relays. We will assume that both relays have the same setting. The time–current curves of the two relays are shown in Figure 7.8. Note again that this is not the standard protection coordination because the current through the two breakers is not the same.

The worst case situation is a fault just downstream of breaker CB2; this occurs when the current through breaker CB1 is highest. The current through breaker CB2 is the maximum value used for protection coordination, indicated as I_{max}. As shown in the figure, the relay will generate a tripping signal after a time t_{min}. For breaker CB1 to be selective, it should have sufficient time margin with CB2. Suppose that the minimum acceptable time margin is Δt, so the minimum tripping time of the relay with CB1 is $t_{min} + \Delta t$. The maximum acceptable fault current contribution from the generator (through CB1) is the current that results in this tripping time: $I_{CB1,max}$ in the figure.

Example 7.1 Consider a feeder equipped with standard inverse overcurrent protection. The relation between tripping time t and current I is as follows [174, p. 76]:

$$t = k \times \frac{0.14}{(I/I_s)^{0.02} - 1} \tag{7.1}$$

where k is the time multiplier setting and I_s is the pickup current setting. Assume that the maximum fault current I_{max} is 20 times the pickup current. Assume also that the time multiplier is set to 1.0. The tripping time is 2.3 s for a fault just downstream of the breaker. Three methods for selecting the minimum margin are compared:

- Short fixed margin: 0.3 s
- Long fixed margin: 0.5 s
- Variable margin: 0.25 s plus 27.5% of the downstream setting

The resulting minimum tripping time of the relay in the nonfaulted feeder is 2.3, 2.6, and 2.8 s, respectively. Using (7.1), this corresponds to current equal to 13.8, 11.4, and 8.5 times the pickup current. When the contribution from the generator exceeds this value, it is time to reconsider the protection setting.

Example 7.2 Instead of the standard inverse curves from Example 7.1, assume the "typical operating curves for inverse-time relays" as presented in Figure 5.8 of Ref. 174. We again assume a maximum current equal to 20 times the pickup current and a time multiplier setting equal to 1.0. The operating time is 0.06 s. Using the same three methods as in Example 7.1 gives the following minimum operating times for the relay in the nonfaulted feeder: 0.36, 0.56, and 0.33 s. In terms of current contribution from the generator, this corresponds to 4.0, 3.3, and 4.2 times the pickup current.

Example 7.3 Consider the worst-case situation in Example 7.2, where the maximum fault current contribution from the generator is 3.3 times the pickup current. What would be the hosting capacity?

Assume that the generator contributes five times its rated current to the fault. This is on the high end of the scale so that the hosting capacity will be on the low end. A fault current contribution of 3.3 times the pickup current corresponds thus to a rated current of 0.66 times the pickup current. The setting of the pickup current is typically somewhat above the maximum load current, say 1.1 times.

The resulting hosting capacity is 0.6 times the maximum load current.

From the above examples, it follows that rather large generator units are needed before unwanted trips occur during upstream faults. Changing the protection settings will most likely be able to prevent unwanted trips. The situation for downstream faults, however, is more severe. A disadvantage of inverse-time overcurrent protection is that the fault-clearing time can become very long for low fault currents. The presence of a generator on the feeder will reduce the fault current and can easily result in unacceptably long fault-clearing times. The actual hosting capacity will strongly depend on the specific situations, but especially for long feeders, the margin may be very small. In this case, a small generator can result in unacceptably long fault-clearing times. The solution in this case is to add an additional breaker somewhere along the feeder—just downstream of the generator location seems the most appropriate location.

7.3 CALCULATING THE FAULT CURRENTS

In this section, we will give expressions for calculating the fault currents through the breaker at the start of a feeder when a generator is connected somewhere along the feeder. These expressions will be used in Section 7.4 to estimate the hosting capacity.

The initial situation considered here is always a well-designed distribution system, but without any generation connected. When designing a distribution system aimed at containing significant amounts of generation, for example, industrial systems with on-site generation, the protection will be from the beginning such that it will remain selective even with generation connected.

7.3.1 Upstream Faults

Consider the network configuration shown in Figure 7.9, which corresponds to Figure 7.1. A distribution network with nominal voltage U_{nom} is supplied via a transformer with rating S_{t} and impedance ϵ_{t} from a higher level network with fault

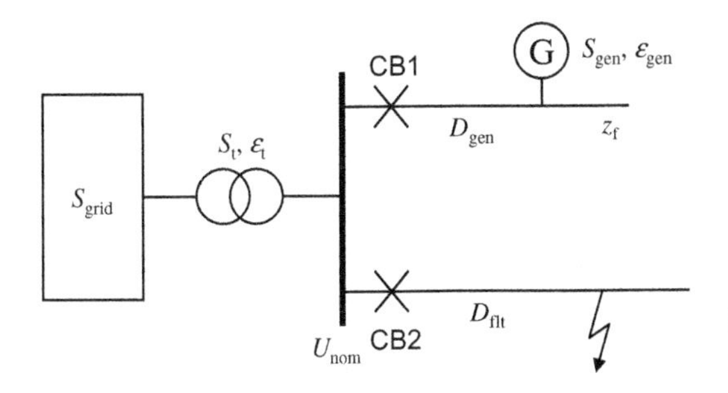

Figure 7.9 Fault on a neighboring feeder: base for calculations.

level S_{grid}. A generator with rating S_{gen} is connected at a distance D_{gen} along a feeder with impedance z_f per unit length. A fault occurs at distance D_{flt} on a neighboring feeder with the same impedance per unit length. The transient impedance of the generator, including any generator transformer, is equal to ϵ_{gen} on a base S_{gen}.

We will start with performing the calculations for a three-phase fault and continue with nonsymmetrical faults. It should be noted that the reason for discussing three-phase faults first is not because these faults are most common or most severe. The calculations for three-phase faults are easier and, therefore, more suitable to illustrate the calculation methods.

7.3.1.1 *Three-Phase Faults*

To calculate the current through breaker CB1 for a three-phase fault, the circuit theory model in Figure 7.10 is used. The source voltages E_{grid} and E_{gen} depend on the prefault power flows. Here we will assume that both are equal to $U_{nom}/\sqrt{3}$, that is, the nominal value of the phase-to-ground voltage.

The impedances in Figure 7.10 are obtained from the system and component parameters in Figure 7.9 by the following expressions:

$$Z_1 = \frac{U_{nom}^2}{S_{grid}} + \epsilon_t \frac{U_{nom}^2}{S_t} \tag{7.2}$$

represents the source impedance at the substation.

$$Z_4 = D_{gen} \times z_f \tag{7.3}$$

represents the feeder impedance between the substation and the generator.

$$Z_5 = \epsilon_{gen} \frac{U_{nom}^2}{S_{gen}} \tag{7.4}$$

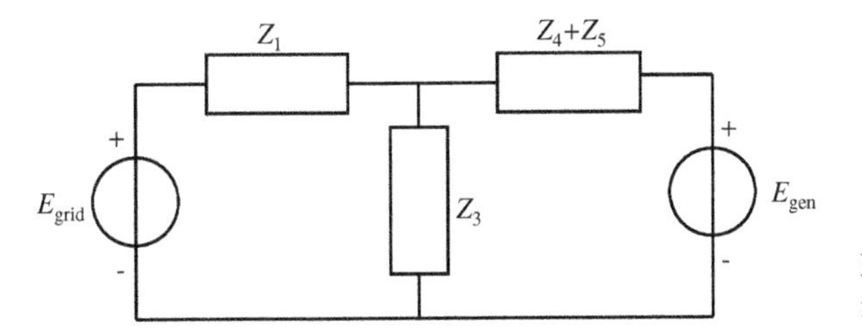

Figure 7.10 Circuit theory model of Figure 7.10.

represents the generator impedance.

$$Z_3 = D_{\mathrm{flt}} \times z_{\mathrm{f}} \tag{7.5}$$

represents the impedance between the substation and the fault. In (7.2)–(7.5), all impedances are referred to the voltage level U_{nom}.

The current through CB1 is found from the following expression:

$$I_{\mathrm{CB1}} = \frac{U_{\mathrm{nom}}/\sqrt{3}}{Z_4 + Z_5 + Z_1 Z_3/(Z_1 + Z_3)} \tag{7.6}$$

The highest current through CB1 occurs for $D_{\mathrm{gen}} = 0$ and $D_{\mathrm{flt}} = 0$

$$I_{\mathrm{CB1}} = \frac{S_{\mathrm{gen}}/\sqrt{3}}{\epsilon_{\mathrm{gen}} \times U_{\mathrm{nom}}} \tag{7.7}$$

Example 7.4 Consider a 50 kVA generator connected to a 400 V low-voltage feeder at 1 km from a 200 kVA transformer. The transformer impedance is 4%; the generator impedance is 20%; the fault level on primary side of the transformer is equal to 39 MVA; the total feeder length is 2 km; and the feeder impedance is equal to 0.28 Ω/km. Calculate the highest current through CB1, in Figure 7.9, for an external fault.

Using expressions (7.2)–(7.4), we get for the location with the highest current:

$$Z_1 = 0.036\ \Omega$$
$$Z_4 + Z_5 = 0.92\ \Omega$$
$$Z_3 = 0$$

The current through CB1 is obtained from (7.6) as

$$I_{\mathrm{CB1}} = 250\ \mathrm{A}$$

Exercise 7.1 What is the current through CB2, the breaker protecting the faulted feeder, for the case in Example 7.4?

As long as the current through CB1 during an upstream fault is less than the maximum load current I_{load}, there is no risk for mal-trip. This results, from (7.7), in the following lower limit for the hosting capacity:

$$S_{\mathrm{gen}} < \epsilon_{\mathrm{gen}} \times \sqrt{3} \times I_{\mathrm{load}} \times U_{\mathrm{nom}} \tag{7.8}$$

This can further be related to the feeder load by realizing that $S_{\mathrm{load}} = \sqrt{3} \times I_{\mathrm{load}} \times U_{\mathrm{nom}}$, so that the inequality becomes

$$S_{\mathrm{gen}} < \epsilon_{\mathrm{gen}} \times S_{\mathrm{load}} \tag{7.9}$$

When this limit is exceeded, it does not imply that mal-trip will occur. After all, the current at which CB1 opens can be set much higher than the highest load current. The highest possible current setting for CB1 is the minimum fault current for a fault on the feeder, $I_{f,min}$. Using this setting results in the following condition:

$$S_{gen} < \epsilon_{gen} \times S_{f,min} \tag{7.10}$$

where $S_{f,min} = I_{f,min} \times U_{nom}$ is the minimum fault level along the feeder. This expression is not suitable for a quick estimation of the hosting capacity as the minimum fault current decreases with increasing amount of distributed generation.

Once the generator size exceeds the limit in (7.10), it is no longer possible to discriminate between a fault on the feeder and an external fault, by current discrimination alone. This again does not mean that alternative (more expensive) protection schemes are needed. Instead, time discrimination could be used where CB1 (the breaker on the feeder with the large DG unit) is delayed compared to the breakers on the other feeders. With reference to Figure 7.2, CB3 is time delayed compared to CB1 and CB2 in a network without generation. For small amounts of generation, this is still possible as long as the limits in (7.9) and (7.10) are not exceeded. Once these limits are exceeded, CB1 should be time delayed compared to CB2, and CB3 should in turn be time delayed compared to CB1. This might, however, make CB3 too slow to protect transformers and cables against damage.

Such a solution is no longer possible when CB1 trips out due to a fault behind CB4. Also, when two feeders from the same bus exceed the hosting capacity, time coordination is no longer possible. Possible solutions to be considered are directional overcurrent protection, distance protection, and differential protection.

7.3.1.2 *Single-Phase Faults*

For nonsymmetrical, single-phase, and two-phase faults, symmetrical component models are needed to calculate the fault currents. The equivalent networks for positive-, negative-, and zero-sequence models are connected at the fault position. The connection depends on the type of fault. For a detailed description of symmetrical component models, we refer to Refs. 12, 42, and 131. Consider again a fault on the neighboring feeder, as shown in Figure 7.9. The positive-, negative-, and zero-sequence models are shown in Figures 7.11–7.13, respectively. To simplify the expressions and the diagrams, we have introduced the following abbreviations, in figures as well as in some of the equations:

$$\begin{aligned} Z_{45p} &= Z_{4p} + Z_{5p} \\ Z_{45n} &= Z_{4n} + Z_{5n} \\ Z_{45z} &= Z_{4z} + Z_{5z} \end{aligned} \tag{7.11}$$

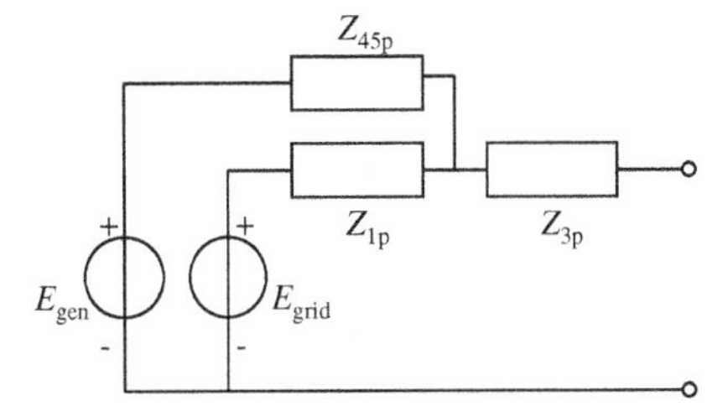

Figure 7.11 Positive-sequence model for a fault on the neighboring feeder.

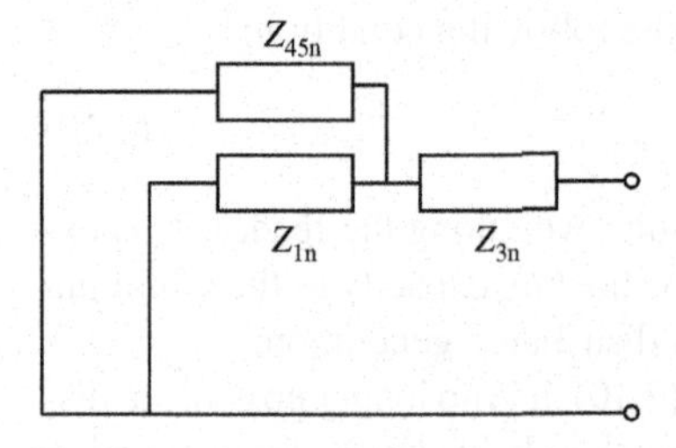

Figure 7.12 Negative-sequence model for a fault on the neighboring feeder.

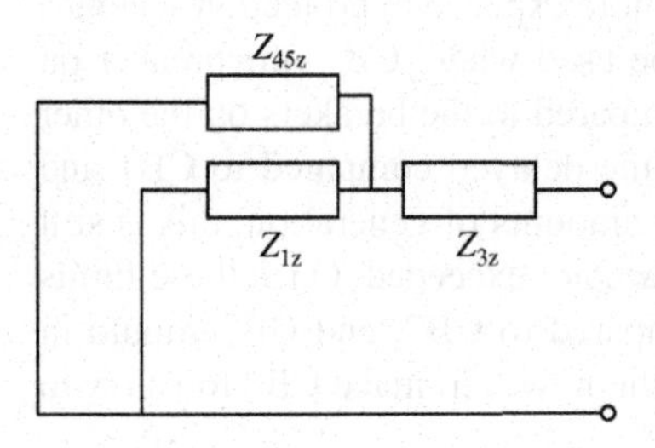

Figure 7.13 Zero-sequence model for a fault on the neighboring feeder.

To calculate the fault currents for a single-phase fault, the three sequence networks are connected in series, as shown in Figure 7.14. The subscripts p, n, and z refer to positive-, negative-, and zero-sequence networks, respectively. The symmetrical components of the current through circuit breaker CB1, the one that may open unwanted, are indicated as I_{p}, I_{n}, and I_{z} in the figure.

The fault current for a single-phase fault is obtained from Figure 7.14 as

$$I_{\mathrm{F}} = \frac{E}{Z_{1\mathrm{p}}Z_{45\mathrm{p}}/(Z_{1\mathrm{p}}+Z_{45\mathrm{p}}) + Z_{1n}Z_{45n}/(Z_{1\mathrm{n}}+Z_{45\mathrm{n}}) + Z_{1\mathrm{z}}Z_{45\mathrm{z}}/(Z_{1\mathrm{z}}+Z_{45\mathrm{z}})} \tag{7.12}$$

Here, like elsewhere, the worst case is considered, that is, a fault just downstream of breaker CB2, for which $Z_{3\mathrm{p}} = Z_{3\mathrm{n}} = Z_{3\mathrm{z}} = 0$.

The positive-, negative-, and zero-sequence components of the current through breaker CB1 are obtained from the following:

$$I_{\mathrm{p}} = \frac{Z_{1\mathrm{p}}}{Z_{1\mathrm{p}}+Z_{4\mathrm{p}}+Z_{5\mathrm{p}}} \times I_{\mathrm{F}} \tag{7.13}$$

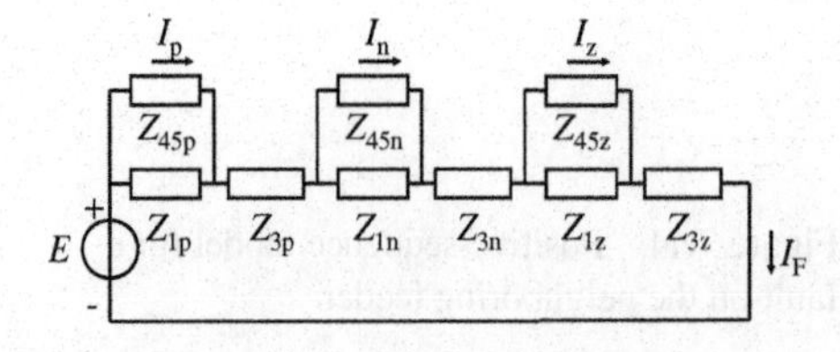

Figure 7.14 Equivalent scheme for a single-phase fault on the neighboring feeder.

$$I_n = \frac{Z_{1n}}{Z_{1n} + Z_{4n} + Z_{5n}} \times I_F \tag{7.14}$$

$$I_z = \frac{Z_{1z}}{Z_{1z} + Z_{4z} + Z_{5z}} \times I_F \tag{7.15}$$

The impedances Z_{1p}, Z_{1n}, and Z_{1z} are the symmetrical component impedances of the grid: overlying grid and transformer impedances. The positive- and negative-sequence impedances are equal. However, the value of the zero-sequence impedance depends on the transformer type and the neutral grounding. For a noneffectively grounded system (common in Europe for medium-voltage feeders), this impedance is high; in Figure 7.14, the impedance Z_{1n} would be much larger than Z_{1p}. The generator connection would have to be such that its zero-sequence impedance, Z_{2n} in Figure 7.14, is also large. The result is that no significant current flows during a single-phase fault, which is one of the purposes of noneffective (high-impedance) grounding. In this section, we will restrict ourselves to effectively grounded systems. In these cases, we can assume the zero-sequence impedance of the source to be equal to or smaller than its positive-sequence equivalent.

The impedances Z_{3p}, Z_{3n}, and Z_{3z} represent the cable or line of the neighboring feeder. Unwanted operation occurs only when this impedance is small. The worst case (maximum current through the breaker) is obtained for zero impedance.

The impedances Z_{5p}, Z_{5n}, and Z_{5z} represent the generator impedance including any generator transformer. The positive-sequence impedance is the one used earlier for the three-phase fault calculations. For rotating machines, the negative-sequence impedance is significantly smaller than the positive-sequence impedance, a factor of 5–10 being typical. In case no generator transformer is used, as may be the case for generators connected to the low-voltage network, a substantial part of the negative-sequence fault current may flow through the nonfaulted feeder with the generator. This could result in unwanted tripping of breaker CB1 for relatively low generator sizes. Note also that the negative-sequence current is even present when there is no positive-sequence contribution to the fault. Unwanted trip during single-phase faults may thus even occur for induction generators or for generators connected through a power electronics converter. The zero-sequence impedance of the generator is most likely to be very high. An exception occurs when a why-delta generator transformer is used, with the why-connected winding on the grid side. This connection is sometimes used in North America and it results in an increase of the fault current for single-phase faults as well as coordination problems for the earth fault protection [123, 124]. Most network operators, therefore, do not allow such a connection.

The phase currents through breaker CB1 are obtained by using the symmetrical component transformation:

$$\begin{aligned} I_a &= I_p + I_n + I_z \\ I_b &= a^2 I_p + a I_n + I_z \\ I_c &= a I_p + a^2 I_n + I_z \end{aligned} \tag{7.16}$$

where $a = -1/2 + 1/2j\sqrt{3}$ is a rotation over 120° in the complex plane. The actual behavior of the breaker will depend on the details of the protection algorithm. It seems reasonable, however, to assume that the breaker will trip when one of the three-phase currents exceeds the threshold. Unwanted trip will, thus, occur when the highest of the currents in (7.16) is higher than the overcurrent setting of breaker CB1.

The diagram in Figure 7.14 is very general. In reality, a number of simplifications can be made based on the physical properties of the system.

- The generator does not contribute any zero-sequence current to the fault.
- Positive- and negative-sequence impedances are equal for all elements apart from the generator.
- The ratio between zero- and positive-sequence impedances of the grid, seen at the low-voltage side of the distribution transformer, is equal to α.
- The ratio between negative- and positive-sequence impedances of the generator, seen at the point of connection, is equal to β.
- The ratio between zero- and positive-sequence impedances of the feeder is equal to γ.

In terms of the impedances in Figure 7.14 and later in Figure 7.22, this translates into the following expressions, where $Z_1 = Z_{1p}$, and so on.

$$\begin{aligned} Z_{1n} &= Z_1 & Z_{1z} &= \alpha Z_1 \\ Z_{3n} &= Z_3 & Z_{3z} &= \gamma Z_3 \\ Z_{4n} &= Z_4 & Z_{4z} &= \gamma Z_4 \\ Z_{5n} &= \beta Z_5 & Z_{5z} &= \infty \\ Z_{6n} &= Z_6 & Z_{6z} &= \gamma Z_6 \end{aligned} \tag{7.17}$$

The impedance Z_6 represents the feeder downstream of the generator location (see Figure 7.22).

Example 7.5 Consider the same configuration as in Example 7.4. For single-phase faults, some additional information is needed, which are the factors introduced in (7.17). The following values are used in this example: $\alpha = 0.60$, $\beta = 0.75$, and $\gamma = 1.80$. Calculate the phase currents through breaker CB1 for the worst-case external single-phase fault. The impedance values used for the calculations are summarized in Table 7.1.

The worst-case external fault is a bus just downstream of the breaker in a neighboring feeder. The fault current is calculated using (7.12). We introduce some additional notation to

TABLE 7.1 Symmetrical Component Impedances Used in Example 7.5

	Z_1	Z_4	Z_5	Z_6
Positive-sequence	36	280	640	280
Negative-sequence	36	280	480	280
Zero-sequence	21.6	504	–	504

All values in mΩ.

simplify the expressions:

$$Z_p = Z_{1p}//(Z_{4p} + Z_{5p}) = 36\,\mathrm{m\Omega}//920\,\mathrm{m\Omega} = 34.6\,\mathrm{m\Omega}$$
$$Z_n = Z_{1n}//(Z_{4n} + Z_{5n}) = 36\,\mathrm{m\Omega}//760\,\mathrm{m\Omega} = 34.4\,\mathrm{m\Omega}$$
$$Z_z = Z_{1z} = 21.6\,\mathrm{m\Omega}$$

The fault current is equal to

$$I_F = \frac{E}{Z_p + Z_n + Z_z} = \frac{231\,\mathrm{V}}{34.6 + 34.4 + 21.6\,\mathrm{m\Omega}} = 2550\,\mathrm{A}$$

Only the part of the current through breaker CB1 is of interest. To calculate this, we use (7.13)–(7.15), resulting in

$$I_p = \frac{Z_{1p}}{Z_{1p} + Z_{4p} + Z_{5p}} \times I_F = 96.0\,\mathrm{A}$$
$$I_n = \frac{Z_{1n}}{Z_{1n} + Z_{4n} + Z_{5n}} \times I_F = 115.3\,\mathrm{A}$$
$$I_z = 0$$

The phase currents through the breaker are obtained from (7.16), resulting in

$$I_a = 211.3\,\mathrm{A}$$
$$I_b = -105.7 + j16.8\,\mathrm{A}$$
$$I_c = -105.7 - j16.8\,\mathrm{A}$$

The magnitude of the current in the nonfaulted phases (absolute value of the complex currents) is equal to 107 A.

7.3.1.3 Phase-to-Phase Faults

The currents during a phase-to-phase fault on the neighboring feeder are obtained by connecting the positive- and negative-sequence networks shown in Figures 7.11 and 7.12. The resulting equivalent scheme is shown in Figure 7.15.

The fault current is obtained from the following expression:

$$I_F = \frac{E}{Z_{1p}Z_{45p}/(Z_{1p} + Z_{45p}) + Z_{3p} + Z_{3n} + Z_{1n}Z_{45n}/(Z_{1n} + Z_{45n})} \tag{7.18}$$

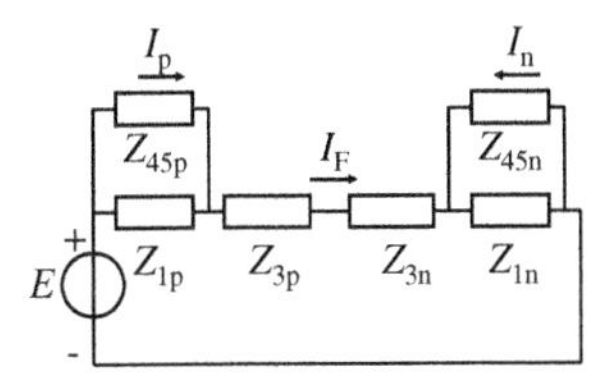

Figure 7.15 Equivalent scheme for a phase-to-phase fault on the neighboring feeder.

Positive- and negative-sequence currents through breaker CB1 are obtained from the following:

$$I_p = \frac{Z_{1p}}{Z_{1p} + Z_{4p} + Z_{5p}} \times I_F \tag{7.19}$$

$$I_n = -\frac{Z_{1n}}{Z_{1n} + Z_{4n} + Z_{5n}} \times I_F \tag{7.20}$$

The rest of the calculations proceeds as before, with the phase currents obtained from the symmetrical component currents using the same expressions:

$$\begin{aligned} I_a &= I_p + I_n \\ I_b &= a^2 I_p + a I_n \\ I_c &= a I_p + a^2 I_n \end{aligned} \tag{7.21}$$

Example 7.6 Consider a phase-to-phase fault on the neighboring feeder in the same system as in Examples 7.4 and 7.5. Calculate the phase currents through breaker CB1 for the worst-case upstream fault.

The impedance values in Table 7.1 can again be used. Using (7.18) results in $I_F =$ 3347.0 A. Combining this with (7.19) and (7.20) gives for the symmetrical component currents through the breaker:

$$\begin{aligned} I_p &= 126.0\,\text{A} \\ I_n &= -151.4\,\text{A} \end{aligned}$$

and for the phase currents:

$$\begin{aligned} I_a &= -25.3\,\text{A} \\ I_b &= 12.7 - j240.2\,\text{A} \\ I_c &= 12.7 + j240.2\,\text{A} \end{aligned}$$

The magnitude of the current in the faulted phases is 240.6 A. The current in the non-faulted phase is small, 25 A.

7.3.1.4 Two-Phase-to-Ground Faults The equivalent scheme for calculating the currents during a two-phase-to-ground fault on a neighboring feeder is shown in Figure 7.16.

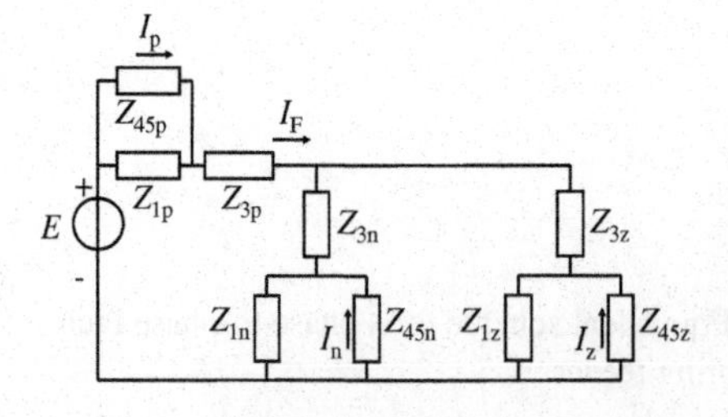

Figure 7.16 Equivalent scheme for a two-phase-to-ground fault on the neighboring feeder.

Assuming that $Z_3 = 0$ (the worst-case upstream fault), the fault current I_F is obtained from

$$I_F = \frac{E}{Z_F} \tag{7.22}$$

where $Z_F = Z_{1p}//Z_{45p} + Z_{1n}//Z_{45n}//Z_{1z}//Z_{45z}$. The symmetrical component currents through breaker CB1 are obtained from

$$I_p = \frac{Z_{1p}}{Z_{1p} + Z_{4p} + Z_{5p}} \times I_F \tag{7.23}$$

$$I_n = -\frac{Z_n}{Z_n + Z_{4n} + Z_{5n}} \times I_F \tag{7.24}$$

where $Z_n = Z_{1n}//Z_{1z}//Z_{45z}$.

$$I_z = -\frac{Z_z}{Z_z + Z_{4z} + Z_{5z}} \times I_F \tag{7.25}$$

where $Z_z = Z_{1n}//Z_{1z}//Z_{45n}$. The phase currents are again calculated from (7.16).

Example 7.7 Considering the same network configuration as in Examples 7.4 and 7.5, calculate symmetrical component currents and phase currents for the worst-case upstream two-phase-to-ground fault.

Using (7.26) with $Z_3 = 0$ results in

$$Z_F = 36//920 + 36//760/21.6\,\mathrm{m\Omega} = 47.9\,\mathrm{m\Omega}$$

and $I_F = 4821$ A. From (7.23)–(7.25) the symmetrical component currents through breaker CB1 are as follows:

$$I_p = \frac{36}{36 + 920} \times 4821 = 181.5\,\mathrm{A}$$

$$I_n = -\frac{13.5}{760 + 13.5} \times 4821 = -84.1\,\mathrm{A}$$

As the generator neutral is not connected to the system neutral, there is no zero-sequence contribution to the fault. The resulting phase currents are as follows:

$$I_a = 97.4\,\mathrm{A}$$
$$I_b = -48.7 - j230.1\,\mathrm{A}$$
$$I_c = -48.7 + j230.1\,\mathrm{A}$$

The magnitude of the current in the faulted phases is equal to 235.2 A. The current in the nonfaulted phase is 97 A.

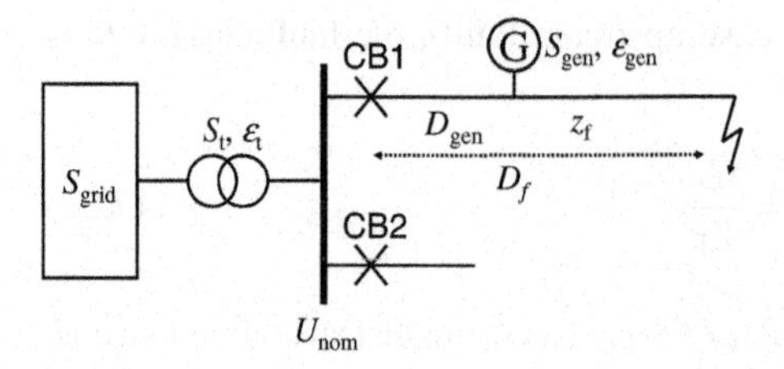

Figure 7.17 Fault downstream of the generator unit: base for calculations.

7.3.2 Downstream Faults

7.3.2.1 Three-Phase Faults For a fault downstream of a generator, the generator contribution to the fault reduces the fault current through the breaker that should detect the fault. The result could be that the overcurrent protection cannot distinguish the fault from a high load situation. The situation is shown in Figure 7.17, corresponding to Figure 7.3.

The corresponding circuit diagram for a three-phase fault is shown in Figure 7.18, where the impedances are related to the parameters in Figure 7.17, through the following expressions:

$$Z_1 = \frac{U_{\text{nom}}^2}{S_{\text{grid}}} + \epsilon_t \frac{U_{\text{nom}}^2}{S_t} \tag{7.26}$$

$$Z_4 = z_f \times D_{\text{gen}} \tag{7.27}$$

$$Z_5 = \epsilon_{\text{gen}} \frac{U_{\text{nom}}^2}{S_{\text{gen}}} \tag{7.28}$$

$$Z_6 = z_f \times \left(D_f - D_{\text{gen}}\right) \tag{7.29}$$

The current through the circuit breaker at the start of the feeder (CB1 in Figure 7.17) is obtained from the circuit in Figure 7.18 as the current through Z_1 and Z_4. Using the approximation

$$E_{\text{grid}} = E_{\text{gen}} = E = U_{\text{nom}}/\sqrt{3} \tag{7.30}$$

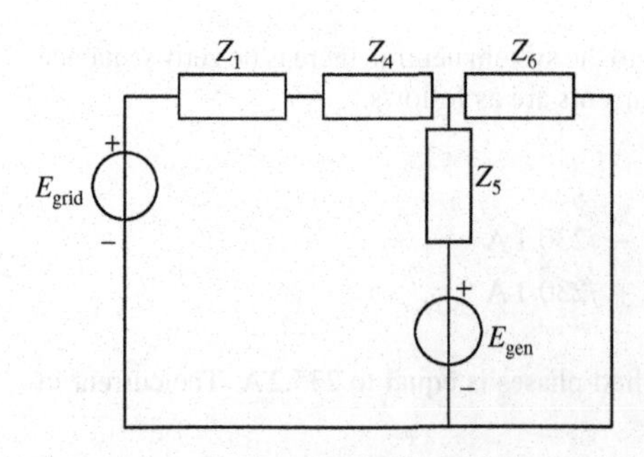

Figure 7.18 Circuit theory model of Figure 7.17.

results in

$$I_{CB1} = \frac{U_{nom}}{\sqrt{3}} \frac{Z_5}{(Z_5 + Z_6)(Z_1 + Z_4) + Z_5 Z_6} \tag{7.31}$$

Example 7.8 Consider a three-phase fault downstream of the generator as in Example 7.4. Calculate the currents through the circuit breaker for the worst-case downstream three-phase fault.

Using (7.26)–(7.31) with the data given in Example 7.4 results in

$$I_{CB1} = 314.5\,\text{A}$$

As this is a three-phase fault, the current magnitude is the same in each phase.

As long as this minimum fault current is higher than the maximum nonfault current, it is possible to choose a current setting such that the protection operates correctly. As we saw before, the maximum nonfault current occurs for a fault on a neighboring feeder. The hosting capacity is found as the generator size S_{gen} for which expression (7.31) becomes equal to (7.6), where $Z_3 = 0$. Rewriting this gives the following quadratic equation in Z_5:

$$Z_5^2 - (Z_1 + Z_6)Z_5 - (Z_1 + Z_4)Z_6 = 0 \tag{7.32}$$

Once Z_5 is known, the hosting capacity is obtained as S_{gen} from (7.28). This expression, however, considers only three-phase faults. For a more accurate estimation, all fault types should be considered. This is discussed further in Section 7.4.

7.3.2.2 Single-Phase Faults Consider next a fault downstream of the generator, as shown in Figure 7.17. The impedance definitions in the equivalent circuit of Figure 7.18 are used to obtain the symmetrical component models shown in Figures 7.19–7.21. The resulting scheme to calculate the fault currents during a single-phase fault is shown in Figure 7.22, where $Z_{14p} = Z_{1p} + Z_{4p}$, and so on.

The fault current is obtained from Figure 7.22 as

$$I_F = \frac{E}{\dfrac{Z_{14p}Z_{5p}}{Z_{14p} + Z_{5p}} + \dfrac{Z_{14n}Z_{5n}}{Z_{14n} + Z_{5n}} + \dfrac{Z_{14z}Z_{5z}}{Z_{14z} + Z_{5z}} + Z_{6p} + Z_{6n} + Z_{6z}} \tag{7.33}$$

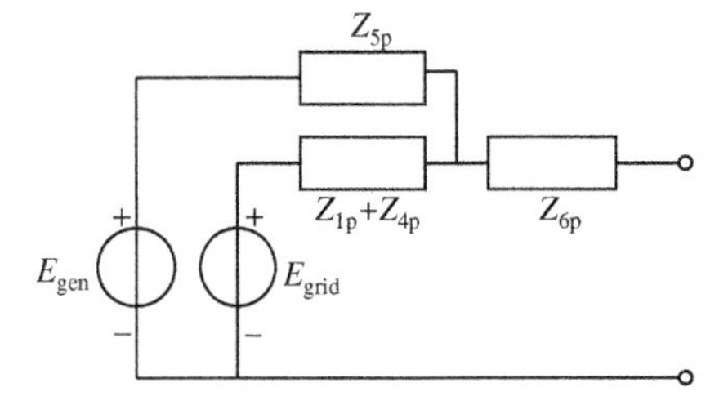

Figure 7.19 Positive-sequence model for a fault downstream of the generator.

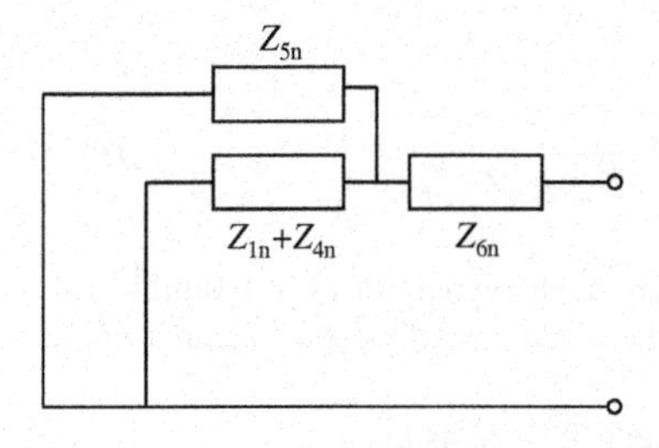

Figure 7.20 Negative-sequence model for a fault downstream of the generator.

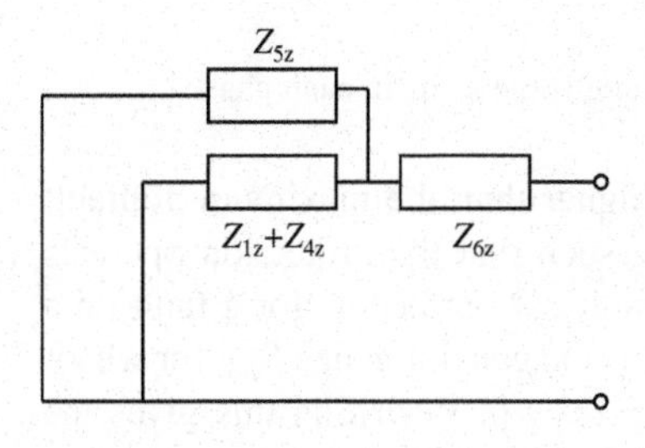

Figure 7.21 Zero-sequence model for a fault downstream of the generator.

The positive-, negative-, and zero-sequence currents through breaker CB1 are obtained from the following current divider expressions:

$$I_{\mathrm{p}} = \frac{Z_{5\mathrm{p}}}{Z_{1\mathrm{p}} + Z_{4\mathrm{p}} + Z_{5\mathrm{p}}} \times I_{\mathrm{F}} \tag{7.34}$$

$$I_{\mathrm{n}} = \frac{Z_{5\mathrm{n}}}{Z_{1\mathrm{n}} + Z_{4\mathrm{n}} + Z_{5\mathrm{n}}} \times I_{\mathrm{F}} \tag{7.35}$$

$$I_{\mathrm{z}} = \frac{Z_{5\mathrm{z}}}{Z_{1\mathrm{z}} + Z_{4\mathrm{z}} + Z_{5\mathrm{z}}} \times I_{\mathrm{F}} \tag{7.36}$$

The phase currents are calculated as before.

Example 7.9 For the same system as in Examples 7.4 and 7.5, calculate the smallest phase currents for a fault on the same feeder as the generator.

As the same system is used, the impedance values in Table 7.1 remain valid. The fault current is obtained from (7.33). We again introduce a new notation, but different from the one

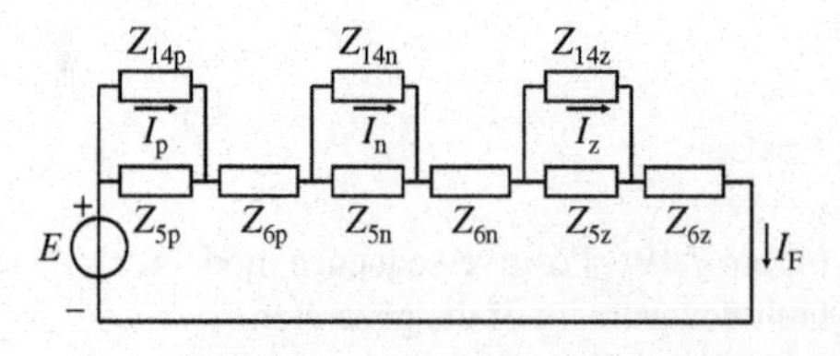

Figure 7.22 Equivalent scheme for a single-phase fault downstream of the generator.

in Example 7.5.

$$Z_p = \frac{(Z_{1p} + Z_{4p})Z_{5p}}{Z_{1p} + Z_{4p} + Z_{5p}} = 211.55\,\text{m}\Omega$$

$$Z_n = \frac{(Z_{1n} + Z_{4n})Z_{5n}}{Z_{1n} + Z_{4n} + Z_{5n}} = 190.55\,\text{m}\Omega$$

$$Z_z = Z_{1z} + Z_{4z} = 525.6\,\text{m}\Omega$$

$$I_F = \frac{231\,\text{V}}{211.55 + 190.55 + 525.6 + 280 + 280 + 504\,\text{m}\Omega} = 115.95\,\text{A}$$

The symmetrical component currents through breaker CB1 are calculated using (7.34)–(7.36):

$$I_p = \frac{640}{36 + 280 + 640} \times 115.95 = 77.6\,\text{A}$$

$$I_n = \frac{480}{36 + 280 + 480} \times 115.95 = 69.9\,\text{A}$$

$$I_z = 115.95\,\text{A}$$

Using (7.16) gives the phase currents:

$$I_a = 263.5\,\text{A}$$

$$I_b = 42.1 - j6.7\,\text{A}$$

$$I_c = 42.1 + j6.7\,\text{A}$$

The current in the faulted phase is 263 A and in the nonfaulted phases 42 A.

7.3.2.3 Phase-to-Phase Faults For a phase-to-phase fault downstream of the generator, the equivalent scheme is obtained by combining (5.53) and (5.54). The result is shown in Figure 7.23.

The expression for the fault current is

$$I_F = \frac{E}{(Z_{1p} + Z_{4p})Z_{5p}/(Z_{1p} + Z_{4p} + Z_{5p}) + Z_{6p} + Z_{6n} + (Z_{1n} + Z_{4n})Z_{5n}/(Z_{1n} + Z_{4n} + Z_{5n})} \tag{7.37}$$

The positive- and negative-sequence currents through breaker CB1 are as follows:

$$I_p = \frac{Z_{5p}}{Z_{1p} + Z_{4p} + Z_{5p}} \times I_F \tag{7.38}$$

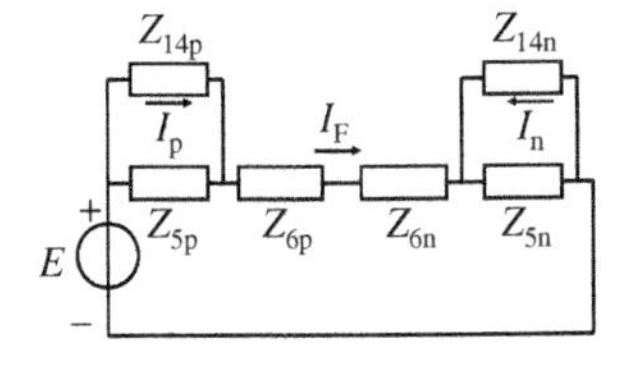

Figure 7.23 Equivalent scheme for a phase-to-phase fault downstream of the generator.

$$I_{\text{n}} = -\frac{Z_{5\text{n}}}{Z_{1\text{n}} + Z_{4\text{n}} + Z_{5\text{n}}} \times I_{\text{F}} \tag{7.39}$$

The phase currents are obtained as before.

Example 7.10 Consider a downstream phase-to-phase fault for the distribution system as in Examples 7.4 and 7.5, with the impedances as in Table 7.1. Calculate the currents at the start of the feeder for the worst-case downstream fault.

From (7.37), the following impedances are calculated:

$$Z_{\text{p}} = (Z_{1\text{p}} + Z_{4\text{p}})//Z_{5\text{p}} = 316//640\,\text{m}\Omega = 211.6\,\text{m}\Omega$$
$$Z_{\text{n}} = (Z_{1\text{n}} + Z_{4\text{n}})//Z_{5\text{n}} = 316//480\,\text{m}\Omega = 190.6\,\text{m}\Omega$$

The fault current for a phase-to-phase fault at the end of the feeder is

$$I_{\text{F}} = \frac{230.9\,\text{V}}{211.5 + 280 + 280 + 190.6\,\text{m}\Omega} = 240\,\text{A}$$

Using (7.38) and (7.39), the positive- and negative-sequence currents through breaker CB1 are

$$I_{\text{p}} = \frac{640}{36 + 280 + 640} \times 240 = 160.7\,\text{A}$$

$$I_{\text{n}} = -\frac{480}{36 + 280 + 480} \times 240 = -144.7\,\text{A}$$

The phase currents through the breaker are as follows:

$$I_{\text{a}} = 16.0\,\text{A}$$
$$I_{\text{b}} = -8.0 - j264.5\,\text{A}$$
$$I_{\text{c}} = -8.0 + j264.5\,\text{A}$$

The fault current is 265 A in the faulted phases and 16 A in the nonfaulted phase.

7.3.2.4 *Two-Phase-to-Ground Faults*

The equivalent scheme to calculate the currents during a downstream two-phase-to-ground fault is shown in Figure 7.24. The scheme is getting more and more complicated, but it is still possible to obtain reasonably simple expressions for the fault currents.

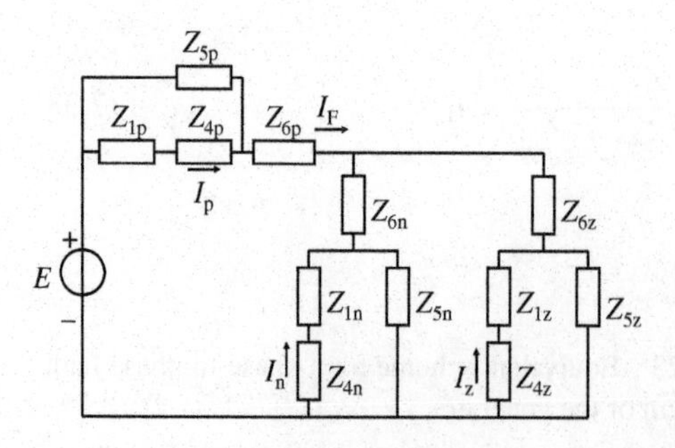

Figure 7.24 Equivalent scheme for a two-phase-to-ground fault downstream of the generator.

The expression for the fault current I_F reads as follows:

$$I_F = \frac{E}{Z_p + (Z_n Z_z/(Z_n + Z_z))} \tag{7.40}$$

with

$$Z_p = Z_{6p} + \frac{Z_{5p}(Z_{1p} + Z_{4p})}{Z_{1p} + Z_{4p} + Z_{5p}} \tag{7.41}$$

$$Z_n = Z_{6n} + \frac{Z_{5n}(Z_{1n} + Z_{4n})}{Z_{1n} + Z_{4n} + Z_{5n}} \tag{7.42}$$

$$Z_z = Z_{6z} + \frac{Z_{5z}(Z_{1z} + Z_{4z})}{Z_{1z} + Z_{4z} + Z_{5z}} \tag{7.43}$$

The symmetrical component currents through breaker CB1 are obtained from the following expressions:

$$I_p = \frac{Z_{5p}}{Z_{1p} + Z_{4p} + Z_{5p}} \times I_F \tag{7.44}$$

$$I_n = -\frac{Z_z}{Z_n + Z_z} \times \frac{Z_{5n}}{Z_{1n} + Z_{4n} + Z_{5n}} \times I_F \tag{7.45}$$

$$I_z = -\frac{Z_n}{Z_n + Z_z} \times \frac{Z_{5z}}{Z_{1z} + Z_{4z} + Z_{5z}} \times I_F \tag{7.46}$$

Exercise 7.2 Using the same case as in Examples 7.4 and 7.5, calculate the symmetrical component currents and the phase currents for the worst-case two-phase-to-ground fault downstream of the generator.

7.3.3 Induction Generators, Power Electronics, and Motor Load

All the calculations in the previous examples are based on synchronous generators. Also, the example calculations in the next section will all be based on the synchronous generator. For the synchronous generator, the contribution to the fault current is largest. However, induction generators are commonly used for wind power installations, so that their total impact may still be larger. It is, therefore, worth to discuss the impact of induction generators on the fault currents. An induction generator does not contribute any zero-sequence current to the fault. In the circuit diagram in Figures 7.13 and 7.21 and in all subsequent schemes and expressions, we have to use $Z_{5z} = \infty$. The contribution of an induction generator to the fault current decays quickly and has disappeared two to three cycles after the start of the fault. When fast protection or current-limiting fuses are used, the same impedance values as for the synchronous generator can be used. The earlier examples and the hosting capacity values in the next section remain valid. For slower protection, where the decision to trip or not is made over a period of 100 ms or longer, the contribution of the induction generator to the positive-sequence current can be neglected. In Figures 7.11 and 7.19 and in all subsequent calculations, we have to insert $Z_{5p} = \infty$.

The contribution of an induction generator to the fault current is the same as the contribution of an induction motor. Induction motors often make up a substantial part of the load. For a complete picture, they have to be considered in the calculations as well. Including induction motor load will indicate that it is no longer possible to use the kind of simplified models used in this chapter. Instead, an appropriate power system analysis package is needed for the protection coordination.

Large generator units with power electronics interface also have the ability to impact the fault current. The contribution strongly depends on the control algorithm used. Many converters do not inject any current when the voltage at their terminals is low. The contribution to the fault current is zero in that case. Others contribute only positive-sequence current or positive- and negative-sequence currents, but limited to a value per phase somewhat above the rated current. In no case, however, is the contribution expected to be much higher than the rated current. This should be compared with a contribution up to six times the rated current for rotating machines. The control algorithm of power electronics converters may inject negative-sequence or even zero-sequence currents during a fault in the grid. Existing small generation units do use very limited control, but increasing requirements on fault ride-through will certainly change that. There are already indications that large wind parks may interfere with the protection in subtransmission systems.

The contribution of double-fed induction generators to the fault current is studied in Ref. 304. A detailed simulation model is used. It is shown that this type of generator can contribute 6–10 times the rated current during the start of a fault. The current decays in 100–200 ms. The peak current is higher and the decay is faster for smaller units. The conclusion from this study should be that double-fed induction generators contribute as much to the make current as that of induction or synchronous generators. Their impact on the break current is, however, limited.

7.4 CALCULATING THE HOSTING CAPACITY

The current setting of the breaker at the start of the feeder should be such that the breaker will not open for any fault outside the feeder. But the breaker should open for any fault on the feeder, also for the one with the lowest fault current. If we do not want to make use of time grading, the current setting should be

- larger than the highest current for an upstream fault, and
- lower than the lowest current for a downstream fault.

Consider, as an example, a 400 V low-voltage network: a 200 kVA, 4% distribution transformer supplied from a 39 MVA medium-voltage network. A 50 kVA generator is connected at 1 km from the distribution transformer on a low-voltage

TABLE 7.2 Fault Currents for Different Types of Upstream and Downstream Faults

	Upstream	Downstream
Three-phase fault	251 A (Example 7.4)	314 A (Example 7.8)
Single-phase fault	212 A (Example 7.5)	263 A (Example 7.9)
Phase-to-phase fault	241 A (Example 7.6)	264 A (Example 7.10)
Two-phase-to-ground fault	236 A (Example 7.7)	293 A (Exercise 7.2)

overhead feeder of 2 km length with an impedance equal to 0.28 Ω/km. The positive-sequence impedance of the generator is equal to 20%. This case has been studied in the previous examples.

The fault currents for the worst-case upstream and downstream faults are given in Table 7.2, including reference to the examples where the calculations are shown. The worst-case upstream fault is a fault on a neighboring feeder, immediately downstream of the breaker protecting that feeder. The worst-case downstream fault is a fault at the end of the feeder to which the generator is connected. The values for the upstream faults have been rounded upward, whereas the one for downstream faults have been rounded downward.

The "setting margin" in this example is between 251 and 263 A. Only current settings within this range will give a correct coordination of the protection. The setting margin is very small in this case and, in practice, it would be wise to look for alternative solutions, like an additional time grading step. The phase currents for different types of faults, both upstream and downstream faults, are shown in Figure 7.25, as a function of the generator size. The decaying curves hold for downstream faults, whereas the increasing curves, all starting at zero, are for upstream faults.

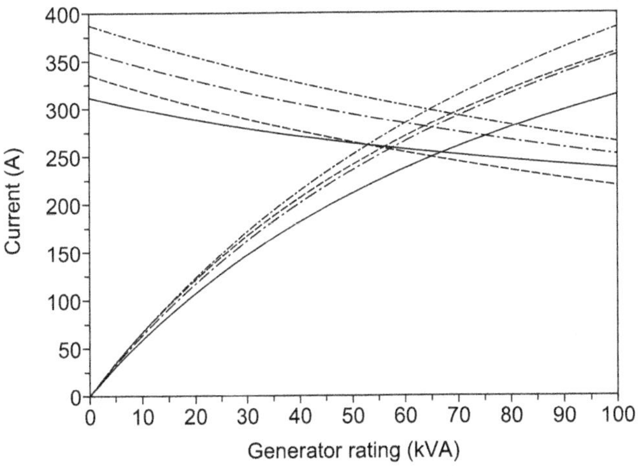

Figure 7.25 Phase currents for upstream and downstream faults as a function of the generator size: single-phase (solid), two and three-phase (dashed).

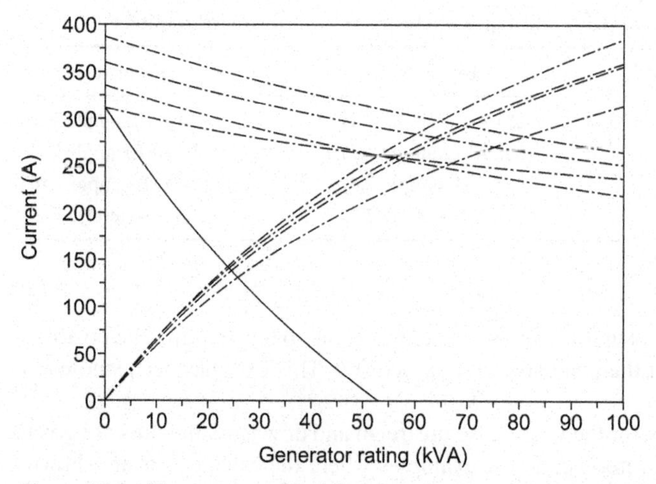

Figure 7.26 Phase current for different faults (dotted lines) and setting margin (solid line) as a function of the generator size.

The setting margin is shown separately in Figure 7.26. For generator sizes exceeding 52 kVA, the setting margin is zero. In this case, it is no longer possible to find a proper current setting of the breaker at the start of the feeder.

The actual hosting capacity depends on the minimum setting margin that the network operator accepts. The larger the required setting margin, the lower the hosting capacity. For a minimum setting margin equal to 50 A, the hosting capacity is 41 kVA; and when the setting margin is not allowed to drop below 100 A, a generator with a capacity of not more than 31 kVA can be connected. A smaller setting margin will increase the risk of incorrect protection operation, which in turn reduces the supply reliability as experienced by the customers.

The hosting capacity has been calculated for different generator and system parameters, using the methodology explained in the previous sections. The hosting capacity has been calculated for zero setting margin (this is an upper limit of the hosting capacity, which in practice would never be acceptable) and for a required setting margin of 100 A. The latter is a more realistic value in practice.

The same low-voltage feeder as before has been used: a generator unit connected at 1 km from a 200 kVA distribution transformer (see Examples 7.4 and 7.5 for a complete description of the case). The hosting capacity has been calculated for different lengths of the feeder; the results are shown in Figure 7.27. The hosting capacity decreases quickly with increasing feeder length. The reason is that the fault current for a fault at the remote end of the feeder becomes smaller with increasing feeder length. The setting margin without generator becomes smaller so that there is less space to add generation. For a 1 km feeder, that margin is almost 600 A; for 2, 3, and 4 km feeder lengths, the margin becomes 310, 210, and 150 A, respectively.

The hosting capacity is smaller when a 100 A margin is required; the difference is about 20 kVA for a 2 km long feeder. For a 4 km long feeder, the hosting capacity becomes very small, less than 10 kVA for 100 A setting margin. This means that the feeder is already close to its limits, even without any generation. Adding a small

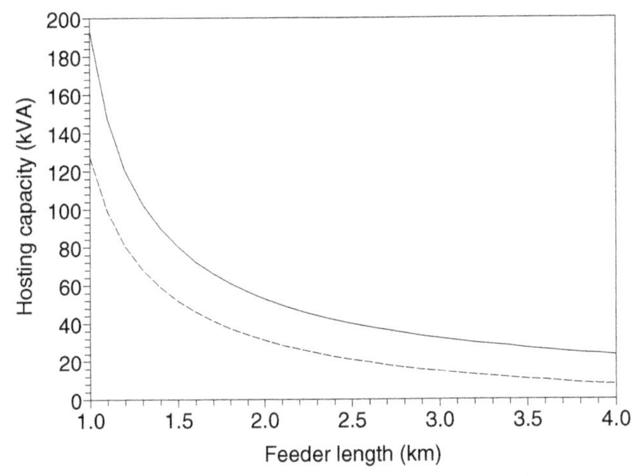

Figure 7.27 Hosting capacity as a function of the length of a low-voltage feeder: zero setting margin (solid line) and 100 A setting margin (dashed line).

amount of generation will shift the performance over the threshold. For a feeder longer than 6 km, the initial setting margin drops below 100 A and no generation can be connected. It should be noted, however, that low-voltage feeders longer than 2 km are very uncommon. Connecting generation to such feeders will very quickly result in problems, not only with protection coordination but also with voltage control and overloads.

The hosting capacity has also been calculated for the generator connected at different locations along the feeder, where the length was again taken as 2 km. As shown in Figure 7.28, the hosting capacity somewhat increases when the generator is connected further away from the main bus. For a generator connected close to

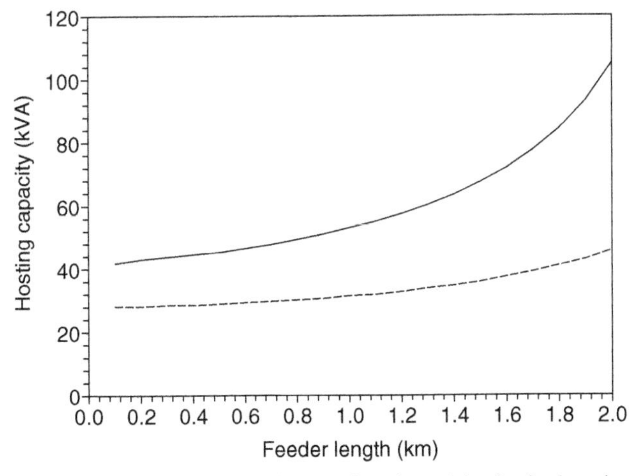

Figure 7.28 Hosting capacity as a function of the feeder location on a low-voltage feeder: zero setting margin (solid line) and 100 A setting margin (dashed line).

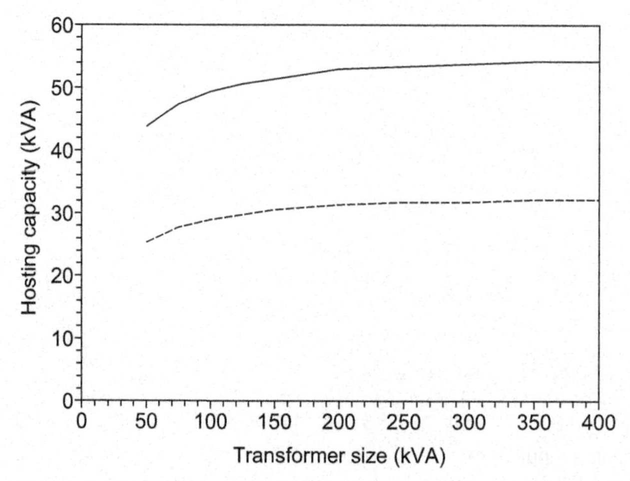

Figure 7.29 Hosting capacity as a function of the transformer size for a low-voltage feeder: zero setting margin (solid line) and 100 A setting margin (dashed line).

the main bus, the hosting capacity is 30–40 kVA, depending on the required setting margin. When the generator is toward the end of the 2 km feeder, the hosting capacity becomes between 40 and 100 kVA.

The transformer size does not have much impact on the hosting capacity. As shown in Figure 7.29, there is only a small increase in hosting capacity with increasing transformer size. The situation looks completely different, however, whenever the generator size is expressed as a percentage of the transformer size. The hosting capacity is between 50% and 90% of the transformer size for a 50 kVA transformer, but only between 8% and 13% for a 400 kVA transformer. Protection coordination may, thus, limit the amount of generation that can be connected to large distribution transformers. In the figure, it has not been taken into account that large transformers are typically used in areas with high load density, where long low-voltage feeders are less common.

The impact of the impedance ratios α, β, and γ has also been studied. It was found from the calculations that their impact on the hosting capacity values is small.

The calculations have been repeated for a generator connected to a 10 kV medium-voltage feeder. The base case consists of a generator connected at 2 km from the main MV bus along a 10 km feeder with positive-sequence impedance 0.35 Ω/km and zero-sequence impedance 1.8 times the positive-sequence impedance. The generator impedance is assumed at 25%, including the generator transformer. The medium-voltage bus is supplied from a 2000 MVA high-voltage network through a 10 MVA, 12% transformer. We assume that the medium-voltage network is noneffectively grounded, so that single-phase and two-phase-to-ground faults do not have to be considered. For single-phase faults, the fault currents are small and dedicated earth fault protection is used instead. For two-phase-to-ground faults, the fault currents are the same as for a phase-to-phase fault. The hosting capacity for this base case is 4.1 and 3.7 MVA for zero and 100 A setting margin.

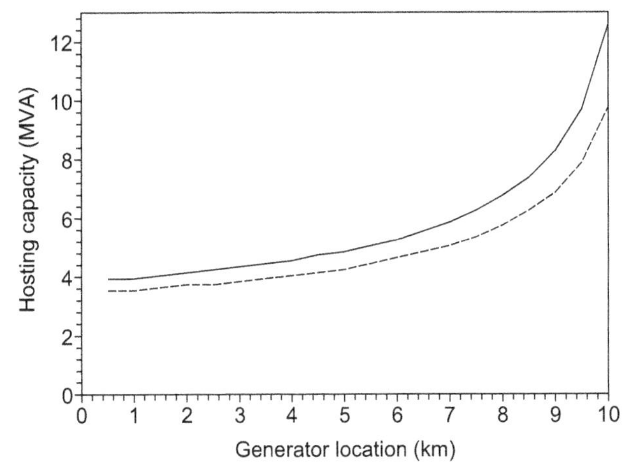

Figure 7.30 Hosting capacity as a function of location of the generator along the medium-voltage feeder: zero setting margin (solid line) and 100 A setting margin (dashed line).

The hosting capacity as a function of the location of the generator along the feeder is shown in Figure 7.30. The contribution of the generator to an upstream fault gets smaller when the generator is connected further away from the main bus. This results in an increase in hosting capacity.

The impact of the feeder length is shown in Figure 7.31. With longer feeders, the current for the worst-case downstream fault decreases, which results in a decrease in the hosting capacity. The decrease is, however, not as severe as for a low-voltage feeder because the transformer impedance is a larger part of the source impedance at the fault location. Even for a 20 km feeder, the hosting capacity is still between 20%

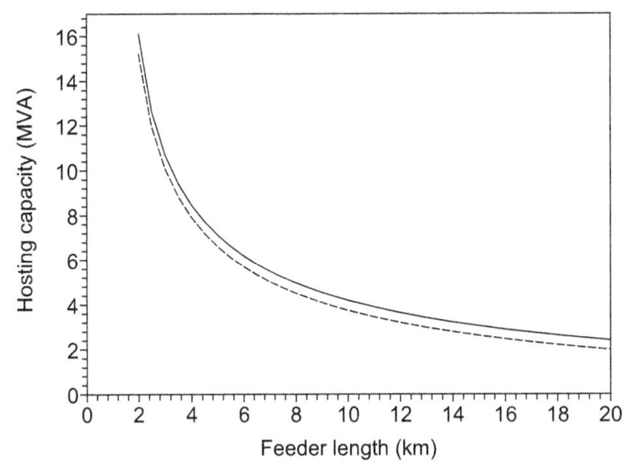

Figure 7.31 Hosting capacity as a function of the length of a medium-voltage feeder: zero setting margin (solid line) and 100 A setting margin (dashed line).

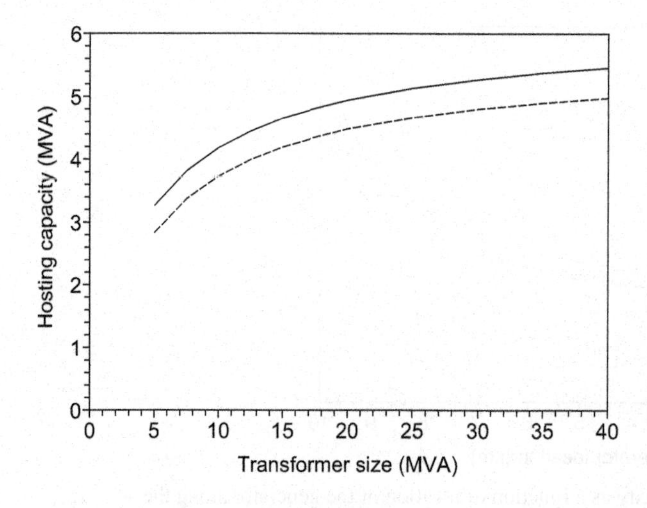

Figure 7.32 Hosting capacity as a function of the HV/MV transformer size: zero setting margin (solid line) and 100 A setting margin (dashed line).

and 25% of the transformer rating. Only for a feeder longer than 40 km (not shown in the figure) does the hosting capacity become less than 10% of the transformer rating.

A larger HV/MV transformer results in a higher short-circuit capacity at the main MV bus. The result is a higher current for a downstream fault and, thus, a higher hosting capacity. This is shown in Figure 7.32, where the hosting capacity is around 3 MVA for a 5 MVA transformer and around 5 MVA for a 40 MVA transformer. Note that the hosting capacity decreases as a percentage of the transformer size from around 60% down to around 12%. In the calculations, the per unit impedance has been kept constant. In practice, the per unit impedance often increases for larger transformers so as to prevent excessive fault currents.

The nominal voltage, Figure 7.33, is one of the parameters that has the biggest impact on the hosting capacity. By increasing the nominal voltage from 10 to 30 kV,

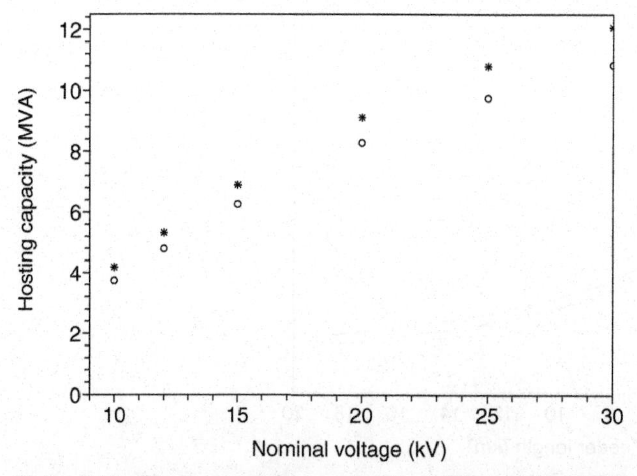

Figure 7.33 Hosting capacity as a function of the nominal voltage: zero setting margin (stars) and 100 A setting margin (circles).

the hosting capacity increases from around 4 to around 11 MVA. The increase is slightly less than by a factor of 3 (between 2.88 and 2.90). It is a good approximation to state that the hosting capacity is proportional to the voltage level. In the calculations, all per unit impedances have been kept the same and so has the feeder impedance in ohm. In practice, such parameters for a 10 kV network are likely to be different from that for a 30 kV feeder.

7.5 BUSBAR PROTECTION

In distribution networks, instead of busbar differential protection, a blocking scheme is sometimes used to speed up the protection for busbar faults [395]. The configuration is shown in Figure 7.34. When a fault occurs downstream of the busbar, one of the breakers CB2, CB3, or CB4 will detect an overcurrent and so will breaker CB1. For a busbar fault, only the breaker CB1 will detect an overcurrent. Normal time-grading overcurrent protection would result in breaker CB1, being typically 250–500 ms slower than the breakers in the outgoing feeders. This could make the protection too slow in case of a busbar fault. Instead, CB1 is given only a short delay compared to the other breakers and instead the other breakers send a blocking signal to CB1 whenever they detect an overcurrent.

The presence of a large synchronous generator interferes with this scheme, as illustrated in Figure 7.35. For a fault on the busbar, there are now two contributions to the fault current: from the grid and from the generator. If the generator is large enough, the fault current from the generator will trigger the overcurrent protection with CB4, which in turn will block the fast opening of CB1. In the same way, as the presence of distributed generation results in an unwanted trip for a fault on a neighboring feeder, it will cause a fail-to-trip for a busbar fault.

The hosting capacity for this fail-to-trip to occur can be calculated in the same way as that for the unwanted trip. In fact, the hosting capacity will normally be similar, unless a different triggering level is used for the blocking signal as for the local trip.

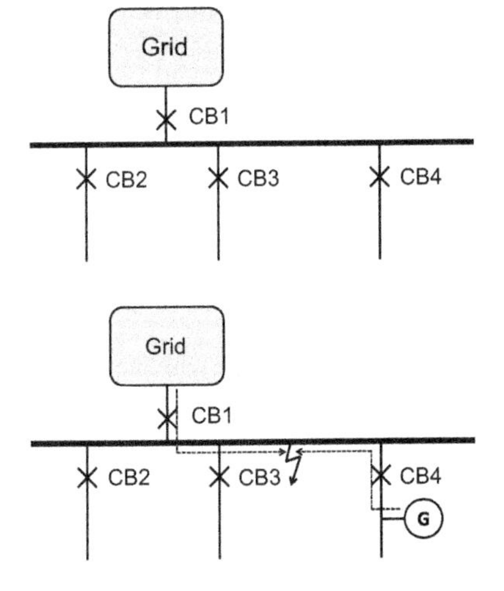

Figure 7.34 Distribution busbar.

Figure 7.35 Distribution busbar with downstream generator.

7.6 EXCESSIVE FAULT CURRENT

Most generators connected to a distribution feeder will result in some increase in the fault current. Here, it is important to distinguish between the "make current" and the "break current." The make current is the peak current that occurs shortly after fault initiation. It is this current that determines the rating of much switchgear, such as busbars and disconnectors. The "break current" is the current magnitude around fault clearing. This determines the required breaking capacity of the circuit breakers. Almost all generators will increase the make current; in fact, other equipment connected to the distribution feeder will also often increase the make current. Also, capacitor banks will contribute to the fault, albeit just a few milliseconds, and they may, therefore, also contribute to the make current.

The increase in break current is very strongly related to the unwanted trip of the protection for an upstream fault. Consider again the configuration shown in Figure 7.1. The current through breaker CB2 is the sum of the contribution from the grid and the contribution from the generator. The same current that results in protection mal-trip gives an increase in the fault current. The steady-state fault current through breaker CB2 can be calculated from the circuit diagram in Figure 7.10, where $Z_3 = 0$ represents the worst case:

$$I_{\mathrm{CB2}} = \frac{E_{\mathrm{grid}}}{Z_1} + \frac{E_{\mathrm{gen}}}{Z_4 + Z_5} \tag{7.47}$$

The closer the generator located to the bus, the larger its contribution to the fault current:

$$\Delta I_{\mathrm{CB2}} = I_{\mathrm{CB1}} = \frac{E_{\mathrm{gen}}}{D_{\mathrm{gen}} \times Z_{\mathrm{f}} + \epsilon_{\mathrm{gen}}(U_{\mathrm{nom}}^2/S_{\mathrm{gen}})} \tag{7.48}$$

Example 7.11 Consider the same medium-voltage case as in Section 7.4. A 2 MVA generator is connected to a 10 kV feeder with a positive-sequence impedance of 0.35 Ω/km. The impedance of the generator is 17%; the impedance of the generator transformer is 5%. This results in the increase in fault current:

$$\Delta I_{\mathrm{CB2}} = \frac{10\,\mathrm{kV}/\sqrt{3}}{2 \times 0.35\,\Omega + (0.17 + 0.05) \times (10\,\mathrm{kV}^2/2\,\mathrm{MVA})} = 493\,\mathrm{A} \tag{7.49}$$

The fault current contribution from the grid depends on the transformer size. For a 10 MVA, 10% transformer, the contribution from the grid is 5774 A. For a 40 MVA, 20% transformer, this is 11.5 kA. The increase in fault current due to the generator is 8.5% and 4.3%, respectively. This should normally be within the safety margin.

Here it is assumed that the contribution from the generator is in phase with the contribution from the grid, so their magnitudes add, which gives an upper limit for the increase in fault current.

The worst-case situation occurs when the generator is located close to the main bus. In such case, its contribution to the fault current is biggest because there is no feeder impedance to limit the current. Assume a generator with rating $S_{\rm gen}$ and impedance $\epsilon_{\rm gen}$ close to a bus supplied through a transformer with rating $S_{\rm tr}$ and impedance $\epsilon_{\rm tr}$. The fault level at the bus without generation is approximately

$$S_{\rm k1} = \frac{S_{\rm tr}}{\epsilon_{\rm tr}} \tag{7.50}$$

After adding the generator, the fault level increases to

$$S_{\rm k2} = \frac{S_{\rm tr}}{\epsilon_{\rm tr}} + \frac{S_{\rm gen}}{\epsilon_{\rm gen}} \tag{7.51}$$

Combining (7.50) and (7.51) gives the following expression for the relative increase in fault level:

$$\frac{S_{\rm k2}}{S_{\rm k1}} = 1 + \frac{S_{\rm gen}}{S_{\rm tr}} \times \frac{\epsilon_{\rm tr}}{\epsilon_{\rm gen}} \tag{7.52}$$

An increase by 10% in fault level is obtained when the second term on the right-hand side equals 0.1, which corresponds to a generator size equal to

$$S_{\rm gen} = 0.1 \times \frac{\epsilon_{\rm gen}}{\epsilon_{\rm tr}} \times S_{\rm tr} \tag{7.53}$$

Example 7.12 Consider a low-voltage network: the impedance of the distribution transformer is 5% and a synchronous generator is directly connected to the low-voltage feeder, $\epsilon_{\rm gen} = 0.17$. What is the hosting capacity when a 10% increase in fault level can be accepted.

Using (7.53) gives

$$S_{\rm gen} = 0.1 \times \frac{0.17}{0.15} \times S_{\rm tr} = 0.34 S_{\rm tr}$$

The hosting capacity is about one-third of the transformer rating.

Example 7.13 Consider a rural medium-voltage network: the impedance of the HV/MV transformer is 10% and the generator is connected through a generator transformer with a total impedance of 22%. An increase up to 10% in fault level is acceptable. Using (7.53) gives for the hosting capacity,

$$S_{\rm gen} = 0.22 S_{\rm tr}$$

or a bit more than one-fifth of the transformer rating.

Example 7.14 Consider an urban medium-voltage network: the impedance of the HV/MV transformer is 20% and the generator is connected through a generator transformer with a total impedance of 22%. An increase up to 10% in fault level is acceptable. Using (7.53) gives for the hosting capacity:

$$S_{\rm gen} = 0.11 S_{\rm tr}$$

or about one-tenth of the transformer rating.

From these three examples, we conclude that the hosting capacity, in relation to the size of the feeding transformer, is highest for low-voltage networks and lowest for urban medium-voltage networks. But even in the latter case, a generator size 11% of the transformer size can be accepted. The transformer size is often twice the load size, so the hosting capacity is about 20% of the total load connected to the HV/MV transformer.

As mentioned before, the increase in break current is mainly due to synchronous machines. For directly grounded low-voltage networks, the highest fault current occurs for single-phase faults. Also, induction machines can somewhat increase the break current in that case. The make current, on the other hand, is increased by almost all types of generation. The biggest increase is for generators with a rotating machine interface. A large induction machine contributes almost as much to the make current as a synchronous machine. For smaller induction machines, the contribution is less because of the high damping of the subtransient current. The contribution from double-fed induction generators is somewhat unclear and seems to strongly depend on the control and protection of the generator. Studies have shown, however, that the initial contribution to the fault current (i.e., the contribution to the make current) is about the same as the contribution of a normal induction generation. The contribution from generators with power electronics converter interface will typically be at most twice the rated current. Increasing amounts of distribution will often go together with more capacitance connected to the distribution system. The initial fault current contribution of a capacitor bank can be very big.

7.7 GENERATOR PROTECTION

7.7.1 General Requirements

Like all other components in the power system, distributed generators are equipped with protection to detect abnormal situations in the device or in the power system that require the immediate removal of the unit from the system. For this reason, there is always a circuit breaker or a fuse at the interface with the grid. This might be at the generator terminals or at the point of connection with the grid. Different countries give different requirements or recommendations for the types of protection and its settings; even different network operators within the same country often place different requirements on the protection of a distributed generator. An overview of the protection requirements, biased toward UK practice and regulations, is given in Ref. 232.

For small generators (50–500 kVA), the requirements are as follows:

- Three-phase time-delayed overcurrent protection. A voltage-dependent relay is an option, but the requirement for a voltage transformer is a disadvantage of such a relay.
- Earth fault time-delayed overcurrent.
- Reverse power.
- A combination of protection relays to detect noncontrolled island operation. These are discussed in further detail in Section 7.7.4.

An overview of protection at the interconnection with a distributed generation is also given in Ref. 193, with reference to industrial installations with on-site generation. The following protection elements are mentioned, grouped by their function:

- **Anti-islanding protection.** Undervoltage, overvoltage, underfrequency, overfrequency, instantaneous overvoltage, and directional power.
- **Shunt fault clearing.** Ground overcurrent, phase overcurrent, ground undervoltage, and overvoltage.
- **Abnormal operating conditions.** Negative-sequence overvoltage, negative-sequence overcurrent, and loss of protection potential.
- **Restoration.** Reconnection timer and synchronization check.

The protection of the generator has two distinctive functions:

- Remove the generator from the grid in case of a fault or another abnormal situation in the generator. The aim of this is to protect the grid against, for example, high currents. It also indirectly protects other customers against abnormal situations in the generator.
- Remove the generator from the grid in case of a fault or another abnormal situation in the grid. The aim of this is not only to protect the generator but also to prevent the generator from making an abnormal situation in the grid worse. The main concern here is "noncontrolled island operation," as will be discussed in detail in Section 7.7.3.

Like with the protection in the grid, the concerns are failure to operate and unwanted operation (see Section 7.1). In addition to the protection of distributed generators, the main concern is failure to operate during faults or other abnormal operations in the grid. Unwanted trips are a concern for industrial CHP installations and possibly also for future domestic CHP. The tripping of the electrical generation will typically also result in tripping of the heat production and thus of the industrial process. Restarting this may take several hours with high resulting costs due to lost production. Most smaller distributed generators (like solar and wind power) will reconnect automatically a few minutes after the supply becomes normal again. The consequences of an unwanted trip, for example, due to a voltage dip, will thus be only a few minutes of lost energy production. Recently, network operators have shown serious concern for the mass tripping of distributed generation. We will come back to this in Section 7.7.6 and in Chapter 8.

Failure to operate during faults and other abnormal situations in the grid will be discussed in more detail in several of the forthcoming sections.

7.7.2 Insufficient Fault Current

In Section 7.6, the situation was discussed in which the generator contributes too much to the fault current. This is especially a concern with synchronous and induction generators. The opposite situation occurs for generation connected through a power electronics converter. The fault current contribution from these generators is too small for the overcurrent protection to detect the fault. With reference to Figure 7.36, the

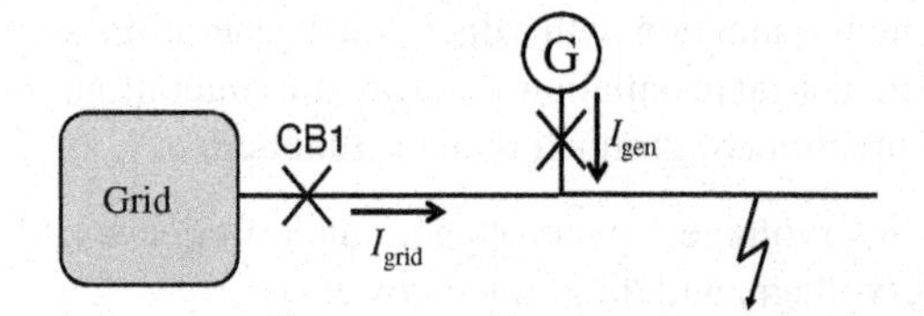

Figure 7.36 Fault contributions from the grid and from a generator connected to a distribution feeder.

fault contribution from the grid I_{grid} is sufficient for the protection to detect the fault. As a result, the breaker CB1 opens. The contribution from the generator I_{gen} is too small to be detected by the overcurrent protection and the generator remains feeding the fault.

The current produced by a converter is typically actively limited to a value somewhat above the rated current. In some cases, the current even drops during the fault, depending on the control algorithm used. The current may either become so small that the fault clears by itself or the generator contribution, although small, may be enough to maintain a fault current. If the fault clears, the generator will go into noncontrolled island operation. The resulting voltage and frequency during island operation depend on the balance between production and consumption (both active and reactive powers) in the island.

When automatic reclosing is used, as is often the case with overhead distribution feeders, the breaker CB1 is closed again after a certain time (ranging from less than 1 s to 1 min, but most typically a few seconds). When the fault has cleared but the generator is still connected, the breaker will close onto a system with an unknown frequency, voltage, and phase. The result could be large currents, damage to equipment in the grid, damage to the generator, and damage to other end-user equipment. It is obvious that such a situation should be avoided.

During the noncontrolled island operation, the feeder will typically not be earthed. Even if the system is normally solidly earthed, the opening of the circuit breaker will remove the system earthing as well. Unless the generator or the generator transformer has an earth connection, the feeder will be isolated and the fault current will be small. This makes it more likely that the fault will extinguish. But if the fault remains, the voltage between the nonfaulted phases and ground will be 1.7 ($\sqrt{3}$) times the nominal phase-to-ground voltage. The frequency will be determined by the generator and most likely deviate from the nominal power system frequency. As a result, the voltage difference over the recloser may be up to 2.7 times the amplitude of the nominal phase-to-ground voltage.

When the generator or the generator transformer has an earthed neutral and when the fault involves multiple phases, the voltages in the islanded part of the feeder will be smaller, so the consequences of reclosing will be less severe.

The recloser will in any case observe almost always a significant current upon reclosing and permanently trip the feeder, even when the fault has cleared or would have cleared without generation present. In this way, the number of long interruptions will increase for the customers. It is difficult to quantify this, but a commonly used value is that 80–90% of faults are transient faults. The presence of

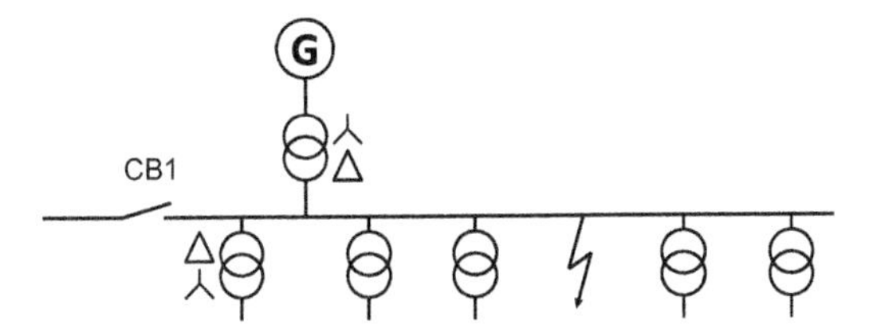

Figure 7.37 Distributed generator connected through a generator transformer.

distributed generation could, thus, increase the number of long interruptions by a factor of 5–10.

In many cases, certainly at medium-voltage feeders, the generators are connected to the feeder through a delta–why-connected generator transformer. The why-connection is typically on the generator side so as to not impact the earth fault protection in the medium-voltage grid. The configuration is shown in Figure 7.37. Because the feeder breaker (CB1) is open, the feeder is isolated.

To calculate the voltages and currents during a single-phase fault, the three symmetrical component circuits should be connected in series, resulting in Figure 7.38. In the diagram, Z_{tr1} and Z_{tr2} are positive- and negative-sequence impedances, respectively, of the generator plus the generator transformer; Z_{L1} and Z_{L2} are positive and negative-sequence impedances, respectively, of the load as seen from the medium-voltage grid, that is, including the impedance of the distribution transformer; and C is the total feeder capacitance.

The load impedance typically has a high-resistive component, but the transformer impedances are normally mainly inductive. These inductances, together with the feeder capacitance, form a resonance circuit that might become excited (see Section 7.7.5). From the diagram in Figure 7.38, it also follows that the generator is not loaded by the presence of the single-phase fault, but that it needs to supply all the positive-sequence load on the feeder during island operation. This case is discussed among others in Refs. 254, 309, and 317.

As already mentioned, the feeder will be nongrounded during island operation. Once the feeder breaker opens, the voltage between nonfaulted phases and ground rises to 1.7 times its normal value. This is not a concern as long as all load is connected through delta–why transformers or similar. In Europe, this is the case for the majority of medium-voltage feeders. As high-impedance earthing is common in Europe for medium-voltage feeders, the load is connected in such a way that the elevated voltage at medium voltage is not transferred to the end-user equipment.

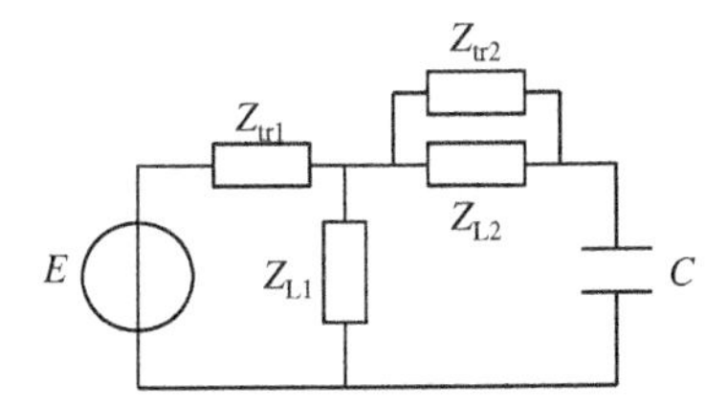

Figure 7.38 Equivalent circuit for the configuration in Figure 7.37 during a single-phase fault.

The situation in North America is rather different. Medium-voltage feeders are solidly earthed and single-phase distribution transformers are rather normal. The phase-to-ground voltages transfer directly from medium-voltage feeders to the equipment terminals. When the fault does not extinguish and the generator does not trip, this overvoltage will remain until the feeder breaker closes again, which can take up to several seconds. Damage to end-user equipment is a likely consequence of this. These overvoltages can be prevented by connecting the generator transformer star-grounded on the grid side, and in this way providing an additional system grounding. This will, however, result in higher currents during earth faults that may interfere with the existing earth fault protection. The various grounding and overvoltage issues for North American systems and the way these are treated by BC Hydro (British Columbia, Canada) are discussed in detail in Ref. 317. Two general connection rules are that a 122% temporary overvoltage during reclosing is accepted, and that the reduction in the sensitivity of the feeder ground protection should not be more than 5%. A grounding impedance is used to comply with this. When this is not possible, a more detailed study is performed to look for other solutions.

When a generator is connected to a noneffectively grounded medium-voltage feeder through a delta–star transformer, as is often the case, it will not be able to detect a single-phase earth fault at medium voltage. Some generators use the zero-sequence voltage on medium-voltage side of the transformer as an input signal and trip when this voltage gets high. This method can, however, not distinguish between a fault on the local feeder and a fault on another feeder, thus resulting in many unnecessary trips.

7.7.3 Noncontrolled Island Operation

The problem of the noncontrolled island is rather complication and we will not be able to discuss all possible aspects of it here. A first, and very important, distinction is between "controlled island operation" and "noncontrolled island operation." Controlled island operation is a method to improve the reliability of supply; one or more generators are equipped with the proper control and protection equipment to guarantee a reliable and safe operation. Controlled island operation is used among others to obtain high reliability in industrial installations, hospitals, and data centers that require a higher reliability than that can be offered by the public supply. In countries or regions with a low reliability of supply, controlled island operation is also used by domestic customers. Most commonly diesel generators are used, but at remote locations battery packs could also be used, being a cheaper solution. The presence of distributed generation enables a wider scale use of controlled island operation. This is often mentioned as an important advantage of distributed generation.

Non-controlled island operation, on the other hand, is a serious concern and should be avoided whenever possible. We use the term noncontrolled island operation when one or more generators power one or more loads, without a galvanic connection to the rest of the grid, and when this situation is unintended. Even when control equipment is available to maintain all parameters within their appropriate range, we still talk about non-controlled island operation whenever this situation is unintended.

There are a number of reasons why noncontrolled island operation should be avoided:

- The voltage magnitude could reach values far outside its normal operating range. Also, unacceptable transient overvoltages or frequencies could occur. This could damage equipment in the network, end-user equipment, and endanger personal safety.
- The protection of the distribution network is not designed for island operation; a fault, therefore, might not be cleared or cleared too slow. This could again lead to equipment damage or injury.
- Opening a breaker in a radial network will no longer guarantee that the downstream network is indeed de-energized. This could endanger the safety of maintenance personnel.
- Automatic reclosing could result in damage to the generator or to the equipment in the network. This is studied with a realistic simulation model in Refs. 254 and 255, where it is shown that it is not unlikely that the generator is still energizing the feeder at the reclosing instant.
- Harmonic resonance or even ferroresonance may occur during the island operation; as the source is weak, a small amount of capacitance can give a low resonance frequency [239].
- The network operator may still be responsible for the voltage delivered to the customers, but without any possibility to control this voltage.

It is also important to distinguish between "short-time island operation" and "long-time island operation" or "sustained island operation." The latter requires a sustained balance between the production and the generation, both for active and reactive powers. Such is not very likely, unless dedicated control systems are used or when control systems installed for other purposes unintentionally take over the control during island operation. Short-time island operation is more likely [255] and it is this that causes the problems with autoreclosing and that might result in dangerous overvoltages. Experimental results presented in Ref. 347 give a 84.8 ms island operation with a 1.7 kW solar panel. Personnel safety is, however, more concerned with the long-time island operation.

The main reasons for anti-islanding detection are personnel safety and the risk of automatic reclosing on a life feeder. The consequences of both can be rather severe (death or injury and serious equipment damage), so most network operators want to prevent noncontrolled island operation at any cost. Any inconveniences of this for the generator should simply be accepted. Therefore, all technical recommendations and regulations for the connection of distributed generation require the presence of anti-islanding protection.

Noncontrolled island operation can occur after any of the following two initiating events:

- A fault in the distribution feeder followed by the opening of the upstream fuse or breaker. The fault current contribution from the generators is too small to detect the fault. Either the fault clears by itself or the generators maintain a small fault current. This case is easier to detect than the next one, because of the presence of the fault. But even here, detection is difficult in some cases, as was discussed in Section 7.6.
- An upstream breaker is opened or a fuse opens without the presence of a fault. This may be an inadvertent operation or the intentional opening at the start of a planned interruption.

The discussion on noncontrolled island operation is still going on. There are two opposing opinions in the discussion. The "safety principle" is that noncontrolled island operation should be absolutely avoided. This is in fact a common principle behind most protection issues: safety overrides everything else. The strict requirements on islanding protection are based on this principle. There is, however, also the risk-based approach, where the principle is that there should be a fair balance of the consequences. In this case, the probability and consequences of undetected island operation should be weighted with the probability and consequences of unnecessary trips. The latter consequences are mainly economical, whereas the former include personnel injury and death. Weighting these has never been easy.

The picture has, however, changed with larger penetration of distributed generation. The consequences of mass tripping of distributed generation have grown, as was shown during two of the latest large-scale blackouts in Europe (see Chapter 8). Transmission system operators will less and less accept unnecessary tripping of distributed generation, which in turn could lead to the need to accept a higher probability of nondetected island operation. In Ref. 68, the probability of balance between the production and the consumption is quantified. Detailed measurements of variations in active and reactive power consumption are obtained. The conclusion from the study is that the probability of noncontrolled island operation is not negligible.

The introduction of controlled island operation will also increase the risk of noncontrolled island operation as it is defined here. Suppose that a number of customers connected to a distribution feeder have the ability to supply their own load sustained during island operation. A tripping of the feeder could result in the generators of these customers picking up the load of the whole feeder and operating the feeder as an island. Although both the voltage and the current might be kept within limits, this is an unwanted situation that should be avoided. An appropriate solution is to block the control of the generator as long as the islanding breaker is closed.

Equipping distributed generation with voltage control was proposed in Chapter 2 to prevent overvoltages. This will, however, also increase the risk of noncontrolled islanding.

7.7.4 Islanding Detection

7.7.4.1 Basic Methods Reliable islanding detection, also called "anti-islanding protection" or simply "islanding protection," is seen by many network operators as one of the main challenges in integrating large amounts of distributed generation.

The term "loss-of-mains protection" is also used. The first line of defense to prevent noncontrolled island operation consists of rather simple relays. Once the voltage magnitude or frequency comes out of its normal range, the generator is tripped. The basic method to detect noncontrolled island operation consists of the following protection devices:

- Undervoltage protection
- Overvoltage protection
- Underfrequency protection
- Overfrequency protection

The overvoltage protection is the most important one, because this is where there is immediate danger to equipment damage. This should preferably be a combination of instantaneous trip for high overvoltage (for example, 1.5 per unit peak voltage [239]) and slower protection at, for example, 110% rms voltage. As we will see later, most countries do not require such a two-stage overvoltage protection, but instead have a relatively fast tripping requirement for relatively minor overvoltages.

The immediate disadvantage with such a scheme is that voltage magnitude and frequency do sometimes reach values outside its normal range, albeit during short periods of time. Voltage dips due to remote faults are the most common example [42]. Also, it is very difficult to accurately measure frequency with a high time resolution, so the fast under- or overfrequency protection would almost certainly result in far more trips than needed. To prevent an excessive number of unnecessary trips, a certain delay is introduced. The European standard for the connection of "microgeneration" (generation with a rated current less than 16 A per phase, connected to the low-voltage network) prescribes the following protection settings [135]:

- Overvoltage protection: 200 ms, 115% of 230 V
- Undervoltage protection: 1.5 s, 85% of 230 V
- Overfrequency: 500 ms, 51 Hz
- Underfrequency: 500 ms, 47 Hz

These are the "maximum range values," that is, the protection should trip the generator within 200 ms, when the voltage rises above 115% of 230 V (the nominal voltage). The standard document further states: "Settings should be as close as possible to the limits to avoid nuisance tripping." It was, however, not possible to obtain agreement between the member states on the protection settings. The above given values are only "default settings" and individual European countries are allowed to require different settings. An informative annex with the document gives the setting prescribed in different European countries. This informative annex illustrates very well the difficulty in finding appropriate settings. The variation in settings for overvoltage and undervoltage is shown in Figure 7.39. The most sensitive overvoltage setting is 106%, 120 ms (in Belgium), whereas Italy accepts overvoltages as high as 120% of nominal and the Netherlands allows the generator to remain connected for as long as 2 s.

The setting of the undervoltage protection shows a similar wide range, with time setting from 150 ms up to 2 s and voltage settings from 50% up to 90%. The trade-off

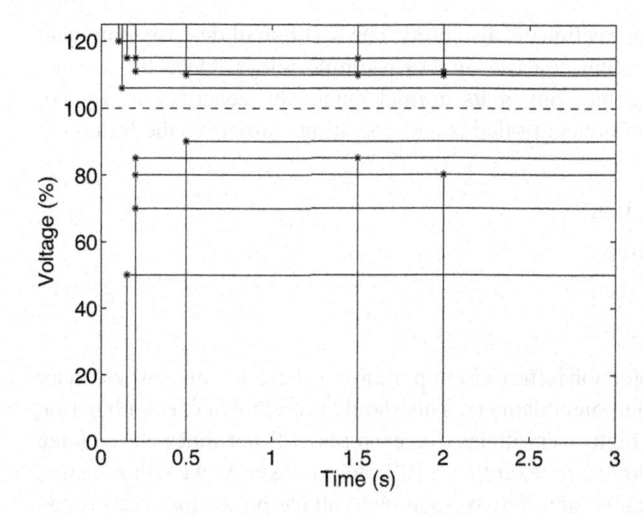

Figure 7.39 Overvoltage and undervoltage protection settings for distributed generation in different European countries.

between preventing noncontrolled island operation and preventing unnecessary (and often also unwanted) tripping apparently falls out completely different in different countries. One of the main problems in setting the protection is that no country has sufficient experience with large-scale integration of microgeneration to be able to estimate the probability and the risk with noncontrolled island operation. As we will see later, other methods for islanding detection are in use as well. Therefore, not all countries place equally high importance on using undervoltage and overvoltage protection to prevent noncontrolled island operation.

A number of countries (notably Denmark, Finland, and Sweden) have two-stage settings with fast tripping for large deviations from normal and slow tripping for smaller deviations. This provides a reasonable amount of protection against excessive voltage magnitudes, while at the same time limiting the probability of unnecessary trips. The requirements for these three countries are reproduced in Figure 7.40. We see fast settings (150–200 ms) when the voltage exceeds 115% and slow settings (up to 60 s) for smaller deviations. The latter settings also prevent overvoltages due to high amounts of production during periods of low consumption, as discussed in detail in Chapter 5.

Also, the undervoltage protection shows a fast stage (150–200 ms) and a slow stage (5–60 s). Note that short-duration undervoltages are much less dangerous than overvoltages. Long-duration undervoltages could, however, result in equipment damage, mainly due to higher currents. The normal consequence of undervoltages of longer duration is that the equipment will trip, which it would have done anyway because the feeder is no longer supplied from the grid.

In Figure 7.40, two curves are shown for Denmark (for undervoltage as well as for overvoltage). While other countries only give "maximum range values," Denmark gives a range of permissible settings. For example, the 200 ms overvoltage setting is allowed to be between 113% and 115% of nominal. Any setting outside this range is not permitted. This is to prevent extremely sensitive settings of the protection, which would result in many unnecessary trips. For undervoltages, the generator is allowed

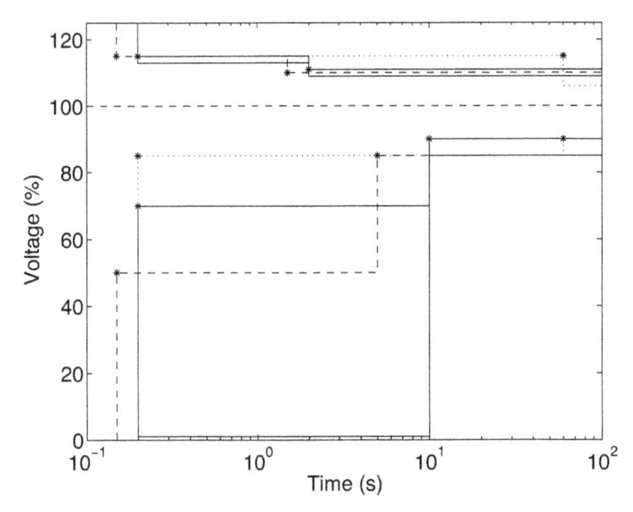

Figure 7.40 Overvoltage and undervoltage protection settings for distributed generation in Denmark (solid), Finland (dashed), and Sweden (dotted).

to trip when the voltage drops below 70% for 200 ms; the generator should trip when the voltage drops below 90% for 10 s.

The IEEE standard for the connection of distributed generation, IEEE Standard 1547-2003 [220], among others includes the following requirements (where DR stands for "distributed resources," and EPS for "electric power system." The "Area EPS" is in most cases synonym for the local feeder.):

- "The DR unit shall cease to energize the Area EPS for faults on the Area EPS circuit to which it is connected."
- "The DR shall not energize the Area EPS when the Area EPS is de-energized."
- "The DR shall cease to energize the Area EPS circuit to which it is connected prior to reclosure by the Area EPS."
- "When any voltage is in a range given [...], the DR shall cease to energize the Area EPS within the clearing time as indicated."
- "When the system frequency is in a range given [...], the DR shall cease to energize the Area EPS within the clearing time as indicated."

The requirements in this standard are a combination of general requirements (the first three) and specific requirements (the last two). The general requirements should be interpreted as input to the design process, whereas the specific requirements give immediate input to the setting of the protection. The maximum clearing times for undervoltage and overvoltage, according to IEEE Standard 1547, are shown in Figure 7.41. A two-stage setting is used, as in Figure 7.40, with the shortest clearing time equal to 160 ms. Long-term operation (beyond 1 or 2 s) is permitted only when the terminal voltage is between 88% and 110% of the nominal voltage.

The choice of the settings according to IEEE Standard 1547 is not discussed in the standard document, but some information can still be obtained on the expected

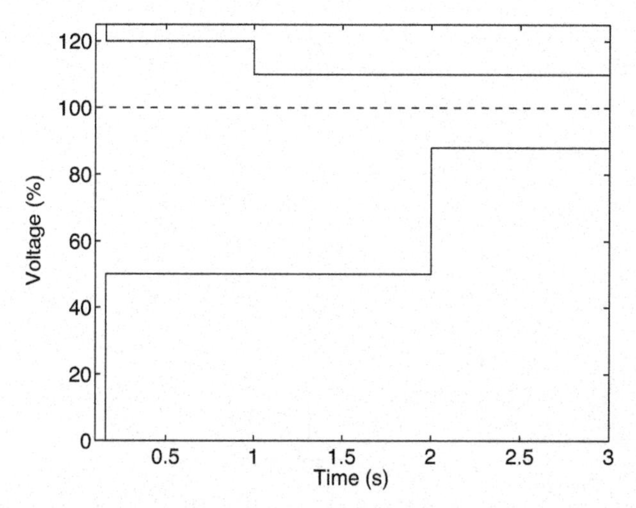

Figure 7.41 Overvoltage and undervoltage protection settings for distributed generation according to IEEE.

number of unnecessary trips. The EPRI Distribution Power Quality survey [120, 184] gives detailed data on the number of voltage dips that can be expected in medium-voltage distribution networks. These data can be used to estimate how often necessary and unnecessary trips on undervoltage are expected to occur. From the data published in Ref. 120, it follows that undervoltage trips are expected to occur about seven times per year. This includes necessary and unnecessary trips. The number of short interruptions (voltage magnitude below 10% of nominal) is 5.1 per year with a duration of 160 ms or longer. We may assume that the majority of these would require the disconnection of the generator. To summarize, we find that a generator at an "average location" during an "average year" will experience five necessary and two unnecessary trips.

Information about the number of voltage dips experienced by a low-voltage customer is obtained from the NPL survey [119, 120]. Using the same reasoning as for medium-voltage distribution networks above results in about 12 necessary and 8 unnecessary trips per year for a generator connected to low-voltage distribution networks.

The underfrequency and overfrequency settings for microgeneration in different countries (as cited in EN 50438 [135]) are shown in Figure 7.42. With the exception of Denmark, these are maximum clearing times. Also for frequency, we see a large range in settings, from less than 200 ms up to 2 s in time, from 49.5 to 47 Hz in underfrequency, and from 50.5 to 52 Hz in overfrequency. Like before, the more sensitive the settings, the smaller the probability of uncontrolled island operation, but the larger the probability of unnecessary tripping.

The concern with unnecessary tripping is especially with underfrequency, hence the low settings (as low as 47 Hz) in some countries. The loss of one or more large generators (or a splitting of the synchronized system) might result in a shortage of

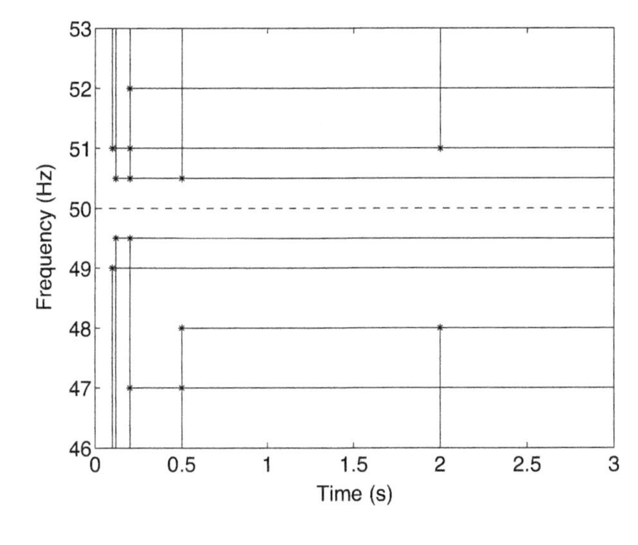

Figure 7.42 Overfrequency and underfrequency protection settings for distributed generation in different European countries.

generation. The loss of all microgeneration due to underfrequency tripping will further deteriorate the situation. We will come back to this in Chapter 8.

The underfrequency and overfrequency requirements according to IEEE Standard 1547-2003 [220] are shown in Figure 7.43. For underfrequency, two curves are given: the solid curve holds for small units (up to 30 kW), whereas the dotted rectangle gives the range of settings for large units (above 30 kW). The actual setting should, according to the standard, be coordinated with the underfrequency settings in the public grid.

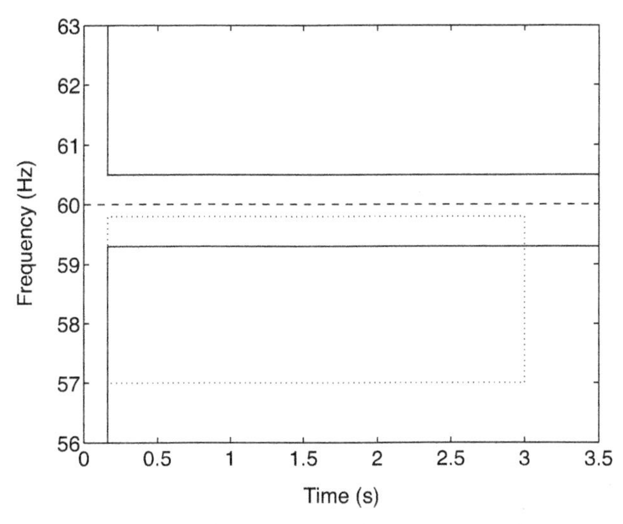

Figure 7.43 Overvoltage and underfrequency protection settings for distributed generation according to IEEE.

7.7.4.2 ROCOF and Vector Shift Two other methods for islanding detecting are in use as well: rate of change of frequency (ROCOF)" and vector shift (also known as "vector surge"). Both have the ability to detect noncontrolled island operation faster than any of the earlier methods, but both are also often more sensitive to other disturbances. This is again the trade-off between fail-to-trip and unnecessary trip. During island operation (either controlled or noncontrolled), the frequency changes with the unbalance between generation P_g and consumption P_c [45, Section 2.1]:

$$\frac{df}{dt} = \frac{f_0}{2H}\left(P_g - P_c\right) \tag{7.54}$$

where f_0 is the power system frequency and H is the inertia constant of the island, that is, the ratio between the total rotational energy and the rated power. Both P_g and P_c are in per unit, with this rated power as a base. A ROCOF relay detects the unbalance between the generated and the consumed active power within the electrical island. The setting of the relay is again a point of discussion. Only for three countries does EN 50438 [135] gives values for the setting of a ROCOF relay with microgeneration:

- In Belgium, the relay should trip within 100 ms (in practice, instantaneously) when the rate of change of frequency exceeds 1 Hz/s.
- In Denmark, the setting should be at 200 ms for a threshold between 2.5 and 3.5 Hz/s.
- In Ireland, the maximum settings are 500 ms and 0.4 Hz/s.

According to Ref. 164, typical settings are 0.125 Hz/s for Great Britain and 0.5 Hz/s for Ireland. The ROCOF relay is notorious for its unnecessary trips. The trade-off between fail-to-trip and unnecessary trips due to ROCOF and vector shift relays is studied in Ref. 126 and 166. According to Ref. 126, the ROCOF relay is better at detecting island operation than the vector surge relay. The sensitivity not only depends on the setting of the relay but also on the type of generator. For a synchronous machine, the sensitivity is roughly proportional to the pickup value of the relay. For a pickup value of 0.4 Hz/s, an unbalance of 5% between production and consumption can be detected. For a DFIG machine, any setting up to 10 Hz/s can cope with unbalance as small as a few percent. It is the loss of the phase-locked loop reference that makes the control system unstable and quickly results in a large shortage of generation. The other side of the trade-off (unnecessary tripping of the relay) is studied in Ref. 126 as well. To ensure stability, settings as high as 10 Hz/s are needed for the synchronous machine and as high as 7.2 Hz/s for the DFIG machine. The main concern are three-phase fault: they cause the highest apparent change in frequency for the relay. When accepting unnecessary trips for all but single-phase faults, a setting of 1 Hz/s would give a stable operation.

A significant source of confusion with ROCOF relays is the difference in performance between different relays for the same disturbance. In order to provide reliable protection against noncontrolled island operation, the relay should detect such a situation quickly. In many cases, the relay is set to trip instantaneously. Within a period of one to several hundreds of milliseconds, the relay should determine the frequency twice and compare the difference between these two values with a threshold value.

Determining the frequency of a signal (in this case, of the voltage) can be done in a number of ways. Most methods are in their essence a time measurement and this can be done very accurately. The simplest and cheapest method is to count the difference in time between a number of zero crossings of the voltage [265]. For a stationary signal, this gives a very accurate frequency measurement. However, the voltage is rarely stationary at the start of noncontrolled island operation, especially not when the initiating event is a feeder fault. The fault often results in a sudden jump in phase angle of the voltage. This is referred to as a "phase-angle jump" in the power quality literature 42, Section 4.5. To prevent unnecessary tripping on this phase-angle jump, most ROCOF relays have an undervoltage blocking, where the trip signal is blocked when the terminal voltage is below, for example, 80% of the nominal voltage.

Even when all using zero-crossing techniques, different relays use different details of the methods, like different time delay and different measuring window [401]. The undervoltage blocking may also be implemented differently: the phase-angle jump occurs both at the start and at the end of the voltage dip. The blocking should remain for some time after the voltage has recovered. This will, however, make the relay slower.

A study was done by National Grid in the United Kingdom to estimate the risk of mass tripping of distributed generation on their ROCOF relay due to the loss of a large generator [319]. The highest rate of change of frequency observed in the grid of Great Britain during a 5-year period was −0.0195 Hz/s. Some of the results of this study are shown in Figures 7.44–7.46. The first figure shows how the loss of more generation results in a faster change in frequency (note that the absolute value of the rate of change in frequency is given). It is interesting to see that the same loss of generation can give a large range of values for the rate of change in frequency. The highest value is, however, more or less linear with the amount of generation lost. Figure 7.45 correlates the rate of change of frequency with the amount of distributed

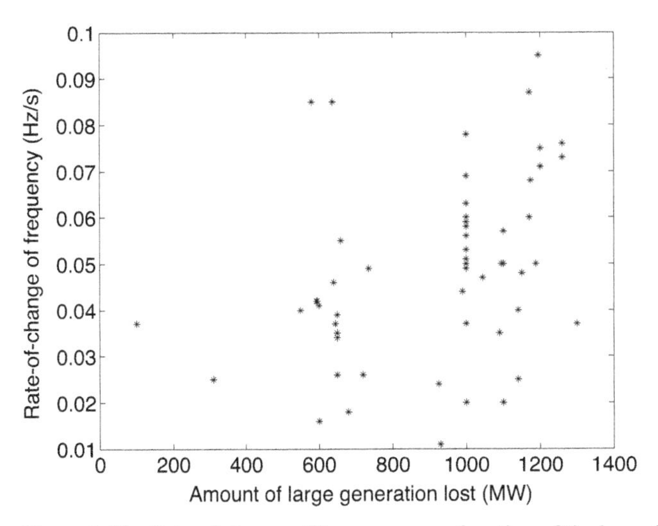

Figure 7.44 Rate of change of frequency as a function of the loss of generation for the Great Britain system.

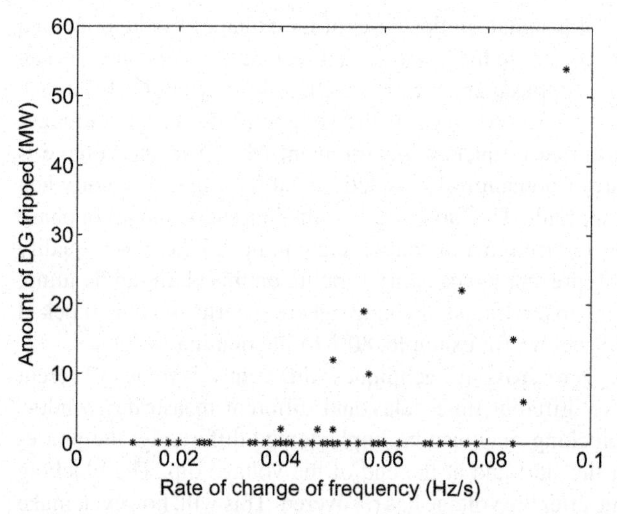

Figure 7.45 Amount of distributed generation tripped as a function of the rate of change of frequency.

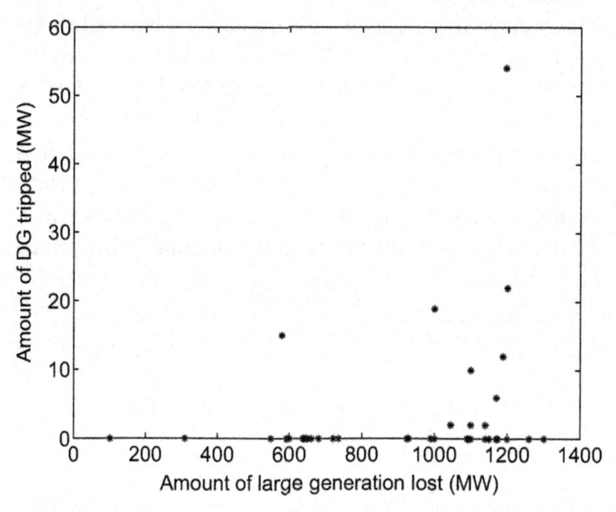

Figure 7.46 Amount of distributed generation tripped as a function of the loss of large generation.

generation that is unnecessary tripped. As mentioned before, a typical setting of the ROCOF relay in Great Britain is at 0.5 Hz/s, so none of the events in the figure should have resulted in a loss of distributed generation. However, once the rate of change of frequency exceeds 0.05 Hz/s, distributed generation starts to trip. Note also that even for higher values of the rate of change of frequency than this, in many cases no distributed generation at all trips. Figure 7.46 directly links the loss of large generation with the loss of distributed generation. With the exception of one event, all loss of distributed generation occurred for 1000 MW or more loss of large generation. The loss of more than about 1300 MW is very uncommon in Great Britain, so it is not possible to say what the loss of distributed generation will be when, for example, 2000 MW of generation is lost. The highest loss of distributed generation, of 55 MW, is however only 4.5% of the amount of large generation lost (1200 MW), so this will not significantly impact the system behavior.

As shown in Section 4.6 of Ref. 42, phase-angle jumps up to 20° can occur in the phase-to-neutral or phase-to-phase voltages, even when the voltage magnitude remains above 80%. The number of faults causing such a large phase-angle jump is still limited however, so let us assume that we want the relay to be immune (i.e., not produce an unnecessary trip) for phase-angle jumps up to 10°. These events have a much bigger potential impact on the ROCOF relay than the real drops in frequency due to the loss of a large generator.

A 10° phase-angle jump will change the time between zero crossings by

$$\frac{10}{360} \times 20\,\text{ms} = 0.56\,\text{ms} \tag{7.55}$$

Let us assume that the distance between zero crossings increases (a "positive phase-angle jump"). If the frequency calculation takes place every cycle, the apparent frequency would jump to

$$\frac{1}{20.56\,\text{ms}} = 48.65\,\text{Hz} \tag{7.56}$$

and the apparent rate of change of frequency would be

$$\frac{50\,\text{Hz} - 48.65\,\text{Hz}}{20\,\text{ms}} = 67.5\,\text{Hz/s} \tag{7.57}$$

This is obviously far beyond any reasonable setting of the relay. Fortunately, the detection time is, in practice, at least a few cycles; according to Ref. 401, the range is 40 ms–2 s. For a 100 ms measurement window, the length of five cycles is measured (in a 50 Hz system). For the same 10° phase-angle jump as before, the length of these five cycles becomes 100.56 ms. This results in an apparent frequency of

$$\frac{5}{100.56\,\text{ms}} = 49.72\,\text{Hz} \tag{7.58}$$

and an apparent rate of change of frequency of

$$\frac{50\,\text{Hz} - 49.72\,\text{Hz}}{100\,\text{ms}} = -2.8\,\text{Hz/s} \tag{7.59}$$

As was mentioned before (and shown in Ref. 126), a setting of 0.4 Hz/s is needed to detect an unbalance of 5% for a synchronous machine. Such a setting would, however, result in unnecessary trips for many voltage dips due to faults.

The vector shift relay (also referred to as "vector surge relay") directly measures the difference in duration between consecutive cycles. It is no longer needed to measure two frequency values and take the difference between them. The vector shift relay is triggered because of the phase-angle jump in the voltage when the local generator takes over the load from the public supply. A drop in frequency also somewhat contributes to the vector shift, but this impact is small as can be seen from the following example.

Example 7.15 Consider an island system where the frequency drops with 1 Hz/s. This corresponds to a drop of 0.02 Hz per cycle. If the previous cycle lasts 20 ms, the next cycle will last 20.008 ms. This extra duration of 0.002 ms corresponds to a vector shift equal to

$$\frac{0.008}{20} \times 360^\circ = 0.144^\circ$$

To trigger the relay for a pickup of 6° would require the frequency to change with 41.7 Hz/s.

A relation between settings for ROCOF relays and for vector shift relays is used in Ref. 166 to be able to compare their performance. The extreme and average values of the permissible setting range are used for this, resulting in the following:

- A setting of 0.1 Hz/s corresponds to a vector shift of 2°.
- A setting of 0.5 Hz/s corresponds to a vector shift of 10°.
- A setting of 1.2 Hz/s corresponds to a vector shift of 20°.

The following settings are given for microgeneration in EN 50438 [135]:

- 7° instantaneous (120 ms at most) for Belgium. The vector shift relay can be blocked when the voltage magnitude is below 80% of nominal.
- In Denmark, the vector shift relay should trip within 200 ms when the vector shift exceeds a threshold value of at least 30°.
- In Ireland, the maximum settings are 500 ms and 6°.

According to Ref. 164, a recommended setting for the United Kingdom is 6°.

The vector shift relay is typically blocked when the voltage drops below a threshold of 80% or 85% of the nominal voltage. This is to prevent unnecessary trips due to voltage dips. As we saw before, phase-angle jumps up to about 10° can theoretically occur due to faults in distribution systems for voltage magnitudes above 80%. A setting value of 5–10° together with undervoltage blocking will prevent most of the unnecessary trips due to faults. However, unnecessary trips by the ROCOF relay or the vector shift relay are continuously reported, even for relatively high settings of the pickup value.

7.7.4.3 *Active Methods*

The disadvantage with any so-called "passive method" is that it can detect island operation only when there is an unbalance between production and consumption in the island. When both the active and the reactive power demand by the load is covered by the generators connected to the island, no passive method can detect the difference between island operation and grid-connected operation. Such a "perfect balance" can be due to the pure coincidence or due to the control system of one or more of generators creating the right conditions. The former may seem unlikely, but it is seen by many network operators and government agencies responsible for safety as an unacceptable risk. The latter may occur more and more often in the future when distributed generation is equipped with control algorithms, for example, to maintain the voltage within acceptable limits.

A number of so-called "active methods" for islanding detection are being proposed. These have the advantage that they can detect any island operation, even when both active and reactive powers are in perfect balance. A number of different methods are mentioned in the literature. A brief overview is given below:

- A small disturbance signal is created in the voltage controller. During grid connected operation, the impact of this signal will be small. However, during island operation, the signal will dominate and the controller may become unstable or an island-detection relay will detect the signal and trip the generator [126].
- A power line carrier signal is generated in the grid. When the signal disappears, the generator is tripped [126].
- Estimating the fault level by switching an inductor or a capacitor [93, 126]. Switching an inductor or a capacitor close to the voltage zero crossing, using a thyristor switch, will only cause a limited voltage distortion. The current amplitude will, however, give information about the local fault level. In most cases, the difference in fault level between grid-connected and island operation is very big, so there is no need for an accurate fault level estimation. The method will not work when large generator units are connected to weak parts of the grid. This may, however, also cause other problems beyond islanding detection, so this should not be seen as a serious limitation to this method.
- Active frequency drift method also known as "Sandia frequency shift" [93, 126, 236, 394]. A converter generates a waveform of a frequency that somewhat deviates from the normal power system frequency, for example, 65 Hz in a 60 Hz system. During grid-connected operation, the grid determines the frequency. During island operation, the frequency is determined by the converter and it will quickly jump to 65 Hz triggering the operation of the over-frequency protection. The standard method of active frequency shift uses a constant but nonnominal frequency. The algorithm as developed by Sandia National Laboratory goes a step further and tries to destabilize the system frequency to achieve a faster islanding detection. This obviously does not work during grid-connected operation as the system is far too strong. The disadvantage of any such method is that it results in harmonic distortion of the current waveform, which could again be a concern for weak systems.
- Inverters could operate as current sources instead of as voltage sources and have a destabilizing signal [239].
- Reactive power export error detection [93]. The control system of the generator either consumes or generates such an amount of reactive power that reactive power balance will not be possible. Island operation can be detected when the reactive power flow suddenly changes; otherwise, the reactive power unbalance will make the island unstable and the generator will trip on undervoltage or overvoltage.

These methods are, however, not immune for unnecessary trips. For example, with the first method, the disturbance signal could dominate the voltage in a weak system. With the second method, the power line carrier signal might temporary disappear that would be incorrectly interpreted as island operation.

7.7.4.4 *Other Methods*

Some authors propose an intertrip scheme where the opening of the feeder breaker will automatically result in the tripping of all generators connected to the feeder [93]. This could be rather expensive when a dedicated communication circuit (typically a metallic wire) is required and the distance between the main substation and the generator is several kilometers. It will also become complicated when many small generators are involved. The recent developments in substation automation, with the communication standard IEC 61850 [216] being the most visible exponent, however, indicate that cheap and reliable intertrip schemes will be within reach soon.

A number of real-life tests of the anti-islanding protection in Spain were presented in Ref. 341. The 20 kV breaker was opened in a 130/20 kV substation, supplying five solar power plants between 900 kW and 5 MW size. The five plants used five different inverter brands, manufactured in three different countries. Two such tests were performed; in both cases, a noncontrolled island resulted, with loads between 600 kW and 2.5 MW. In one of the two cases, uncontrolled islanding took place for 13 min after which it was actively terminated. Although all inverters were equipped with anti-islanding detecting, each of them would probably have individually detected the island, they collectively did not detect the island operation.

In Ref. 281 it is shown by simulations how different active methods for islanding detecting could compensate for each other. This could confirm the behavior observed in a realistic field test by Ref. 341.

BC Hydro (British Columbia, Canada) uses the basic rule that an island is unstable when the load exceeds twice the generator size [317]. In that case, a noncontrolled island will quickly collapse, so that no additional measures are needed beyond the standard voltage and frequency protection. In fact, the underfrequency protection will trip the generators well within 1 s even for insensitive settings.

When no sufficiently fast and reliable anti-islanding protection is available, two relatively cheap solutions are as follows:

- Automatically checking the voltage immediately downstream of the recloser before making a reclosing attempt. The reclosing will be blocked when the feeder is still energized, that is, when it is operating as an island. The disadvantage is that noncontrolled islanding, even short-time, will result in a long interruption.
- Delay the reclosing to allow the anti-islanding protection to intervene and for the voltage to disappear. This will increase the duration of the short interruption.

7.7.5 Harmonic Resonance During Island Operation

During island operation, controlled or noncontrolled, the harmonic impedance of the source changes. This results in new harmonic resonance frequency. Consider the island operation of a medium-voltage feeder. A harmonic source located at low voltage will observe the following impedance at angular frequency ω:

$$Z(\omega) = j\omega L_{tr} + \frac{j\omega L_{gen}}{1 - \omega^2 L_{gen} C} \tag{7.60}$$

where L_{tr} is the inductance of the distribution transformer, L_{gen} is the inductance of the generator transformer plus the generator, and C is the capacitance of the medium-voltage feeder. The harmonic impedance is highest, that is, a resonance occurs, for

$$1 - \omega^2 L_{gen} C = 0 \tag{7.61}$$

which results in the following expression for the resonance frequency:

$$f_{res} = \frac{1}{2\pi\sqrt{L_{gen} C}} \tag{7.62}$$

The lower the resonance frequency, the higher the probability that there is a significant source of harmonic distortion. The main concern in areas with dominantly commercial and domestic loads is with harmonic orders 5 and 7, that is, resonance frequencies of 250 and 350 Hz. Low resonance frequencies are obtained for high values of C and high values of L_{gen}. The latter means small generators, the former means underground cables. The capacitance of underground cables strongly depends on their construction and on the voltage levels. Wherever possible, data should be obtained from the cable manufacturer. Based on the tables given in Ref. 1, we conclude that the following typical values may be used when no manufacturer data are available:

- For 12 kV cables with cross sections of 185 and 240 mm^2, the capacitance varies between 350 and 600 nF/km. These are the kind of cables that can be expected in an urban environment. The maximum cable length would be a few kilometers. Lower capacity (smaller cross section) cables typically have a lower capacitance.
- For 24 kV cables with cross sections of 50 and 70 mm^2, the capacitance varies between 150 and 300 nF/km. These would be used in rural networks, with cable length up to a few tens of kilometers.

For a generator with rating S_{gen} and per unit impedance ϵ_{gen}, the inductance is found from

$$L_{gen} = \frac{1}{2\pi\omega_0}\epsilon_{gen} \times \frac{U_{nom}^2}{S_{gen}} \tag{7.63}$$

Example 7.16 Consider a 12 kV urban feeder of 3 km length with capacitance 450 nF/km. A 50 Hz 1 MW generator with 25% impedance is supplying the feeder in island. What is the harmonic resonance frequency seen for a low-voltage customer?

The generator inductance according to (7.63) is

$$L_{gen} = \frac{1}{2\pi 50\,\text{Hz}} \times 0.25 \times \frac{12\,\text{kV}^2}{1\,\text{MW}} = 115\,\text{mH}$$

Using (7.62) gives for the resonance frequency:

$$f_{\text{res}} = \frac{1}{2\pi\sqrt{115\,\text{mH} \times 1350\,\text{nF}}} = 403.9\,\text{Hz}$$

This is a frequency at which the harmonic emission is normally low, so no excessive harmonic voltages are expected.

From (7.62) and (7.63), it follows easily that the resonance frequency is proportional to the square root of the generator size. Thus, a four times as large generator will double the resonance frequency.

Example 7.17 Consider again the generator and feeder from Example 7.16. Which generator size results in a resonance frequency of 350 Hz?

A generator size of 1 MW results in a resonance frequency of 403.9 Hz; the resonance frequency is proportional to the square root of the generator size. This implies that a generator of size

$$\left(\frac{350\,\text{Hz}}{403.9\,\text{Hz}}\right)^2 = 751\,\text{kW}$$

will result in a resonance frequency of 350 Hz. A generator of 750 kW size would be connected to the medium-voltage network, so that this is not an unrealistic case. According to [Table 1.5] in Ref. 232, units larger than 250 kW would be connected to medium voltage.

Exercise 7.3 Calculate the generator sizes that will result in resonance frequencies of 250, 550, and 650 Hz? In which cases would the resonance frequencies be a concern?

Exercise 7.4 Repeat the calculations for a 24 kV rural feeder with a length of 18 km and a capacitance of 230 nF/km. A 60 Hz, 1 MW generator with 27% impedance is connected to the feeder. Calculate the resonance frequency. For which generator size will the resonance frequency occur at the 5th, 7th, 11th, or 13th harmonic in a 60 Hz system?

For the urban case (Examples 7.16 and 7.17), a resonance frequency below the fifth harmonic is rather unlikely. But for the rural case (Exercise 7.4), lower resonance frequencies are possible due to the higher voltage level and the longer feeders. For the rural example, a 240 kW unit would give a resonance frequency of 60 Hz.

Ferroresonance is a specific type of resonance: like the harmonic resonance discussed before, ferroresonance involves a capacitance and an inductance. The ferroresonance, however, involves a nonlinear inductance, in most cases the magnetizing inductance of a transformer. The nonlinear character of the inductance indicates that the resonance frequency depends on the magnitude of the voltage. The magnitude will adjust itself to a value such that the resonance frequency becomes equal to the frequency at which most energy is available, which is the power system frequency (50 or 60 Hz). During the noncontrolled island operation, the frequency may also change to obtain a balance between production and consumption in the island. Ferroresonance is known for its high voltage magnitudes (3 per unit is non uncommon) and for its highly distorted and irregular voltage and current waveforms. The following necessary conditions for ferroresonance are given in Refs. 239 and 309:

- The generator must be in island operation.
- The generator must be capable of supplying the island load. According to Ref. 308, the load should be less than three times the generator rating.
- Sufficient capacitance must be available: 30–400% of the generator rating. The range in capacitance resulting in ferroresonance is, according to Ref. 308, 25–500% of the generator rating.
- A transformer must be present to serve as a nonlinear inductance.

Both synchronous and induction generators can be involved in ferroresonance. It is stated that a combination of the two can make things worse [239]. Self-excitation of induction generators is a related phenomenon that can result in overvoltages of 1.5–2 per unit [309].

7.7.6 Protection Coordination

The previous section has mainly discussed protection requirements, with the main aim being to protect the network from noncontrolled island operation. Besides, many system operators place requirements on the immunity of distributed generation during voltage dips and large frequency deviations. The different requirements on the setting of the protection are summarized in Figure 7.45. The actual protection of the generator should be such that on the one hand it will not trip for any voltage dip above the "immunity curve," and on the other hand it will certainly trip for any dip below the "protection curve." The actual setting of the protection (which is what determines the immunity in practice) can only be in the area between the two curves.

The protection and immunity curves should be chosen such that the immunity curve is always to the left and above the protection curve. The curves in Figure 7.47 are an example of correct coordination. This is, however, not always the case and an example of incorrect coordination is shown in Figure 7.48.

From the discussion about islanding detection in the previous section, it is obvious that it is not always simple to coordinate between the protection and immunity

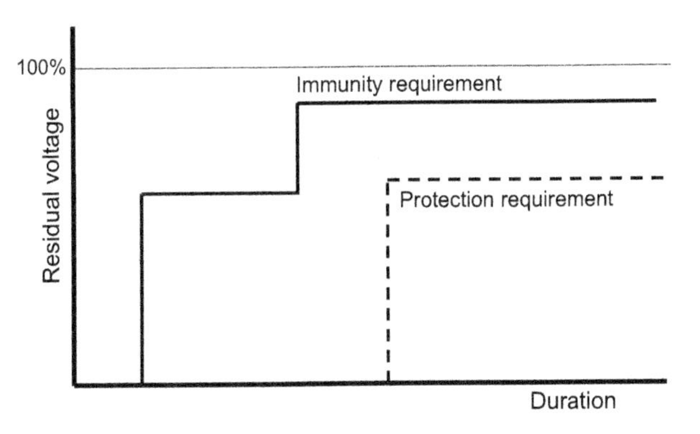

Figure 7.47 Protection and immunity curves for distributed generation.

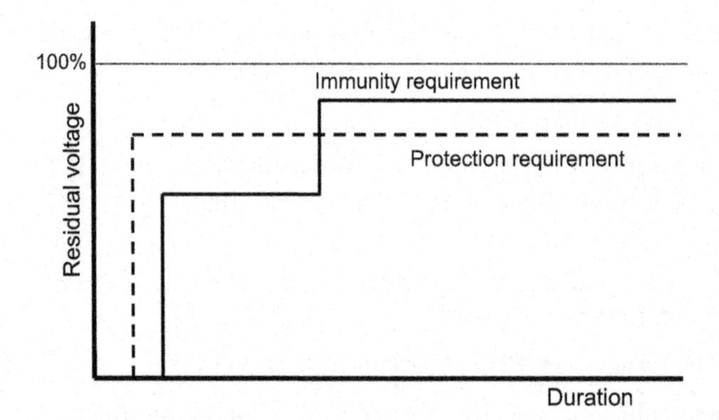

Figure 7.48 Incorrect coordination between protection and immunity curves for distributed generation.

curves. The undervoltage protection of the generator can, however, be easily coordinated with the immunity requirements. For the ROCOF and vector shift protection, it is less obvious as there is no direct link between the drop in voltage and the apparent frequency or phase-angle shift. The coordination is further complicated because immunity and protection requirements are typically set by different entities. The protection requirements are dictated by the distribution network operator, who wants the distributed generation to trip as soon as possible to prevent the noncontrolled island operation and also the other interferences with the protection of the distribution network. The immunity requirements are set by the transmission system operator who wants to keep the distributed generation to remain connected as long as possible to the network. We will discuss this in further detail in Chapter 8.

7.8 INCREASING THE HOSTING CAPACITY

Once the hosting capacity, as discussed in the previous section, is exceeded, mitigation actions are needed in the distribution network to prevent incorrect operation of the protection. Note that the ultimate impact of protection maloperation is a deterioration in the reliability of supply as experienced by the end users. For a complete assessment, the impact of other phenomena on the supply reliability should also be considered. When the incorrect operation of the protection after connection of a generator unit remains a rare event compared with other causes of supply interruptions, it may be decided to allow the connection without taking any additional measures. This is a decision that cannot be taken by the protection engineers alone, it rather requires a broader discussion, possibly including even the energy regulator. In this section, we however consider that the protection maloperation is an unacceptable event and that mitigation action is needed.

A general recommendation is that the setting of the protection in the distribution system should be recalculated when a large generator unit is connected. A detailed systematic method for this is presented in Refs. 290 and 291.

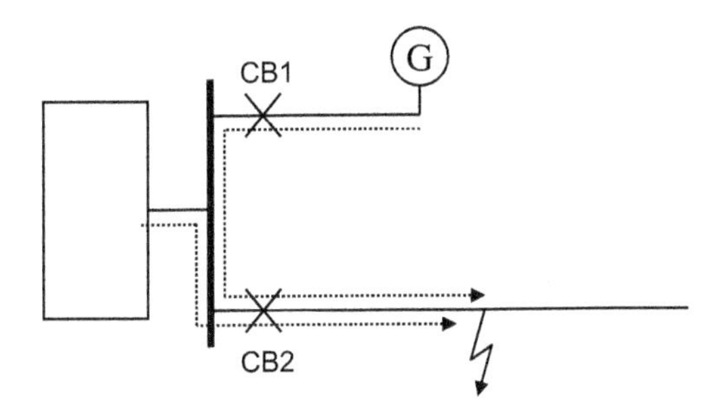

Figure 7.49 Dedicated feeder for a generator unit

7.8.1 Dedicated Feeder

A common solution against adverse impact of a generator unit on the distribution network is to connect it through a dedicated feeder to a sufficiently strong location in the network. As shown in Figure 7.49, there is no load connected to the same feeder. As the feeder stops at the generator unit, there is no risk of a fail-to-trip for a fault behind the generator. On the other hand, the risk of an incorrect trip due to a fault on another feeder remains. Such a trip will, however, only impact the generator and no other customers. Furthermore, as shown in some of the previous examples, the hosting capacity is high for a feeder when the generator is connected at the end of the feeder.

Using the same notation as in the earlier examples, the hosting capacity is reached when the maximum current for an external fault, according to (7.6), becomes equal to the minimum current for an internal fault, according to (7.31). For a dedicated feeder, there are no faults downstream of the generator, so $Z_6 = 0$. From (7.32) or from comparing the two expressions, it is easy to find the requirement $Z_1 = Z_5$. Using (7.28) and (7.2) gives the following expression for the hosting capacity:

$$S_{gen} = \frac{\epsilon_{gen}}{(1/S_{grid}) + (\epsilon_t/S_t}) \approx \frac{\epsilon_{gen}}{\epsilon_t} S_t \qquad (7.64)$$

The generator size has to be a substantial part of the transformer size; the contribution of the generator to a fault on grid side of the generator transformer should be the same as the contribution of the grid to a fault on secondary side of the grid transformer.

The use of a dedicated feeder will also solve a lot of problems with non-controlled island operation. As there is no load connected to the feeder, with the exception of a small amount of load needed for the auxiliary supply of the generator, balance between production and consumption is unlikely. Furthermore, the reclosing time can be coordinated with the protection of the generator, or automatic reclosing can be disabled completely.

The disadvantage of a dedicated feeder is, however, the high costs, especially when the generator is at a remote location. Also, such a solution is not practical for many small generators connected to the low-voltage network, as it could be possible with many micro-CHP units or rooftop solar panels. But these will most likely not be synchronous machine units, so their contribution to the fault current will be small.

7.8.2 Increased Generator Impedance

Incorrect operation of the protection is directly related to the high fault current contribution from the generator to the feeder. Limiting this fault current will result immediately in an increase in the hosting capacity. In fact, it can rather easily be shown that doubling the impedance of the generator will double the hosting capacity. Increasing the generator impedance by adding a series reactance is proposed in Ref. 436 as a solution for low penetration. A high impedance will, however, make it even more difficult to detect the fault, thus increasing the risk of noncontrolled island operation. It will also increase the risk of both the harmonic resonances at low frequencies during island operation and the transient or voltage instability. The installation of fault current limiters is also proposed in Ref. 129.

7.8.3 Generator Tripping

Another commonly used approach to prevent generator units from interfering with the distribution system protection is to equip them with very sensitive overcurrent and undervoltage protection. The result of such a setting is that the generator will always trip first. The setting of the rest of the protection no longer needs to take into account the contribution from the generator unit.

Consider the case in Figure 7.50. The instantaneous trip of CB5 does not always solve the incorrect operation of the protection. Fail-to-trip behind the generator is mitigated now, although the fault-clearing time may be somewhat longer. The increase in fault-clearing time is equal to the time it takes for CB5 to open. Where it concerns with the unwanted trip of CB1, this could still happen if CB1 is also set to instantaneous trip. In this case, breaker CB1 should be delayed compared to CB5. At the same time, CB3 and CB6 should be delayed compared to CB1. The delay in the tripping of CB6 could be a concern when the infeed is from an MV/LV transformer. Fast tripping of that breaker is required as an additional safety measure in case of a short circuit when personnel is around.

Another disadvantage is the regular tripping of the generator unit during faults elsewhere in the grid. This holds especially in case the tripping occurs on undervoltage. However, using overcurrent protection, it may be possible to limit the number of generator trips. The generator unit should trip for any fault that would result in unwanted operation of CB1. This implies that its current setting should be lower than

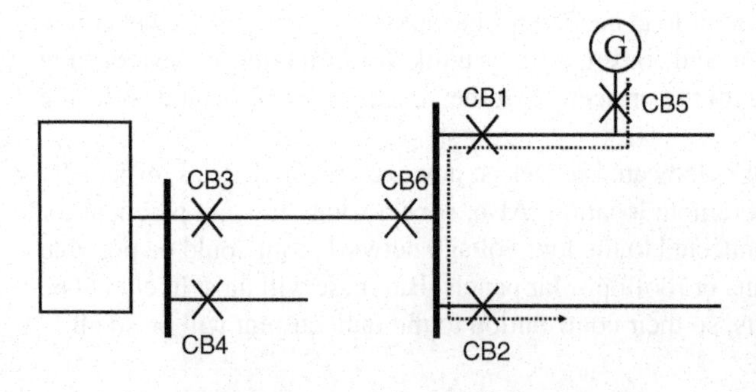

Figure 7.50 The need for protection coordination on a distribution feeder with generation.

the current setting for CB1. This setting is in turn less than the minimum fault current. The generator unit should also trip for a fault at the end of the feeder, thus allowing CB1 to clear that fault. The current setting of the generator unit should thus be less than the generator contribution to a fault at the end of the feeder.

Many distribution network operators require generator units to be equipped with sensitive protection against islanding. This protection will typically result in instantaneous trip for any fault on the feeder or on a neighboring feeder. The protection problem is immediately solved in this way. The transmission system operators are, however, not very happy with mass tripping of generator units connected to the distribution system. With large penetration of distributed generation, an alternative for the islanding detection needs to be developed.

7.8.4 Time-Current Setting

A more sophisticated method of protection coordination would attempt to limit the number of unnecessary trips of the generator unit. Consider again the network shown in Figure 7.50. There should be a time-grading coordination sequence for each possible fault current path. For a fault downstream of breaker CB2, the fault current path is CB2–CB6–CB3. Thus, CB6 and CB3 should be delayed compared to CB2. In the same way, CB6 and CB3 should be delayed compared to CB1. The presence of the generator unit creates a new fault current path for a fault downstream of breaker CB2: CB2–CB1–CB5. This requires that CB1 is delayed compared to CB2 and CB5 is delayed compared to CB1.

- Assume that CB2 is equipped with instantaneous trip.
- CB1 should be delayed by one time margin.
- CB3, CB6, and CB5 should be delayed by two time margins.

The setting of the islanding detecting is also a matter of protection coordination. The undervoltage curve, as shown in Figure 7.51, should be coordinated with the

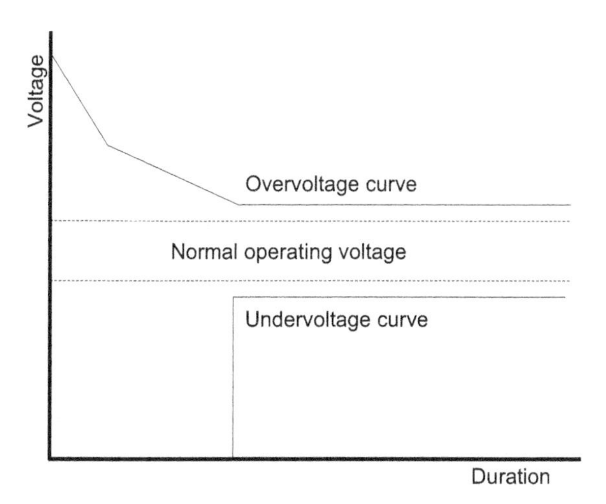

Figure 7.51 Overvoltage and undervoltage settings for islanding detection.

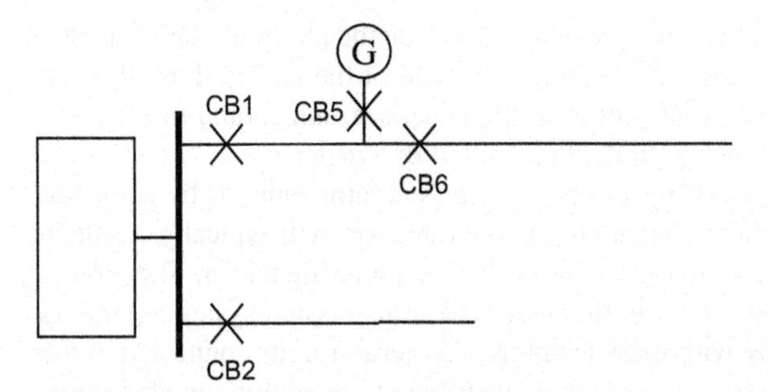

Figure 7.52 Additional circuit breaker downstream of the generator unit.

protection. Any requirements set by the transmission system operator should also be considered. These requirements are in fact protection coordination requirements. The overvoltage curve should be such that no equipment damage due to overvoltages occurs.

7.8.5 Adding an Additional Circuit Breaker

To prevent a fail-to-trip for a fault downstream of the generator, an additional circuit breaker is needed. There is a range of possible locations for this breaker, but immediately downstream of the generator would be an appropriate location. The new breaker (CB6 in Figure 7.52) clears faults toward the end of the feeder that are not seen by CB1. The setting of breaker CB1 can be increased to prevent mal-trip for external faults.

There is no need to change any of the settings of CB1, but delaying CB1 compared to CB6 (when possible) would increase the reliability for all customers connected between CB1 and CB6, including the generator unit.

The risk with having multiple breakers in cascade is that the failure of the downstream breaker will not be detected, or with a long delay, by the upstream breaker. Using inverse-time characteristics could give very long fault-clearing times. An intertrip from the downstream to the upstream breaker could solve this problem [239], but would obviously further increase the costs.

7.8.6 Directional Protection

When the amount of generation is high for two neighboring feeders, it is no longer possible to coordinate the protection using only overcurrent protection. With reference to Figure 7.53, breaker CB1 should be delayed compared to CB2; but at the same time, breaker CB2 should be delayed compared to CB1. A directional overcurrent relay at one of the two locations is a possible solution. If, for example, CB2 is equipped with directional detection, it will not trip for a fault behind CB1. Therefore, CB2 can be equipped with instantaneous trip. If breaker CB1 is next delayed compared to CB2, selectivity is again achieved.

Directional overcurrent relays have been used for many years in subtransmission systems and in meshed distribution systems. In addition to these "classical directional protection algorithms," a number of advanced algorithms have been proposed. References 83 and 315 propose a wavelet-based scheme for directional detection. Protection devices are located at different locations in the distribution system. The

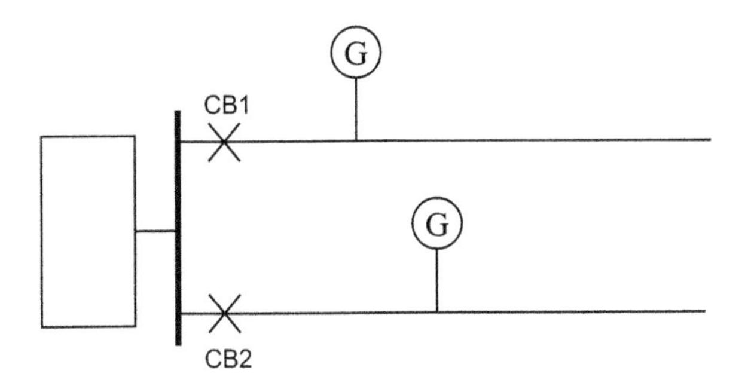

Figure 7.53 Distribution grid with large amount of generation on two neighboring feeders.

information from the individual devices is used to identify the faulted section. A master-slave directional scheme is proposed in Ref. 436. The directional protection algorithms proposed in Ref. 159 use the high frequency part of voltages and currents to determine the direction to a fault. This algorithm, like the others proposed, has a wider application range than that for feeders with distributed generation. In Ref. 122, an algorithm based on negative-sequence voltage and current is proposed.

7.8.7 Differential or Distance Protection

Differential protection could also solve many of the protection coordination problems associated with the integration of distributed generation. In Ref. 230, the advantages of differential protection are studied. An important conclusion from the study is that high-impedance differential protection should be preferred above impedance differential protection.

In Refs. 90 and 380, the use of distance protection is propagated for distribution feeders with generation. The application appears to be mainly feeders with large individual generators or with wind parks connected. To save costs, the voltage input to the relay is not obtained from a voltage transformer at 11 kV. Instead, the voltage is measured at the low-voltage side of the distribution transformer. Using some additional information, like the transformer size, the error in distance estimation is kept within reasonable limits. It should be noted that the aim of a distance relay is not to accurately determine the distance to a fault, but to give a reasonable indication of whether the fault is before the next substation. An error of 10% is, therefore, not really a problem. The study presented in Ref. 90 also showed that the feeder resistance should be considered in the setting of distance protection at 11 kV feeders.

Distance protection as a possible solution is studied in Ref. 294. It is shown that some of the differences between transmission and distribution systems indicate that the error in the distance estimation can be much larger for a distribution feeder than that at transmission level. The error is highest (up to 50% of the actual distance) for low load and a large generator connected to the feeder.

7.8.8 Advanced Protection Schemes

The so-called "adaptive protection" concept is studied by several researchers as a method for solving protection coordination problems due to distributed generation.

The basic philosophy behind adaptive protection is that the relay settings are continuously adjusted to the present system state. Whenever a generator becomes online or goes off-line, or when any other switching action is made, the optimal protection settings are calculated centrally and communicated to the individual protection relays. Whenever a fault occurs, the decision about whether to trip or not is, however, still made by each unit individually without communication between the relays. This approach has the advantage that the protection settings no longer need to cover all operational states, which will make it easier to obtain selectivity.

An outline for a "multiagent system" is proposed in Ref. 467, where relays or relay agents embedded in other intelligent devices exchange information to obtain optimal settings. The method proposed especially aims at detecting high-impedance faults. The disadvantage of the method proposed in Ref. 467 is that it requires measurements at many locations. Once substation automation with standardized communication protocols becomes widely applied, this should not be a problem, but applying methods like this in existing systems would be far too expensive. In Ref. 88, it is further proposed to change the protection settings with load variations. It does not become clear from the paper how practical this is. One of the purposes of the protection is to be able to distinguish between faults and load variations. A kind of blocking scheme could be used as a basis, assuming that the actual load currents are available to the protection.

An interesting scheme using conventional equipment is proposed in Ref. 170. Under this scheme, all fuses on a distribution feeder are replaced by reclosers with proper relays. The coordination between the various reclosers and the generator protection is discussed in detail in the paper. The disadvantage of the scheme is obviously the high costs associated with the installation of a recloser with each lateral.

In Ref. 62 it is proposed to temporary operate parts of the distribution in island after a fault. A central protection device is proposed that monitors voltages and currents at several locations along the feeder. Upon detecting and locating a fault, this device sends the proper clearing signals to the interrupting devices (breakers and reclosers). The opening of these devices will take place before any of the fuses reaches its minimum melting time. In this way, any impact of the fuses is prevented. The central protection device also trips all generator units connected to the faulted part of the feeder. An attempt at reclosing is made next by closing only one of the breakers to the faulted part of the network. This can be done without synchronizing because all generation has been tripped. If the fault has cleared, the remaining breakers are closed one by one, using synchronizing check. If the fault is permanent, maintenance personnel will be informed of the fault location. A scheme like this has many advantages, and has no limits on the amount of generation that can be connected to the feeder. The proposed scheme is in fact an automatic version of the way in which transmission systems are restored after an outage. The disadvantages of the proposed scheme are that it relies completely on one central device, in need of real-time communication to make a decision; that at least one generator in each zone should be equipped with facilities for controlled island operation; and that there may be strong opposition to automatic restoration schemes because it may endanger the safety of maintenance personnel.

The solution proposed in Ref. 33 combines a central relay in the substation taking information from all outgoing feeders and the relays along the feeder using only local information. In this way, the communication is limited to the substation. The central relay is based on directional protection, thus requiring voltage and current input.

In Ref. 239, a central device (the term "transfer trip scheme" is used) is proposed that monitors the status of all breakers on the feeder. When one of them is opened, this is detected and all generators downstream of the breaker receive a trip signal. In this way' the island operation is prevented. The acceptance of a scheme like this will depend on the costs for communication and, possibly even more important, on the perception of the risk that there will be a noncontrolled but stable island occurring despite this scheme. For this to happen, the transfer trip scheme should fail just at the moment that there is a close-enough balance between production and consumption. As both are small probabilities, and as they are independent, the risk seems to be sufficiently small.

7.8.9 Islanding Protection

The increased probability of noncontrolled island operation, according to many distribution network operators, is one of the biggest concerns with a large penetration of distributed generation. Although the risk is already significantly reduced with under- and overvoltages and frequency protection, this is not sufficient for many network operators. The probability of island operation can be further reduced with ROCOF and vector shift protection, but no passive method can completely rule out the possibility of noncontrolled island operation. Both latter types of island detection are, however, extremely sensitive and will result in many unwanted trips.

When considering the need for advanced islanding protection, it is very important to distinguish between the short-term and long-term island operations. Short-term island operation is reasonably likely and does not have to be stable to create dangerous voltages. In fact, high transient overvoltages could occur immediately after the opening of the feeder breaker, which cannot be prevented by any of the existing anti-islanding schemes. The potential impact of short-term islanding is mainly equipment damage and the risk of out-of-phase reclosing with fast autoreclosing schemes. Long-term noncontrolled island operation does not seem very likely, but cannot be ruled out. Its potential impact is especially on the safety of maintenance personnel and on the strange legal situation that the network operator is responsible for the operation of the island but has absolutely no control about it.

Active methods for islanding protection reduce the probability of noncontrolled island operation, but at this stage there is not enough experience with any of the methods to decide about this. Using an intertrip signal or the disappearance of a power line carrier signal appear to be the most promising methods. The former would be a blocking scheme, whereas the latter would be permitting scheme where the generator trips as soon as the power line carrier signal disappears.

The issue of unwanted trips due to the islanding protection is being discussed a lot, because of its potentially big impact on transmission system operation. Supporting the transmission system during a serious disturbance would require accepting a certain risk of noncontrolled island operation or the use of a suitable active method for islanding protection.

Finally, some additional measures are needed, as noncontrolled island operation can never be completely ruled out. New maintenance procedures are developed to protect maintenance personnel against unexpectedly energized distribution networks. Suitable overvoltage protection (in the form of surge arresters or similar) should be developed to prevent equipment damage during the short-term island operation. Reclosers should be equipped with a check to prevent them from switching onto a life feeder. Once these measures are in place, a certain probability of noncontrolled island operation may become acceptable. This could in turn solve many of the problems due to unwanted trips.

CHAPTER 8

TRANSMISSION SYSTEM OPERATION

Before starting this chapter, a note about the nomenclature for the different synchronous systems in Europe is pertinent. The transmission system in Europe has traditionally been divided into a number of different synchronous systems. The divisions were partly geographical and partly political. Each synchronous system has its own operational and planning rules. Two of these systems were known as the UCTE system and the Nordel system, named after the organizations that coordinated the operation of the system. Recently, the European Commission has decided that the operation of all transmission systems should be ruled by one body: ENTSO-E (the European Network for Transmission System Operators of Electricity) (http://www.entsoe.eu). The earlier bodies, such as UCTE and Nordel, no longer exist. The systems that were operated by Nordel and UCTE are now referred to as "regional group Nordic" and "regional group continental Europe," respectively. In this book, we will refer to the former as the "Nordic (synchronous) system" and to the latter as the "European (synchronous) system".

8.1 IMPACT OF DISTRIBUTED GENERATION

Generation connected to the distribution system (i.e., distributed generation) will reduce the power flow through the transmission system. The distributed generation compensates part of the consumption in the same distribution network and results in smaller flows through the transmission transformer. This is discussed in Chapter 4: reduced risk of overload and reduced losses are a consequence. Reduced power flows through the transmission system will also result in less risk of instability. Overall, the introduction of distributed generation will thus be a good thing for the transmission system. There are, however, a number of concerns with transmission system operation: some are minor concerns and some are serious concerns. We will summarize these concerns in this section and go into further detail in several of the following sections. An overview of the various solutions will again be given in the final section of the chapter.

In this chapter, we will go beyond the commonly used definition of distributed generation and also include large wind parks connected to the transmission system.

Integration of Distributed Generation in the Power System, First Edition. Math Bollen and Fainan Hassan.

They impact the transmission system operation in ways similar to wind power connected to the distribution system. The same will hold for large solar power installations that may appear within a few years' time. Such large wind parks or solar power installations are often found at remote locations. The first large offshore wind parks have appeared and many more are planned. This requires investments in the transmission system to transport the power to the consumption centers. This is a problem very similar to the situation with many hydropower resources. These are also typically located far away from the consumption. The same holds to a lesser extent for (especially) nuclear power and large coal-fired power stations as there is often strong opposition to building these close to large cities. New large thermal generation will also more likely be built close to the coast or near large waterways, where cooling water is available. This may again not be close to the load centers. Recently, plans were published for building huge solar power installations in Northern Africa and transporting the power to Europe [163]. This will obviously require an enforcement of the European transmission grid. In the same way, enforcements of the North American grid are needed to transport wind power from the Midwest to the east coast of the United States [103, 266]. The same situation will arise in China, where the wind power resources are being developed in the western part of the country, far away from the main centers of consumption on the east coast [156]. In Europe, much of the existing wind power development is closer to the most densely populated parts of Europe (United Kingdom, Germany, France, Benelux), which limits the need for transport over long distances. The occurrence of bottlenecks is mainly due to the production being more concentrated in areas with high wind speeds [154]. Future developments of offshore wind, solar power, and balancing with the help of hydropower imply that serious investments in the transmission grid remain high on the agenda [155].

The introduction of distributed generation will also result in different patterns of power flow. Traditionally, the power flow between transmission and distribution would vary between two states, "high load" and "low load." These were the ones that had to be considered, with the load variations showing distinctive and predictable daily, weekly, and seasonal patterns. Next to these, some loads were temperature dependent. Electric heating and air conditioning load are both strongly dependent on the temperature. Peak consumption occurs in Northern Europe, where electric heating is very common, during the coldest winter days. In hot climates, with large amounts of air conditioning loads, the peak occurs during the hottest summer days. Because of this, maintenance on generation is preferably not scheduled during winter in a cold climate or during summer in a hot climate. In some countries with moderate climate, a double peak has started to appear. This makes the scheduling of maintenance more difficult.

The change in existing generation patterns with the introduction of new generation is not restricted to distributed generation. When new (cheap) generation is introduced close to existing expensive generation, the new generation will simply replace the existing generation and there will be no impact on the power flows. However, when the new generation appears far from the most expensive generation, there will be a new flow from the location of the new generation to the (earlier) locations of the most expensive generation, superimposed on the existing power flows.

This new flow may result in an increase or a decrease in power flow. This holds for all new types of generation and also for reduced electricity consumption.

From the above, it follows that the introduction of weather-dependent generation (solar, wind, combined heat-and-power) will not so much introduce completely new phenomena but it will create more complexity in existing phenomena. Instead of only temperature dependent, the power flows also become dependent on wind speed and cloud cover. The power flow patterns will become more complicated. Let us have a look at the different types of generations.

The production by combined heat-and-power (CHP) purely depends on the heat demand. For industrial installations, the heat demand is rather constant, so it does not influence the power flow pattern. CHP for space heating shows a strong temperature dependence and some dependence on wind speed and insolation, as well as a daily and a weekly pattern. The daily and weekly patterns mainly occur in domestic and commercial applications where different temperature settings are used for different times of the day and the week. CHP for space heating will be used only in colder climates and during colder periods in moderate climates. The production by CHP will thus reduce the peak load, which is again a good thing for the transmission system. The combination of CHP and electric heating will also make the power flows through the system less sensitive to errors in the temperature prediction.

For solar power, we can distinguish a daily and a seasonal pattern. The daily pattern, with maximum production at noon, shows a reasonable correlation with the normal consumption pattern experienced at transmission level. However, daily peak consumption often occurs in the morning and in the early evening. Solar power might somewhat reduce the morning peak, but it will not be able to contribute much to supply the evening peak. The seasonal pattern, with maximum production in summer, is correlated with the consumption by air conditioning, but opposite to the consumption by electric heating. The highest temperatures, and thus the highest consumption by air conditioning, do not occur at noon but several hours later. In the same way, the hottest days occur several weeks after the sun reaches its most northerly or most southerly position in the sky. Despite this, solar power is still expected to result in some reduction in peak load in hot climates. Solar power in combination with energy storage, for example, in the form of molten salt, will be able to contribute to the evening peak. To be able to contribute (more) to the morning peak, energy storage should be available to cover the whole night. This may be beyond the capability of molten salt storage. Battery storage could be a solution here. However, no amount of storage will be able to shift the annual peak in insolation to the annual peak in consumption.

Wind power shows a mild seasonal variation and at some locations a daily pattern as well. This is however not very pronounced and the variations are dominated by passing weather systems (high-and low-pressure systems). It should be noted here that the main experience with wind power comes from regions with moderate climate where the passing of weather patterns is a normal part of the climate. In tropical regions, this may be less common and more constant wind power production could occur. Both wind power and solar power will result in more variations in the power flows through the transmission system, with only weak daily and seasonal variations. This certainly will make life more difficult for the transmission system operator.

The term "intermittent generation" is often used to refer to this strong variation in production.

Even more difficult for the operation of the transmission system is the issue of predictability of the power flows. To predict the production by weather-dependent sources, prediction of wind speed, temperature, and cloud cover is needed. The main concern at the moment is the prediction of wind speed. For the future, prediction of cloud cover may become equally important. An error in prediction makes it difficult for the transmission system operator to schedule the conventional generation units and the required reserves. We will discuss prediction in more detail in Section 8.4. The impact of prediction errors on the operation is discussed in Section 8.3.5.

The operation of the transmission system should always be such that there is sufficient reserve available to tolerate the loss of any single component: the so-called $(N-1)$-criterion. The behavior of the system due to the loss of a single component (typically a fault) should therefore be well known and understood by the transmission system operator. The introduction of distributed generation will impact this behavior in a number of ways:

- Different pre-event power flows will result in different stability margins. The system might be stable for a certain event when the production is as predicted, but instable when the production deviates a lot from prediction. Transmission system operators do not have any guidance on how much deviation from prediction should be considered in deciding whether the system is stable when a certain event occurs. When insufficient margin is scheduled, this could result in a blackout, but there is a high cost associated with maintaining high margins all the time.
- Increased production by distributed generation will result in less conventional generators being in operation, often expressed as "weakening of the transmission system." This will, in many cases, result in a less stable system; voltage, frequency, and angular stability could be impacted. It will also make it more difficult to schedule the required amount of active and reactive power reserves. The reduction in fault level and the adverse impact of this on the power quality are discussed in Chapter 6.
- Distributed generation has a dynamic behavior different from load or conventional generation. Detailed dynamic models are needed for distributed generation and large wind parks. There is a lot of emphasis on the latter at the moment. Reactive power flows during and after a fault may have a lot of influence on the short-term voltage stability. The main concern at the moment is the unnecessary tripping of generation due to a fault or a frequency deviation. This could result in the sudden loss of large amounts of generation. This will be discussed in more detail in Section 8.10.

The increase in uncertainty due to both prediction errors and uncertain dynamic behavior during faults requires more reserves. At the same time, the reduction in the amount of conventional generation being in operation brings down the amount of reserves available. This explains the often expressed fear that less and less conventional generators have to provide more and more reserves. The situation is however not that bleak in reality. The transmission system operator will schedule additional conventional generators to ensure that sufficient reserve is available. The main limitations are economical: the transmission system operator has to pay for the running of production units that do not result from the market settlement. Also, some of the existing operational rules set limits to the amount of distributed generation that can be integrated, for example, the rule that each transmission system operator should schedule its own reserves.

The remaining of this chapter will start with an overview of the basic principles of transmission system operation. Section 8.2 will introduce the basics, including the $(N-1)$ criterion and different types of operational reserve. Frequency control, balancing, and the different types of reserves will be discussed in more detail in Section 8.3. Difference will be made between primary, secondary, and tertiary reserves. Predictions of consumption and production by the new types of generation and their impact on the reserves are discussed in Section 8.4. Section 8.5 gives a brief overview of the impact of distributed generation on the restoration of the supply after a blackout. The impact of distributed generation on different types of stability are discussed in the following sections: short-term voltage stability in Section 8.6.1, long-term voltage stability in Section 8.6.2, frequency stability in Section 8.8, and angular stability in Section 8.9.

The amount of kinetic energy present in the system is an important parameter for both frequency and angular stability. The possible reduction in the amount of kinetic energy with replacement of convention generation by distributed generation is treated separately in Section 8.7. Another parameter that has a strong impact on the stability of the transmission system is the behavior of distributed generation during large disturbances in the power system. The so-called "fault ride-through" is discussed in Section 8.10: a brief overview of both the need for fault ride-through requirements and the kind of requirements set by transmission system operators. Some thoughts about storage, a subject getting more attention recently, are presented in Section 8.11. Different methods for increasing the amount of distributed generation that can be integrated in the transmission system are discussed in Section 8.13, including HVDC, alternative ways of keeping reserves, and stochastic methods.

8.2 FUNDAMENTALS OF TRANSMISSION SYSTEM OPERATION

The operation of the transmission system is fundamentally different from that of the distribution system. The distribution system is operated in a passive way: once it is built, there is limited intervention. The transmission system is operated actively, with both automatic and manual interventions. For those not familiar with the way in which the transmission system is operated, we have summarized some of the basics in this section. Note that we describe the existing way of operation; massive deployment of distributed generation will require changes in this, which will be discussed later. What we will present next is a general description of the operational principles; the detailed operational rules vary between countries.

8.2.1 Operational Reserve and ($N-1$) Criterion

The operation of the power system is based on the balance between production and consumption at all levels and locations. At most locations, the balance is kept by importing any shortage from elsewhere and exporting any surplus to somewhere else. For example, a distribution transformer will import from a higher voltage level such an amount of power that exactly covers the demand from all customers located at the low-voltage side and all the losses. This holds for both the active and the reactive power. This may sound complicated, but it is nothing more than Kirchhoff's laws. Applying the law of conservation of energy would lead to the same conclusion.

The main active control takes place at the transmission level: automatic control is done almost exclusively by the large power stations and manual control is done by control personal at different timescales. To maintain the balance and compensate for any unexpected unbalance, it is important to maintain sufficient reserve to be available at different timescales. The rules for the amount of reserve to be available vary between system operators, but a general rule is that the reserve should be enough to cover the loss of any single component as well as the likely loss of multiple components.

Reactive power control and voltage control also take place in distribution systems but to a limited extent. The HV/MV transformers are often equipped with automatic on-load tap changers. Switched capacitor banks are sometimes found at distribution levels and some industrial installations are even equipped with a static var compensator (SVC).

Planning of the production and operational reserves takes place over a range of timescales, from less than 1 h to several months. The latter concerns the seasonal variations, where (in cold climates) the planning for the winter season takes place several months ahead so as to be able to take measures when it turns out that there is insufficient reserve available to cope with a cold winter. In fact, at the time of working on this chapter (in December 2009), the winter was not only colder than on average but it was especially earlier than normal. As two of the nuclear power units were still in maintenance by the end of December, the electricity price went very high. This in turn resulted in (unfounded) accusations that the owners of the nuclear power plants intentionally kept those units out of operation to drive up the electricity price [328].

Planning of production and reserves further takes place on a weekly basis and on a daily basis. Production planning takes at first place through the electricity market, where producers and consumers agree on the electricity price. The electricity market functions such that the total predicted consumption equals the total scheduled production plus the predicted production outside the market (like most distributed generation). The term "day-ahead market" is often used to distinguish it from any "intraday markets." Next to the electricity market (with many buyers and sellers), a market for reserve capacity exists, with the transmission system operator as the only buyer. After the market settlement, the transmission system operator takes over. The transmission system operator is allowed to intervene in the market when the resulting scheduling of production units does not result in a secure operational state. Even during the day of trading, the transmission system operator can change the amount of reserves or even the scheduling of production units. This may, for example, be needed after the failure of a large production unit or an important transmission line.

Also, prediction errors in consumption or in production outside the market may require rescheduling. At the shortest timescale, typically 15 min and less, the balance is kept by automatic control systems using the power system frequency as the control variable. More about this can be found in Section 8.3. The term "balancing" or also "buyback" is used for the intervention by the transmission-system operator in the market. Sometimes, the term "intraday balancing" is used for any intervention during the day of trading. The term balancing is sometimes even used for the automatic frequency control. In most countries, functioning balancing markets exist, with again the transmission system operator as the sole buyer.

The fundamental rule in the scheduling of reserves is that no failure of a single component should result in any loss of load. The service to the customers should not be affected at all by such a failure. This holds for all generator units, for all transmission lines and cables, for all HVDC links, for all capacitor banks, and for all other components in the transmission system. This is commonly known as the $(N - 1)$ criterion. As far as (active power) production is concerned, this implies that the system should be able to cope with the largest likely loss of production. This is called the "dimensioning failure." In the Nordic system, this is 1200 MW, corresponding to the loss of the biggest production unit. In the European system, this is 3000 MW, corresponding to the loss of the two biggest production units shortly after each other. The European system is much larger than the Nordic system; this makes it easier to have 3000 MW reserve, and also the probability of two big units failing shortly after each other is larger in a system with more large units. The $(N - 1)$ criterion is not only the basis for the amount of reserves in production, but also sets the requirements for reserves in reactive power and for secure transfer capacity.

8.2.2 Different Types of Reserve

The reserves in production capacity (active power reserves) consist of different types, depending on the speed at which they can be made available. The fastest reserves are made available automatically, with frequency or voltage as a control signal. A recent report by the organization of European transmission system operators [150] distinguishes between the three reserve categories:

- *Frequency containment reserves* that come available in 15–30 s when the power system frequency deviates significantly from its nominal value. These are the reserves that keep the system stable after the loss of a large production unit. The size of these reserves is determined by the size of the largest production unit.
- *Frequency restoration reserves* that come available in 0.5–15 min whenever the frequency deviates from its nominal value. These reserves are needed to maintain the normal operational balance between production and consumption and to compensate for prediction errors. The size of these reserves is determined by the uncertainty in prediction.
- *Replacement reserves* that are needed to replace the first two types of reserves whenever they are depleted. The delivery timescale is 15 min or longer.

The terminology used for the different types of reserves is different for different synchronous systems. For a secure operation of a synchronous transmission system, it is important to agree on the amount of reserves needed and who is responsible for maintaining them. Therefore, clear rules exist for this in each synchronous system; however, the rules and terminology are not the same everywhere. This is partly for historical reasons and partly because of different control methods in use. Some examples will be given next.

The grid code for the Nordic synchronous system distinguishes between the following types of active power reserve [323]:

- **Frequency-controlled normal operating reserve**: This automatically controlled reserve is used to maintain the balance between production and consumption during normal operation. Normal operation is considered to keep the frequency between 49.9 and 50.1 Hz, so that the reserve can be completely activated at 49.9 or 50.1 Hz. As this is a normal operation, no sudden changes in production or consumption take place, and the control of the reserve can be rather slow. A time constant of a few minutes is acceptable according to the grid code. Under the above-mentioned ETSO classification, this would classify as "frequency restoration reserve".
- **Frequency-controlled disturbance reserve**: This automatically controlled reserve is activated when the frequency drops below 49.9 Hz. It should be sufficient to maintain the system frequency above 49.5 Hz for the so-called "dimensioning fault" of 1200 MW. Half of the reserve should be available within 5 s and the other half within 30 s. It is this reserve that is used to ensure that the $(N-1)$ criterion holds. This would classify as frequency containment reserve.
- **Fast active disturbance reserve**: This is controlled manually and should be available within 15 min. This reserve is used to replete the frequency-controlled normal operating reserve and the frequency-controlled disturbance reserve after a disturbance. The timescale of 15 min is important in the operation of the Nordic grid. A situation with insufficient reserve (the so-called "disturbed state") should be ended within 15 min; that is, within 15 min after a severe disturbance, the $(N-1)$ criterion should hold again (the so-called "normal state"). This would classify as "replacement reserve," but available within 15 min.
- **Slow active disturbance reserve**: This is also controlled manually and comes available after 15 min. This is used, for example, for balancing and to compensate for prediction errors. This would also classify as replacement reserve.

In the European synchronous system, the following distinction is made according to Ref. 416:

- *Primary reserve* (also known as "seconds reserve") is available almost instantaneously, that is, at a timescale between several seconds and about 1 min. The primary reserve is spread throughout the system and all primary reserves over the whole system should at least be 3000 MW. Half of the primary reserves should be available in 15 s and the rest within 30 s. The primary reserve must be available during at least 15 min. The 3000 MW of primary reserve is spread over production units throughout the synchronous system; each transmission system operator is required to contribute with a certain amount to the total primary reserve. According to the ETSO classification, this would class as frequency containment reserve.

- The *secondary reserve* or "minutes reserve" is part of the automatic control needed to restore the balance in each control area. The secondary control also returns the frequency to its nominal value after the loss of a large amount of production. The secondary control results in the primary reserve being restored in all control areas, apart from the one that suffered the loss of production. The secondary control should start its action latest by 30 s after the loss of production. The control action should be completed within 15 min from the start of the disturbance. The amount of secondary reserve is decided for each control area by the responsible transmission system operator. There are no general requirements, but the operational handbook does give detailed guidance on this. The secondary reserve is a combination of frequency restoration reserve and replacement reserve.
- *Tertiary reserve* concerns all reserves that require more than 15 min to be made available. Tertiary reserve is typically made available manually, although some slower reserves are sometimes under the control of the secondary controller. This would classify as replacement reserve.

In the literature, a distinction is often made between "load following reserves" and "contingency reserves." The former is used continuously to compensate small unbalance between production and consumption. Its function is similar to the above-mentioned "frequency response reserves," although the term is also used for reserves needed to follow the daily load curve. Contingency reserves (also called "operating reserves," "reliability reserves," and some other terms) is synonym to frequency containment reserves.

8.2.3 Automatic or Manual Secondary Control

A comparison of the different types of reserves in the Nordic and the European system shows that the main difference is in the minutes reserve. This is made available in the European system through automatic control, whereas it is manually controlled in the Nordic system. The difference points to a more fundamental difference in the way balances are maintained, as illustrated schematically in Figures 8.1 and 8.2. In the European synchronous system, Figure 8.1, the balance is kept automatically for each

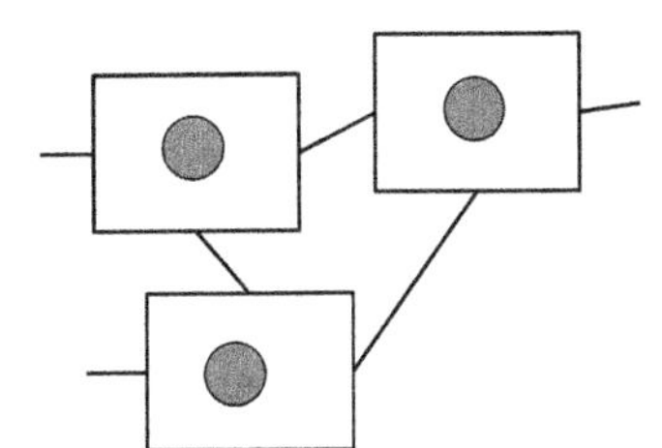

Figure 8.1 Basic principle of operational security used within the European synchronous system.

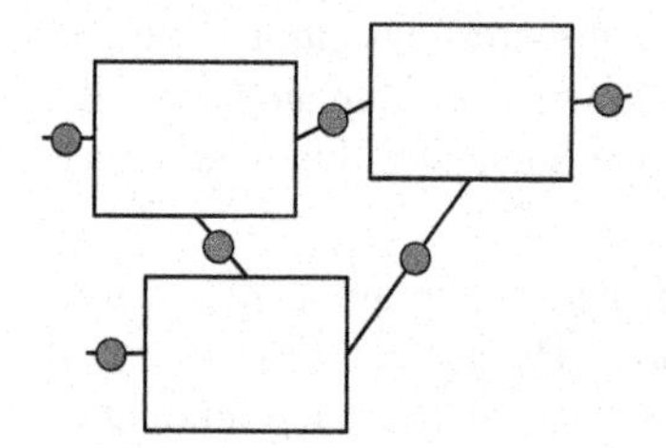

Figure 8.2 Basic principle of operational security used within the Nordic synchronous system.

control area. In this way, the power exchange between individual countries is kept the same as scheduled. Deviations from the scheduled exchange are accepted for only up to 15 min, or in case of emergencies. The control algorithm used for this is discussed in detail in Section 8.3.3.

The operational method prescribed by the North American Electric Reliability Council (NERC) is also based on keeping the balance within each control area. The so-called "area control error" should be kept within strict limits according to NERC's regulations.

The control method used in the Nordic system (Figure 8.2), is not based on keeping balance within different control areas, but instead it is based on limiting the exchange between the areas. A number of "cuts" are defined together with maximum power transfer through these cuts. The transmission system operators use manual control of production to keep the power transfer below these limits.

8.3 FREQUENCY CONTROL, BALANCING, AND RESERVES

8.3.1 The Need for Reserves

In the previous section, different types of reserves where introduced. In this section, we will explain how those reserves are used during the operation of the transmission system. We will thereby use the distinction between primary, secondary, and tertiary reserves. The authors are aware that these terms are not very well defined and that different transmission system operators use different terminologies. The terms primary, secondary, and tertiary reserves are used here preliminarily to refer to the different timescales at which the reserves can be made available.

As long as the actual consumption and production are as predicted, there is no need for the transmission system operator to intervene. There are two distinctive reasons for differences between prediction and actual values:

- The forced outage of a large generator unit or the forced outage of an HVDC link with another synchronous system
- Prediction errors for consumption or production by distributed generation

The first one has traditionally been the main reason for operating with reserves, where the loss of the largest generator unit has been setting the requirements for the primary reserve. The second reason for unbalance used to be a minor issue. Some "load following reserve" is allocated for taking care of this. Prediction errors in, for

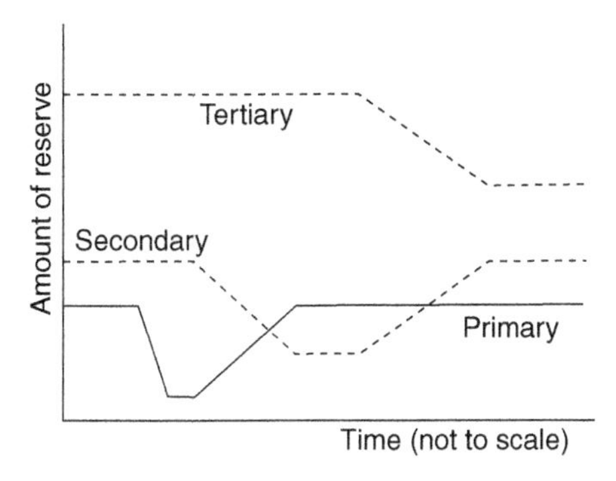

Figure 8.3 Changes in primary, secondary, and tertiary reserves after the loss of a large generation unit.

example, wind power production increase this second reason for unbalance. We will come back to this later, but first will have a closer look at what happens with the reserves upon the loss of a large generator unit. The reserves as a function of time are shown in Figure 8.3.

The loss of the large generator results in a reduction of the primary reserve (the reserves become actual production). This is made up, automatically or manually, from the secondary reserves. The reduction in secondary reserves is next made up, manually, from the tertiary reserves. Note that all this does not matter directly to the customer. The customer is directly impacted only when the primary reserves are depleted. In that case, the system becomes unstable, leading to a loss of load through automatic load shedding or in the form of a large blackout. Once the primary reserve becomes less than the size of the largest production unit, the system is "insecure": the operation no longer complies with the $(N - 1)$ criterion. This is also a situation that rarely occurs and when it occurs, the primary reserves are very quickly replenished with the secondary and tertiary reserves. But the transmission system operator is obliged to maintain a certain amount of primary and secondary reserves. Using the terminology introduced in Chapter 3, maintaining reserves is a secondary aim (it does not directly impact the customer), whereas preventing loss of load is the primary aim.

8.3.2 Primary Control and Reserves

The principle of primary power-frequency control for conventional generators is shown in Figure 8.4 [45]. The input to the speed governor is a corrected power setting (corrected for the deviation of the frequency from its setting). The speed governor delivers a signal to the steam valves with a thermal power station, or the water inlet with a hydropower station, to regulate the amount of steam or water that reaches the turbine. The turbine reacts to this, with a delay of several seconds, by changing the amount of power delivered to the electrical machines and thus to the grid. In the figure, f_{act} is the actual frequency, f_{set} is the frequency set point (in most cases, the nominal frequency of 50 or 60 Hz), P_{sched} is the power setting, and P_{act} is the actually produced active power. The droop setting R determines how much the steady-state frequency drops with an increase in consumption. Having a nonzero droop allows the controllers to operate independently, while still sharing the burden of any load variations.

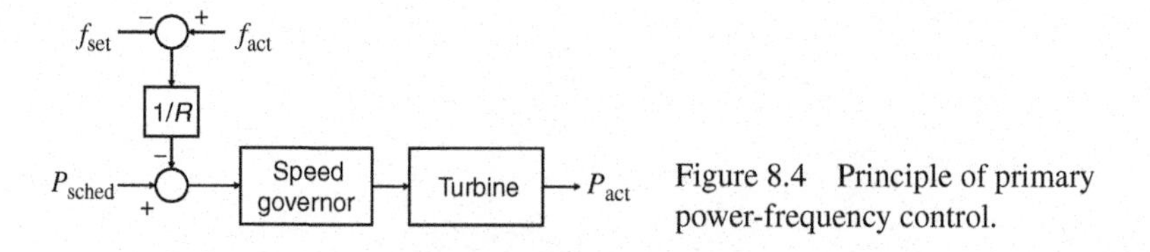

Figure 8.4 Principle of primary power-frequency control.

The same principle can be used when distributed generation or curtailable load is used as primary reserve. In that case, the governor becomes an electronic controller that determines the actual production or consumption.

In steady state, the produced power, using the scheme in Figure 8.4, is the same as the input signal to the governor:

$$P_{act} = P_{sched} - \frac{1}{R}(f_{act} - f_{set}) \tag{8.1}$$

This can be rewritten to give the frequency drop due to a power unbalance:

$$f_{act} = f_{set} - R(P_{act} - P_{sched}) \tag{8.2}$$

A shortage in production in the synchronous system (e.g., due to the loss of another generator) will immediately result in a drop in frequency (see Section 8.8). The primary control reacts to this by increasing the production. In the new steady state, each generator equipped with frequency control produces more power than scheduled, whereas the frequency is lower than the set value.

The choice of the power setting P_{sched} for each generator is, in principle, the amount of power the generator owner agreed to produce according to its bid on the electricity market. The sum of the settings over all generators is equal to the predicted consumption. When the production would be exactly equal to the prediction, the frequency would be exactly equal to the nominal frequency (actually to the frequency setting). We will come back to this later when discussing secondary and tertiary control, where the settings can be changed.

The primary control operates at timescales up to about 1 min. The presence of distributed generation does not add much additional variations in power at this timescale. Any local variations in wind speed or irradiation are averaged over the synchronous area. Only for small isolated systems (such as small islands) has distributed generation the ability to impact the primary control. Apart from those systems, the direct impact of distributed generation on the primary control is perceived as small.

An indirect, and much more discussed, consequence of the introduction of large amounts of distributed generation is that there will be less conventional generation available to provide the primary reserve and the primary control. This will not be so much of a technical problem, unless we talk about very large penetration, but merely an economic problem.

Example 8.1 The Northern European synchronous system (Norway, Finland, Sweden, and part of Denmark) has a minimum consumption of about 30,000 MW according to statistics published on the web sites of the national system operators of these countries. The primary reserve should be at least 1200 MW to cope with the loss of the largest unit, which is 4% of the minimum consumption. Replacing 15,000 MW of conventional generation with distributed generation that does not contribute to the primary reserve would result in the spinning reserve being at most 8% of the installed conventional generation.

Primary reserve can be present in a number of forms. In most cases, the majority is in the form of "spinning reserve" with large thermal or hydro units. The units operate in such a way that the production can be increased to a certain amount within several seconds. There is a cost associated with having spinning reserve available. With thermal units, the efficiency becomes somewhat less, and with hydropower no direct costs are associated with having spinning reserve. However, spinning reserve also means that the generator owner does not produce all the energy that could be produced and therefore does not sell all the energy that could be sold. When the electricity price is higher than the marginal cost for the unit (which is often the case for hydropower units), this becomes a loss of profit for the generator owner. With reservoir-based hydropower, the energy can still be sold later, but that may be at a lower price. Despite this, reservoir-based hydropower units are seen as the best choice for providing spinning reserve. Nuclear power units are by some system operators not used for providing spinning reserve. See, for example, Ref. 437 for further discussion on the costs of keeping spinning reserves.

Primary reserve can also be provided by HVDC links with other synchronous systems. In that way, one makes use of the primary reserve of the neighboring system. Note that this does not typically increase the spinning reserve requirements for the neighboring system because the loss of the largest unit in both systems, within a short period, is small and rarely considered in the scheduling of primary reserve.

More recent discussions concern the use of large wind power installations for providing primary reserve (a large wind park in Denmark is occasionally used for this purpose) and using load reduction as primary reserve [244, 245]. Distributed generation and load may actually have advantages compared to large conventional power stations for maintaining primary reserve. Small units could be much faster in increasing their production than large units, and their efficiency is in general not strongly dependent on the actual production. However, the drawback is that the production should be kept at a level less than the maximum possible amount, for sources such as solar, wind, wave, or run of river. This would mean a loss of income for the owner of the unit, unless there is a suitable compensation for this, for example, through the primary reserve market. An important drawback for society is that the power kept in reserve has to be produced elsewhere. As long as the total energy consumption is more than the amount of renewable energy available (and that will be the case for a long time to come), requiring renewable sources to provide primary reserve will result in more energy being produced from nonrenewable sources. The term "spilled wind" is sometimes used to describe this for wind power.

Using load reduction as a form of primary reserve has very small operational costs. There may however be some costs, or other inconvenience, for the customer when the spinning reserve is needed. Instead of operating the large power stations such that they can increase their production by, for example, 1000 MW within a few seconds, the consumption can be reduced by 1000 MW in the same amount of time. The simplest implementation would be to equip a number of large consumers with minimum frequency relays that operate for a frequency drop of, for example, 0.2 Hz. This could result in a surplus of production, but that can be easily compensated by the existing large production units. The disadvantage would obviously be that the disruption for those consumers would be severe. However, one may think of applications where large customers are willing to suffer the loss of part of their consumption a few times per year or less in return for a certain payment. This payment may be continuous or only in case the reserve is actually used.

Another method of keeping primary reserves being developed is through storage such as flywheels or batteries. According to Ref. 367, at least three system operators in the United States have opened their spinning reserve markets for fast response systems, such as flywheels and battery storage. At this stage, no information is available on the feasibility of this.

Also discussed a lot over the last few years is to gradually reduce consumption with decreasing frequency in the same way as the production is gradually increased with decreasing frequency. The details of this are still under discussion, but the potential appears to be huge. Air conditioning load is seen as suitable for this. According to Ref. 244, air condition can be turned off completely for 10 min without incurring any costs. A study in a medium-sized hotel showed that load could be curtailed for 15–30 min on a regular basis and for 60 min occasionally. The disadvantage is the need for large investments because a large number of devices are needed to make a substantial contribution, as can be concluded from Example 8.2. The investment costs would be rather large as every unit would have to be equipped with control equipment.

Large HVAC installations in office buildings, schools, or shopping centers could provide a cheaper solution, but it would still require thousands of buildings to participate before a serious impact can be made. Primary reserve requirements vary somewhat between synchronous systems, but 1000 MW is a typical size, so at least 100 MW would be needed to make an impact. It is also important to note that the primary reserve requirements, under the existing operational rules, are constant throughout the year. The availability of air conditioning load varies, however, strongly through the year.

Example 8.2 Suppose that 900 MW of primary reserve should be achieved by reduction in the consumption at device level. Air conditioning units form a suitable type of device for this because the temperature increase in a building is rather slow and an increase of a few degrees is acceptable for a certain time. If we assume a standard size of 750 W per unit, this would require 1.2 million units. However, a single unit is only operating part of the time, so we need more units. When each unit is operating 60% of the time (this would be on a rather hot day), it would require 2 million units, where it is assumed that the unit can be turned off completely.

Example 8.3 Experiments performed on a 162-room hotel in the United States showed that curtailment of the consumption, for primary reserve purposes, by 22–37% was possible. The average power taken by the hotel varied between 180 and 320 kW. Taking the middle values of 30% and 250 kW gives a primary reserve of 75 kW. To replace 900 MW of spinning reserve, a total of 12,000 such hotels are needed. However, when we consider 25% of 200 kW per hotel, 18,000 hotels would be needed.

TABLE 8.1 Example of Settings for Underfrequency Load Shedding [165]

Step	Frequency (Hz)	Delay (s)	Amount of load (%)
A	59.7	0.28	9
B	59.4	0.28	7
C	59.1	0.28	7
D	58.8	0.28	6
E	58.5	0.28	5
F	58.2	0.28	7
L	59.4	10.0	5
M	59.7	12.0	5
N	59.1	8.0	5

One type of curtailment of the consumption is actually implemented already on a wide scale, but this is very rarely activated. What we refer to here is the so-called "under-frequency load shedding." A significant part of the load, starting at about 10%, is automatically tripped when the shortage in production is so large that it cannot be compensated by the primary reserve in the generators. In a typical implementation, one or more feeders in a number of medium-voltage substations are equipped with underfrequency relays that trip the feeder instantaneously when the frequency drops below a set value. Two examples of settings for underfrequency load shedding are shown in Tables 8.1 and 8.2. The former one is standard for Florida [165] and the latter is the recommendation made by ENTSO-E for the European synchronous system. Note the difference in approach between a large number of steps very close in frequency and a small number of steps.

With increasing amounts of distributed generation, the total costs of keeping primary reserve will probably not change much but the costs would be carried by less production units. These units will however have more market power in a primary reserve market, which could result in an increase in the price for primary reserve. Also, the market settlement could result in conventional production being in operation for which it is more expensive to keep primary reserve. Another possible consequence of having the primary reserve concentrated in less units is that the loss of a large unit would induce larger power flows because the spinning reserve has to be transported further. Transfer capacity has to be held available to transport this, so that the secure transfer capacity will become less.

Increasing amounts of distributed generation could also impact the under-frequency load shedding. Traditionally, a distribution feeder would not contain any

TABLE 8.2 Example of Settings for Underfrequency Load Shedding

Frequency (Hz)	Amount of load (%)
49.0	10–20
48.7	10–15
48.5	10–15

generation, so tripping this feeder would always result in a load reduction. Nowadays, and even more so in the future, the feeder may contain so much distributed generation that it becomes a net producer. Tripping the feeder would actually increase the load. It is unclear how severe this impact is and whether it will endanger the effectiveness of the load shedding scheme. With large amounts of distributed generation, it is important to check the feeders involved in the load shedding scheme to see if tripping these would indeed result in the expected load reduction.

8.3.3 Secondary Control and Reserves

Secondary control operates at a timescale of minutes. Its aim is to bring the power system frequency back to its nominal value of 50 or 60 Hz, to restore the primary reserve after the loss of a large production unit, and to reduce the power flows between areas after the loss of a primary unit. As seen from Example 8.4, the amount of primary reserve is larger for large control areas, but the amount of secondary reserve is the same for each control area.

Example 8.4 The loss of a 1000 MW production unit somewhere in the European synchronous system, for example, in Belgium, is compensated by an increase in production by generators throughout the system. The contribution from a number of countries is as follows [105]: France 238 MW, Germany 275 MW, Spain and Portugal 105 MW, the Netherlands 37 MW, and Belgium 37 MW. The secondary control should next return the balance between production and consumption within each country. For most countries, the production can be reduced to its pre-event values. However, in Belgium, the country where the loss of production took place, the production should be increased by 963 MW (it was increased by 37 MW already by the primary control). This is taken care of by the secondary control.

The secondary control typically uses automatic mechanisms to keep the difference between production and consumption constant for each control area. The principle is illustrated in Figure 8.5. The interconnected system is divided into so-called "control areas." In Europe, most control areas correspond to individual countries, Germany being the main exception. In the United States, the control area correspond to the individual transmission system operators. The secondary control obtains information from the active power flows between the control area and the neighboring control area. The total exchange with all the other control areas is maintained constant.

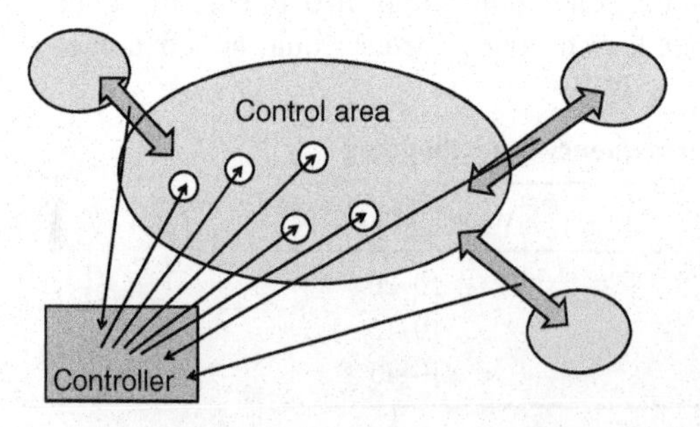

Figure 8.5 Principle for secondary control in the European interconnected system.

The control function of the secondary control, that is, the function that the controller aims to maintain as close to zero as possible, is given by the following expression [231]:

$$G = (P_{act} - P_{sched}) + K_{sec}(f_{act} - f_{set}) \tag{8.3}$$

where P_{act} and P_{sched} are the actual and the scheduled total export out of the control area, f_{act} and f_{set} are the measured frequency and the set value of the frequency, K_{sec} is the setting of the controller. The set value of the frequency is in almost all cases equal to the nominal frequency, 50 or 60 Hz. The value of (8.3) is referred to as the "area control error," or ACE. In the United States, the regulatory authority NERC sets limits on the integrated ACE over different timescales.

For a synchronous system, (8.3) is kept close to zero for each control area. The sum of all area control errors is given by the following expression:

$$\sum_i G_i = \sum_i P_{act,i} - \sum_i P_{sched,i} + (f_{act} - f_{set}) \sum_i K_{sec,i} \tag{8.4}$$

where the frequency is assumed to be the same throughout the system. In an interconnected system, the total K factor is the sum of the K factor for each of the control areas. In the European interconnected system, the total K factor is equal to 26,680 MW/Hz [416]. The sum of all exports from the individual control areas is zero, both scheduled and actual values. The first two terms on the right-hand side of (8.4) are thus zero. As long as the control error is zero for each control area, the third term is also zero. The secondary controllers thus not only maintain the same difference between production and consumption in each control area, but also ensure that the frequency remains close to its set value.

When there is an unbalance in one of the control areas, that is, the total export or import deviates from the scheduled value, the frequency will deviate from its nominal value. Consider a synchronous system consisting of N control areas; in control area 1, the export deviates from the scheduled value due to a lack of secondary reserve. The area control error for area 1 is

$$G_1 = \Delta P_1 + K_1 \Delta f \tag{8.5}$$

where $\Delta P_i = P_{act,i} - P_{sched,i}$ is the surplus of generation in control area i and $\Delta f = f_{act} - f_{set}$ is the overfrequency in the system. The area control errors for the other areas are zero, so their sum is also equal to zero:

$$0 = \sum_{i=2}^{N} G_i = \sum_{i=2}^{N} \Delta P_i + \Delta f \sum_{i=2}^{N} K_i \tag{8.6}$$

The sum of all power exchanges is equal to zero, so

$$\sum_{i=2}^{N} \Delta P_i = -\Delta P_1 \tag{8.7}$$

Inserting (8.7) into (8.6) gives the frequency deviation in the system as a function of the unbalance in one control area:

$$f_{\rm act} = f_{\rm set} + \frac{\Delta P_1}{\sum_{i=2}^{N} K_i} \tag{8.8}$$

A production shortage ($\Delta P < 0$) in one control area will result in a reduced frequency throughout the synchronous system. A situation such as this could occur when there is insufficient secondary reserve in a control area after the loss of a large generator in that area. The automatic secondary control will then use the reserves in other control areas, but at the expense of a drop in frequency. The transmission system operators will correct this through bilateral agreements as part of the tertiary control.

Note that the frequency in (8.8) is the steady-state frequency at a timescale of several minutes. Immediately after the loss of the large generator and during the action of the primary control, much larger frequency deviations will occur.

Example 8.5 Consider a large synchronous system with a total control strength $\sum K_i$ of 25,500 MW/Hz. A shortage of 1350 MW in production occurs in a control area with a control strength of 2500 MW/Hz. Using (8.8) gives for the system frequency

$$f_{\rm act} = 50\,{\rm Hz} - \frac{1350\,{\rm MW}}{23{,}000\,{\rm MW/Hz}} = 49.94\,{\rm Hz} \tag{8.9}$$

where 23,000 MW is the control strength of the rest of the system. It shows that even a fairly large power unbalance in one control area does not result in a serious drop in frequency.

Exercise 8.1 What would be the frequency deviation if this area was not part of a synchronous system? Assume the same control strength, 2500 MW/Hz, as in Example 8.5.

Exercise 8.2 The size of the dimensioning failure is 3000 MW for the synchronous system in Example 8.5. What would be the resulting frequency error when there is an unbalance of 3000 MW in the same area as in Example 8.5?

Exercise 8.3 What has a bigger impact on the system frequency: an unbalance of 1000 MW in a small control area or the same unbalance in a large control area?

The direct impact of distributed generation on secondary control is similar to its impact on primary control. The variation in production by distributed generation on a timescale of a few minutes, averaged over a whole control area, is rather small. Several studies have been performed, in different countries, to estimate the need for additional reserves due to the integration of large amounts of distributed generation. The main emphasis has been on wind power, but other types of generation have also been studied. Most studies consider timescales from about 1 min to about 1 h, that is, roughly corresponding to the secondary reserves.

An overview of different studies is presented in Ref. 161. The results cannot be directly compared and the details of the studies vary a lot, but the overview still gives a good impression of the impact of wind power on the need for reserves. Secondary reserve as discussed in these studies is the frequency restoration reserve also called "load following reserve." The secondary reserve also contains a replacement reserve part that is almost equal to the size of the largest unit in the control area. This can be of the order of 500–1500 MW, which is much larger than the need for reserves we discuss here.

- Going from zero to 25% wind power increases the need for reserves from 137 to 157 MW.
- Going from zero to 50% wind power increases the need for reserves from 50 to 100 MW.
- An additional 2200 MW in a system with 8000 MW peak consumption will not noticeably increase the need for reserves.
- Going from zero to 15% wind power increases the need for reserves from 60 to 69 MW.
- 15,000 MW wind would require an additional 10% of reserve.
- 20% (30%) wind increases the need for 1 min reserve by 3% (7%). 30% wind increases the need for 5 min reserves by 8%.
- 20% wind power will increase the ramping rate requirements from 15 to 25 MW/min.

Overall the impact of even large amounts of wind power on the need for secondary reserve is limited.

A study on the need for reserves in the Netherlands is presented in Ref. 168 for two different scenarios. For a scenario with 8% biomass and 19% wind power, the need for reserves would increase from 250 to 350 MW both upward and downward. For a scenario with 8% biomass, 15% micro-CHP, 4% photovoltaics, and 7% wind, the need for reserves would increase to 750 MW.

A basic rule for the automatic secondary control is that each control area has its own reserves available. This could cause a problem for small control areas with large amounts of distributed generation. With weather-dependent production, a small geographical extent implies that all production is exposed to about the same weather. An example of large short-term variations in wind power production in a relatively small control area is discussed in Ref. 180. Southwestern Public Service in Amarillo, TX, has a peak load of 5500 MW and an expected installed wind power capacity of 8300 MW. The 10 min changes in wind power production were shown to be up to about 25% of installed capacity; the hourly changes were about 50%. Other studies [103] indicate that the variations are biggest at 50% production. In this case, that would imply that for a wind power production of 4000 MW, 1000 MW reserves (25%) should be available within 10 min and 2000 MW within 1 h. This is a large amount of reserve for a network operator with a peak load of 5500 MW.

The indirect effect of distributed generation on the secondary control appears to be much bigger than the need for increased reserves. When the control area is exporting more than agreed, or importing less than agreed, the area controller will reduce the setting of the frequency controller of the generators within the control area. If the export is too small or the import too large, the set points for the primary control in Figure 8.4 will be set higher. In steady state, the difference between production and

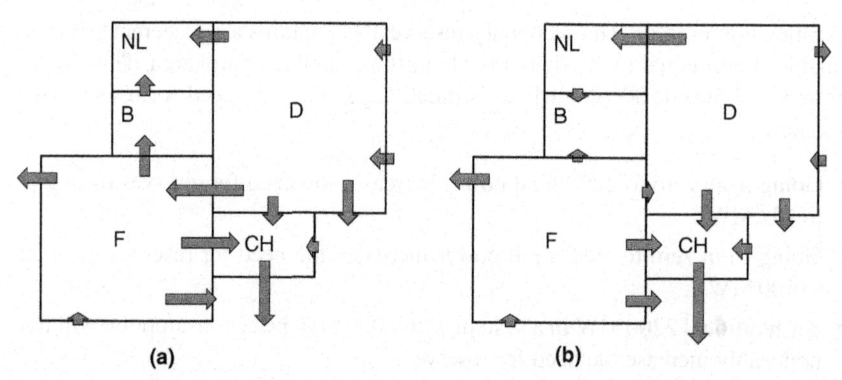

Figure 8.6 Scheduled exchange (a) and actual power flows (b) for Saturday evening November 4, 2006.

consumption is, for each control area, the same as scheduled. However, this does not mean that the power flows between countries are the same as scheduled. An example of how much difference there can be between actual and scheduled power flows is shown in Example 8.6. The scheduling is based only on the difference between the production and the consumption per control area. No consideration is given to where in the control area the power is produced or consumed. Consumption will not likely change location, but with unpredictable generation such as wind power, the production could take place at a completely different location than expected. As a result, the power flows between the control areas could become completely different than expected. Maintaining the same difference between production and consumption per control area no longer gives any guarantee for the cross-border power flows.

Example 8.6 The scheduled exchanges and the power flows in the northwestern part of the European interconnected network are shown in Figure 8.6 for Saturday November 4, 2006, 22:09 [415]. The length of the arrows is proportional to the amount of power, where the scheduled exchange between Germany and the Netherlands is close to 2000 MW. Both Germany and France are export countries. As shown in Table 8.3, Germany is the main export country, with 7649 MW net export; Germany also transfers 1781 MW from Poland and the Czech Republic to the west and the south. France has a net export of 5172 MW, but with an import of 3781 MW from Germany and a small import (317 MW) from Spain. The Netherlands, Switzerland, and Belgium are import countries. However, Belgium also serves as a transport country between France and the Netherlands; Switzerland transports from Germany and France to Italy.

TABLE 8.3 Net Balance Between Import and Export, Saturday Evening November 4, 2006

Country	Scheduled	Actual
The Netherlands	+2875 MW	+3217 MW
Belgium	+879 MW	+1043 MW
Germany	−7649 MW	−7594 MW
France	−5172 MW	−5724 MW
Switzerland	+1434 MW	+1456 MW

The exchange program shown in the figure is set up the day before based on market transactions between consumers and producers in different countries. The exchange program is used to determine if the expected flows require any interventions in the market. The net export and import have not changed much compared to the planned exchange programs (see Table 8.3), but the physical power flows are completely different from the ones scheduled the day before.

The large power flow from Germany to France has disappeared and instead there is a small flow from France to Germany. The transfer from Germany to the Netherlands has doubled and so has the transfer from Germany to Switzerland. This means that in the border regions between Germany and the Netherlands and between Germany and Switzerland, the power flow is 1500–2000 MW more than that scheduled the day before. It should be noted that the number of transmission lines between countries is limited. The $(N-1)$ criterion implies that an actual overload is not very likely. However, an unscheduled 2000 MW flow could easily result in the $(N-1)$ criterion no longer holding.

This is exactly what was the case in the European network this Saturday evening. The manual disconnection of a double-circuit line resulted in the splitting of the European grid into three parts. The western part suffered from a huge shortage of generation, resulting in the shedding of about 17,000 MW of load [415].

When the principles for secondary control were set up, it was assumed that each country would keep its own balance between production and consumption, so that cross-border transfers would be small. Only temporarily, for example, due to the outage of a large production unit, would cross-border transfers occur. This is no longer valid in any case; the transfer of power between countries is large and the cross-border transfer capacity often sets the limit for economic transactions. The international connections are often most heavily loaded on the night between Saturday and Sunday, when the consumption is lowest. The low consumption results in a low electricity price, which implies that only low-cost production will be running. This just turns out to be more concentrated in Europe than high-cost production.

To illustrate the unpredictability of cross-border power flows, the transfer between Energie Baden-Württemberg (EnBW), Austria, Switzerland, and France is shown in Figure 8.7 for 1 week around Christmas 2009 (December 21, 20:15 through December 28, 20:00). EnBW is the network operator for Baden-Württemberg, the German State in the southwestern part of Germany, with Stuttgart as its main city. It is obvious from the figure that the scheduled power flows have no relation at all with

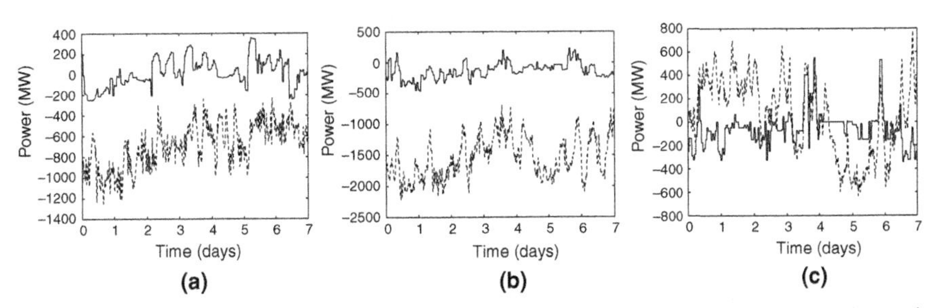

Figure 8.7 Scheduled (solid) and physical (dashed) power flow between EnBW and Austria (a), Switzerland (b), and France (c).

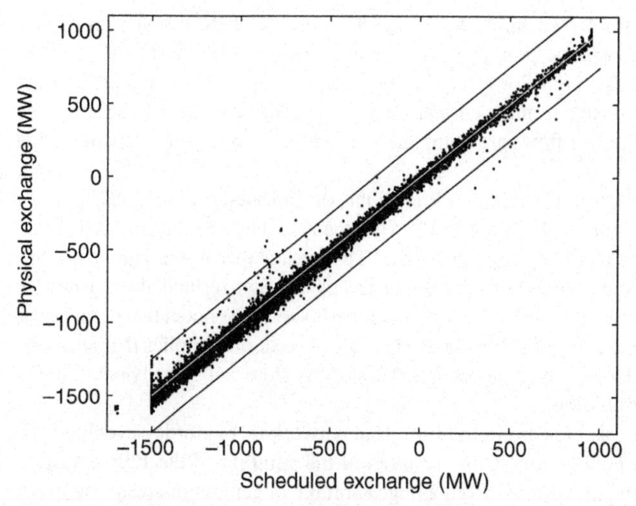

Figure 8.8 Scheduled and physical exchanges between Denmark and Germany.

the actual power flows. The total wind power production in Germany reached a value of around 20,000 MW on day 4 of the period shown in the figure, but it was less than 500 MW for day 1 through day 3.

The western part of Denmark has a very high concentration of distributed generation. During 2008, 23.3% of the electrical energy was produced by combined-heat-and-power and 24.6% by wind power. This control area is part of the European synchronous system; the only connection within this system is with Germany. The automatic secondary control will thus aim to maintain the exchange with Germany equal to its scheduled value. The correlation between the scheduled and the physical exchange between Denmark and Germany, for each hour of the year 2008, is illustrated in Figure 8.8. The white line is the diagonal where scheduled and physical exchanges are the same. The two black lines indicate 250 MW deviation from the scheduled exchange.

The physical exchange is within about 100 MW from the scheduled exchange most of the time, but occasional deviations of more than 250 MW occur. The highest deviation during the year was 540 MW. This may appear rather large deviations; however, the loss of a large power station in Denmark would result in several hundred MW being imported from Germany as spinning reserve. The system should be able to cope with this, so a 500 MW deviation is within the capabilities of the system. Obviously, these deviations should not occur too often, otherwise the risk becomes too high that the transfer capacity is not sufficient upon the loss of a large power station in Denmark.

The reason that the exchange with Germany remains rather close to the scheduled values, despite the huge amount of distributed generation, is the presence of HVDC links with the Nordic system. Western Denmark is connected through HVDC with both Norway and Sweden. Any intraday balancing, to compensate for prediction errors, takes place through these HVDC links. The AC connection with Germany is needed to only compensate for uncertainties at shorter timescales. In the central parts of Europe, as in Southern Germany (see Figure 8.7), there are no possibilities to use HVDC links, or other means of active power flow control, to perform intraday balancing on power flows. Several transmission system operators are considering the installation of phase shifting transformers (also known as "quadrature boosters") to have a better control of the power flow. Such devices are already in operation on the border between the Netherlands and Germany [248, 421]. HVDC links are further considered between Germany and Belgium [101] and between France and Spain.

8.3.4 Tertiary Control and Reserves

The tertiary control considers timescales of 15 min and longer. The transmission system operator starts new production units or changes the set points of units already in operation to accommodate for any depletion of primary and secondary reserves. As discussed previously, any loss of primary reserves should be made up by the secondary reserves. The reduction in secondary reserves is next made up from the tertiary reserves. The tertiary reserves cover all timescales from 15 min onward.

At a timescale of hours, we see the largest variations in production due to distributed generation. Temperature, irradiance, and wind speed show large variations mainly on this timescale. What matters for the operation of the transmission system is not so much the variation in production but the unpredictability of the production. We will discuss prediction in more detail in Section 8.4, but we can mention here that the main concern is the prediction of a change in weather (i.e., the movement of a weather pattern). A change in weather will result in serious changes in production and is difficult to predict accurately. It is often not too difficult to predict the change in weather or the occurrence of extreme weather, but the exact location and timing are difficult to predict. Here, it should be kept in mind that the scheduling of production takes place 12–36 h ahead of time (from noon the day ahead). During the intraday balancing, corrections in scheduling can still be made, but this can easily impact the availability of reserves.

In the existing system, the biggest change in consumption occurs during the morning and afternoon. This is when production units have to be started fastest and prediction errors have the biggest impact. During the morning pickup, reserve production should be available to cope with this. Wind power production can increase or decrease at a high rate at any time of day. This will make the scheduling of reserve power more difficult. The concern is not so much with the frequency control of synchronous systems. Variations in wind power will cancel out over a large geographical area, whereas the daily variations in consumption will occur at the same time within the same time zone. The concern is mainly for the transport of power. Locally, a shortage or surplus of power may occur that could result in transmission system overloads.

Some examples of the impact of distributed generation on the rise and drop in net consumption have been obtained for Western Denmark (data from Ref. 142). The hour by hour change in the actual consumption for the year 2008 is shown in Figure 8.9 as a function of time of the week. The daily and weekly patterns are clearly visible and do not show much variation throughout the year.

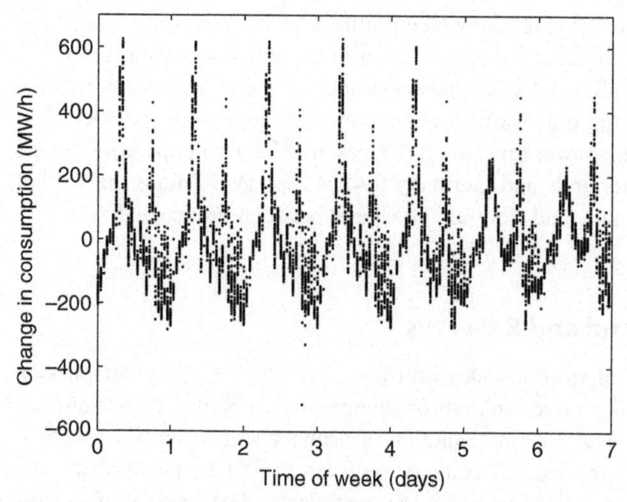

Figure 8.9 Hourly change in consumption, as a function of the day of the week.

The biggest increase is about 600 MW/h and occurs every weekday around 8 a.m. This is the well-known morning pickup of the consumption. A second peak of around 400 MW/h occurs every day around 6 p.m. It is slightly bigger on Friday through Sunday than the other days. The biggest decrease in consumption, about 250 MW/h, takes place around midnight every day. The one outlier is a reduction of over 500 MW/h between 6 p.m. and 7 p.m. on Christmas eve, December 24, when many industries are closed for Christmas. Important to note here is that the increase and decrease of the consumption is rather predictable. The system operator ensures that sufficient reserve is available every weekday morning and evening to cope with the increase in consumption.

When adding combined heat-and-power to the generation mix, the demand on the conventional generation becomes the consumption minus the production by combined heat-and-power. The hourly changes in this demand versus the time of the week are shown in Figure 8.10. The highest increases in demand have actually somewhat decreased. Adding combined heat-and-power to the energy mix reduces the need for reserve power during the morning pickup. Also, the evening pickup becomes somewhat lower, but not as significant as the morning pickup.

Adding wind power results in the changes shown in Figure 8.11. The general pattern is the same as before, but the peaks become less predictable. The highest increase only moderately increases, from 520 to 575 MW/h, but changes of 200–250 MW/h can occur any time of the day.

To illustrate the potential impact of wind power, production and consumption during a 5-day period are plotted in Figure 8.12. The consumption and the CHP production show their standard daily patterns; the wind power production shows an irregular pattern with periods of one to several days. On the morning of day 37, the wind power production drops within a few hours from almost 1800 MW to less than 800 MW, at the same time as the consumption shows its fast morning increase. The

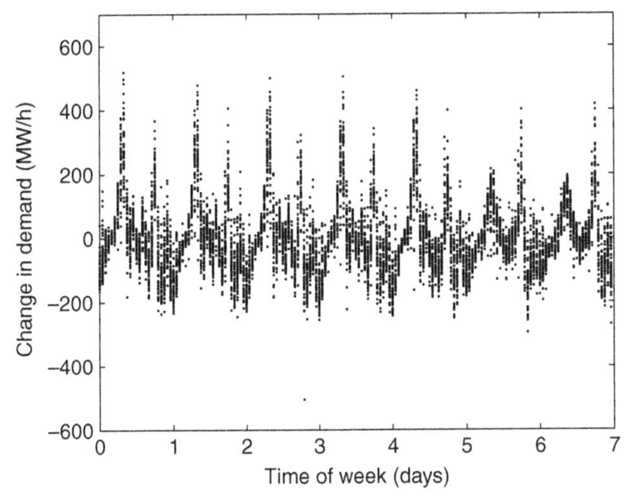

Figure 8.10 Hourly change in consumption minus CHP production, as a function of the day of the week.

impact on the system is limited in this case because the CHP consumption increases at the same time, compensating a large part of the loss in wind power production.

The fastest changes, for different combinations of production and consumption, are summarized in Table 8.4. As the changes in wind power have no correlation with the changes in consumption and CHP, the system operator has to be prepared for an additional 350 MW/h increase in demand on top of the 520 MW/h for the system without wind power (but with CHP).

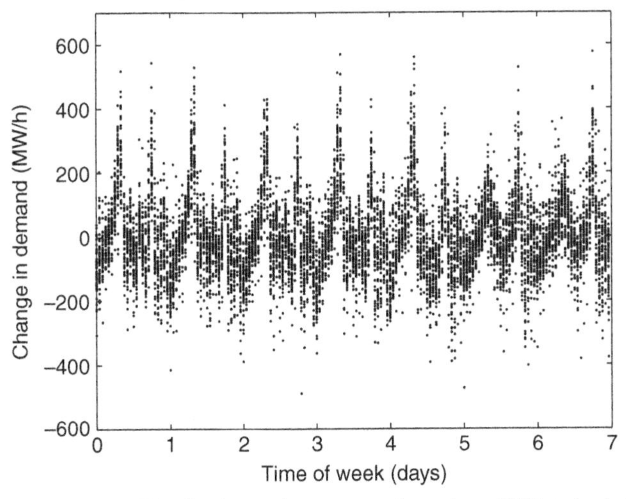

Figure 8.11 Hourly change in consumption minus CHP and wind power production, as a function of the day of the week.

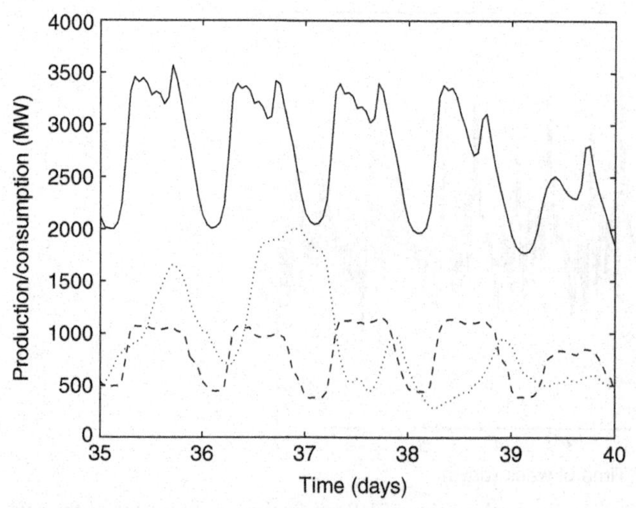

Figure 8.12 Variation in consumption (solid line), CHP (dashed), and wind power (dotted) during a 5-day period.

There is very limited experience with the scheduling of tertiary reserve, for example, in the form of fast-start units, in systems with large amounts of distributed generation. The worst-case situation would be the production due to distributed generation dropping to zero owing to a change in weather. This would require an amount of tertiary reserve equal to the production by distributed generation, where we assume that the drop takes place at a timescale beyond the secondary control. This does not seem to be realistic and will certainly result in too much tertiary reserves. Transmission system operators with large amounts of distributed generation in their system use ad hoc methods for this. The impact of tertiary reserves on the supply reliability is actually rather small, as both primary and secondary reserves have to be depleted as well before the customer is in any way impacted. For an individual transmission system operator, there is however a cost associated with a shortage of reserves within the control area. Reserves have to be bought elsewhere, which will require reserve in transfer capacity as well. Probabilistic and economic studies are needed to determine what are reasonable amounts of tertiary reserve in systems with large amounts of

TABLE 8.4 Highest Increases in Consumption and Production

	Highest increase (MW/h)	Highest decrease (MW/h)
Consumption	+627	−516
CHP	+466	−289
Wind power	+342	−341
Consumption – wind power	+862	−529
Consumption – CHP	+520	−505
Consumption – wind – CHP	+575	−489

distributed generation. Instead of fast-start units, any of the alternative methods for primary reserve mentioned in Section 8.3.2 could be used for tertiary reserve, where it should be noted that tertiary reserves will rarely be used as actual production. This implies, for example, that tertiary reserve should preferably not be scheduled with renewable sources of energy. However, having potential curtailment of consumption as tertiary reserve would for the customer being involved give only a very small risk of actually being curtailed.

8.3.5 Impact of Decay in Production on Reserves

The role of the reserves after the loss of a large production unit, as illustrated in Figure 8.3, will not be impacted by wind power or any other type of distributed generation. Even the amount of reserves is not directly impacted by the introduction of wind power. The tripping of a large wind park, when it is producing a lot of power, will have a similar impact as the tripping of a large conventional generator. When wind parks become larger than the largest conventional units, measures should be taken to prevent the tripping of the whole park at once. This is relatively simple from a technical viewpoint, although it will most likely require additional transmission lines and substations.

The main challenge with large amounts of distributed generation is an unpredicted change in weather resulting in a large reduction in production. This could be a drop in wind speed, an increase in cloud cover, or a rise in temperature. These variations take place at timescales of several minutes or longer. The shortage of production will initially result in the primary reserve being used and the frequency dropping. This will however be quickly replenished by the secondary reserve. The transmission system operator will next use the tertiary reserve to replenish the secondary reserve. The result is that the drop in production from distributed generation will at first result in a drop in tertiary reserve. Once the tertiary reserve is used up, the secondary reserve will reduce. Once the secondary reserve is used up, the primary reserve will be used. Only when all reserves are finished will there be a shortage of production that impacts the customers.

The reduction in reserves due to an unpredicted decrease in wind power production is shown schematically in Figure 8.13. For simplicity, it is assumed that there

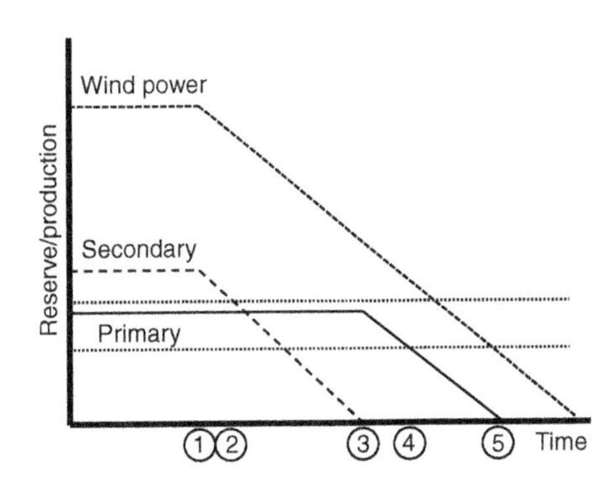

Figure 8.13 Drop in primary and secondary reserves due to an unexpected decrease in wind power production. The horizontal dotted lines indicate the minimum requirements for primary reserve (bottom line) and secondary reserve (top line).

are no more tertiary reserves that can be made available fast enough. In practice, there will always be a certain amount of tertiary reserve available. The drop in secondary reserve will in that case start only when this tertiary reserve is depleted.

A number of relevant time instants can be distinguished in the figure (labeled 1–5):

1. The production from wind power unexpectedly starts to drop. The reduction in wind power is compensated by a reduction in the secondary reserves.
2. The secondary reserves become less than the requirements. The transmission system operator has to initiate making more secondary reserves available. Here, we assume that this is not possible fast enough. The transmission operator will not be able to fulfill requirements, but the customer is not impacted by that. Also, the operational security is not endangered. However, if two large power stations would fail before the secondary reserve is replenished, this could result in a shortage of production.
3. The amount of secondary reserve becomes zero and the primary reserve takes over. From this moment, the frequency in the grid will drop.
4. The amount of primary reserve becomes insufficient. Tripping of a large generator unit could result in a shortage of production. From this moment, the system is no longer secure.
5. Also, the primary reserve is depleted and thc system suffers a shortage of production. Without any additional countermeasures, the frequency will drop quickly and the underfrequency load shedding will reduce the consumption.

Example 8.7 According to the grid code for the Nordic synchronous system [323], the following reserves should be available:

- Frequency-controlled normal operation reserve ("load following reserve"): 600 MW
- Frequency-controlled emergency reserve ("primary reserve"): 1000 MW
- Fast acting disturbance reserve ("secondary reserve"):
  - in Finland: 1000 MW
  - in Sweden: 1200 MW
  - in Eastern Denmark: 600 MW
  - in Norway: 1600 MW

An unexpected continuous decrease in production by distributed generation will result in some drop in the load following reserve and some drop in the primary reserve, but mainly in a continuous decrease in the secondary reserve as the transmission system operators will replenished the drop in primary and load-following reserves. If we assume that all four countries contribute to this, there is a margin of 4400 MW before the frequency-controlled reserve becomes insufficient. Next, the production can drop by another 600 MW before the frequency starts to drop. After another 1000 MW decrease in production, a shortage of production will occur. The total unexpected drop in production that the system can cope with is 6000 MW.

From the example, is follows that only a very large drop in wind power production will result in an actual shortage of production. This is very unlikely to happen, especially because the availability of tertiary reserves has not been included in the above calculations. It is however the task of the transmission system operator to maintain sufficient operational reserves. Although the probability of an actual shortage of generation is small, there is a high probability that the amount of secondary reserves will be insufficient. According to operational rules, this is an unacceptable situation. The maximum drop in production that can be allowed is therefore not determined by the total amount of reserves but by the speed with which new reserves can be made available.

In discussions it is often stated that the power system is able to cope with the rise in consumption in the morning. The system would therefore also be able to cope with a similar drop in production from distributed generation. This is true as long as the drop in production is predicted long before scheduling the starting of generators. For an unexpected drop in production, there may not be enough time to maintain sufficient secondary reserve. The maximum unexpected drop in production with which the system can cope depends completely on the mix of production. In Sweden and Norway, the large amount of hydropower generation will allow a fast increase in production. This will be able to compensate for almost any decrease in wind power production. Denmark, on the other hand, has mainly oil- and coal-fired power stations where possibilities for a fast increase in production are rather limited. Fortunately, the HVDC links with Norway and Sweden offer plenty of tertiary reserve instead.

The above discussion assumed that all secondary reserves can be made available to compensate for the drop in production. When the interconnected system consists of a number of regions linked by weak interconnectors, the situation becomes somewhat different. The transport of reserves throughout the interconnected system is often a serious limitation to the planning of reserves. Primary reserves are always planned for the interconnected system as a whole, but secondary and tertiary reserves are typically the responsibility of every single transmission system operator. Consider one control area connected by a tie line to the rest of the interconnected system. This situation is shown schematically in Figure 8.14. The production in the area consists of wind power, P_{wind}, and conventional generation, P_{conv}. The total consumption is equal to P_{cons} and the balance is made up of import P_{tie} over a tie line with capacity S_{tie}.

$$P_{\text{cons}} = P_{\text{wind}} + P_{\text{conv}} + P_{\text{tie}} \tag{8.10}$$

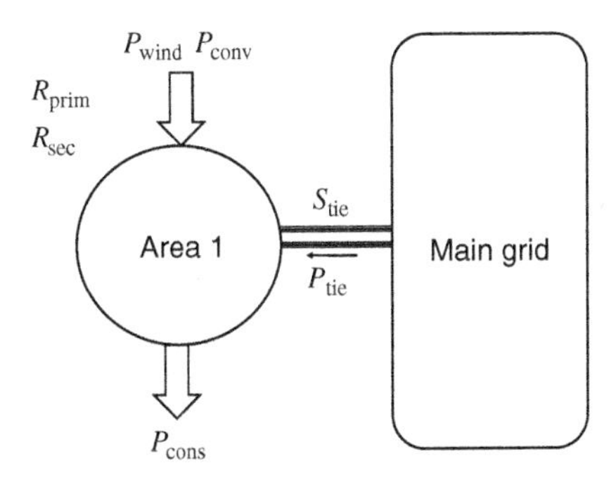

Figure 8.14 Single area as part of a large interconnected system.

The transmission system operator for the area further disposes of primary reserve R_{prim} and secondary reserve R_{sec}. We assume that there is automatic secondary control, so that production and consumption are balanced within the control area. The secondary controller will ensure that the power flow over the tie line is kept constant at a timescale of minutes or longer, as long as there is secondary reserve left in the area.

An unexpected decrease in wind power production will result in an increase in the production by conventional generation, thereby depleting the secondary reserve. If the secondary reserve should at least be equal to $R_{sec,min}$, the maximum acceptable decrease in wind power production is equal to $R_{sec} - R_{sec,min}$. Note again that we did not consider the availability of any tertiary reserve that can be turned into secondary reserve. We are looking here at a worst-case scenario, where the tertiary reserve is either depleted or cannot be made available fast enough. Further decrease of the wind power production will simply result in a depletion of the secondary reserve. The operational security is not endangered when the secondary reserve drops below its minimum permissible level. The operational security is endangered only when the primary reserve drops below its minimum permissible level.

Example 8.8 Consider an area with 6000 MW consumption, 4000 MW production from wind power and 2000 MW from conventional generation. Primary and secondary reserves are 300 and 700 MW, respectively. The total transfer capacity from the rest of the system is 650 MW. In the initial state, there is no power flow over the tie line. The wind power production unexpectedly starts to decrease.

The secondary reserve, initially 700 MW, is depleted when the wind power production has dropped by 700–3300 MW. The conventional production has at the same time increased by 700–2700 MW. The power flow over the tie line remains zero.

When the secondary reserve in a certain control area is depleted, the conventional generation can no longer increase its production fast enough to compensate for the decrease in wind power. This results in a drop in frequency and more import into the control area. As discussed in Section 8.3.3, the shortage of production in the control area is balanced by the secondary reserve in the rest of the synchronous system. In a large synchronous system, the total amount of secondary reserve is very big. For each control area, the secondary reserve should be at least equal to the size of the largest generator unit in that area. It is unlikely that all of this will be depleted by a shortage in one control area. The limitation is instead the import capacity into the area.

An important distinction has to be made between the transfer capacity of the tie lines and the secure import into the area. As mentioned previously, the most important operational rule is the $(N - 1)$ criterion, which states that the loss of any single component shall not result in any loss of load. The sudden loss of a large amount of production within the control area results in primary reserve throughout the synchronous system being used. If this control area is a small part of the synchronous system, which is typically the case, the loss of a large production unit will result in a surge of power into the area. A margin should be available to prevent overload of the tie lines during this. A large margin is needed only with import into the area. With export a much smaller margin is needed, only to cope with the contribution from the primary reserve in the control area to a loss of production elsewhere. Next to this, margin should be available for the loss of a tie line. We will not take the latter limitation into account in the below examples.

Example 8.9 Consider the same system as in Example 8.8. The control area with the decreasing amount of wind power is 15% of the size of the total synchronous system. This means that 85% of the primary reserve is imported after the loss of a large production unit. Assume that the largest unit has a size of 500 MW. The loss of this unit will result in an increase in the import by 425 MW (85% of 500 MW).

Assume that there was no power exchange with the rest of the system as long as there was secondary reserve available. We saw in Example 8.8 that the secondary reserve becomes depleted once the wind power production has dropped by 700 MW. Any further drop will be covered by the secondary reserve elsewhere in the system and result in import into the area. The tie lines have a transfer capacity of 650 MW, of which 425 MW should be kept as reserve to maintain the operational security. This means that the operational security is no longer maintained once the import is more than 225 MW. This takes place when the wind power production has dropped by 925 MW to a value of 3075 MW.

Once the operational security is no longer maintained, the network operator should take action to fulfill the criterion within typically 10–15 min. Using the primary reserve to increase production, that is, relying on primary reserve outside the control area, is not a solution. The reduction in local primary reserve will require additional reserve to be available in the transfer. The reduction in secure transfer capacity is exactly the same as the reduction in local primary reserve, so the system will not become more secure.

For further reduction in wind power, the transmission system operator will have to use manual load shedding to maintain operational security. But let us assume that nonsecure operation is permitted. In that case, the final limit is the transfer capacity of the tie line. Once the import exceeds the transfer capacity for one of the lines, this line will trip and other lines will trip one after the other. The result is island operation of the control area. As there is a large shortage of production and no reserve at all, the frequency will drop very quickly. This will end either with a blackout or with underfrequency load shedding of a large part of the load.

Example 8.10 For the scenario from the previous examples, the tie line will become overloaded and trip once the wind power production has dropped to 2350 MW. A summary of the results from the examples is given in Table 8.5.

TABLE 8.5 Reduction in Wind Power Production According to the Scenario in the Examples

	Initial state (MW)	Example 8.8 (MW)	Example 8.9 (MW)	Example 8.10 (MW)
Wind power	4000	3300	3075	2350
Conventional	2000	2700	2700	3000
Import	0	0	225	650
Secondary reserve	700	0	0	0
Primary reserve	300	300	300	0
Consumption	6000	6000	6000	6000

Exercise 8.4 Recalculate the values in Table 8.5 when the initial export is 500 MW. Explain the differences!

The conclusions from the above discussion and examples can be summarized as follows:

- The reduction in wind power can be as much as the secondary reserves plus the difference between the initial import and the secure import limit, before the operational security requirement is violated.
- Without failure of any important component, the reduction in wind power can be as much as the secondary reserves, the primary reserves, and the difference between the initial import and the total transfer capacity of the tie lines, before a blackout or serious load shedding occurs.

8.4 PREDICTION OF PRODUCTION AND CONSUMPTION

The earlier mentioned day-ahead scheduling of the electricity market makes use of prediction of consumption and production. The sum of all scheduled production is equal to the predicted consumption minus the predicted consumption outside the electricity market. The latter may contain more than just distributed generation, but here we will just consider distributed generation. The details of the balance between scheduled and predicted production and consumption depend on the specific rules for the electricity market, but in all cases the market settlement is dependent on predictions. When the actual production and consumption deviate from their predictions, an unbalance occurs that is taken care of by the secondary and tertiary control as explained in Section 8.3. Large prediction errors could deplete the reserves, and in the worst case this could result in loss of load, as illustrated in Section 8.3.5.

Traditionally, the main source of error was in the prediction of the consumption, as the amount of distributed generation was too small to make a difference. Despite many years of development, there remain errors in the prediction of the consumption. In Figure 8.15, a comparison is made between the prediction and the actual consumption for a Monday in Sweden during the spring of 2009. The first prediction is made by the network operator about 2 days ahead; the second prediction is made by the balance responsible parties about 1 day ahead. As shown in the figure, the two predictions are close, but both deviate up to about 950 MWh/h from the actual consumption. The fact that this was the first working day after the introduction of summer time ("daylight saving time") might have contributed to the prediction error.

The example shown in Figure 8.15 is for a country where the power consumption has shown little growth over the last 10–20 years. In countries with a very fast growing electricity consumption, as in many developing countries, load prediction becomes much more difficult as there is no past experience that can be used. Future changes in consumption patterns will also make it more difficult to predict the load in industrialized countries. The possible introduction of electric cars is often used as an example, but also a change from gas fired to electric heating or the other way around will introduce new errors in prediction.

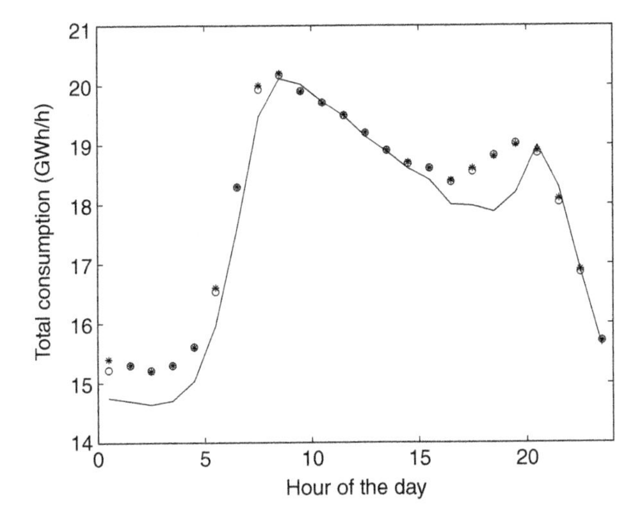

Figure 8.15 Electricity consumption in Sweden on March 30, 2009 (solid line), together with the prediction by the network operator (stars) and by the balance responsible parties (circles).

The introduction of distributed generation introduces an additional source of error. In principle, any type of production outside the electricity market (the term "nondispatchabe generation" is often used for this by system operators) is a possible source of error, but the main source is due to weather-dependent generation such as combined heat-and-power, solar power, and wind power. In the last few years, the emphasis has been almost exclusively on prediction errors for wind power.

As an example, the predicted and actual production rates are presented in Figure 8.16 for the northeastern part of Germany. The data have been obtained as 15-min averages from Ref. 177. The prediction data are the day-ahead data obtained from the weather office. The actual wind power production was estimated from measurements at a number of selected installations.

The figure not only shows that the prediction gives a good impression of the actual production, but also shows that deviations of several hundred MW are not

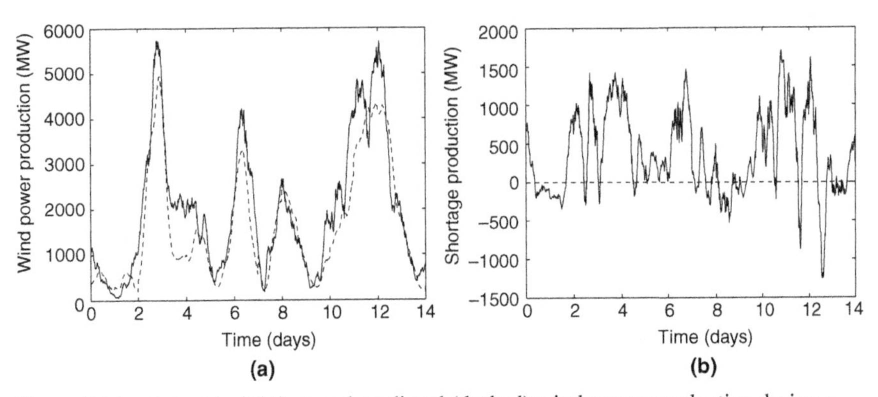

Figure 8.16 Actual (solid line) and predicted (dashed) wind power production during a 2-week period in January 2009 (a): difference between the actual and the predicted production (b).

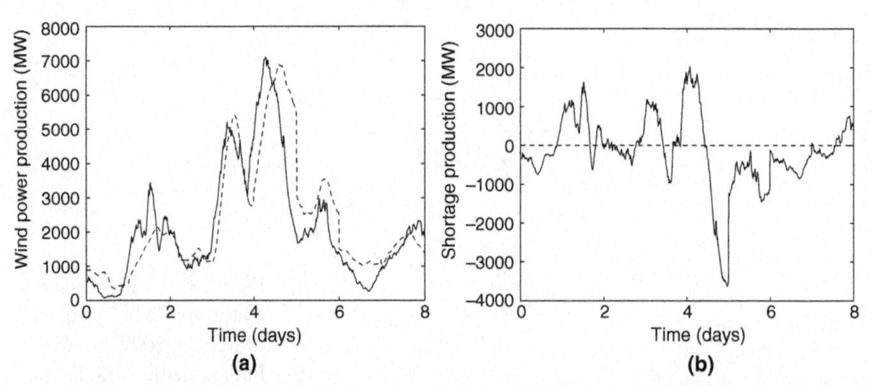

Figure 8.17 Actual (solid line) and predicted (dashed) wind power production during a 1-week period in February 2009 (a): difference between the actual and the predicted production (b).

uncommon. The plot on the right shows the difference between the actual and the predicted production, that is, the prediction error. When the value is positive, that is, the actual wind power production is larger than the predicted value, there is a surplus of production. This requires a reduction in the amount of production by conventional power stations or a curtailment of the wind power production. As long as the system security is not compromised, the conventional generation is reduced. A more serious situation occurs when the actual wind power production is less than the predicted value. In that case, there is a shortage in production, which can only be mitigated by increasing the production by conventional generation. This requires the presence of sufficient reserves. During the period shown in Figure 8.16, the shortage was at most 500 MW during the first several days. However, at day 12, a shortage of 1200 MW occurred. Even larger shortages are shown in Figure 8.17: the plots in (a) show that a weather system arrives several hours earlier than predicted. Especially the earlier passing of the weather system (i.e., the earlier than predicted drop in wind speed) resulted in a serious shortage in production and in this case about 3600 MW. Such a difference between the predicted and the actual production will at first result in a depletion of tertiary and secondary reserves, as explained in detail in Section 8.3.5.

The correlation between the predicted and the actual production is shown in Figure 8.18 again for the northeastern part of Germany, but now over a 1-year period (2009). The middle diagonal line corresponds to a zero prediction error; the upper diagonal line corresponds to an actual production 1500 MW more than predicted; the lower diagonal line corresponds to an actual production 1500 MW less than predicted. From the figure, one concludes immediately that a reserve of 1500–2000 MW is needed to accommodate for prediction errors. As mentioned previously, a prediction of a too low value is less of a concern for the operational security than the prediction of a too high value. From an economic viewpoint, both too high and too low predictions result in additional costs.

The highest negative prediction error (shortage of production) during 2009 was about 3600 MW, the case shown in Figure 8.17. Using the worst-case approach, an additional reserve equal to 3600 MW should be available in the control area to prevent

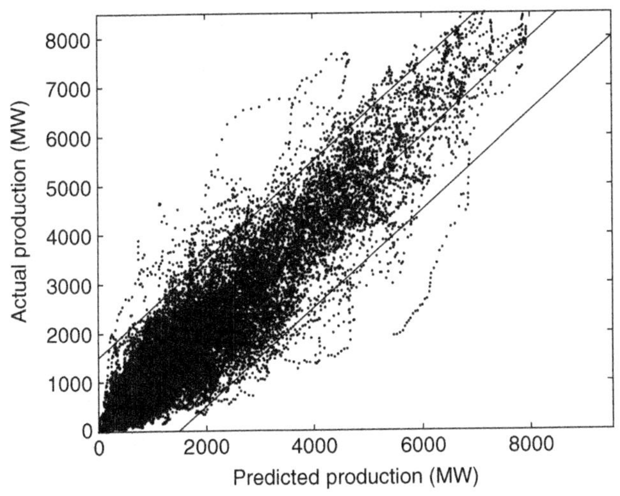

Figure 8.18 Predicted production and actual production for the northeastern part of Germany during 2009.

a shortage of production. Considering probabilities, a different picture is obtained. The probability distribution function of the prediction error is shown in Figure 8.19, where a negative value again corresponds to a shortage in production compared to the prediction. The detailed picture in Figure 8.19a shows that the probability of more than 1000 MW shortage is about 3%; the probability of more than 2000 MW shortage is about 0.3%.

A number of studies have been performed in which the additional operational costs due to the uncertainty in wind power have been estimated. The operational costs assuming perfect prediction have been compared with the operational costs for state-of-the-art prediction and with the operational costs when no prediction is used. The results of some of these studies, performed in the states of New York, Texas,

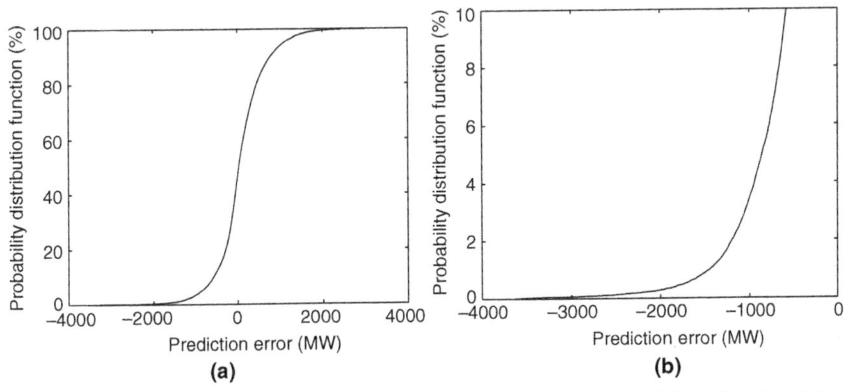

Figure 8.19 Probability distribution function of the prediction error: full scale (a) and details for production shortage (b).

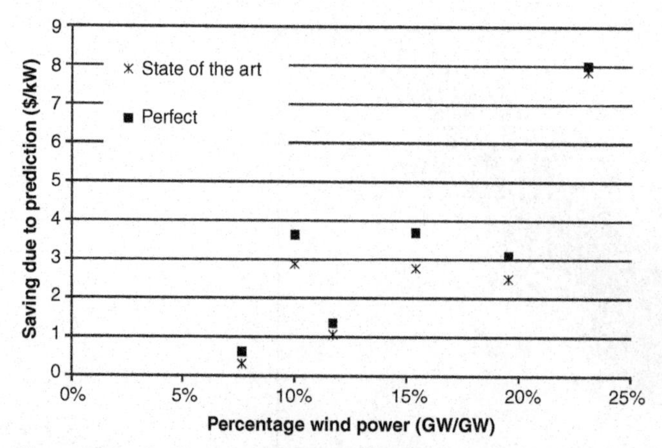

Figure 8.20 Savings that can be obtained by using state-of-the-art prediction (stars) and perfect prediction (squares) as a function of the amount of wind power in the system for six studies performed in three different systems.

and California, are presented in Ref. 103, which in turn is the base for Figure 8.20. The savings are expressed in U.S. dollar per kW of peak load. The contribution of one household to the peak load is somewhere between 1 and 4 kW, varying between different countries. This value thus gives a good indication of the additional costs per household due to the unpredictability of wind power. The horizontal scale gives the amount of wind power in the system, expressed as the ratio between the installed capacity and the peak load.

The figure shows that the costs due to the uncertainty of wind power is negligible as long as the installed capacity is less than about 5% of the peak load. For higher percentages, the costs can no longer be neglected and they increase with the amount of wind power installed. However, even for 23% of wind power, the costs per household will not exceed some tens of dollars per year. This is only a small part of the annual electricity bill. In other words, the unpredictability of wind power does not seem to be an economic issue. A study performed in Sweden [80, 288] considered the situation with 4000 MW wind power in a system with about 25,000 MW peak load. Although the need for reserves would double under that scenario, this would increase the total price of electricity for the end user (per kWh consumption) by only 1–2%.

The preliminary results from a large integration study for the eastern part of the United States show that the costs for additional reserves due to the variability dominate over the costs due to prediction errors [103]. In other words, the costs due to variability are higher than the costs due to uncertainty. The rules for scheduling reserves will thus have an important impact on the costs of integration. It would be fair to compare the different rules for their ability to maintain the operational security. Probabilistic methods should be developed to make such a comparison.

Studies after the impact of prediction errors consider this as an economic issue: the transmission system operator has to balance production and consumption and make additional reserves available. There are costs associated with this and it is the transmission system operator that has to carry these costs, which in the end will result in an increase in the use-of-system charge. The costs will obviously depend strongly on the details of the electricity market, the reserve market, and on the methods used for scheduling of reserves. This makes it difficult to compare the results from different studies.

An interesting overview of methods for wind power prediction is given in Ref. 180. Strong emphasis is placed on the need for specific information on so-called "wind events": a fast increase or decrease in the wind speed. A fast increase could always be mitigated by curtailment of the wind power production, but a fast decrease requires in all cases the availability of reserves. Instead of predicting a value and an uncertainty as a function of time, it is proposed to give more specific information about the occurrence of events. An example would be the expected value and the uncertainty in the arrival time of the event.

In most electricity markets, the main scheduling of production takes place 1 day before (the so-called "day-ahead market"). Wind power predictions 1 day ahead are rather inaccurate and large errors often occur. Some information on prediction errors in the Spanish electricity market is presented in Ref. 6. A prediction tool is used that gives an hourly value of wind generation with an 85% confidence interval. Based on this, reserves are scheduled, with on average 630 MW of additional reserves being required. In the 15% of cases that the reserves are insufficient, intraday rescheduling of the reserves is used to maintain system security.

The prediction for the day-ahead market is based on weather predictions obtained from a weather office. These predictions are the results from numerical calculations based on weather observations performed the day before that. As a result, the scheduling of production and reserves is based on the observed weather 2–3 days before. A better prediction is obtained when predictions from different sources are combined. This will even give some indication about the uncertainty in prediction.

Progress is still being made in the accuracy of weather predictions, but to obtain a significant improvement in the scheduling of production and reserves, the lead time has to be brought down from a few days to a few hours or less. This will still allow the transmission system operator to schedule the generation and it will at the same time give much more accurate scheduling of production and reserves. Several methods are under development for predicting wind power production several hours into the future. The methods are typically based on measurements of wind speed and wind power production during several hours over a large geographical area. Such methods could give transmission system operators a few hours advance warning of large changes in production. Overviews of recent developments on wind power prediction are given in Refs. 6 and 148.

8.5 RESTORATION AFTER A BLACKOUT

A complete collapse of a large part of the transmission system (often call a "blackout") is rare, but if it happens, it is important to be able to restore the system as soon as possible. Some large blackouts last just a few hours, whereas in other cases, it takes a few days before all customers have their electricity back. The time it takes to get the power back is to a large extent determined by the start-up time of the generators.

TABLE 8.6 Start-Up Time of Different Types of Generation

Generation type	Start-up time
Conventional thermal plants	2 – 4 h
Nuclear power plants	20 – 30 h
Combined cycle gas turbines	20 – 40 min
Hydropower units	2 – 5 min
Wind power	Almost instantaneous
Fuel cells	Almost instantaneous
Photovoltaic	Almost instantaneous
Combined heat-and-power	Depending on the process

The start-up time of generators varies strongly, as shown in Table 8.6, with the values obtained from Ref. 105.

Next to the start-up time, the black-start capability and the ability to operate in island are important factors in the restoration of the supply. Large thermal plants often require a large amount of power from the grid to start. The so-called "auxiliary power" is 10–20% of the production capacity of the unit. Some large thermal plants are able to operate in island, that is, producing only their auxiliary power while not being connected to the grid. Nuclear power stations often are equipped with back-up generators (such as gas turbines) to prevent an uncontrolled shutdown upon loss of the grid. These can be used for black start of the generators as well. Unfortunately, starting a nuclear power plants takes a day or longer, so it is important that the unit has the ability of island operation.

Hydropower is more easy to be equipped with black-start capability because the need for auxiliary power is less. Also, a hydropower plant can start producing within a few minutes, whereas this would take hours for thermal plants. But even here a backup generator is needed, where diesel units with their own local fuel tanks are seen as the best solution. The costs of installing and maintaining the black-start capability, and often the need to have personnel on-site to take care of the black start, imply that the "black-start service" has to be enforced in the connection agreement or bought by the system operator. In countries without sufficient hydropower, the black-start service has to be provided by thermal units, despite the higher costs for this.

The presence of distributed generation will impact the restoration process in a number of ways. As long as some large units remain, they can be equipped with black-start capability. Only a small number of units require black-start capability, although the long start-up time of large thermal units would make it worthwhile to have several of them equipped with black-start capability. Note that even with high penetration of distributed generation, there would still be large conventional units needed, although they would be in operation less of the time.

Some types of distributed generation also have black-start capability, which could make them suitable for restoring the power system starting at the distribution level. Wind turbines with induction machine interface do not have black-start capacity, but units with power electronics interface could be used for this. A small diesel generator or a battery block would be needed, but that is already present in most larger installations for other reasons. This same holds for photovoltaics, where also a small amount of energy is needed to power the protection and control equipment. An important advantage of these installations is that they take only a short time to start, at most a few minutes. The disadvantages are the need for communication that cannot always be relied on during a blackout, and the fact that the black-start ability depends on the availability of wind or sun.

Industrial CHP is relatively large (often in the 1–30 MW range) and has synchronous machines as interface. The installations are often equipped with island operation capability to improve the reliability. Synchronization with the public grid after a blackout is relatively simple in that case. However, complete restart might take a longer time because the electricity production is often strongly linked to the heat demand in the industrial process, which could take many hours to come to full production again.

Distributed generation can contribute to the restoration of the supply in a positive way. At local level, a generator can maintain a part of the load during the blackout and thus reduce the duration of the interruption for those customers. This ability to operate as a controlled island does not come for free: it will require investment in protection and control equipment.

For customers, to supply their own load during an interruption of the public grid should not be restricted by the authorities. There may however be safety issues involved that could place restrictions on this. Operating part of the distribution grid as an island independent of the rest of the grid is in most countries not allowed. There are however no technical limitations to island operation (industrial installations, small islands, remote villages, all operate like this and with a reasonable reliability) and allowing it could reduce the duration of a blackout for many customers. It would also allow restoration of the grid, starting with the lowest voltage levels. The Danish transmission system operator is looking into methods for allowing controlled island operation of medium-voltage networks during a blackout of the transmission system or even to prevent a blackout of the transmission system.

There are however economic limitations to this. A large penetration of distributed generation is needed, so that a significant amount of consumption can be covered. Either the production capacity should exceed the peak consumption or the distribution network should be equipped with load shedding capabilities. The protection and voltage regulation should further be able to function correctly during grid-connected operation and island operation. At least one of the generators in the island should be equipped with frequency control. Voltage control should be present as well, either in the generators or in the form of switched capacitor banks or SVCs. Finally, there should be synchronization equipment to connect several of these islands and build up the power system from the bottom-up.

8.6 VOLTAGE STABILITY

Voltage stability concerns the ability of the system to maintain the voltage magnitude while transporting active and reactive powers. In most cases, the concern is with the transport of reactive power, but even for active power, there is a limit to what the system can transport. A distinction should be made between "short-term voltage stability" at timescales up to a few seconds and "long-term voltage stability" at timescales up to several minutes. Most of the discussion is about long-term voltage stability, where the term "voltage collapse" refers to the situation where the system is no longer able to maintain the voltage. Short-term voltage collapse has traditionally been a concern in industrial distribution systems with large amounts of induction motor load, but different trends have also resulted in it potentially becoming a concern for public transmission and distribution systems. We will discuss both short-term and long-term voltage stabilities in the following sections, including the influence of distributed generation.

8.6.1 Short-Term Voltage Stability

Short-term voltage stability involves loads that require increased active and reactive power at timescales of seconds or less. The situation that leads to voltage collapse is when the loading of the system increases with decreasing voltage magnitude. This will give a positive feedback: a reduction in voltage magnitude will increase the loading that in turn will cause further reduction in voltage magnitude. There are indications that a growing amount of small customer equipment (computers, lighting) takes a constant power from the supply, independent of the voltage. Measurements on individual devices confirm this, but such equipment does not seem to be in widespread use yet and no measurements could be produced to show the impact of this at a higher voltage level. Seen from the transmission system, a large part of the consumption currently is constant power at a timescale of minutes due to the actions of the automatic tap changers. The wide-scale introduction of the type of electronic load mentioned would cause the power to remain constant at much shorter timescales. The voltage stability limit would not be impacted much, but the collapse would happen much faster, making it harder for the transmission system operator to take measures.

The main reason for short-term voltage collapse in the existing system remains the presence of large amounts of motor load. Induction machines, the main type of motor load, take a higher current when the voltage magnitude drops. The main increase is in the reactive part of the current, which is the part that causes the main voltage drop in the transmission system. Already during normal operation, the presence of large amounts of induction motor load could result in voltage collapse. However, the main concern is the reactive power consumed by induction motor load after a fault. The situation becomes worse when the fault results in the loss of one of the transformers or lines feeding the motor load. The actual stability limit depends strongly on the motor parameters, the loading, and the properties of the mechanical load. It is however possible to estimate the stability limit using some rough approximations. For a more in-depth treatment of short-term voltage stability, the reader is referred to the several books on power system stability, for example, Refs. 256, 400, and 417.

A short-circuit in the grid results in a reduction in voltage magnitude, a "voltage dip," at the terminals of the machine. During this dip, the air-gap magnetic field is removed from the rotating machine. After the fault, this field needs to be built up again. Without the presence of air-gap magnetic field, the motor behaves in the same way as a short-circuited transformer taking a current of about six times the rated current, with mainly a reactive component. This high reactive current results in a reduced voltage at the terminals of the motor. The reduced voltage gives a reduction in electrical torque: the electrical torque is proportional to the square of the voltage for the same rotational speed. When the electrical torque is less than the mechanical torque (the load torque), the motor will not be able to accelerate back to its normal speed: the motor will stall. The operational torque of a motor is typically around 50% of the maximum torque (the pull-out torque), so a voltage drop down to 70% will prevent the motor from coming back.

Consider a motor load of rated power S_{M} connected to a system with short-circuit capacity S_k. Immediately after the fault, the motor load consumes reactive power $6 \times S_{\mathrm{M}}$ resulting in a voltage drop

$$\Delta U = \frac{6S_{\mathrm{M}}}{S_k} \tag{8.11}$$

The voltage drop should at most be 0.3 pu for the system to be stable, so that

$$\frac{6S_{\mathrm{M}}}{S_k} < 0.3 \tag{8.12}$$

Rewriting this gives the following requirement for the short-circuit power, which is a rule of thumb in the design of industrial distribution systems with large amounts of induction motor load:

$$S_k > 20 \times S_{\mathrm{M}} \tag{8.13}$$

When the short-circuit capacity is more than 20 times the rated power of the motor load, the system is stable.

The behavior of induction generators is very similar to that of induction motors, as discussed, for example, in Refs. 11, 268, and 293. Both induction motors and induction generators require a large amount of reactive power after a fault. The same reasoning as given before for induction motors also applies to induction generators, so the rule of thumb for the minimum system strength will hold for both. Whenever the total amount of induction machines (generators and motors) is more than about 5% of the short-circuit capacity, short-term voltage collapse could become an issue.

Example 8.11 Consider a 132 kV substation with 2316 MVA fault level. (This is the median value for all 132 kV buses in the United Kingdom; see Table 6.9.) The peak load is 150 MW. The amount of induction motor load is estimated at 25% with 0.85 power factor.

The total rating of induction motor load supplied from this bus is

$$\frac{0.25 \times 150}{0.85} = 44.1\,\text{MVA}$$

The rule of thumb gives us a maximum amount of induction motor load equal to 5% of the fault level, which is 115.8 MVA in this case. Beyond the 44.1 MVA already present on consumption side, at most 71 MVA can be added as induction generators.

Exercise 8.5 How sensitive is the result on the assumptions made for the induction motor load?

Note that the short-circuit capacity required is the one available after the fault. This is important when considering redundancy or compliance with the $(N - 1)$ criterion. For example, when a group of induction generators is connected to the grid via two cables, so at to improve the reliability, the short-circuit capacity should be 20 times the rating of the machines, with one of the cables not available. The same holds for industrial installations with large amounts of induction motor load. One may of course decide not to have any redundancy. In that case, the factor of 20 should hold for both cables in operation, but a fault in one of the cables could result in short-term voltage collapse and loss of the group of induction generators.

The calculations performed in Example 8.11 give an indication of the limitations posed by the postfault inrush of induction generators. This is obviously a worst-case situation, where peak consumption and maximum production are assumed to occur together, and exactly at this moment a three-phase fault occurs close to the substation. This may seem a high set of coincidences, but it is this worst-case approach that has resulted in the high level of reliability that we have come to expect from the transmission system. A consequence of this approach is obviously that it will make it more expensive to connect large amounts of distributed generation or large wind parks.

The operation of the transmission system should be such that no single fault results in instability. When the fault level varies, the system should be stable, even for the lowest fault level. Curtailment of production could be used during operational states with low fault levels. The curtailment of production should consist of taking induction generators out of operation, not just of reducing the production of individual generators. In this case, there is no need for an automatic disconnection after, for example, the loss of a transmission line (such as in an "intertrip" or "special protection" scheme). Under the rules for the operation of the transmission system, the system is allowed to be insecure for up to 10–30 min. This allows manual curtailment of production.

It may however be possible in some case to improve the stability by using intertrip schemes. This will strongly depend on the specific properties of the system. When short-term voltage stability is expected for a limited number of fault locations, the wind park causing the voltage collapse can be disconnected quickly whenever any of these faults occur. This can be achieved through intertrips from the breakers clearing those faults to the breakers that will disconnect the wind park. More advanced methods might be possible where only some of the turbines are disconnected and where the amount of turbines disconnected depends on the amount of production immediately before the fault.

When the voltage collapse would impact only one customer and the customer would accept being disconnected occasionally, it is allowed to operate without reserve. Such a situation could occur not only with the connection of a large wind park but also with the connection of an industrial installation. The customer would in that case accept a deterioration of the reliability against a reduction in the connection fee.

For a more accurate estimation of the hosting capacity, a detailed dynamic model of both the consumption and the generation is needed. Such a model needs to include the consumption and generation connected to nearby substations as well. Detailed models of wind parks (the main source of induction generators) are typically available.

The main uncertainty is in the modeling of the dynamic behavior of the consumption. Some models are recommended for stability studies, but these are in most cases based on experience gained many years ago and possibly no longer valid for modern types of consumption. The change in the consumption for, especially, domestic and commercial customers implies that the results obtained from such models may no longer be valid. Should short-term voltage stability become a serious barrier to the connection of wind power, a renewed modeling effort is needed. The most suitable way seems to be the widespread use of disturbance recorders to obtain data on the actual dynamic behavior of the load connected to the transmission system.

Short-term voltage stability is a concern only with induction generators. Synchronous generators do not suffer from this issue (they suffer from angular stability instead), neither do generators with full converter interface. The behavior of double-fed induction generators depends on the details of their control and protection. Although it is possible to prevent short-term voltage collapse with such generators, most of them connected to the grid at the moment will deteriorate the short-term voltage stability. The by far most important application of induction generators is for wind power. Here, the last few years have seen a shift first to double-fed induction generators and then to generators with full converter interface. This implies that short-term voltage stability will likely become less of an issue. Large numbers of smaller wind turbines connected to the distribution system, using cheap induction machines, could however bring the subject back on the agenda.

A number of mitigation methods are available to reduce the risk of short-term voltage collapse. Full converter and synchronous machine interfaces were already mentioned as an option. For induction machine interface, the main solution is to make large amounts of reactive power available. Switched capacitor banks, static var compensators, and STATCOMS all help to improve the stability. The reactive power consumption after a fault is about six times the rated power, so every MVA to be added above the limit requires 6 Mvar in reactive power.

Example 8.12 Continuing from Example 8.11, a 50 Mvar switched capacitor bank is added to provide voltage support after a fault. The hosting capacity for induction generators increases by $50/6 = 8.3$ MVA to 80 MVA, an increase by about 10%.

There is however a limit to the amount of capacitor banks that can be connected to a subtransmission system. Harmonic resonances occur between the capacitor bank and the source. Also, transients will occur when the bank is energized and the fault is cleared. For large banks, harmonic filters are probably needed, which will further increase the costs.

Exercise 8.6 Assume that the largest possible capacitor bank is the one that gives a resonance at harmonic 5.5 (see Section 6.4.5). What will be the increase in hosting capacity?

Tripping the induction generators on undervoltage is a way of protecting the system against short-term voltage collapse. Care has to be taken that not all units are reconnected at the same time when the voltage has recovered. Alternatively soft starters can be used to limit the inrush current.

Tripping of the induction generation has become a problem by itself because it might result in a sudden large loss of production due to a fault in the transmission system. The fault ride-through requirements posed by transmission system operators make that tripping on undervoltage to prevent voltage collapse is no longer an appropriate solution. This will be discussed further in Section 8.10.

A study after the short-term voltage stability of the Danish transmission system is presented in Ref. 8. During periods with high wind and low consumption, the production from wind power covers a large part or sometimes even the whole of the consumption in the western part of Denmark. Operating wind power would expose the system to a high risk of short-term voltage collapse after a fault. Especially faults near the AC interconnection with Germany would result in voltage collapse. The cause is a combination of the reactive power drawn by the induction machines after the fault and the lack of reactive power reserve when the conventional power stations are not in operation. To prevent this, a number of conventional generators, equipped with reactive power reserve, have to be in operation. As these generators have a minimum production, this results in a surplus of active power in the Western Denmark system. This is typically exported via DC links to Sweden and Norway. The minimum number of conventional units that should be in operation depends on the number of AC links with Germany that are in operation. The DC links to Norway and Sweden do not provide any reactive power reserve. It is shown in Ref. 8 that the system recovers, even after the worst-case fault, when all links to Germany and at least two conventional generators are in operation. As the start-up time of these generators is rather long (up to several hours), a third generator has to be in operation to provide reserve in case of a forced outage. When one of the links to Germany is out of operation, at least five conventional generators should be in operation to provide sufficient stability and reserve: four for stability and one as reserve. Using the same reasoning as before, at least four generators would have to be in operation even when both links with Germany are available. Whether three or four conventional generators are in operation depends on the perceived probability of one of the lines failing and a severe fault occurring before additional generation can be started.

This can also be described in terms of primary and secondary reserves, but now for reactive power. The primary reserve is the reactive power needed for the induction machines after the fault. This reactive power is self-recovering (the demand drops after a few seconds), so there is no need for secondary reserves to replenish the primary reserve. There is however secondary reserve needed to replenish the primary reserve after the loss of a conventional generator or after the loss of one of the transmission lines to Germany. In the first case, the system loses primary reserves together with the loss of the generator. In the second case, the primary reserve requirements increase. The secondary reserve should be made available in such a time that the probability of a severe fault occurring is sufficiently small.

8.6.2 Long-Term Voltage Stability

Long-term voltage stability is a very complicated subject that requires many more than a few pages to explain in all detail. Several good books are available that give an in-depth treatment of the subject, for example, Refs. 256, 400, and 417. In this section,

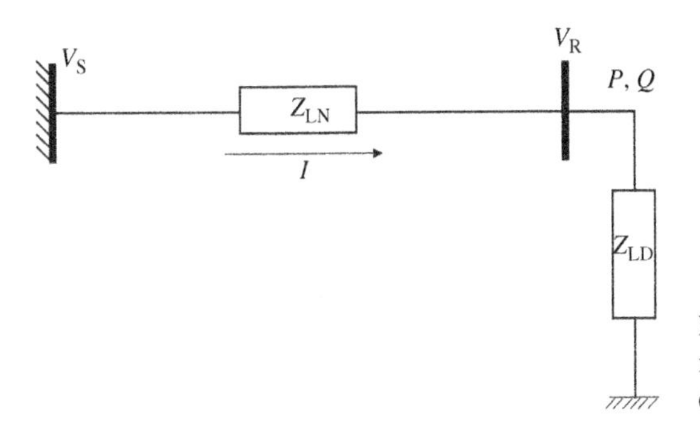

Figure 8.21 Transmission line model for explaining voltage collapse.

we will merely give a very brief introduction of the physics behind long-term voltage stability and address some of the ways in which this is impacted by the presence of distributed generation.

Long-term voltage stability is a typical transmission system issue related to the transfer of large amounts of power over long distances. Consider the circuit in Figure 8.21. A load with impedance Z_{LD} is supplied through a line with impedance Z_{LN}. The supply busbar is infinite; that is, voltage V_S remains constant. The current I is given by

$$I = \frac{V_S}{Z_{LD} + Z_{LN}} \tag{8.14}$$

where $Z_{LD} = Z_{LD} \cos\phi + jZ_{LD} \sin\phi$ and $Z_{LN} = Z_{LN} \cos\theta + jZ_{LN} \sin\theta$.

The receiving end voltage is given by

$$V_R = \frac{Z_{LD}}{Z_{LD} + Z_{LN}} \tag{8.15}$$

The magnitude of the receiving end voltage is given by

$$V_R = Z_{LD} I = \frac{1}{\sqrt{1 + (Z_{LD}/Z_{LN})^2 + 2(Z_{LD}/Z_{LN})\cos(\theta - \phi)}} \frac{Z_{LD}}{Z_{LN}} V_S \tag{8.16}$$

The complex power supplied to the load is obtained from the complex voltage and current:

$$S_R = V_R I^* \tag{8.17}$$

$$= \frac{V_S V_S^*}{(Z_{LD} + Z_{LN})(Z_{LD} + Z_{LN})^*} Z_{LD} \tag{8.18}$$

The active power supplied to the load is the real part of the complex power:

$$P_R = \frac{V_S^*}{(Z_{LD} + Z_{LN})(Z_{LD} + Z_{LN})^*} R_{LD} \tag{8.19}$$

$$= \frac{Z_{LD}}{1 + (Z_{LD}/Z_{LN})^2 + 2(Z_{LD}/Z_{LN})\cos(\theta - \phi)} \left(\frac{V_S}{Z_{LN}}\right)^2 \cos\phi \tag{8.20}$$

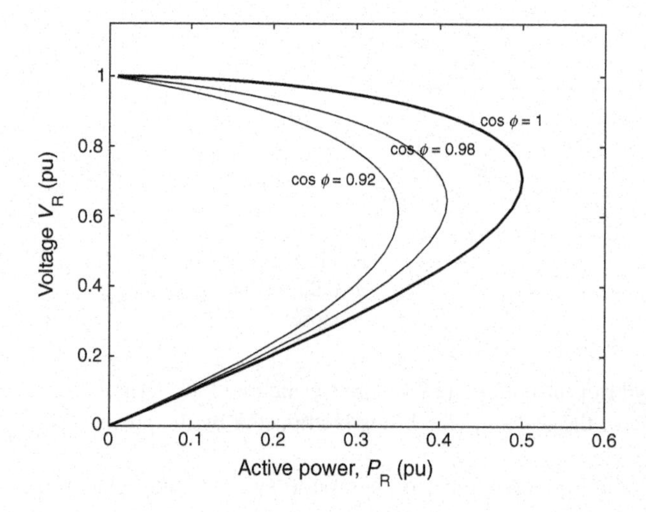

Figure 8.22 Relation between active power transported and receiving end voltage for a transmission line.

Figure 8.22 shows the relation between active power P and voltage V_R for the system in Figure 8.21. The relation is given for different values of the power factor of the load ($\cos\phi$). The losses of the transmission line are neglected ($\sin\theta = 1$).

Note that two values of voltage exist for each value of the transported power: one for a high voltage (small voltage drop) and a low current and the other for a low voltage and a high current. There are two operational regions on these curves: the part of the curve above the knee, where small changes in power cause small changes in voltage and the part below the knee where small changes in load cause large voltage changes and *voltage instability* will have occurred. For a given line, there is a maximum amount of power that can be transported. This value goes down very quickly for decreasing power factor, as shown by comparing the three curves.

For unity power factor and neglecting the losses, the maximum amount of active power that can be transported over a transmission line is

$$P_{\max} = 0.5\frac{U^2}{X} \tag{8.21}$$

where U is the voltage at the sending end and X the line reactance.

Exercise 8.7 Prove this!

Expressions for the maximum power transfer with decreasing power factor (increasing amount of reactive power) are given in Table 8.7. The first column gives the amount of reactive power load as a fraction of the active power, the second column gives the load power factor, and the third column gives the expression for the maximum amount of active power that can be transported, with reference to (8.21).

When the transmission system moves toward its voltage collapse limit, the action of tap changing transformers is interesting. If the receiving end transformer tap ratio changes to keep the load voltage constant, the line current increases, hence

TABLE 8.7 Effect of Reactive Power Load on the Transport Capacity of a Transmission Line

Q/P	$\cos\phi$	P_{max}
0%	1.000	$0.5\frac{U^2}{X}$
10%	0.995	$0.45\frac{U^2}{X}$
20%	0.98	$0.41\frac{U^2}{X}$
30%	0.96	$0.37\frac{U^2}{X}$
40%	0.93	$0.34\frac{U^2}{X}$

causing the voltage to decrease further. Some network operators block the operation of tap changers as a last resort when the system moves toward voltage collapse.

From Figure 8.22 and Table 8.7, it is evident that the power factor is a critical parameter for voltage stability. For low power factors, voltage collapse is more probable. That is why, it is justified for long transmission lines to keep the power factor close to unity with the use of capacitors, synchronous compensators, and so on.

Example 8.13 Consider a transmission system where the load power factor is 0.93 or higher. The transport capacity of a transmission line, according to Table 8.7, is then

$$P_{max} = 0.34\frac{U^2}{X}$$

Consider a 400 kV transmission line with an impedance of 300 mΩ/km. The transfer capacity according to (8.22) is 1800 MW for a 100 km line, 360 MW for a 500 km line, and only 180 MW for a 1000 km line. In most cases, the power factor would be higher (0.93 was the lower limit after all), so the transport capacity is typically higher. However, the load power factor is not predictable, so the lowest value (worst case) has to be considered for design of the system.

Suppose that the system contains a corridor consisting of eight such transmission lines in parallel, each with a length of 500 km. The total amount of power that can be transported over this transmission corridor is 8×360 MW = 2880 MW.

However, the loss of one line would lead to voltage collapse in such a case. Operating under the $(N-1)$ criterion gives a secure transfer capacity of 7×360 MW = 2520 MW.

Using perfect reactive power compensation (e.g., with a number of static var compensators) would increase the secure transfer capacity to 3720 MW.

If, next to perfect reactive power compensation, the operating voltage is increased to 430 kV, the secure transfer capacity increases to 4300 MW. Without building extra lines, the secure transfer capacity has increased from 2520 to 4300 MW, that is, an increase by 70%.

A very simplified example system, to illustrate the impact of distributed generation on voltage stability, is shown in Figure 8.23. The system consists of two geographically separated areas, connected through two transmission lines ("tie lines" in the figure). Within each of the two areas, the system is strong enough; the limitation in transfer capacity is the voltage stability limit of the corridor between the two areas.

The concern with voltage stability is never when all lines are in operation. The $(N-1)$ criterion implies that it is not allowed to operate a transmission system for

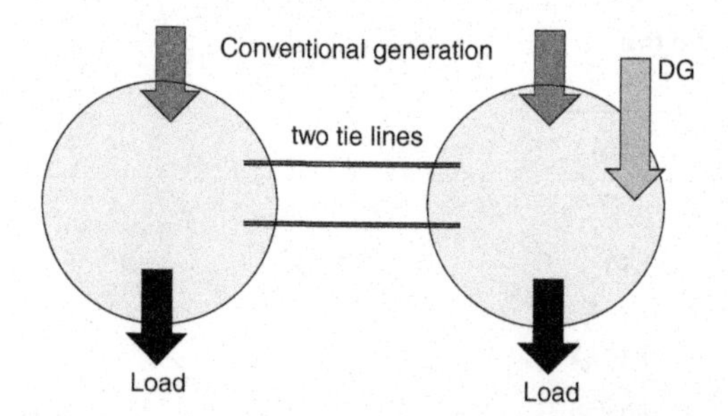

Figure 8.23 Example system with two main load-generation areas.

a longer time close to the voltage stability limit. In that case, the transfer should be reduced. The concern is what happens when one of the lines of the corridor is lost. The flow of power through the corridor should be less than the transfer capacity of the corridor. The flow of power is equal to the unbalance between the two regions. Using the notation in Figure 8.24, for the power flow through the corridor we get:

$$P_{\text{corr}} = (P_1 - L_1) = (P_2 + P_{\text{DG}} - L_2) \tag{8.22}$$

According to the $(N-1)$ criterion, the system should be stable for all individual events ("contingencies") that might occur in the system. The loss of the largest generator is often the worst-case event. The amount of power transport through the corridor should therefore always be less than the transfer capacity minus the amount of primary reserve that has to be imported upon loss of the largest generator on the receiving size. The shift from large conventional stations to distributed generation may result in the largest units being in operation less often. If the loss of the largest unit is the worst case, the shift to smaller units will increase the amount of power that can be transported. The secure transfer capacity should also consider any transfer of primary reserve after the loss of a large production unit.

Let P_{VSL} be the voltage stability limit for the transfer through the corridor and $P_{\text{gen,large}}$ the import upon loss of the largest generator on the receiving end. The secure transfer capacity is in that case equal to

$$P_{\text{corr,max}} = P_{\text{VSL}} - P_{\text{gen,large}} \tag{8.23}$$

For an important corridor, the worst case may no longer be the loss of the largest power station, but the loss of one of the lines through the corridor. Assuming that the

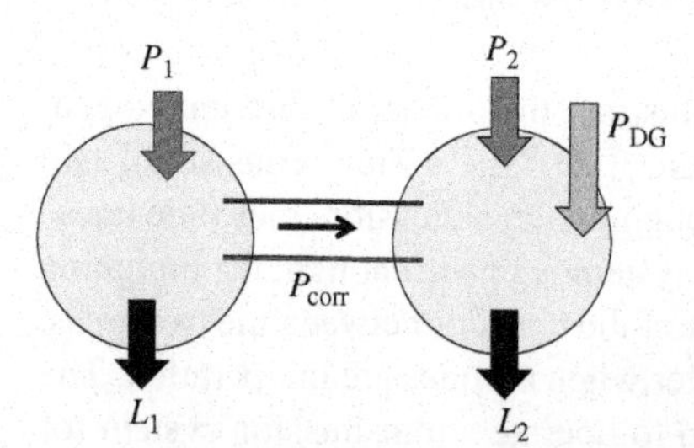

Figure 8.24 Power flows in the example system from Figure 8.23.

corridor consists of N_L parallel lines, the secure transfer capacity is

$$P_{\text{corr,max}} = \left(1 - \frac{1}{N_L}\right) P_{\text{VSL}} \tag{8.24}$$

Here, it is assumed that the transfer capacity is proportional to the number of lines. In reality, not all lines will have exactly the same impact on the transfer capacity and also here the worst case is typically taken.

The loss of a line becomes the worst case when the secure transfer capacity according to (8.24) is less than the one according to (8.23). The following inequality results from this

$$P_{\text{VSL}} > N_L \times P_{\text{gen,large}} \tag{8.25}$$

In other words, the loss of a line is the worst case when the voltage stability limit divided by the number of lines is larger than the import upon loss of the largest generator.

Example 8.14 A country consists of four geographical areas connected by three transmission corridors. These corridors (1, 2, and 3) have a voltage stability limit of 5000, 7000, and 6000 MW, respectively, and consist of 4, 8, and 3 lines, respectively. The import upon loss of the largest generator unit is 1200 MW in each of the regions. What is the secure transfer capacity through each of the corridors?

For corridor 1, we get from (8.22)

$$P_{\text{corr,max}} = 5000 - 1200 = 3800\,\text{MW}$$

and from (8.23)

$$P_{\text{corr,max}} = \frac{3}{4} \times 5000 = 3750\,\text{MW}$$

The loss of one of the lines is the worst case, giving a secure transfer capacity of 3750 MW. In the same way, we find that, for corridor 2, the loss of the largest generator is the worst case, with a secure transfer capacity of 5200 MW. For corridor 3, the secure transfer capacity is 4000 MW, and the loss of a line is the worst case.

Exercise 8.8 If the cost of building an additional line is the same for each corridor, which is the most effective corridor to build a new transmission line? Assume that the voltage stability limit increases linearly with the number of lines.

The above discussion has been, intentionally, kept somewhat simplified. In reality, the situation is much more complex. With the loss of the largest generator, the other generators in the interconnected system will take over the production using the primary reserve. This redistribution of production will occur within about 1 min. Long-term voltage collapse is a rather slow process, taking several minutes. The transmission system operator may therefore consider the situation after redistribution of the production as the reference for determining whether the system is stable or not. The result of this depends to a large extent on where the primary reserve is located. When all primary reserve is located on the receiving end of the transmission corridor, the transfer over the corridor will be the same after redistribution as before the loss of the generator. The loss of the generator would in that case not impact the long-term power flow through the corridor. The opposite effect occurs when the primary reserve is located on the sending end of the corridor. In that case, the corridor will have to be able to transport the primary reserve. In most systems, the primary reserve is distributed through the system, so only part of it has to be transported.

Another issue not considered here is the reactive power balance. Large power stations not only produce active power, but also support the reactive power balance. After the loss of a large power station, a temporary shortage of reactive power could occur in the area. Reactive power transfer causes a serious drop in the voltage stability limit. When there is not enough reactive power reserve within a certain region, the effective transfer capacity to this region is significantly reduced because part of the capacity should be kept in reserve to transfer reactive power in case of an emergency.

The immunity or fault ride-through of distributed generation is another important factor to the voltage stability. Consider again the situation shown in Figure 8.24. A fault somewhere in the area could result in the loss of a large amount of distributed generation. If we assume that the consumption L_2 in the area is not impacted (in reality, the consumption will also be reduced by the fault), this loss of production has to be covered by increased import into the area. This gives a third upper limit to the secure transfer capacity into the area:

$$P_{\text{corr,max}} = P_{\text{VSL}} - P_{\text{DG}} \tag{8.26}$$

where the worst case is taken: loss of all distributed generation and no loss of consumption. This combination in rather unlikely and a more accurate estimation is needed in case (8.26) sets the limit.

Under this reasoning, distributed generation sets a limit to the transfer capacity when the maximum amount of distributed generation that trips due to a fault becomes more than the import upon loss of the largest generator and more than the transfer capacity per line.

The tripping of distributed generation reduces the secure transfer capacity as well when it takes place due to a fault that also results in the loss of one of the lines in the corridor or the loss of the largest generator. In that case, the secure transfer capacity is reduced by the maximum amount of distributed generation that may trip due to such a fault.

The presence of distributed generation in the distributed system will reduce the active power flows, which in turn improves the voltage stability. The reduced current also reduces the reactive power losses in the transmission and distribution transformers. A detailed theoretical study [34] looking into the different combinations shows that the presence of generation in the distribution system during periods of high consumption reduces active and reactive power demand on the transmission system. This holds for all types of generation, even for induction generators. The conclusion from this study is thus that any type of distributed generation will initially improve the voltage stability.

8.7 KINETIC ENERGY AND INERTIA CONSTANT

Both transient stability and frequency stability are strongly impacted by the amount of kinetic energy present in the system. This amount is in turn directly related to the inertia constant of individual production units. A reduction in kinetic energy will make the system less stable. Transient stability will be discussed in more detail in Section 8.9 and frequency stability in Section 8.8, whereas this section will treat the impact of distributed generation on the amount of kinetic energy present in the system.

The kinetic energy of a mass with moment of inertia J rotating at an angular speed ω is equal to

$$\mathcal{E}_{\text{kin}} = \frac{1}{2} J\omega^2 \tag{8.27}$$

In power system studies, the kinetic energy of a production unit is often expressed through the "inertia constant." The inertia constant H is the ratio of the kinetic energy at nominal speed (i.e., corresponding to 50 or 60 Hz frequency in the system) and the rated power S_{nom} of the unit:

$$H = \frac{(1/2)J\omega_0}{S_{\text{nom}}} \tag{8.28}$$

where $\omega_0 = 2\pi f_0$ and f_0 is the nominal frequency. The inertia constant for an electric motor can be defined in the same way. The total amount of kinetic energy present in a synchronous system is the sum of the kinetic energy of all individual machines:

$$\mathcal{E}_{\text{tot}} = \sum_i \mathcal{E}_i = \sum_i S_i H_i \tag{8.29}$$

The right-hand side of the expression, strictly speaking, only holds when the actual system frequency is equal to the nominal frequency. Any deviation from the nominal frequency results in an increase or decrease in the total amount of kinetic energy present. This is one of the stabilizing factors of large synchronous systems. However, the frequency never deviates more than a few percent from the nominal frequency, so (8.29) is a good approximation.

In system studies, an inertia constant for the system as a whole is sometimes used. It is defined in the same way as for an individual machine:

$$H_{\text{tot}} = \frac{\mathcal{E}_{\text{tot}}}{S_{\text{tot}}} = \frac{\sum_{\text{i}} S_i H_i}{S_{\text{tot}}} \tag{8.30}$$

The choice of the base power S_{tot} plays an important role here. When the base power is taken as the sum of the rated power of all individual machines, a much lower value is obtained than that in case the total consumption is used as the base power. It is important to consider this when discussing the impact of distributed generation on the inertia constant.

Replacing large generators connected to the transmission by distributed generation will change the amount of kinetic energy present in the system, which can have

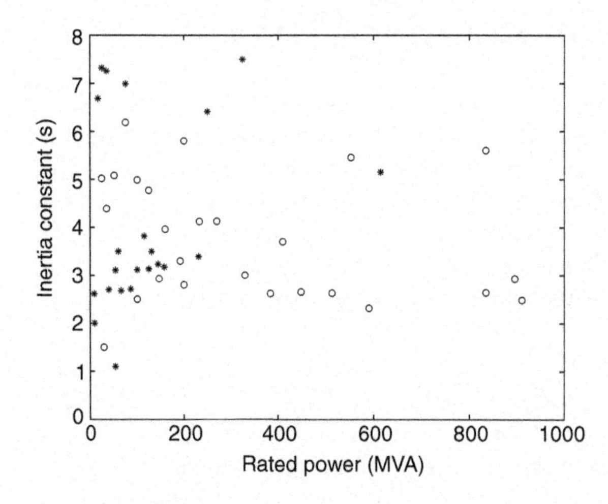

Figure 8.25 Inertia constant of hydro units (stars) and steam units (circles) over a range of rated power from different sources.

a big impact on the stability. Before looking at this in more detail, we will first present some data on the inertia constant of different generators.

Data from the literature on the inertia constant of conventional steam and hydro-generators have been used to obtain a relation between the size of the generator unit and the inertia constant of the unit. Most of the data are obtained from the classical list of typical data in Appendix A in Ref. 14. Further data have been obtained from Refs. 77, 166, 252, 313, 318, 331, and 440. The data have been used to create Figure 8.25. There is no clear consistency in the inertia constants: they vary between 2.5 and 7.5 s for the majority of generators. The inertia constant depends strongly on the construction of the generator. Hydropower units appear to cluster around 3 s and 7 s. Some smaller units (less than 100 MW) show a very small inertia constant, down to 1 s, but overall there does not seem to be a relationship between inertia constant and generator size.

Some more information on the inertia constant for large generator units, from different sources in the literature, is summarized in Table 8.8. The table confirms the large range of values that was also observed in Figure 8.25. However, several sources refer to larger values of the inertia constant for large thermal units with four-pole synchronous machines (running at 1500 or 1800 rpm, depending on the system frequency). Inertia constants up to 10 s may occur for some units, but these are certainly not typical values. The values up to 15 s mentioned in Ref. 194 should be seen as very exceptional. An exception is the value given for the Japanese 60 Hz system, which is higher than the value for any of the generator units. This is due to the choice of the base power. For generator units, the rated apparent power is used. However, for a large system, the actual active power load is a more appropriate value. The sum of the rated power of all generators connected to the grid will obviously be larger than the actual active power load. Also, the inertia constant will be further increased by the kinetic energy of all motor loads connected to the system.

Some data on small hydro units in Alaska [99] is shown in Figure 8.26, together with data on wind turbines [171, 271, 369] and small synchronous generators

TABLE 8.8 Published Values for the Inertia Constant of Large Generators

Generator type	Size	Inertia constant (s)	Sources
Japanese 60 Hz system	95,000 MW	14–18	223
Thermal unit (2-pole)	Up to 500 MW	3–6	183
Thermal unit (2-pole)		2.5–6	256
Thermal unit (2-pole)		4–7	179, 428
Thermal unit (4-pole)	Up to 100 MW	6–10	183
Thermal unit (4-pole)		4–10	256
Thermal unit (4-pole)		6–9	179, 428
Large thermal unit		3.5	72
Large turbogenerator		4	440
Turbogenerators	50 MW class	2–5	351
Turbogenerator		2.8	91
Turbogenerator (2-pole)		3–5	14
Turbogenerator (4-pole)		5–8	14
Noncondensing unit		3–4	179, 428
Thermal unit, forced cooling		7–10	194
Thermal unit, natural cooling		10–15	194
Combined cycle gas turbine		7–8	72
Hydro unit		2–4	179, 183, 256, 428
Hydro unit		6–8	194
Large hydro unit		4	72
Large hydro unit		3.0–5.5	14
Synchronous condenser		1–1.25	179, 428

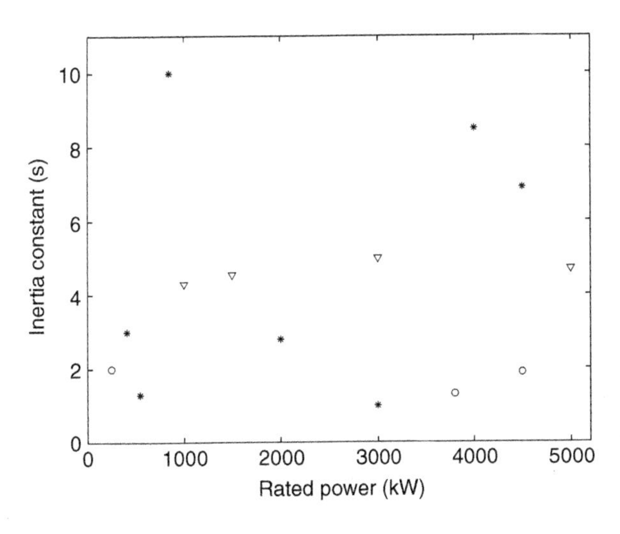

Figure 8.26 Inertia constant of small hydropower units (stars), synchronous generators (circles), and wind turbines (triangles).

TABLE 8.9 Published Values for the Inertia Constant of Motors and Small Generators

Generator type	Size	Inertia constant (s)	Sources
Synchronous motor		2	440
Synchronous motor		3–5	194
Synchronous motor		1–5	179
Induction motor		3	3
Induction motor	4.3 MVA	2.0	388
Induction motor	94 kVA–4.7 MVA	0.45–1.98	388
Synchronous generators	1.5–5 MVA	0.5–2	108
Small diesel generators		0.8–1.5	247
Induction generators		3–5	164
Landfill unit		1.0	72
Waste unit		3	72
Biomass unit		3	72
Small hydro unit		1.5	72
Small hydro unit		1.5–4.0	14
Wind turbine	2 MW	2.4–6.8	249
Wind turbine		2.2	313
Wind turbine		3	72
Wind turbine		4.3	69
CHP with combined cycle gas turbine		4–6	72
CHP with heat recovery steam turbine		3.5	72
CHP with gas turbine		6	72
CHP with internal combustion engine		1.7	72
CHP with steam turbine		3	72

[226, 388]. The figure shows a wide range in inertia constants for the hydropower units, without any correlation with the generator size.

The inertia constant of small generators is, as far as the limited amount of data available shows, somewhat smaller than the inertia constant of large conventional generators (Table 8.9). The inertia constant of wind turbines is of the same range as that of large conventional generator units. The inertia constant of small thermal generators, such as the ones used in combined heat-and-power installations is only about half the value for large units. During periods of low wind, the amount of connected units per MW consumption will be higher for wind power than for conventional generation. As a result, there is also more kinetic energy available per MW consumption. For high wind speed throughout the system, a situation could occur in which the amount of kinetic energy becomes somewhat less. The values in the above tables and figures however indicate that the reduction will not be spectacular. The inertia on the consumption side also contributes to the system inertia and this one is not impacted by the introduction of distributed generation.

Distributed generators connected through a power electronics interface do not contribute at all to the inertia constant of the system. When conventional generators are replaced by such distributed generation, the amount of kinetic energy in the system will decrease. It is this reduction that most concerns the system operators, as it could result in frequency as well as angular instability.

Example 8.15 Consider a region with 10,000 MVA installed generation in the form of high-inertia thermal units. Assume that the average inertia constant equals

$$\frac{\sum S_i H_i}{S_b} = 7\,\mathrm{s}$$

with $S_b = 10,000\,\mathrm{MVA}$. The total kinetic energy in the system is 70,000 MWs. Connecting 5000 MW of wind power would allow, on a windy day, to reduce the amount of thermal generation by up to 6000 MVA. If we neglect the inertia of the load and assume that wind power is connected by a power electronics interface not contributing to the system inertia, the total kinetic energy goes down to 28,000 MWs, a reduction by a factor of 2.5. Here, we assumed that all large thermal units have the same inertia constant.

Exercise 8.9 What will be the percentage reduction in kinetic energy when the amount of kinetic energy in the load is half the amount in the generation?

The issue of inertia of modern wind turbines is discussed clearly in Section 5.3.7 of Ref. 164. It is explained that what matters is not the actual amount of kinetic energy available in the rotating mass, but the speed at which this energy can be made available to the power system. For wind turbines based on induction generators, the energy is made available when the generator speed follows the power system frequency. In fact, during a fast drop in frequency, the generator slip will increase (compared to the actual frequency) and it will quickly produce more electrical power at the expense of its kinetic energy, thus slowing down in speed. The effective inertia constant is between 3 and 5 s according to theoretical studies and measurements [207, 278] as quoted by Ref. 164.

No such natural transfer of energy from the rotating mass to the grid takes place when a power electronics converter is used, as is the case for double-fed induction generators, full power converters, and HVDC connections to wind parks. These types of generation would not contribute to the kinetic energy available to the system. It is however possible to build "electronic inertia," for example, by using the rate of change of frequency (ROCOF) as an input to the torque control of the actual turbine [164, p. 190; 305]. With all implementations of electronic inertia, it is important that there is some kind of additional energy transfer from the generation side to the grid side of the converter once the system frequency drops. A simple implementation, for example for microturbines, would be to let the turbine speed be proportional to the system frequency. A reduction in system frequency would result in a reduction in turbine speed and the difference in kinetic energy would become available.

With nonrotating distributed generation (such as solar panels), such an implementation would require some energy storage or some reserve. The latter would result in a reduction in the production of renewable energy. The loss of renewable energy may be limited by having the electronic inertia only activated when there is not enough kinetic energy connected to the system. Such a reasoning could naturally result in a "kinetic energy market."

Using electronic inertia, alternative solutions are possible, some of which might be better than the classical solution. The amount of energy made available (i.e., the reduction in rotational speed) could, for example, depend on the rate of change of frequency. In that case, more energy would be made available when the system stability is endangered more. Control schemes could also be developed where a drop in frequency results in a higher effective inertia constant than a rise in frequency. The same generator could help in preventing a drop in frequency without slowing down the frequency recovery. With microturbines, a combination of underfrequency protection and electronic inertia could be installed where the turbine speed is reduced very quickly once the frequency drops below a certain threshold. The reader may notice that such schemes introduce a gray zone including both system inertia and primary reserve. In fact, some authors consider the kinetic energy as a form of primary reserve.

8.8 FREQUENCY STABILITY

The frequency stability in a power system is determined by the balance between all production and consumption in the whole synchronous system. Any unbalance will result in a change in frequency, according to

$$\frac{\mathrm{d}f}{\mathrm{d}t} = \frac{1}{2} f_0 \frac{P_{\mathrm{gen}} - P_{\mathrm{load}}}{SH} \tag{8.31}$$

where f_0 is the nominal power system frequency, P_{gen} the total production, P_{load} the total consumption, and SH the total kinetic energy of the rotating mass connected to the system. See the discussions on inertia and kinetic energy in Section 8.7.

The reaction of the production units to a sudden loss of production is shown schematically in Figure 8.27. The change in frequency is determined, according to (8.31), by the difference between production and consumption. Note that the figure is not to scale. For the frequency to return to its original, pre-event, value, the area below the dashed line should be equal to the area above the curve. The frequency disturbance starts with the sudden loss of a large production unit or an HVDC

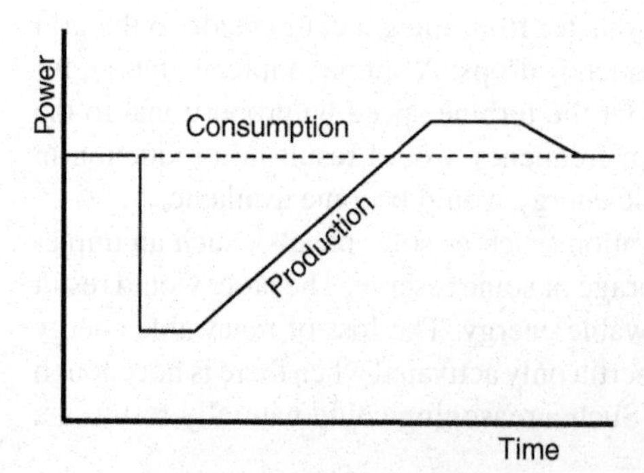

Figure 8.27 Recovery of the total production in a synchronous area due to a sudden shortage of production.

link importing power from another synchronous area. Initially, that is, before the control system of the remaining generation reacts, the frequency decays linearly with time:

$$\frac{\mathrm{d}f}{\mathrm{d}t} = \frac{1}{2} f_0 \frac{\Delta P}{SH} \tag{8.32}$$

After a few seconds, the primary reserve becomes available. This is typically in the form of "spinning reserve," where the large production units are able to very quickly increase their production by a small amount. However, HVDC links could also provide this and some of the links between the Nordic and the European synchronous systems are equipped with emergency control that quickly changes the active power setting of the link in case of a large frequency drop in one of the systems. The total amount of primary reserve present in a synchronous system should be sufficient to cover the loss of the largest unit. Typically, there are also requirements on the spread of the spinning reserve through the system. This is to prevent a shortage of spinning reserve in part of the system after splitting of the system and also to prevent overloads occurring in the transmission system when the spinning reserve is too much concentrated.

To return the frequency to a value close to its original value, the production should be higher than the consumption for a while. In terms of energy, the kinetic energy that was used to limit the drop in frequency at the start of the event should be supplied back again. Some examples of the resulting frequency events are shown in Figure 8.28, all measured in the Nordic synchronous system [45].

The presence of large amounts of distributed generation will impact the frequency stability in two ways. The replacement of large conventional generation by distributed generation may reduce the amount of kinetic energy available to the system, as discussed in Section 8.7. If this is the case, the initial drop in frequency will

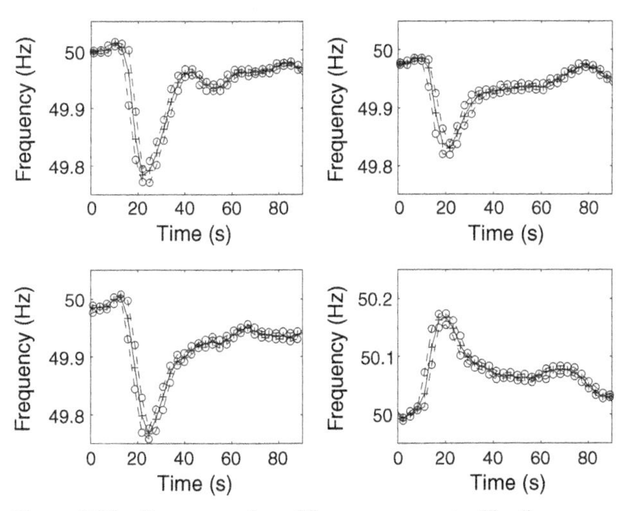

Figure 8.28 Four examples of frequency events. The three curves in each figure give the highest, lowest, and average frequencies over each 3 s time interval.

be larger. A larger drop in frequency is not a concern by itself, as end-user equipment is rather insensitive to the frequency well beyond the values that occur in a large synchronous system. A too large drop in frequency may however result in the unwanted operation of underfrequency protection. The underfrequency protection may be part of the islanding detection for distributed generators (see Chapter 7) or part of the underfrequency load shedding scheme.

A smaller system inertia results in larger frequency variations during normal operation as well as larger frequency drops due to the loss of a large production unit. Both increase the probability of unwanted operation of the underfrequency protection. The frequency variations in the existing synchronous systems are however very small, both during normal operation and during the loss of a large production unit. The margin for reduction of the system inertia is large before the probability of unwanted operation becomes serious. Any reduction of the inertia constant for the whole system due to distributed generation is thus not a concern for the time being. Only with large to very large penetration of distributed generation could the lack of inertia potentially become a concern. This is illustrated in the Example 8.16.

Example 8.16 In a large synchronous system, the frequency varies between 49.85 and 50.15 Hz during normal operation. Owing to the loss of a large production unit, the frequency can drop by 0.25 Hz before the primary reserve is activated. (The values are based on measurements by one of the authors in a number of such systems.) The lowest frequency that can occur due to the loss of a large production unit is 49.6 Hz; that is, when the unit is lost at the moment the normal operating frequency is at its lowest value. Assume that the first stage underfrequency load shedding starts at 49.1 Hz and that we would want some safety margin so that a drop down to 49.2 Hz is the limit. This allows a doubling of the maximum drop compared to the existing system. At first approximation, this corresponds to the total kinetic energy being allowed to drop to half its existing value.

If we neglect the kinetic energy in the load and assume that only the large production units contribute to the kinetic energy, we can replace half of these units by noninertia distributed generation before the frequency drop becomes too large. According to Ref. 164, one-third of the kinetic energy is held by the load, which would allow 75% of the large production units to be replaced.

The impact of wind power on the frequency stability in the Northern Ireland system is studied in Ref. 69. This system is rather small with a peak load around 7000 MW and a minimum load around 2300 MW. Impact on both the rate of change of frequency and the minimum frequency has been studied. During minimum load, the ROCOF due to the loss of the biggest unit is around 0.5 Hz/s without wind power. Up to 1500 MW wind power, there is no impact on the ROCOF, but for 2400 MW wind power, the ROCOF increases to 1.2 Hz/s. During peak load, the impact is small: the ROCOF increases from 0.11 Hz/s for zero wind to 0.12 Hz/s for 2400 MW wind. The minimum frequency during minimum load goes down from 49.60 Hz for zero wind to 49.34 Hz for 2400 MW wind. In all cases, it has been assumed that wind power does not contribute anything to the system inertia. The study shows that wind power has a measurable impact, but even for a high wind power penetration in such a small system, the impact remains limited.

8.9 ANGULAR STABILITY

Next to the thermal limit and voltage stability, the transfer capacity in the transmission system is limited by the "angular stability," also known as "transient stability." Transfer of active power in a system with low losses (as is the transmission system) is possible only when there are sufficient angular differences between parts of the system. However, when the angular difference becomes too large, the system becomes unstable. There are well-defined limits for steady-state operation and during faults. It is the latter that limits the transfer capacity in a transmission system. The steady-state limit is rarely a concern because the $(N-1)$ criterion requires that the system is also stable after the loss of any single component, and thus certainly in normal operation.

As before, we will not go into detail of angular stability. Instead, we will just explain the general principles of angular stability and how the stability is impacted by increasing amounts of distributed generation. For more details, the reader is referred to some of the books on this subject, for example, Refs. 14, 243, 256, 334, and 338.

8.9.1 One Area Against the Infinite Grid

Consider an area of the power system connected to the rest of the system through a transmission line. There is a direct relation between the amount of active power P transported over the transmission line and the angular difference θ between the voltages on both sides of that line. Neglecting the resistive losses, this relation reads as

$$P = \frac{U_1 U_2}{X} \sin\theta \tag{8.33}$$

where U_1 and U_2 are the voltages on both sides of the line and X is the reactance of the line. The relation is shown graphically in Figure 8.29. During normal steady-state operation, the generators operate at constant speed and the mechanical input to the turbine is the same as the electrical output plus the losses. The steady-state operating point is indicated as "1" in the figure.

Upon the occurrence of a fault on the line or somewhere in the neighborhood of the line, the voltage drops on both sides of the line and the amount of electrical

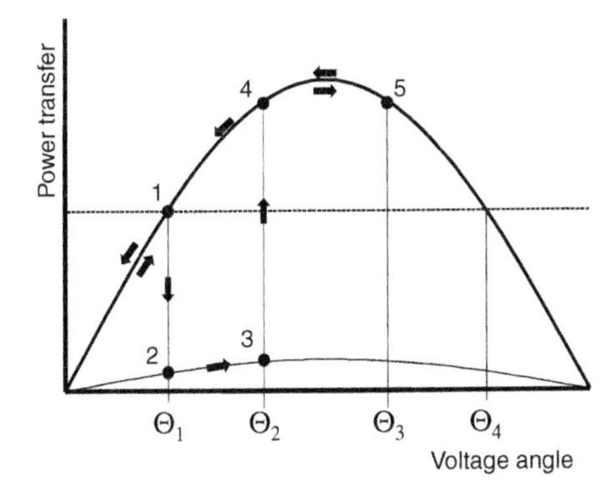

Figure 8.29 Relationship between transferred power and voltage angle: changes in power and voltage during and after a fault.

power that can be transported over the line goes down in accordance with (8.33). This corresponds to point "2" in the figure. The result of the drop in electrical power is an unbalance between electrical and mechanical power. The mechanical power remains constant at the timescales discussed here. The rotating machine increases in speed because it gets more energy from the mechanical side than it can deliver to the electrical side. The angular difference thus increases according to the so-called "swing equation":

$$\frac{d^2\theta}{dt^2} = \frac{1}{2}\omega_0 \frac{\Delta P}{SH} \tag{8.34}$$

where $\Delta P = P_{mech} - P$ is the difference between the mechanical power P_{mech} being provided to the rotating machine and the electrical power P being delivered to the power system. The mechanical power is typically assumed constant and equal to the electrical power transfer before the fault, ω_0 is the rated angular speed (2π times the rated frequency), and SH is the kinetic energy of the rotating machine at nominal speed. Twice integrating (8.34), assuming both ΔP and SH to be constant, gives the following expression for the angular difference as a function of time:

$$\theta(t) = \theta_1 + \frac{1}{4}\omega_0 \frac{\Delta P}{SH} t^2 \tag{8.35}$$

After a certain period, the fault is removed and the voltage recovers to its original value. In the figure, this corresponds to the jump from point "3" just before fault clearing to point "4" just after fault clearing. From this moment, the electrical power being transferred is more than the total mechanical power being delivered to the machines. The result is a reduction in speed. The machines are however still rotating faster than that during steady-state operation, so the angular difference keeps on increasing. At point "5" in Figure 8.29, the frequency has come back to its nominal value (50 or 60 Hz) and the angular difference has reached its maximum value. Next, the angular difference will swing back to its normal operating value at point "1" in the figure.

At the point with the maximum angular difference in the figure (point "5"), the electrical power being transferred is more than the mechanical power being delivered, so the machine still slows down. If, however, the angular difference would increase more, soon a point would be reached where the electrical power becomes less than the mechanical power. Once that point is reached, the generators will keep on accelerating and the system will become unstable. The borderline case is shown in Figure 8.30. It can be shown mathematically that the two areas (A and B in the figure) are exactly equal in size. This rule is referred to as the "equal-area criterion" and it is especially handy in evaluating the stability of simple network configurations. For more complex systems, similar expressions can be derived under certain assumptions [339]. However, in almost all practical cases, the equations are solved numerically using a suitable power system analysis package.

Even in the rather simple case of one machine against an infinite grid, there are no closed expressions for the stability limit. It is however possible to determine which parameters impact the critical fault-clearing time. The critical angle, shown in

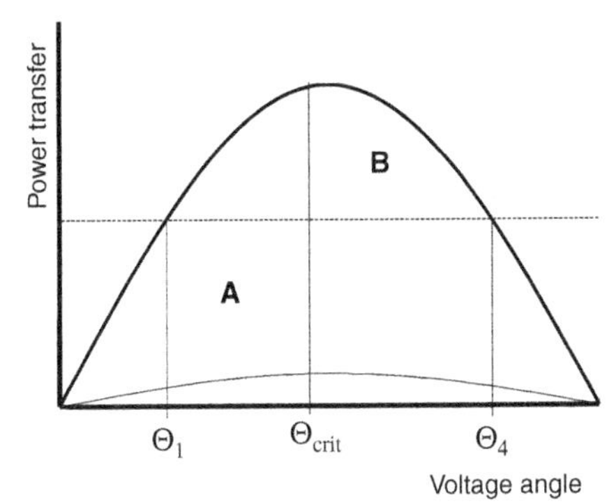

Figure 8.30 Equal-area criterion.

Figure 8.30 depends on the power flow before the fault, the power flow during the fault, and the system strength after the fault.

- The larger the power flow over the transmission line, the larger the prefault angular difference, and the shorter the critical fault-clearing time. Increasing the prefault transfer over the line will make the system less stable.
- The larger the active power taken by the loads and transported over the line during the fault, the less the machines will accelerate and the longer the critical fault-clearing time. Increasing the power consumption during the fault will make the system more stable.
- The stronger the system after the fault, the more power is transferred once the voltage recovers, and the longer the critical fault-clearing time. Increasing the fault level after the fault will make the system more stable.
- The higher the inertia, the slower the machines will accelerate, and the longer the critical fault-clearing time. Increasing the system inertia will make the system more stable.

In the end what matters is whether the actual fault clearing time is longer or shorter than the critical fault-clearing time. The actual fault-clearing time depends to a large extent on the type of protection used. For important connections, the protection used is almost exclusively differential protection or distance protection using intertrip. There is however a minor increase in fault-clearing time with decreasing fault current, thus with decreasing system strength.

The limitations set by transient stability on the power transfer through a transmission line or corridor are explained using the system and notation in Figure 8.31. In normal operation, the production by conventional generation in the area is P_1, the production by distributed generation is P_{DG}, and the local consumption ("load") is equal to L_1. The power export to the main grid through the transmission corridor is equal to

$$P_{corr} = P_1 + P_{DG} - L_1 \tag{8.36}$$

During normal steady-state operation, we assume that all generators in the area have the same rotor angle compared to the infinite grid. In other words, the angular

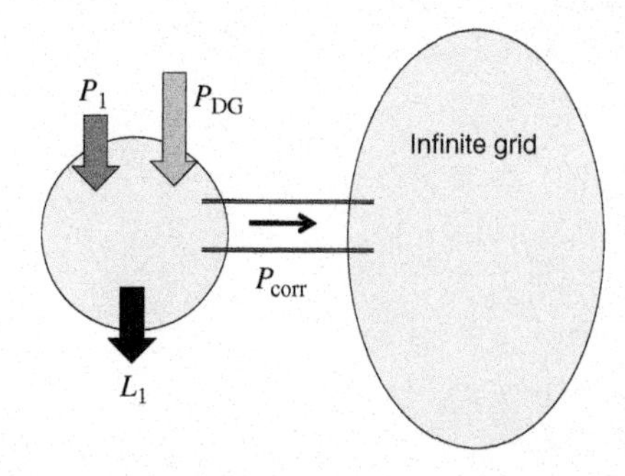

Figure 8.31 Area of the transmission system connected through a transmission corridor to the rest of the system.

differences within the area are small compared to the angular difference between the area and the rest of the grid. We use the relationship between transported power and angle from (8.33) and Figure 8.29. Assuming the voltage on both sides of the transmission corridor to be equal to 1 per unit, we get for the steady-state angular difference between the area and the grid

$$\sin \theta_1 = X_{corr} P_{corr} \tag{8.37}$$

where X_{corr} is the per unit impedance between the area and the grid on the same base as P_{corr}. During a fault, the production and consumption are no longer in balance, for a number of reasons:

- The voltages on both sides of the transmission corridor drop, resulting in an instantaneous drop in the power transfer. The normal procedures for transmission system operation consider the worst case, which is in most cases that the power transfer through the corridor drops to zero.
- The drop in voltage in the area means that the consumption drops. It is difficult to give a general value for this, but the smaller the area, the more the average drop in voltage, and the smaller the consumption during the fault. The fault current contribution by the generators implies that the losses within the area increase. This should be considered in the calculations, as the losses during the fault can become a substantial part of the production.
- The production by distributed generation is also impacted. Typically, the production will drop or even become zero during the fault. Note that the fault ride-through requirements, discussed among others in Section 8.10, concern only the power production shortly after the fault. There are, as yet, no requirements on the power production by distributed generation during a fault.
- The mechanical power delivered to the turbines of the conventional generators can be assumed constant during the fault.

Before the fault, the mechanical power delivered to the turbines is the same as the electrical power produced by the conventional generators, P_1 in (8.36):

$$P_{mech} = P_{corr} + L_1 - P_{DG} \tag{8.38}$$

During the fault, we assume the power through the corridor to drop to zero, whereas the consumption and the distributed generation drop to L_1^* and $P_{\rm DG}^*$, respectively. The electrical power produced by the conventional generators during the fault is equal to

$$P_{\rm e}^* = L_1^* - P_{\rm DG}^* \tag{8.39}$$

The power unbalance, which causes the angular difference between the grid and the region to change, is obtained by subtracting (8.38) from (8.39):

$$\Delta P = P_{\rm corr} + (L_1 - L_1^*) - (P_{\rm DG} - P_{\rm DG}^*) \tag{8.40}$$

The same energy balance holds for the area as for a single machine, and hence (8.34) still holds. The only difference is that the kinetic energy is the total kinetic energy in the area:

$$SH = \sum_i S_i H_i \tag{8.41}$$

From (8.35), this gives the increase in angular difference with time:

$$\theta(t) = \theta_1 + \frac{1}{4}\omega_0 \frac{P_{\rm corr} + (L_1 - L_1^*) - (P_{\rm DG} - P_{\rm DG}^*)}{SH} \times t^2 \tag{8.42}$$

8.9.2 Impact of Distributed Generation: Before the Fault

The angular difference between the region and the infinite grid during steady-state operation, that is, before the fault, is determined by the impedance of the transmission corridor and the amount of power transferred over the corridor, according to

$$\sin\theta_1 = \frac{P_{\rm corr} X_{\rm corr}}{U_1 U_2} \tag{8.43}$$

The larger the angle θ_1, the less the margin for increase in angle during the fault, that is, the less the critical fault-clearing time. The voltages U_1 and U_2 at both sides of the transmission corridor are not impacted by the presence of distributed generation as the voltage is controlled by the (remaining) conventional generators. The power transferred over the corridor is impacted by the amount of distributed generation:

$$P_{\rm corr} = P_1 + P_{\rm DG} - L_1 \tag{8.44}$$

The consumption L_1 will not be impacted by the presence of distributed generation, but the production by the conventional generators, P_1, might be impacted. This depends on the rules of the electricity market and on the marginal costs of production for the local generation compared to generation elsewhere. In a system with a free electricity market, the introduction of distributed generation will result in the most expensive sources of production no longer being operated. If the most expensive production is located locally (i.e., on the left of the corridor in Figure 8.31), P_1 will be reduced together with the increase in $P_{\rm DG}$ and the transfer over the transmission corridor is not impacted. If, however, the most expensive generation is located elsewhere in the grid, the local production will not be reduced and the corridor will be more heavily loaded.

8.9.3 Impact of Distributed Generation: During the Fault

During the fault, production and consumption of power in the area are no longer balanced. As a result, the angular difference between the area and the main grid increases, with constant acceleration (i.e., with the square of the time elapsed since the occurrence of the fault). The increase with time is given by (8.42): a larger power unbalance increases the acceleration; a larger inertia decreases it. Both the power unbalance and the inertia may be impacted by the presence of distributed generation. We will consider the power unbalance first and return to the inertia later.

The power unbalance during the fault is given by (8.40); the impact of distributed generation on each of the three terms in this expression is discussed separately below.

- As discussed in Section 8.9.2 the presence of distributed generation may result in a larger power transfer through the corridor (P_{corr}), and this would increase the angular acceleration. However, for the same power transfer, the presence of distributed generation does not have any impact on this term. It was also discussed previously that an increase in transfer through the corridor is not so much an impact of the distributed generation but a consequence of the electricity market favoring cheap generation, no matter where it is located.
- The second term in the expression, the drop in consumption during the fault, could be indirectly impacted by the presence of distributed generation close to the consumption. If the distributed generation remains connected and contributes to the fault current, this will raise the voltage for the end-user equipment, so the consumption drops less. The less the drop in consumption, $L_1 - L_1^*$, the less the power unbalance. As discussed in Chapter 6, the injection of reactive power is most advantageous for raising the voltage. Not only this improves the power quality in the distribution system, but it also could improve the transient stability.
- The third term is the drop in active power production by the distributed generation during the fault. The more this production drops, the less the power unbalance, and the more stable the system. Tripping the units during the fault would actually improve the stability. The most advantageous for the transient stability would be when the distributed generation during the fault would inject maximum current but purely reactive.

We see that the power balance could actually be positively impacted by the presence of distributed generation. It should be noted that we have considered an area with an export of power. When the area imports power, the impact of distributed generation on the power balance is different. We will discuss this in Section 8.9.6.

The impact of distributed generation on the inertia depends on whether it replaces existing generation or it is added to the existing generation. When distributed generation is added to the existing generation, the total inertia $\sum S_i H_i$ will increase or remain constant. What matters here is the type of generation and whether it remains connected during the fault or not. When the distributed generation trips early during the fault, it will not contribute to the inertia. Also, generation with a power electronics interface will not contribute to the inertia; methods are under development to create "electronic inertia," as discussed in Section 8.7.

A large reduction in the amount of kinetic energy in the area would increase the speed with which the angular difference with the rest of the grid increases during a fault, so that the critical angle will be reached faster. The critical fault-clearing time is proportional to the square root of the amount of kinetic energy (SH). Halving the kinetic energy in the area will reduce the critical fault-clearing time to 70% of its original value.

8.9.4 Impact of Distributed Generation: Critical Fault-Clearing Time

Consider a corridor consisting of a number of transmission lines. We consider the worst-case situation in which a three-phase fault occurs on one of the lines, close to the generation area. At first, we assume that there is no load in the area, so the power produced by the generators drops to zero during the fault. The prefault angle Θ_1 between the area and the main grid is found from

$$\frac{\sin \Theta_1}{X_{\text{pre}}} = P_{\text{corr}} \tag{8.45}$$

where X_{pre} is the impedance of the corridor before the fault, that is, with all lines in operation. The maximum first-swing angle Θ_4 is obtained from the following expression:

$$\frac{\sin \Theta_4}{X_{\text{post}}} = P_{\text{corr}} \tag{8.46}$$

where X_{post} is the impedance after the loss of one line. Also, $\Theta_1 < \pi/2$ and $\Theta_4 > \pi/2$.

The equal-area criteria is illustrated in Figure 8.33, where A is the accelerating area and B the decelerating area.

For this case, the equal-area criterion becomes:

$$(\Theta_{\text{crit}} - \Theta_1)P_{\text{corr}} = \int_{\Theta_{\text{crit}}}^{\Theta_4} \left(\frac{\sin \Theta}{X_{\text{post}}} - P_{\text{corr}} \right) \mathrm{d}\Theta \tag{8.47}$$

with the following solution for the critical fault-clearing angle (*hint:* use A + C = B + C, with C the rectangular area below B to the right of A):

$$\cos \Theta_{\text{crit}} = \cos \Theta_4 + (\Theta_4 - \Theta_1) X_{\text{post}} P_{\text{corr}} \tag{8.48}$$

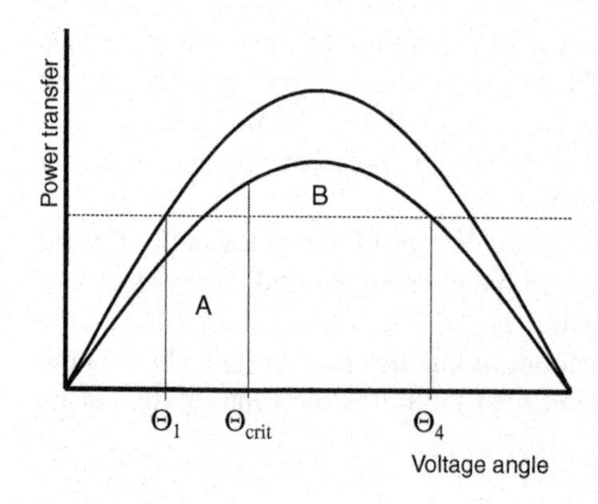

Figure 8.32 Equal-area criterion, no during fault power, and different postfault and prefault impedances.

The critical fault-clearing time is obtained from

$$t_{\text{crit}} = 2\sqrt{\frac{\Theta_{\text{crit}} - \Theta_1}{\omega_0} \times \frac{SH}{P_{\text{corr}}}} \tag{8.49}$$

where SH is the total kinetic energy in the area (sum over all generators of the product of generator rating and inertia constant).

Example 8.17 An area is connected to the main grid through three 400 kV transmission lines of 280 km length. The line impedance is 0.35 ohm/km. For a power transfer of 1000 MW and a kinetic energy in the area equal to 5000 MWs, we get

- $\Theta_1 = 11.8°$
- $\Theta_4 = 162.2°$
- $\Theta_{\text{crit}} = 98.5°$
- $t_{\text{crit}} = 310\,\text{ms}$

Increasing the transferred power to 2000 MW and the kinetic energy to 10,000 MWs (we assume that double the number of generators is in operation) gives a critical fault-clearing time of 205 ms. When we assume that the kinetic energy is proportional to the produced power, the transfer capacity of the corridor is 2400 MW for a fault-clearing time of 140 ms. Keeping the kinetic energy at 5000 MWs would have resulted in a transfer capacity of 2000 MW.

Next, we consider consumption and distributed generation in the region (the situation shown in Figure 8.31). The calculation of the prefault angle and the maximum first-swing angle proceeds the same as before, as in (8.45) and (8.46).

The equal-area criterion gives the following expression, with reference to Figure 8.33:

$$(\Theta_{\text{crit}} - \Theta_1)\Delta P = \int_{\Theta_{\text{crit}}}^{\Theta_4} \left(\frac{\sin\Theta}{X_{\text{post}}} - P_{\text{corr}} \right) d\Theta \tag{8.50}$$

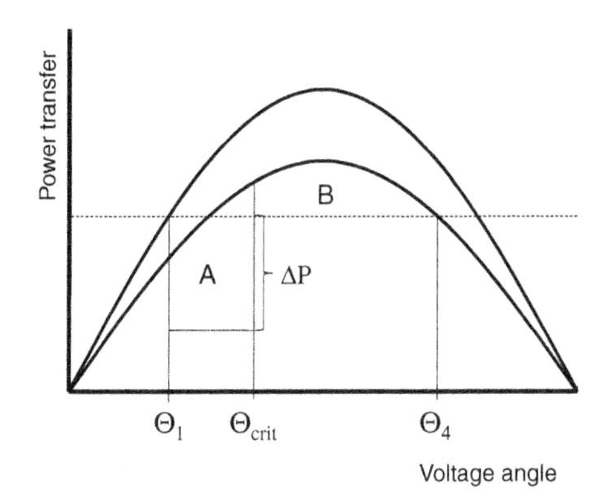

Figure 8.33 Equal-area criterion, considering load within the area, different postfault and prefault impedances.

where ΔP is given by (8.40). The expression can be simplified to the following form:

$$\cos \Theta_{crit} + X_{post}(P_{corr} - \Delta P)\Theta_{crit} = \cos \Theta_4 + \Theta_4 X_{post} P_{corr} - \Theta_1 X_{post} \Delta P \tag{8.51}$$

which cannot be solved analytically. Instead, numerical methods are needed to obtain the critical fault-clearing angle. Once this angle is known, the critical fault-clearing time can again be obtained analytically using (8.42).

Example 8.18 Consider the same corridor as in Example 8.17. The generation area now also contains 3000 MW consumption; during the fault, the consumption drops to 1500 MW. The power transfer over the corridor is 1000 MW, so the region contains 4000 MW of generation. We assume that the inertia constant of the generators is 5 s as before and that the inertia constant of the load is 1 s (using the active power as a base). That gives a total of 23,000 MWs of kinetic energy in the region. Using the above expressions results in a critical fault-clearing time equal to 550 ms. The critical fault-clearing time reduces quickly with increasing power transfer over the corridor. For 2000 MW transfer, the critical fault-clearing time is only 274 ms, assuming an increase in total kinetic energy in relation to the increase in local power generation. For a fault-clearing time of 140 ms, the transfer capacity is 2500 MW, versus 2400 MW in Example 8.17.

The system in Example 8.18 is used to illustrate the impact of distributed generation on the critical fault-clearing time. First, we assume the worst case where distributed generation replaces existing generation in the region, the distributed generation does not contribute to the inertia of the system, the distributed generation maintains its prefault production during the fault, and the distributed generation does not contribute to a raise in voltage during the fault near the consumption. The impact of the distributed generation is in this case only the reduction of the total amount of kinetic energy in the system. Using the same values as in the examples, the introduction of each 100 MW distributed generation will result in 100 MW less conventional generation being present, reducing the kinetic energy by 500 MWs. The critical fault-clearing time as a function of the amount of distributed generation is shown in Figure 8.34 for three values of the power transfer. Increasing amounts of distributed generation

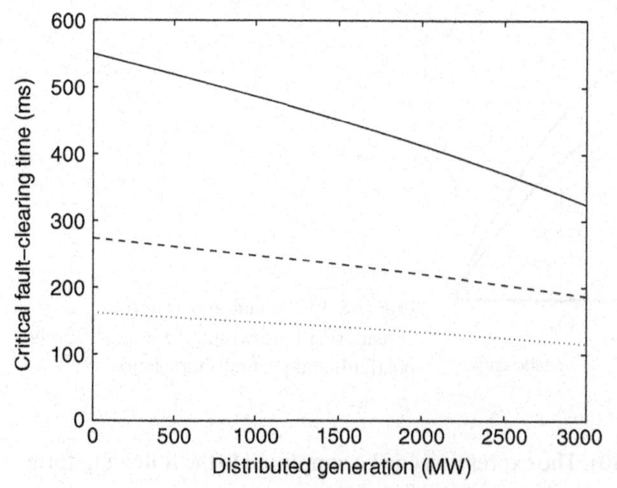

Figure 8.34 Critical fault-clearing time as a function of the amount of distributed generation, worst case: 1000 MW (solid), 2000 MW (dashed), and 2500 MW power transfer.

do result in a reduction in the critical fault-clearing time and thus in a less stable system. However, the reduction is moderate keeping into consideration that 3000 MW corresponds to the whole local load being supplied by distributed generation.

Consider next the case where the distributed generation produces less active power during the fault than before. This helps the conventional generation in maintaining the speed and helps the system remaining stable. The curves in Figure 8.35 are calculated assuming that the production by the distributed generation drops to 25% of its prefault value. Comparing with Figure 8.34 shows that the impact of distributed generation on the angular stability has become much less. For large power transfer, the impact of distributed generation can even be neglected (in this specific example).

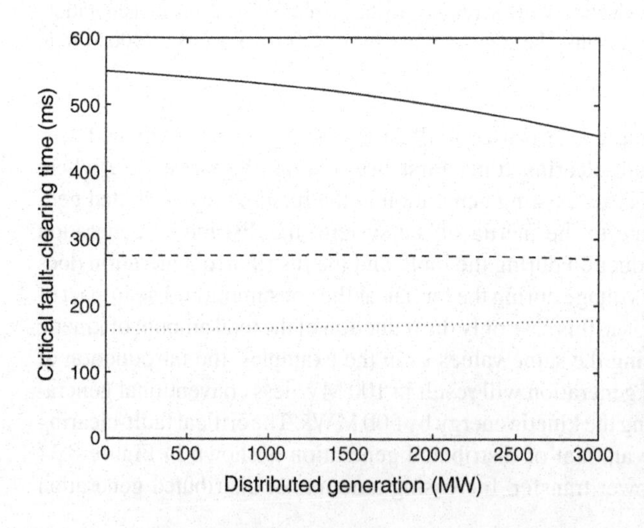

Figure 8.35 Critical fault-clearing time as a function of the amount of distributed generation, distributed generation drops to 25% of its prefault value: 1000 MW (solid), 2000 MW (dashed), and 2500 MW (dotted) power transfer.

TABLE 8.10 Reduction in Transfer Capacity with Increasing Amounts of Distributed Generation, Worst Case

Distributed generation (MW)	Transfer capacity (MW)
0	2590
500	2560
1000	2530
1500	2500
2000	2460
2500	2400
3000	2340

Example 8.19 When all distributed generation would completely drop its production to zero during the fault, the situation could even occur in which the generation decreases speed during the fault instead of increase. Consider the situation with 2600 MW distributed generation and 1000 MW export. The local consumption is 3000 MW as before, so 1400 MW is locally produced by the conventional generation. During the fault, the export becomes zero, the distributed generation becomes zero, and the local production drops to 1500 MW. The electrical power demand on the conventional generators becomes 1500 MW, which is an increase compared to the prefault demand.

The impact of distributed generation on the transfer capacity is small, as shown in Table 8.10. The transfer capacity is defined here as the amount of transferred power that gives a critical fault-clearing time of 140 ms. Even when all local loads are supplied by distributed generation, the transfer capacity drops by only 10% for 3000 MW distributed generation. Here, it is assumed that the distributed generation does not reduce its production during the fault, which is the worst-case situation.

It is important to note that these are just examples and it is not possible to draw any general conclusions from this. The impact of distributed generation on the angular stability may be more severe or less severe than what is shown here. A new calculation, using the appropriate models, is needed for every specific case.

8.9.5 Impact of Distributed Generation: After the Fault

In all the previous calculations, it was assumed that the production and consumption after the fault are exactly the same as before the fault. This is typically not the case. Equipment may have tripped during the fault (due to the voltage dip) and take a long time to come back (i.e., longer than the timescales up to several seconds discussed here). Other equipment remains connected during the fault and takes a higher active power when the voltage recovers. The latter holds especially for motor load. Motor load typically also takes a very high reactive power after the fault, which will reduce the voltage and limit the power transfer. All this is rather independent of the presence of distributed generation, so we will not further discuss it here. In any case, detailed modeling and calculations are needed to include all this in the stability assessment.

The impact of distributed generation, after the fault, has again to do with whether the generation remains connected to the grid during the fault. If the distributed generation recovers immediately upon fault clearing, the system behavior after the fault will be pretty much the same as without distributed generation. If the distributed generation will take a long time to recover (again, longer than several seconds), the area will experience a shortage of generation or at least less of a surplus. As the area experienced a surplus during the fault (loss of the transfer over the corridor, reduction in the local consumption), this additional shortage will improve the stability. The tripping of the distributed generation is however not necessarily a positive thing, as we will see in Section 8.10. Most transmission system operators would prefer the distributed generation to recover their production immediately after a fault. In some cases, delayed recovery would be better for the stability, but there are at the moment no methods for determining during the actual event whether it is better for a distributed generator to remain connected.

8.9.6 Impact of Distributed Generation: Importing Area

When an area of the system is an importing area, the impact of distributed generation becomes different and, in some cases, exactly the opposite of what it was for the exporting area discussed previously. We will again distinguish between the impact before, during and after the fault, but we will not go into the same level of detail as before.

The prefault power flows are shown in Figure 8.36. The difference from Figure 8.31 is only in the direction of the power flow between the area and the main grid. The expression for the angular difference between the region and the main grid is the same as before, but the angle is of opposite sign due to the opposite direction of the power flow. The impact of the distributed generation on the stability depends, such as for the exporting area, on whether the new generation is added to the existing generation or replacing the existing generation. When the distributed generation is added to the existing generation, the power shortage in the area will become less and the power transfer over the corridor becomes less. This will reduce the steady-state angular difference and make the system more stable. Replacing existing local generation with distributed generation will not change the power transfer.

For the during fault impact, we again distinguish between the power unbalance and the amount of kinetic energy. The power unbalance during the fault is given by

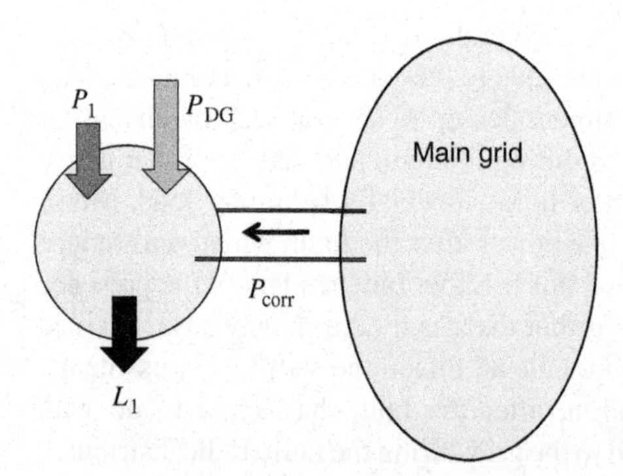

Figure 8.36 Importing area connected to the main grid through a transmission corridor.

the following expression:

$$\Delta P = P_{\text{corr}} - (L_1 - L_1^*) + (P_{\text{DG}} - P_{\text{DG}}^*) \tag{8.52}$$

Instead of the surplus being the concern for the exporting area, it is the shortage of power in the area that is a problem for the importing area. A surplus of power during a fault would cause the angular difference to become less, which is obviously not a problem for the stability. Expression (8.52) gives the shortage of power in the area during the fault. The bigger the shortage, the faster the increase in angular difference between the region and the main grid.

The presence of distributed generation potentially impacts the second and third terms in (8.52) as before. But the impact of this on the stability is just opposite as before. The drop in production by the distributed generation during the fault will increase the value of the third term, which will increase the shortage of power during the fault. Raising the voltage close to the consumption during the fault will result in less drop in consumption during the fault. This will again increase the shortage. When comparing this with the impact for an exporting area, we see that the same behavior of the distributed generation is positive for an exporting area but negative for an importing area. As the same part of the power system may be importing at one time and exporting at another time, it is not possible to give general requirements for the behavior of the distributed generation during the fault. With larger installations, such as wind parks, the settings could be changed depending on the operational state of the system. This would require the development of new infrastructure (among others control and communication) and new ways of looking at transmission system operation.

The impact of distributed generation on the amount of kinetic energy in the region is the same as for an exporting area. There is however less generation connected in an importing area, so a reduction in kinetic energy is noted faster.

After the fault, distributed generation should be available immediately for an importing area to limit the shortage in power and allow the angular difference to return to its prefault value as soon as possible. Tripping of distributed generation during the fault should thus be avoided for an importing area. Note that for the exporting area, tripping of the distributed generation actually improves the angular stability.

8.10 FAULT RIDE-THROUGH

8.10.1 Background

As mentioned previously, the main operational rule for transmission systems is the $(N-1)$criterion. Under this criterion, no single event shall result in any loss of load. The transmission system operator will schedule sufficient reserves to ensure that the $(N-1)$ criterion is maintained. For the scheduling of reserve, it is important to know the behavior of both consumption and production during and after the event. To illustrate this, consider the system shown in Figure 8.37. An area with production and consumption is connected to the main grid by two transmission lines. The difference

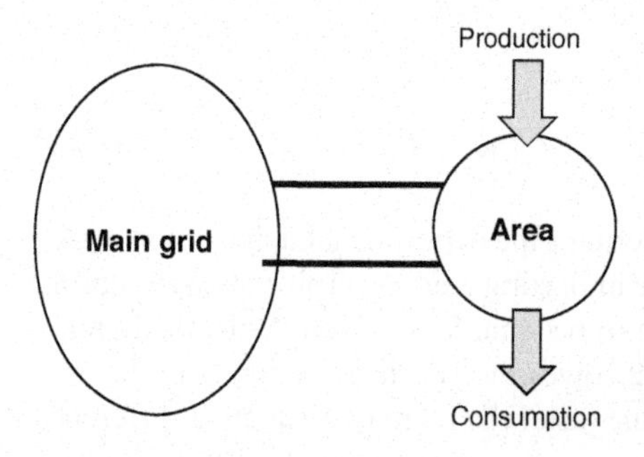

Figure 8.37 Local area with production and consumption connected to the main grid by two transmission lines.

between production and consumption within the area should be less than twice the transfer capacity of each line.

The secure transfer capacity of the transmission corridor is however only once the transfer capacity per line. The loss of one line should not result in any loss of load. The difference between production and consumption within the area should thus be less than the transfer capacity of one line. The scheduling of the production in the region should be such that this difference is not exceeded.

Example 8.20 Assume that the transfer capacity is 60 MW per transmission line and the consumption in the area is 100 MW. The secure transfer capacity of the corridor is 60 MW, so at least 40 MW of production should be available within the area.

Exercise 8.10 What is the secure transfer capacity of the corridor and the minimum amount of production in the region when the corridor in Example 8.20 consists of three lines of 40 MW each?

In practice, the balance between production and consumption is kept by means of actual production. In the case of Example 8.20, there will be 40 MW of production in operation in the area. In theory, it is possible to schedule reserve production that can be made available before the lines become overloaded. This would require a communication infrastructure to make this reserve available fast enough upon the loss of one of the lines. Note that the reserve can be made available not only as an increase in production but also as a reduction in consumption.

Exercise 8.11 Discuss whether this reserve should be considered as primary reserve or as secondary reserve. How could this reserve be made available?

In the above reasoning about the maximum difference between production and consumption in the area, it is assumed that consumption and production are not impacted by the loss of the transmission line. Here, it should be realized that the loss of a transmission line is in most cases due to a short circuit or earth fault. This fault is experienced by both production and consumption as a "voltage dip," a short reduction in voltage magnitude. Such voltage dips are known to result in tripping of large industrial installations [42, 97]. The voltage dips may equally result in tripping of distributed generation. This is partly due to the inherent sensitivity of electrical machines and especially power electronics converters and partly due to the protection requirements. An important cause for the tripping of distributed generation due to voltage dips is the anti-islanding detection. The trade-off between fault ride-through and anti-islanding protection is discussed in Section 7.7.4.

To prevent overloading of the second line after the loss of the other one, extra production should be scheduled to accommodate for tripping of distributed generation due to the fault. In the worst case, that is, when not relying on any fault ride-through for the distributed generation, the amount of conventional production needed is not impacted by the amount of distributed generation. This would make the system more secure as the transport through the transmission corridor would become less. It would however result in an increase in the operational costs, unless the electricity market would result in sufficient production within the region in any case.

Most transmission system operators put strict requirements on the fault ride-through of production units. Currently, this is limited to larger units, especially large wind parks, but there is also a growing trend toward putting fault ride-through requirements on smaller units. The fault ride-through requirements cover immunity against undervoltages, as well as against underfrequency or overfrequency. Sometimes requirements on overvoltage are also in place. Some examples of existing fault ride-through requirements will be given in Section 8.10.3.

A study after the impact of 3500 and 7000 MW wind power on the Norwegian transmission system is presented in Ref. 463. Based on existing requirements for fault ride-through, the amount of wind power expected to trip is calculated for a large number of faults in the transmission system. For 3500 MW of wind power, up to 800 MW is expected to trip. For 7000 MW, the loss of production due to tripping of wind power can be up to 1623 MW. This exceeds the size of the primary reserve in the Nordic synchronous system (1200 MW), so a large blackout could result. It was shown in the same study that the amount of wind power tripping would be reduced to at most 1050 MW when the wind parks would contribute to the short-circuit current. It should be further noted that the study considered an exceptional operational state and that the amount of primary reserve present in the Nordic system is typically more than the minimum requirement of 1200 MW because of the presence of large amounts of hydropower.

8.10.2 Historical Cases

Although the emphasis in the different discussions is very strongly on "undervoltage ride-through," the actual cases of massive tripping are related to underfrequency and overfrequency. Industrial installations in Great Britain have for years been reporting tripping of their CHP installations due to the loss of the HVDC link with France. The rate of change of frequency due to this is sufficient to make the ROCOF relay trip the generator. This was seen as an inconvenience for the industrial installation, similar to the tripping due to voltage dips. It was not seen as a concern for the transmission system operator because the amount of production involved was small. No such problems were reported in the Netherlands or in Denmark, two countries with large amounts of CHP in their agricultural section. The reason for this is that the frequency deviations in the European synchronous system are much smaller than that in the British system.

The earliest reports on the adverse impact of tripping of wind power installations come from small island systems. During a 3-year period, 33 cases were recorded in which wind power installations on Crete tripped on undervoltage. A blackout was caused by the tripping of many wind power installations on underfrequency [336].

The first well-documented case of massive tripping occurred in association with the Italian blackout of September 28, 2003. The blackout was initiated by the loss of two transmission lines between Italy and Switzerland. This resulted in overloading of the other international connections followed by the island operation of the complete Italian system. As this system contained a severe shortage of generation, the frequency started to drop quickly. When the frequency reached 49 Hz, a total of 1700 MW of distributed generation tripped on underfrequency. This further increased the shortage of production. The frequency dropped further and almost the whole country experienced a blackout [414]. This case has been discussed a lot as an example of the unwanted tripping of distributed generation making things worse for the transmission system operator. However, in this case, the blackout would likely still have occurred, even if the distributed generation would have remained online. Also, the distributed generation did perform exactly as required for the anti-islanding protection. Several conventional power stations did trip before they were allowed to trip.

A second case took place a few years later, on November 4, 2006. The overloading of a transmission line in Northern Germany resulted in a cascade effect that ended with the splitting of the European synchronous system in three parts. This case is also discussed in Example 8.6. Owing to the splitting of the system the northern part had a surplus of generation, resulting in a fast rise in frequency. As a result, the wind power and other distributed generation in Northern Germany and Denmark tripped on overfrequency. This contributed to the stabilization of the frequency in this part of the system. However, when these production units reconnected, a new local surplus occurred that almost resulted a further splitting of the system.

At the same time, there was a serious shortage of production in the south western part of the system, resulting in a fast drop in frequency. As during the Italian blackout, all distributed generation tripped when the frequency reached 49 Hz. (Tripping of industrial CHP resulted in several long production stoppages.) The total amount of production tripped was 10,800 MW. The initial shortage of production in this part of the grid was about 9000 MW; however, a total of 17,000 MW of load was tripped by the underfrequency load shedding, corresponding to about 15 million households [415]. It is not possible to exactly contribute load shedding to either the system splitting or the loss of production by distributed generation. But a quick look at the figures justifies the rough estimation that about half of the 15 million households lost their electricity due to the tripping of distributed generation. Here, it should again be clearly emphasized that this is not by itself a consequence of the presence of distributed generation. It is due to a protection setting, as required by the distribution network operator, that has negative consequences at transmission level.

8.10.3 Immunity Requirements

To prevent the mass tripping of production units during events in the transmission system, most transmission system operators put strict requirements on the immunity

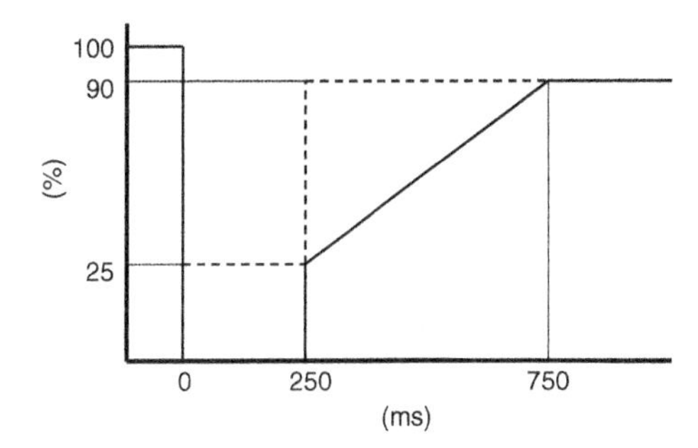

Figure 8.38 Undervoltage requirements in Sweden for production units larger that 100 MW (solid line) and for production units between 1.5 and 100 MW (dashed line).

of production units against frequencies and voltages that deviate from normal. In most cases, the requirements hold for large installations, but smaller and smaller production units are exposed to these requirements. We will give some examples of these requirements below, as they are in place in different countries. The reader should be aware, however, that these are merely examples of the different types of requirements that are in place. The text is not aimed as a benchmarking between countries, nor to allow the readers to know the requirements for their country. The reader should also be aware that the rules are very much under development and that they may well have changed between the time of writing and the time of publication of this book. In all cases, information should be obtained from the transmission system operator when the exact fault ride-through requirements are needed.

The undervoltage immunity requirements in Sweden are shown in Figure 8.38. For units larger than 100 MW, zero voltage should be tolerated for at least 250 ms. Smaller units (1.5–100 MW) only have to tolerate 25% voltage during 250 ms. The slope between 25%, 250 ms and 90%, 750 ms represents the slow voltage recovery after a fault due to dynamic effects. Strictly speaking, this is not the same kind of curve as the voltage-dip immunity curves used in power quality studies. In the power quality field, the immunity is given by the "voltage-tolerance curve" or "voltage dip immunity curves" [97, 218]. This curve connects rectangular dips with different duration and magnitude. The immunity curves used by the Swedish transmission system operator gives the worst-case (nonrectangular or rectangular) voltage dip for which the generator should not trip.

Placing less strict requirements on smaller units than on larger units is common. However, some transmission system operators place different requirements on units depending on the voltage level at which they are connected, for example, in Norway, as shown in Figure 8.39. Above 200 kV, the production unit has to tolerate zero voltage for 200 ms, whereas at lower voltage levels, the unit is allowed to trip when the voltage drops below 15%. Units connected at lower voltage levels, however, have to tolerate longer but more shallow dips. This reflects the typically longer fault-clearing time at lower voltage levels compared to higher voltage levels.

The Danish transmission system operator places different requirements on different types and sizes of production units. The following categories are distinguished:

- All generators with a size up to 11 kW [137]
- Thermal generators larger than 11 kW but smaller than 1.5 MW [139]

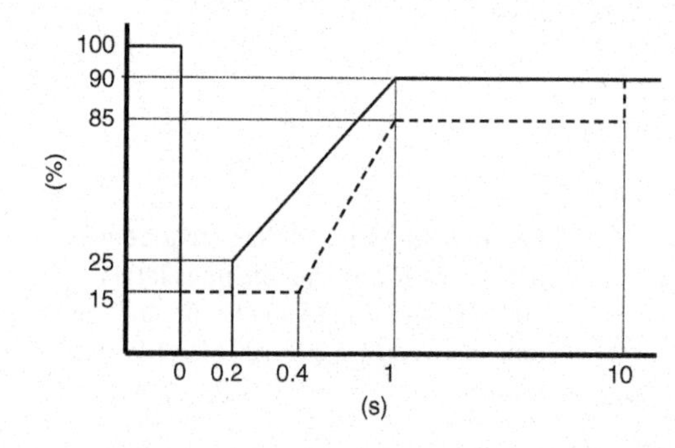

Figure 8.39 Undervoltage requirements in Norway for production units connected at voltage levels above 200 kV (solid line) and for production units connected at voltage levels below 200 kV (dashed line).

- Thermal generators with a size of 1.5 MW or above [138]
- Wind turbines connected to a voltage level less than 100 kV, larger than 11 kW but smaller than 1.5 MW [141]
- Wind turbines connected to a voltage level less than 100 kV, with a size of 1.5 MW or above [141]
- Wind turbines connected to a voltage level of 100 kV or above [140].

For thermal units larger than 200 kW, the generator should remain connected for a voltage drop down to 50% in three phases during 1 s, and a drop down to zero of one phase-to-ground voltage during 1 s [139]. For frequency deviations, any thermal unit larger than 11 kW should remain connected for frequencies between 49.0 and 50.5 Hz, and for 30 min when the frequency is between 47.5 and 51.0 Hz. Some reduction in production is allowed for frequencies below 49.0 Hz. The unit further should remain connected for rate of change in frequency between −2.5 and +2.5 Hz/s.

For voltages above 100 kV, different requirements are in place for single-phase and three-phase faults. The immunity curves for three-phase and single-phase faults are shown in Figures 8.40 and 8.41, respectively. In both cases, the initial voltage can be anywhere in the normal operating range of the voltage magnitude and the voltage after the dip can be on the lower limit of the normal operating range.

For three-phase faults, different requirements are in place in the east of the country (which is part of the Nordic synchronous system) and the west of the country

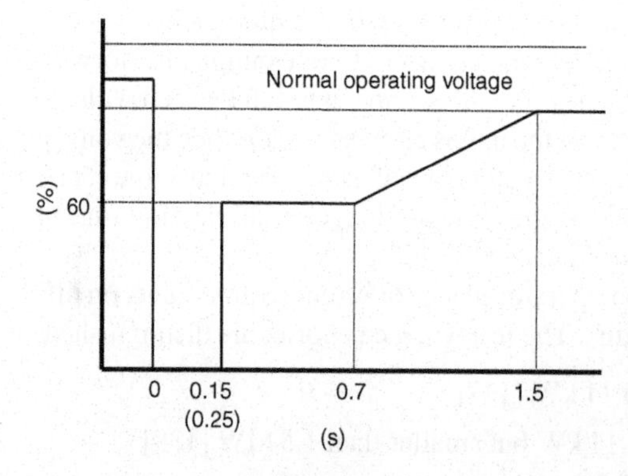

Figure 8.40 Undervoltage requirements in Denmark for thermal power stations of 1.5 MW or larger and connected above 100 kV: three-phase faults.

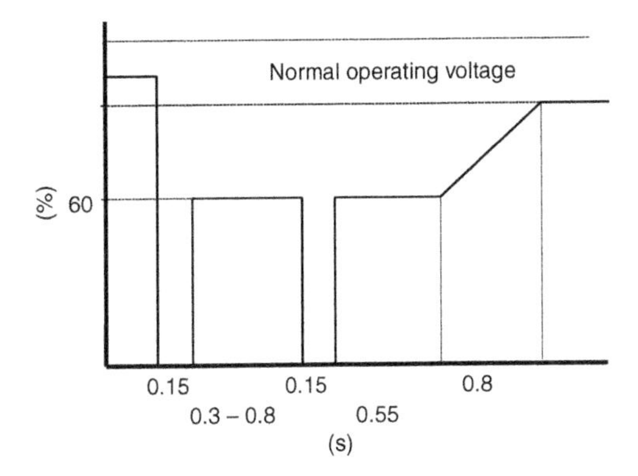

Figure 8.41 Undervoltage requirements in Denmark for thermal power stations of 1.5 MW or larger and connected above 100 kV: single-phase faults.

(which is part of the European synchronous system): in the east of the country, the generator should be able to remain connected when the voltage is zero for 250 ms; in the west of the country, a ride-through of 150 ms is sufficient. For three-phase faults, the generator has to tolerate only one dip; for single-phase faults, two dips 300–800 ms separated in time, should be tolerated.

For thermal generators connected at voltage levels below 100 kV, the following requirements hold: 50% voltage during 1 s in all three phases and zero voltage in one phase during 1 s.

For wind turbines connected to a voltage level less than 100 kV, the requirements as given in Figure 8.42 hold. The figure shows both the immunity requirements (for which the turbine should remain connected) and the protection requirements (for which the turbine should be disconnected). There is an area between region I and region II for which the turbine may trip or remain connected. There are no rules in this region. For overvoltages, there is no region in between region I and region II: the turbine has to trip or remain connected.

A concern often expressed by manufacturers of distributed generation is the wide range of rules and requirements between countries. Different rules between countries regularly cause misunderstandings and result in differences in interpretation of the rules. All this may seriously delay the connection of the installation. The main concern for the manufacturer of generator equipment is with smaller production units,

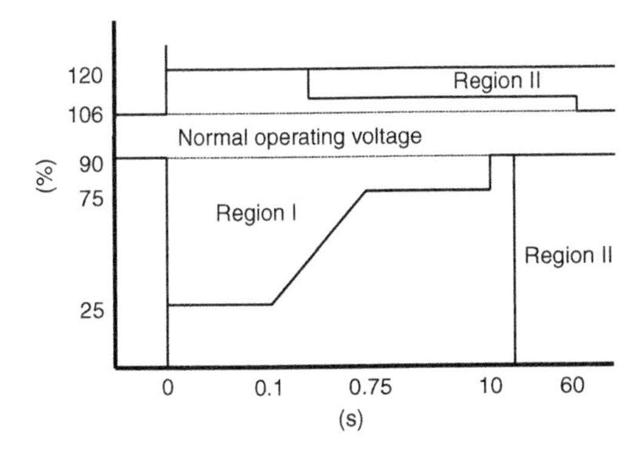

Figure 8.42 Immunity and protection requirements for wind turbines in Denmark. Region I: should not disconnect; region II: should disconnect.

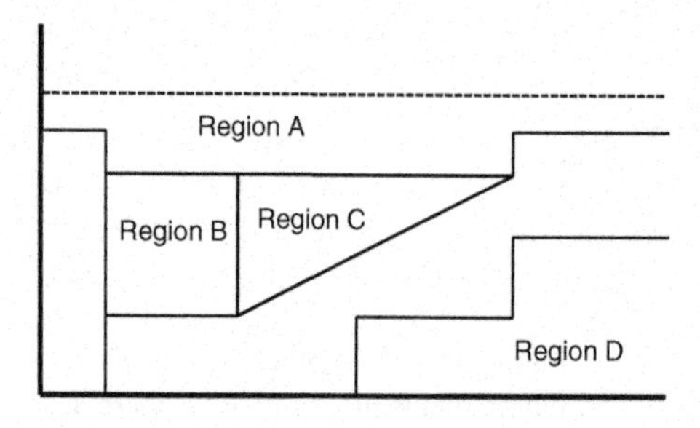

Figure 8.43 Proposed set of immunity and protection requirements by DERlab [324].

where mass production is important to keep the price low. Not only are the rules different between countries, but they also change with time, typically become more strict with time. This concern is expressed among others in the white book published by the DERlab consortium, a group of research laboratories that aim to set up harmonized standards and testing methods for distributed generation [324]. As yet there are no immunity requirements for such small units, but there are protection requirements. To ensure harmonization of protection and future immunity requirements, it is proposed in Ref. 324 to define a number of regions in the voltage-duration plot, as shown in Figure 8.43.

- Region A is the normal operation range, where the generator performance should not be impacted.
- For dips in region B, the generator should remain connected, but there are no requirements on active or reactive power flow.
- For dips in region C, the generator should remain connected and supply a defined current. Requirements might be needed both for the active and reactive parts of the current.
- Region D is the range where the generator should be actively removed from the system as part of the protection requirements.

As mentioned previously, the only well-documented cases of mass tripping of distributed generation were due to frequency deviations. The requirements on immunity against frequency deviations are, however, less complex than for voltage dips. Some of the rules imposed by the Danish transmission system operator are shown in Figure 8.44. The requirements are given in a frequency–voltage diagram; for different combinations of voltage and frequency, the installation must remain connected for different lengths of time.

The following regions are distinguished in Figure 8.44:

- Region A is the normal operation region: the installation should operate as intended whenever the frequency is between 49 and 51 Hz and the voltage between 90% and 106% of nominal.
- In region B, underfrequency between 48 and 49 Hz and normal voltage, the installation should remain connected for 25 min. This is when a serious shortage of production occurs and when underfrequency load shedding is initiated to

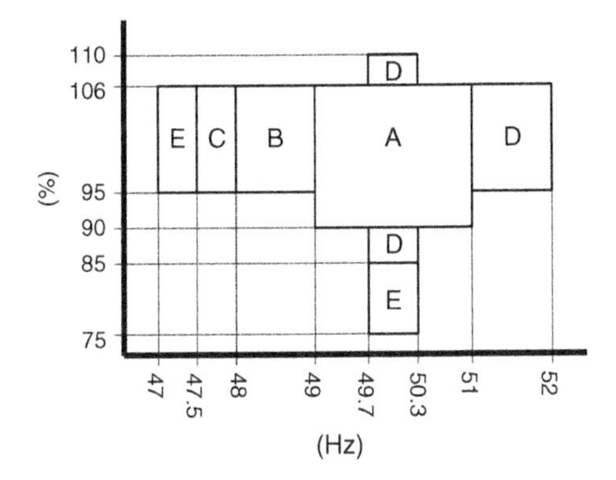

Figure 8.44 Requirements for frequency and voltage ride-through: wind power installations connected at voltages below 100 kV in Denmark.

bring back the balance between production and consumption; loss of additional production would cause additional consumption to be disconnected.

- Region C, underfrequency between 48 and 47.5 Hz, should be tolerated for 5 min. This is a very serious situation that should rarely occur and last only for a short period.
- Region D consists of three parts: overfrequency for normal voltages, mild undervoltages, and mild overvoltages for normal frequency. Here, the requirement is that the installation should remain connected for 1 min.
- In region E, the requirement is to stay connected for only 10 s. This is severe underfrequency at normal voltage or severe undervoltage at normal frequency. In both cases, the system is heavily disturbed and operation of this will not be possible for longer than a few seconds.

There are no requirements for severe deviations in voltage and frequency at the same time. This corresponds to the underlying philosophy of the $(N-1)$ criterion, where not more than one serious event is assumed to occur at the same time.

8.10.4 Achieving Fault Ride-Through

Achieving fault ride-through has very much become a task of the owner of the generation facility, with the transmission system operator setting the requirements. Most developments are directed toward achieving the required immunity for each individual generator unit. This is however not the only possible solution. System-based solutions can be used to maintain the voltage close to the generator units during a fault in the transmission grid. It should be noted here that the immunity requirements should hold for the connection point to the transmission system. The sudden loss of large amounts of generation is a transmission system issue. A fault at distribution level, electrically much closer to the generators, will impact only a small number of generators. Their tripping will not have any impact on the operation of the transmission system. What matters is the behavior of distributed generators for a voltage dip occurring at the connection point with the transmission system. Also, for large wind parks connected to the transmission system, the voltage dip at the connection point should be considered. It is not obvious which voltage level should be considered as the connection point for the purpose of fault ride-through: just the highest voltage level (400 kV in Europe), also subtransmission levels such as 110, 130, or 150 kV, or even the higher distribution voltages such as 50 or 70 kV. What matters here is the impact that the loss of the distributed generation will have on the operation of the transmission or subtransmission system. For the frequency stability (balance between production and consumption), it is unlikely that a fault at subtransmission level would make a serious impact. Note that the primary reserve should at least be equal to the size of the largest production unit, over 1000 MW in most systems. Any loss of production less than the size of the largest production unit will not endanger the frequency stability. However, there may be reasons, such as maintaining reserve of transport capacity at subtransmission level, why massive tripping at lower levels due to a fault in the subtransmission grid is not acceptable.

When a fault occurs in the subtransmission grid, for example at 132 kV, there are at least two transformers between the fault and a distributed generator unit. The voltage at the terminals of the generator is determined by the contribution of the generator to the fault current and by the impedance between the generator and the fault. As discussed in Chapter 7, the contribution of the generator to the fault depends strongly on the type of generator. The contribution is biggest for synchronous machines.

Synchronous machines keep up the voltage in the distribution system during a fault in the transmission system, as discussed in Sections 6.6.1 and 6.6.2. As discussed in these sections, the impact can be significant for large penetration of distributed generation, that is, when it matters most. This is clearly an advantage of using synchronous machines. The drawback is however that synchronous machines could become unstable after the fault; it is the angular stability that matters here. The longer the fault and the lower the voltage, the higher the risk of instability. It also strongly depends on the inertia constant of the synchronous machine and the reactive power control by the machine. Conflicting information on this is given in the literature and it is not possible to draw any general conclusions. However, one conclusion can be drawn: when the machine becomes (angular) unstable, it is not possible to keep it connected to the grid. Either the machine will be damaged or it will be tripped by its protection. The latter will occur sooner, so the fault ride-through limit will always be less than the actual stability limit.

The situation with induction machines is completely different. Induction machines do maintain the voltage during a three-phase fault only for one or two cycles after fault initiation. For nonsymmetrical faults, an induction machine damps the negative-sequence voltage, thus improving the voltage somewhat (see Section 6.6.3 for more details). After the fault, the induction machine can reconnect whenever the system is sufficiently strong. Here, we enter the realm of short-term voltage stability: the machine takes a high amount of reactive power after the fault, which pulls down the voltage so much that the machine cannot transfer sufficient energy from the mechanical to the electrical side of the machine (see Section 8.6.1 for details). The behavior of induction machines is important because of their widespread use in wind power installations. Especially many of the smaller wind turbines contain induction machines. Solutions to improve fault ride-through with induction machines are all related to generating the reactive power close to the machines so that it does not result in voltage drops. In Refs. 267 and 268, a comparison is made between a switched capacitor bank, an SVC, and a STATCOM for their ability to provide fault ride-through for a wind park consisting of induction machines only. So what matters is not the speed with which the reactive power is made available, but only the amount of reactive power. It is shown that a switched capacitor bank is doing equally well as an SVC or a statcom of the same rating. Any overloading capacity of the SVC or STATCOM has not been considered so as to have a fair comparison. There are other advantages to an SVC or a STATCOM, such as their smaller size (which is an issue for offshore wind parks) and their ability to provide voltage and reactive power control during normal operation as well.

The fault ride-through of double-fed induction generators (DFIG) depends strongly on their protection and control. The commonly used way of protecting the power electronics in the rotor circuit is by means of a crowbar switch that instantly results in a normal induction machine operating at high slip. This gives a high reactive power flow that causes the protection of the machine to disconnect it. The fault ride-through of such machines is very limited. More recently built machines use alternative methods of protection and control that allow better fault ride-through. When the turbine remains connected, the converter part can be used to provide reactive power support to the grid after the fault.

An increasing part of the wind turbines being installed are equipped with a full power converter. The immunity issues of these are very similar to the immunity issues of converters used for powering end-user equipment, such as adjustable-speed drives: when the voltage is too low, the power electronic circuits stop functioning properly; during a low voltage, the current has to increase to maintain the power flow; unbalanced dips can cause a high ripple in the DC bus voltage; upon voltage recovery, overcurrents and overvoltages can occur damaging the power electronics. There are two ways of preventing damage to the power electronics: tripping the device on undervoltage or overcurrent, or overrating the components so that they are not damaged. The former approach is most typically used for end-user equipment where only very mild immunity requirements exist. Using overcurrent protection gives better fault ride-through than using undervoltage protection, but the latter provides more safety against equipment damage. For wind turbines where fault ride-through is an issue, overrating of the power electronic components is the method of choice. Next to that, many different power electronic control methods are being developed to improve fault ride-through. Also, the large mechanical forces during and after the fault should be kept under control.

The current research trend is clearly toward improving fault ride-through for individual turbines. There are however possibilities to limit the voltage drop at the terminals of the generator by injecting reactive power at strategic locations in the distribution system and in the collection grid of a wind park.

8.11 STORAGE

Most discussions on power system operation start with the statement that electrical energy cannot be stored. This is however a simplification of the actual physical situation. As discussed in Section 8.7, the energy stored as kinetic energy of all rotating machines connected to the grid is equal to the electricity consumption for several seconds. Strictly speaking, this is not electrical energy, but it becomes available as electrical energy and it plays an essential role in keeping the grid stable, as observed in the discussions on angular and frequency stability in Sections 8.8 and 8.9. Energy is also stored in small batteries, which enables off-grid applications such as mobile phones and laptops. One may say that even the water in the reservoirs with hydropower generators is a form of energy storage. This is not typically seen as storage, although "pumped hydro" is referred to as storage.

Recently, the discussion on storage has been taken up again, as can be seen, for example, from a special issue of the IEEE Power and Energy Magazine (July/August 2009) [161, 326, 367]. To enable a transparent discussion on storage, from the viewpoint of a distribution network or transmission system operator, it is important to distinguish between different types of "storage." Earlier in this chapter, we discussed about different types of "reserve," and storage is in fact one of the forms of reserve. What matters to the system operator is not the form in which the energy is stored, but the amount of energy that can be made available, how fast it can be made available, and for how long it can be made available. For the system operator, it actually does not matter whether the energy is stored in an oil tank, in a fast rotating flywheel, or as water behind a dam.

Although there is no difference from a power system operation viewpoint, there may be difference from an economic viewpoint and from an environmental viewpoint. By temporarily storing electrical energy in another form and later transforming it back again, money and emission can be saved. By temporarily storing a surplus of wind or solar power, more energy from renewable sources can be integrated in the power system. By using energy stored in flywheels as spinning reserve, renewable sources can be used to their full capacity and thermal units can operated at their highest efficiency. Storing energy close to the consumption for peak shaving could be a cheap alternative against upgrading a distribution line.

A number of storage technologies are seriously discussed for power system applications. In the transmission system, the only existing technologies that are of sufficient size to be considered are pumped hydro and compressed air. Pumped hydroelectric motors are used to pump energy from a lower reservoir to a higher reservoir. At a later moment in time, electricity is generated in the same way as in a conventional hydropower installation. Pumped hydro has been around for a very long time, but it has seen a significant development over the last few decades. Modern installations have a single machine that can operate as generator or motor/pump. Also, adjustable-speed installations are being used that allow even more flexibility including frequency regulation when operating as a pump. A list of pumped storage installations around the world presented in Ref. 449 shows that installations exceeding 1000 MW in size are not uncommon. The largest pumped storage installation in the list is a 2700 MW unit built in Japan in 2005. The efficiency of pumped storage installations is between 70% and 85% [130, 449], which means that 15–30% of the energy is lost during the two conversion processes.

Pumped hydro can be made available very fast and for several hours. For example, the Dinorwig pumped storage installation in Wales, UK, can supply about

2000 MW for 5 h [225]. Another large installation, the 1500 MW Raccoon Mountain installation in Tennessee, has energy available for 28 h [402]. According to Ref. 367, pumped storage installations are available in the range from 500 to 3000 MW and their power is available for 10–100 h.

When running idle, the production of the Dinorwig plant can be increased from zero to 1320 MW in 12 s [225], which implies that this installation alone can provide a large part of the primary reserve of a synchronous system.

Exercise 8.12 Would it, under the $(N - 1)$ criterion, be allowed to have all primary reserves concentrated in one unit?

Compressed air storage is an alternative method for temporarily storing energy. By adding compressed air to a gas turbine, its output power can be doubled for the same amount of gas use. The energy stored in the compressed air accounts for the increase in output power. Installations of 290 and 110 MW were built in 1978 in Germany and 1991 in Alabama, United States, respectively. According to Ref. 367, EPRI is developing an installation that can provide 150–400 MW for up to 10 h using underground storage at 100 bar. A 2700 MW installation under development is mentioned in Refs. 161 and 446. Development is reportedly also underway for smaller units, up to 15 MW with the compressed air being stored in pipes. According to Ref. 367, a compressed air plant has a starting time of about 15 min. This would make it not suitable as a primary reserve but very good to restore the reserve after a large outage, where 15 min is often used as an acceptable timescale. Also, for balancing purposes, this is an acceptable timescale. The main problem with compressed air as it is used at the moment is its rather low efficiency. During the compression of the air, it becomes too hot to be stored. The air should be cooled and this energy should be added again later to allow it to be used in a gas turbine. The Alabama installation has a somewhat higher efficiency than the German installation because some of the heat is reused. Currently, research is taking place toward the so-called "adiabatic compression," which has in theory an efficiency close to 100.

The need for fossil fuel, in the form of natural gas, to reheat the compressed air is seen as a disadvantage of compressed air energy storage. In the existing situation, where not all energy can be supplied from renewable sources, the net gain is however clear. The energy will be stored when there is a surplus of wind energy and used when there is a shortage of wind energy. The use of temporary storage will increase the total energy produced by existing wind power installations and make it more economically attractive to build new installations.

Example 8.21 A 600 MW wind park at a location with an average wind speed of 7.5 m/s has the capacity to produce about 1.4 TWh/year (2800 h). But because of limited transfer capacity, 500 h of maximum production cannot be used, resulting in 50,000 MWh of energy not being used. Storing this as compressed air, and adding 10,000 MWh equivalent of natural gas, will produce 40,000 MWh of electricity. As this electricity is produced during periods of low wind speed, the saving in natural gas is 30,000 MWh/year, so the spilled wind has been reduced from 50,000 to 20,000 MWh/year.

This may seem a large amount of energy. However, the storage installation increases the total production of the wind park by only 2.1%. If the park consists of 3 MW turbines, 2.1% would correspond to four turbines. Adding four turbines would not increase the production by 2.1% because the spilled wind would also increase, but adding, for example, six might do. The costs of a compressed air storage installation will have to be compared with the costs of six additional wind turbines. There are obviously other advantages with storage, such as more efficient participation in the electricity market, more accurate prediction of the net power injection into the grid, and the ability to offer reserve power, which might make the investment worthwhile. However, purely from an emission viewpoint, storage might not be worth the investment.

Two other types of storage are discussed a lot recently: battery storage and flywheels. Both have been used for many years as a backup supply for important loads in case of interruptions or voltage dips in the supply. However, developments in both and the need for reserves may make their application area wider. The main applications under development are in the transport sector, for example, for regenerative braking of trains. High-speed flywheels are also being considered as a source of primary reserve. A 1 MW installation consisting of 10 flywheels is being built and is expected to come in operation soon. The flywheels will be able to supply power to the grid for 15 min [326, 367]. The start-up time of such an installation is very short, probably of the order of seconds. However, a certain amount of energy will always be needed for the flywheel to keep on running, so that it is most efficient for applications with a short storage duration. Also, the rating of the installations is on the small size to make them attractive as a source of primary reserve. Several hundreds of the 1 MW installations would be needed to make a significant contribution.

Battery storage is the most commonly used method for temporarily storing electrical energy. The current trend is toward smaller batteries with a higher energy density. Modern lithium-ion batteries store three to five times as much energy per kilogram as the classical lead-acid batteries. Within power engineering, the trend is toward larger installations, as backup for the supply to industrial installations, as a temporary storage of energy from wind power installations, and as a source of primary reserve. The main technology being developed for large power system is the sodium-sulfur (NaS) flow battery. This is a high-temperature battery that makes it less suitable for applications with domestic commercial customers. Also, installations based on nickel-cadmium (NiCd) batteries are under development.

An installation based on NiCd batteries operating in Fairbanks, Alaska, is able to supply 26 MW during 15 min or 40 MW during 7 min [2, 367] and a 34 MW, 245 MWh installation for wind power stabilization is in operation in Northern Japan [367]. Other large installations are a 12 MW, 120 MWh installation in the United Kingdom, a 1 MW, 7.2 MWh installation using NaS batteries by American Electric Power (AEP), and three installations of 2 MW, 14.4 MWh at three different locations, also by AEP [326]. According to Ref. 326 the NaS installation is about 80% efficient at the point of connection.

A recent development mentioned in Ref. 368 is the zinc-bromide (ZnB) flow battery. A 500 kW, 1000 kWh and a 2.8 MWh installation are mentioned. The advantage of this type of installation would be that it can be relatively easily transported, so it can be placed at the location where it has the biggest impact. Congestion at medium-voltage levels seems to be the application considering the sizes mentioned.

This short overview shows that installations supplying tens of magawatts of power during a few hours are technically possible. These kind of installations can become available in a matter of seconds, so they can serve as primary reserve. But as they can be available for several hours, they can replace even gas turbines. No clear information is available on the economics of these installations, but it obviously depends strongly on the details of the electricity market and the differences between the highest and lowest electricity prices.

Thermal storage at consumption side is also discussed as a reserve. "Modular ice-storage systems can produce ice during off-peak periods to power air conditioning during several hours" [367]. "In Europe, a very large thermal storage system (up to 10,000 MWh) is being proposed" [367].

The impact of a large penetration of renewable energy (30% wind and 5% solar) on the transmission system of Arizona, Colarado, Nevada, New Mexico, and Wyoming was studied. Some of the results are presented in Ref. 103. It was, among others, shown that pumped storage use increases with increased penetration of renewable energy, but even for 35% penetration, no need for additional pumped storage could be shown.

8.12 HVDC AND FACTS

Power electronic controllers can be used to impact the flow of active and reactive powers in the transmission system. Several controllers are available for this. The most commonly used ones are the synchronous condenser, the phase shifting transformer, and the static var compensator. The synchronous condenser and the static var compensator inject reactive power and can, for example, be used to enable more power transfer without causing voltage collapse. The phase shifting transformer can be directly used to control active power flows and thus prevent overloading of lines. It does not change the transfer capacity of any specific line, but it allows the power to be more equally spread between different lines. Phase shifting transformers have been used in Great Britain for many years now to control the power flow from Scotland and the north of England to the demand centers in the south. Phase shifting transformers are also being discussed on the border between Germany and the Netherlands to control the transport of wind power from Northern Germany to the west.

When it comes to transferring active power in a controlled way, there is nothing that beats the HVDC link. The power flow through the links can be fully controlled independent of the flows elsewhere. The main disadvantages of HVDC have been its high costs and the lack of experience of transmission system operators with its integration in the AC system. The majority of HVDC lines in operation connect two synchronous systems. For example, the Nordic system is connected to the European system by means of eight links [447], with more links being discussed. Several links are also being discussed between Northern Africa and Southern Europe. HVDC links embedded in an AC system are less common. The 3100 MW Pacific intertie between Oregon and Los Angeles is a classical example [448]. The two examples in Europe are the link between Sweden and Finland and the link between Italy and Greece. Both provide a shortcut through the sea between locations that are weakly linked through the AC grid. A fully embedded three-terminal link is planned between the south of Sweden, the middle of Sweden, and the south of Norway. Two-terminal links are studied between Belgium and Germany, between France and Spain, and between Scotland and England. The Norwegian transmission system operator Statnett is studying a multiterminal HVDC link connecting a number of oil platforms and the onshore transmission system. Future offshore wind parks could connect to the DC nodes at the oil platforms.

The economics of HVDC and AC connections to offshore wind parks are compared in Refs. 65 and 66. The conclusion is that HVDC is more economical for connecting a 100 MW wind park when the distance is more than about 70 km.

The use of HVDC transmission is proposed by several studies as an appropriate technology for integrating more wind power into the power system. It allows the power to be transported in large amounts over long distances, without endangering the stability of the system.

According to a study covering the eastern part of the United States, HVDC connections in combination with large AC "collector grids" are needed to transfer the energy from wind power in the Midwest to the cities on the east coast [103]. In this scenario, over 200,000 MW would be installed, more than half of this in the Midwest.

In Ref. 237, the collection grid is integrated with the HVDC link to obtain an HVDC grid. Groups of four turbines or 2 MW each are connected to one AC–DC converter.

The term "HVDC" refers to a system dedicated to transfer electric power between two AC systems through a DC transmission link by using two power electronic converters, one at each end of the link. It is basically implemented for power trading between AC systems with different frequencies or voltage levels, or asynchronous grids of the same nominal frequency. Also, the technology has found more advantages over the more simple AC cable connection for similar AC systems: there is theoretically no limit on cable length due to the cable capacitance [100]. Hence, HVDC is usually promoted for long-distance transmission, where a curve as shown in Figure 8.45 is

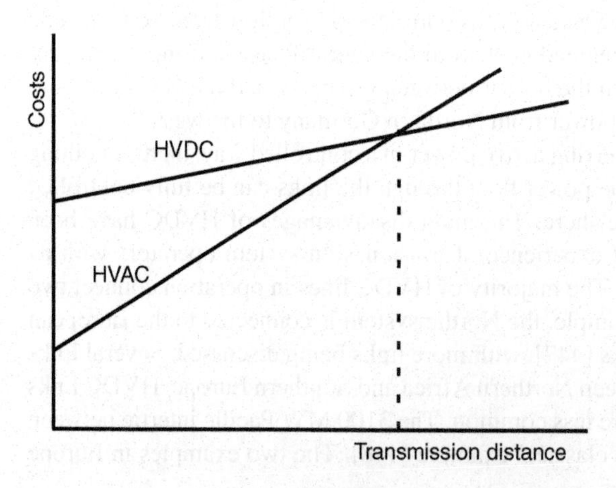

Figure 8.45 Comparison between HVDC and HVAC regarding costs and transmission distance.

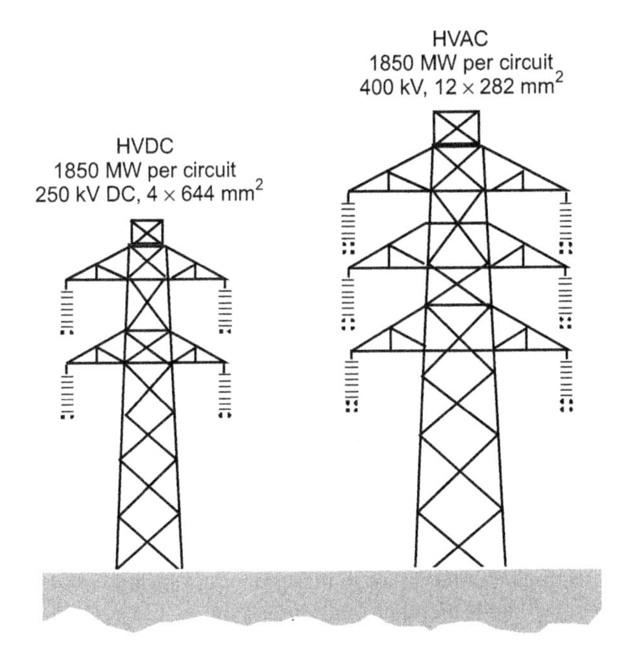

Figure 8.46 Comparison between the towers for DC and AC transmission.

usually used to compare the two solutions for a specific system and determine the break-even distance. The break-even distance is usually calculated as about 600 km for overhead lines and about 40 km for submarine and underground cable links. Moreover, if a monopole configuration is used, the number of wires will be reduced from three to two compared to HVAC transmission [162], as shown in Figure 8.46.

Two different technologies of HVDC are in use: classical HVDC using line commutated converters (LCC-HVDC) and VSC-HVDC using self-commutated converters. The LCC-HVDC technology utilizes thyristors as shown in Figure 8.52. The thyristors use the AC voltage for turn-on but do not have a turnoff capability. As shown in Figure 8.47, if the gate of the thyristor receives a firing pulse at an angle α (measured from the zero crossing of the AC voltage), the thyristor starts conducting. In the figure, a resistive load is assumed, and hence the current zero crossing coincides with the voltage zero crossing and the thyristor accordingly stops conducting at the zero crossing of the voltage. It is also obvious from the figure that the operation of line commutated thyristors inherently produces lagging power factor, since the current starts flowing at α and hence draws reactive power from the grid. The average output voltage of the thyristor is controlled through varying the angle α. Thyristors are

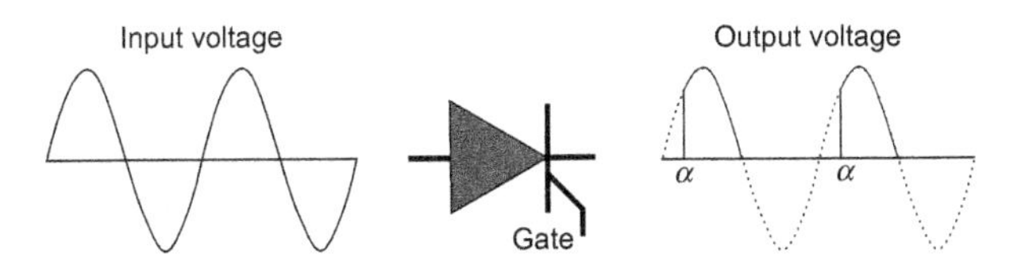

Figure 8.47 Thyristor basic operation: current is assumed in phase with voltage.

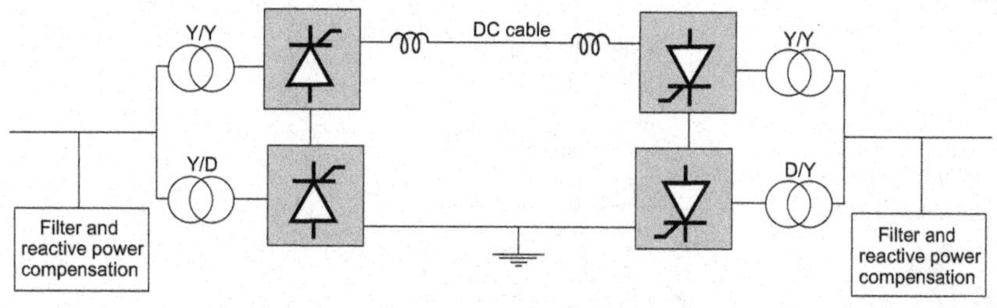

Figure 8.48 12 pulse LCC-HVDC (monopole).

normally turned off at zero current, and hence extra circuits (commutating circuits) may be incorporated to control the zero crossing of the current.

The LCC-HVDC is usually implemented in a 12-pulse configuration, as shown in Figure 8.48, in order to reduce the current harmonics generated at the AC side and the DC voltage ripples. Also, filters and reactive power compensation units are utilized at the AC sides to improve the power quality. The use of thyristors for LCC-HVDC provides the capability of implementation for higher power and voltage levels and also lower losses. On the other hand, the controllability of the LCC-HVDC is limited due to its dependence on the grid voltage.

The VSC-HVDC technology, on the other hand, utilizes IGBT devices as the main switching element, as shown in Figure 8.51. With thc capability of IGBT devices to turn-on and turn off on current regardless of the grid AC voltage, they inherently provide more operational flexibility. Hence, the VSC provides the independent controllability of both active and reactive powers. A typical curve that indicates the control area of the VSC-HVDC is shown in Figure 8.49, where the main limitation is imposed by the amplitude of the DC link voltage compared to the AC terminal voltage. Increased terminal voltage may cause the converter to saturate, and hence the output voltage of the converter should not increase accordingly in order to protect the converter from saturation, which means limiting the reactive power capability in this case.

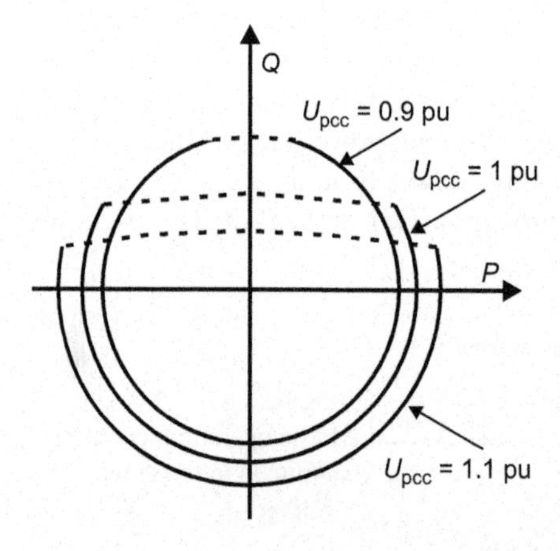

Figure 8.49 A typical *PQ* curve for VSC-HVDC, where the dashed line represents the DC link voltage limitation (also referred to as saturation limit) and U_{pcc} is the AC terminal voltage.

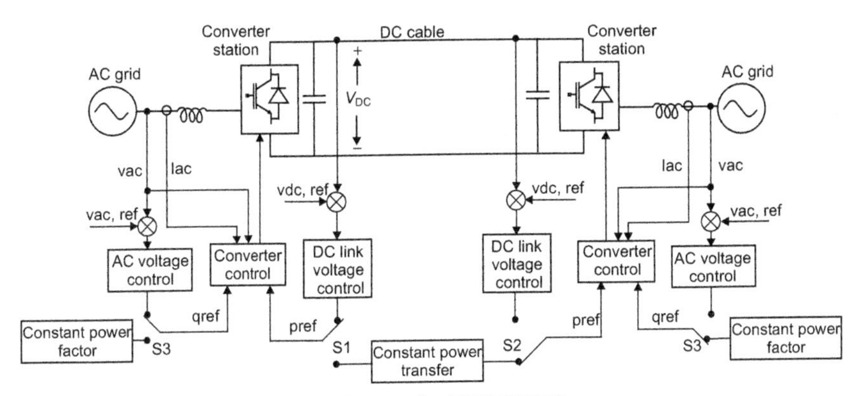

Figure 8.50 Main control block diagram for VSC-HVDC.

The overall control of the VSC-HVDC is shown in Figure 8.50. The switch S3 has basically two positions to achieve either AC voltage control or constant power factor operation at either AC side of the link. The switches S1 and S2 have opposite positions to achieve DC link voltage control using the converter station on one side and the converter station on the other side to control the power flow. The main converter controller at either station is a typical current controller as explained in Chapter 2.

Currently, the highest ratings for LCC-HVDC are 3000 MW transmitted power and ± 600 kV DC, whereas for VSC-HVDC are 550 MW transmitted power and ± 150 kV DC. These maximum ratings are likely to grow in the near future. The main differences, advantages, and disadvantages between the two technologies and their comparison with the HVAC transmission are summarized in Table 8.11 [100, 147, 162, 251, 298, 306, 357]. As the VSC technology for DC transmission is motivated by its wide range of controllability, black-start capability, and overall better dynamic performance, it is currently the main development focus for the industry. The main challenges are to reduce the size and cost, improve the efficiency, and increase the transmission capability (Figures 8.51 and 8.52).

The first commercial installation of VSC-HVDC took place in 1999 when two 70 km, 80 kV, and 50 MW HVDC underground cables were ploughed close to each other to connect the wind power plants in southern Gotland and the power station in Visby, Sweden. During the evaluation of the project, the SVC-HVAC (AC cable with static reactive power compensation at its connection end to the transmission grid) alternative was considered. It was found that the VSC solution was more economical, in addition to a more significant power quality improvement that was going to be introduced over the entire Gotland AC network.

The first offshore version of VSC-HVDC went into operation in the North Sea in 2005 connecting Statoil's gas platform to the grid [321]. In the same year, a double circuit, two times 40 MW subsea HVDC cable installation from Kollsnes to the Troll A platform, 67 km offshore outside Bergen in Norway, was commissioned to feed two medium-voltage compressors.

A VSC-HVDC system for Valhall will replace offshore gas turbines and deliver up to 78 MW power to run the complete complex consisting of five bridge-linked

TABLE 8.11 Comparison Between Different Transmission Options

Requirement	SVC-AC	LCC-DC	VSC-DC
System simplicity	Advantageous	Complicated hardware design	Simple
Voltage control at the connection bus	Limited	Limited	Advantageous
Independent control of active and reactive powers	Only reactive power control	2-quadrant	Advantageous (4-quadrant)
Dependence on AC system	Dependent	Dependent	Independent
Black-start capability	Yes	No	Advantageous
Multiterminal DC grid	Not applicable	Possible	Advantageous
Cable length for the same transmission capacity	Limited	Unlimited	Unlimited
Reactive power demand	Reactive losses	Commutation	Advantageous
Connection to a weak or isolated system	No	Yes, with additional hardware	Advantageous
Investment costs with same voltage level, same capacity, and rather long distance	More investment costs	Less investment costs	Advantageous
Contribution to short-circuit power	Yes	No	No
Critical fault-clearing time	Shorter	Longer	Longer
Sensitivity to DC link faults	N/A	Lower	Higher
Postfault performance	Long recovery	Shorter recovery	Advantageous
Transmission losses	Advantageous	Low losses	High losses

SVC-AC, AC transmission with SVC for reactive power compensation; LCC-DC, DC transmission using line commutated converters technology; VSC-DC, DC transmission using voltage source converters technology.

platforms [321]. This will make Valhall one of the most environment friendly fields off Norways shores. Also, this is the first time a complete offshore power system for an entire field will be fed with electric power from the mainland using DC transmission. It is evident that the use of HVDC systems is a key enabling technology for large integration of renewables without impacting the stability of the grid.

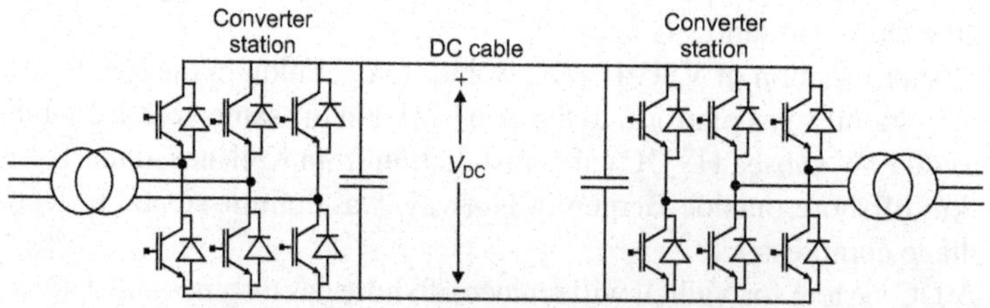

Figure 8.51 Voltage source converter (VSC) based HVDC.

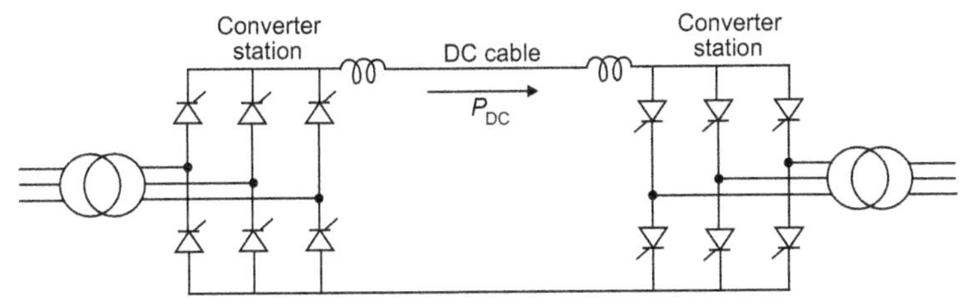

Figure 8.52 Line commutated converters (LCC) based HVDC.

8.13 INCREASING THE HOSTING CAPACITY

The shift from conventional generation of distributed generation may result in a shortage of reserves and ancillary services, while at the same time the uncertainty in prediction increases the need for reserves. Most countries have functioning reserve and balancing markets with sufficient conventional generation available for the transmission system operator to be able to buy the required reserves and other ancillary services. This will obviously increase the operational costs and thus the electricity cost for the consumers. The examples shown previously do however indicate that the increase in cost is limited even for wind power penetration of 20–30%. This does not mean that there are no serious technical challenges to be expected. For example, large prediction errors could result in an insecure or even unstable system. Massive tripping of distributed generation could trigger a blackout. Distributed generation may also otherwise behave in an unexpected way during a disturbance. In this section, we will go through some of the methods discussed or already in use for increasing the amount of distributed generation in the transmission system without endangering the quality and reliability for the end customer and without resulting in excessive costs. Here, it should be mentioned again that we also consider large wind parks connected to the transmission or subtransmission system.

Integration of distributed generation is not the only reason for investments in transmission systems. The open electricity market has also resulted in power transport over longer distances being less predictable. At the same time, the building of new power stations is no longer as uniformly spread through a country or continent as it used to be. Large power stations are being built in areas with access to, for example, cooling water and not necessarily close to the consumption centers. Finally, the growth in consumption is not followed by an equal growth in transfer capacity in the transmission system in all countries. Many of the improvements discussed next, are in many cases also a help against the other challenges that face the transmission system operator.

8.13.1 Alternative Scheduling of Reserves

The existing methods of scheduling operational reserves in synchronous systems are all based on the same fundamental rule: primary reserve is spread over the system; secondary and tertiary reserves are maintained regionally (typically by each transmission system operator). The relatively small geographical spread of some transmission systems makes them very susceptible to prediction errors, which requires large amounts of secondary and tertiary reserves.

Sharing secondary and tertiary reserves over a larger geographical area will save reserves and make the system less susceptible to prediction errors. Not only do random errors cancel when the area increases, but also will different regions experience different weather conditions. Sharing of reserves between neighboring system operators is often referred to as "enlarging the balancing area."

8.13.2 Increasing the Transfer Capacity

Spreading the reserves over a larger geographical area requires that they can be transported to the location where they are needed. This calls for an enforcement of the transmission system's transfer capacity. This can be done by building more transmission lines or installing power electronic controllers that increase the transfer capacity. In fact, one of the reasons that every country had to provide its own reserves was exactly the limited transfer capacity of the transmission system.

The TradeWind study looked at increased transfer capacity between European countries toward the integration of 300,000 MW of wind power into the European power system. This amounts to about 25% of the electricity demand. It was shown that increased transfer capacity would reduce the operational costs by 1.5 billion euros per year, justifying about 20 billion euros in investment costs [103]. It was shown in the same study that the capacity value of wind power increased from 8% for individual countries to 14% for the whole of Europe assuming sufficient transmission capacity.

Looking at this the other way around, the limited transfer capacity of the international connections results in an additional cost of wind power integration equal to about 1.5 billion euros per year. These values are often used in discussions to show that wind power will be extremely expensive. However, considering the size of the European population of about 500 million (European Union only), this amounts to about 3 euros per person per year.

In Ref. 386, reference is made to a "vision scenario" with 20% wind power penetration in the United States. Part of this scenario is a 765 kV grid covering most of the country to transport wind power from the production areas to the consumption areas. This would, according to Ref. 386, require 30,000 km of new 765 kV transmission line with a total cost of $60 billion. This is a large amount of money, but spread over the whole U.S. population (300 million), and assuming a 10-year investment period, this corresponds to $20 per person per year.

8.13.3 Large-Scale Energy Storage

The variability and unpredictability of renewable sources of energy are often presented as a serious barrier to their integration in the power transmission system. Being able to store large amounts of energy would remove this barrier. The only large-scale storage method currently available is pumped storage. Compressed air to increase the efficiency of gas turbines could become a competing technology in the future. All other methods, including hydrogen storage, are far from large-scale application. For local applications, some of the alternative methods may be feasible. Examples are mitigation of voltage variations and overloads at distribution level and to be able to place more secure bids on the electricity market.

In Europe, pumped storage already plays an important role in the operation of the transmission system. It enables a more efficient scheduling of production, so that, for example, large thermal units can be in operation for longer periods of time. At the same time, it limits the need for building additional gas turbines to cope with peaks in consumption. The widespread use of pumped storage does, however, require a strong transmission system connecting the hydro resource with both the production and the load.

In Ref. 66, a case is made for using the pumped hydro resources in Europe to balance offshore wind power. The pumped hydro resources in Europe are, according to Ref. 66, over 37,000 MW. Only the region around the Alps already contains over 20,000 MW of pumped hydro. This region is centrally located for the main consumption in Europe and also for the wind power resources to the north and west and for solar power to the south.

8.13.4 Distributed Generation as Reserve

The use of distributed generation, especially large wind parks, as a source of primary, secondary, or tertiary reserves is discussed at many occasions. There are no real technical limitations to this. In that case, the generator will produce somewhat less than its available capacity. Upon receiving a command, the production can be increased. The command can be generated automatically or manually. Automatic commands are needed for primary reserve and secondary reserve. When using the distributed generation as tertiary reserve, manual commands will be issued by the transmission system operator. Equipping large amounts of small production units with communication and control will require large investment costs that might not be worth the effort. A lot of research and development is going on at the moment to find out appropriate solutions.

Next to this economic barrier, there is a more fundamental objection to using distributed generation as reserve. For a given amount of consumption, requiring a certain generator to contribute to the reserves will simply mean that this amount of power will have to be produced elsewhere. When, for example, 5000 MW of wind power has to provide 5% of reserve, conventional generation units have to produce an additional 250 MW. In almost all industrialized countries, the marginal production is provided by thermal units, gas or coal fired. The 250 MW reserve with wind power will thus result in 250 MW of power being produced by thermal units instead of by wind power. From an environmental viewpoint, this is not a good idea and wherever possible the operational reserves should be placed with conventional units.

This does not mean that the occasional use of distributed generation as a reserve should never be allowed. When, during certain operational states, a wind park can contribute to the reserves, more wind power can be integrated, which will increase the amount of energy coming from renewable sources. The Horns Rev wind park off the Danish west coast is occasionally used for primary or secondary control. This is not so much for technical reasons but because the operational rules in the European synchronous system require each country to contribute with a certain amount of reserves.

The above discussion concerns the use of distributed generation as a reserve when there is a shortage of production. Reserves are also needed for when there is a surplus of production. Using distributed generation as a reserve for such purpose does not have any adverse impacts as long as the reserve is not used. Occasional use has only limited impact on the total energy production. Curtailment of production from distributed generation is a tool that transmission system operators can use in most countries. There are however some priorities where distributed generation is curtailed only when no other options are available. For large wind parks or large CHP installations, this can be easily implemented. For small-scale distributed generation, the costs of communication and control could become a barrier.

8.13.5 Consumption as Reserve

We mentioned previously the use of large-scale energy storage as a reserve and the use of distributed generation as reserve. However, both have their disadvantages: energy storage is always associated with losses; using distributed generation as reserve will increase the use of fossil fuels. From the viewpoint of the transmission system operator, there is no difference between an increase in production or a reduction in consumption. Instead of having 100 MW reserve available in the form of a fast-starting gas turbine, the transmission operator could have a contract with a large industrial customer to reduce the load by 100 MW. The Swedish transmission system operator has access to about 2000 MW of so-called strategic reserves that can be used when the electricity market does not result in sufficient reserves. These reserves consist partly of production units that can be started and partly of industrial installations that can be curtailed.

Much of the ongoing research and development is toward the use of control of consumption as a reserve. Some of the applications are for distribution systems and transmission system applications are also envisaged. Individual domestic customers could offer load reduction as a reserve to the transmission system operator. The probability of secondary and especially tertiary reserve being actually used is not very big, but having it available in the form of possible reduction on the consumption side would save the operation of expensive generators that are used only occasionally. In principle, all consumption could be offered as reserve, but the main interest is in consumption that has an inherent storage. The most obvious examples are heating, cooling, and ventilation. Changing the temperature setting of all airconditioning units in a country could result in a huge reduction in the consumption. When the change is sufficiently large (more than a few degrees), the consumption will drop to almost zero, while slowly recovering to a lower value than before the change. The recovery time of the load is related to the thermal time constant of the building, which is of the order of a few minutes to several hours. What matters is how fast the room cools down or heats up. A good thermal insulation not only reduces the energy consumption for heating or cooling, but also makes more reserves available for the transmission system operator.

It will not be practical for individual customers to directly become involved in the reserve or balancing markets. Several aggregation schemes are being proposed, with technical as well as economic aggregation being studied.

It is important to note that the vision is to have a large amount of reserves available in the form of a temporary reduction in consumption where the impact on the customer is limited. It will not result in any energy savings, but shift the consumption to a later moment in time. In this way, the transmission system can be operated with less reserves on the generation side or in the network. This allows for more power to be transported while accepting the risk of an occasional "noninvasive load shedding."

8.13.6 Requirements on Distributed Generation

The method currently most used by transmission system operators is to put strict requirements on production units, both large ones and small ones. Some of those requirements, especially concerning fault ride-through, are discussed in Section 8.10. Next to fault ride-through, transmission system operators place a lot more requirements, including reactive power exchange with the grid, providing information on the status of individual turbines, harmonic emission and flicker, the ability for the transmission system operator to remotely curtail the production, protection against faults in the collection grid, and limits on the increase in production with time. From a transmission system operator viewpoint, it makes good sense to set these requirements, but it does increase the costs for the wind park owner. With large wind parks, the additional costs are in most cases a limited part of total costs. There is however a trend toward setting requirements for ever smaller units, possibly including microgeneration. The costs of meeting those requirements could become a substantial part of the costs for such a unit, which in turn could form a barrier to their introduction.

An alternative approach would be to introduce markets for all ancillary services. In that case, it would be up to the owner of a production unit to offer a certain service. In Spain, wind power installations that comply with the fault ride-through requirements get payed a higher price for the produced power than the ones that do not comply. Also, units that do not comply are more likely to be curtailed when the security of the transmission system is endangered. However, the disadvantage of any market mechanism is that it could result in a shortage of one of the ancillary services. In other words, this means that other customers could suffer interruptions.

The requirements set by the transmission system operator should not be confused with safety and protection requirements set by a national safety agency or by the distribution system operator. These requirements are needed both for the safety of the owner of the distributed generator and for the safety of other customers. We did discuss some of these in detail in Chapter 7.

8.13.7 Reactive Power Control

One of the important ancillary services in the transmission system is the reactive power. This is usually provided by large production units. These production units are designed such that they can produce or absorb reactive power even during maximum (active power) production. One of the ways of achieving this is by overrating the synchronous machine and the generator transformer with respect to the maximal mechanical power. When these units are replaced by distributed generation, it is no longer obvious that the same amount of reactive power is available to the transmission system operator. Some possible solutions to a shortage of reactive power reserve are as follows:

- Using sources of reactive power independent of conventional power stations. The synchronous condenser is the traditional solution, although a switched capacitor bank will provide reactive power. More advanced solutions include the static var compensator, the regulated reactance [35], and the static compensator (STATCOM). Also, VSC-based HVDC terminals can control the reactive power flow. The use of any of these solutions implies that it is no longer needed to run a production unit only because of the need for reactive power or reactive power reserve. Also, the equipment can be installed at the location where the reactive power is most needed. As always, the scheduling of reserves is very important in the transmission system. The disadvantage of these solutions is that they result in additional costs for the transmission system operator that will one way or the other end up in the increase in electricity price for the consumer.
- At this moment, several transmission system operators require large wind parks connected to the transmission system to have the same reactive power capabilities as a large conventional generator. In practice, this will result in one or more of the above-mentioned solutions being installed in the wind park. The regulated shunt reactance is proposed in Ref. 35 to control the reactive power flow between an offshore wind park and the grid. These solutions increase the costs for the wind park owner and may not result in the reactive power being made available where it is most needed. It may thus still be needed for the transmission system operator to install similar equipment elsewhere.
- When distributed generation comes in the form of many small units, they may be asked to contribute to the reactive power needed by the transmission system operator. The cell controller, developed by the Danish transmission system operator, has this as a basic principle [282, 283]. The transmission system operator will communicate a requirement for the reactive power exchange to the cell controller, for example, with a 130/20 kV transformer. The cell controller uses this to calculate set points for the reactive power exchange with distributed generator units, capacitor banks, and so on in the underlying distribution system. The main disadvantage of such a system is the potentially high costs when large numbers of small units have to get involved. These costs will likely have to be carried by the owner of the generator unit. This could again become a barrier for the introduction of small production units. On the other hand, when market principles are applied, this could also be part of the business model for distributed generation. A technical disadvantage is the need to transport reactive power over long electrical distances. This results in high reactive power losses, but it could also result in increased ohmic losses, overloading, and overvoltages or undervoltages.

8.13.8 Probabilistic Methods

The existing methods for scheduling of reserves are based on deterministic rules, with fixed values for the different kind of reserves, often independent of the time of year and the amount of consumption. The scheduling is to a large extent based on worst-case scenarios. The $(N-1)$ criterion and the scheduling of reserves have contributed a lot to the very high reliability that we have come to expect from transmission systems. It is the task of the transmission system operator to maintain this reliability level, hence their objection to any change in methodology.

The introduction of large amounts of distributed generation, as well as a number of other developments, implies that there is a need for alternative methods instead of or next to the $(N-1)$ criterion [54, 433]. A completely different approach to the scheduling of reserves would be to set an upper limit to the probability that there will be insufficient reserves. Insufficient reserves could mean, for example, that the secondary reserve is depleted, the operational security is no longer maintained, or the primary reserve is depleted. The latter would result in (automatic or manual) load shedding or loss of load.

For every operational state, a number of probabilities are to be calculated, for example,

- The probability that the secondary reserve within a control area becomes depleted. This could be due to the loss of conventional generation, forced outages of tie lines, reduction in the production by distributed generation, or unexpected increase in consumption.
- The probability that a control area is no longer secure, that is, there is at least one single contingency that would result in loss of load for the area.
- The probability of loss of load for the area.

All these probabilities are dependent on the time window considered, that is, the "lead time." The longer the lead time, the higher the probability. If any of the resulting probabilities are too high, additional reserves should be made available.

The additional uncertainty due to distributed generation will make in almost unavoidable to introduce probabilistic methods in the planning and operation of the transmission system. This call for probabilistic methods is repeated regularly, but at the moment, there are no clear developments underway in this direction. Even the amount of research on this subject is very limited.

A report was commissioned by the North American Electric Reliability Corporation (NERC) in 2009 to investigate the impact of 20% wind power on the power system. One of the specific recommendations of the report was the development of probabilistic methods. Specific reference was made to methods for transmission planning, determination of the capacity value, and resource adequacy [227, 387].

The need for stochastic planning and operating tools is also stated by Ref. 103: "Stochastic planning and operating tools and methods are needed to better incorporate the variability and uncertainty associated with the variable output of renewable resources." With the development of probabilistic methods for transmission-system operation, it is also important to distinguish between variability and uncertainty [103]. The former can be modeled in a deterministic way; it is the uncertainty that requires probabilistic methods.

In Ref. 118, methods are proposed for calculating the operational risk from the probability of generator outage and the uncertainty in demand and wind power production. The amount of reserves is next calculated by setting a maximum value for the expected number of load shedding incidents per year. These methods are used in Ref. 182 to further develop wind power prediction and operational scheduling tools.

8.13.9 Development of Standard Models for Distributed Generation

The development of models will by itself not give any improvement in the operation of transmission systems. However, with models that are better trusted by the transmission system operator, a source of uncertainty is removed. This in turn reduces the need for reserves and the cost of operating the transmission system. It is not clear how much additional reserve is needed to accommodate for model uncertainty. It is however clear that there is a significant drive toward the development of standard models, especially for wind parks connected to higher voltage levels. Many transmission system operators require detailed models from the wind park owners as input in their own simulations.

A discussion on the development of standard models is presented in Ref. 348. The need for standard models, often referred to as "generic models," is very strongly emphasized. The validation of the models is very important, as oversimplification could result in unexpected behavior of the system during a disturbance. Simplification is however needed as it will not be possible for a transmission system operator to use a model including a detailed model of every single wind turbine. According to Ref. 348, it is common to model a wind park through one equivalent turbine. However, the work presented in Refs. 267 and 268 shows that incorrect values for the reactive power flow after a fault are obtained when only one equivalent turbine is used. A two-turbine or even five-turbine model would be more appropriate. The modeling becomes even more complicated when a number of turbines may have tripped due to the fault. This will also impact the postfault behavior of the park and should thus be included in the model.

CHAPTER 9

CONCLUSIONS

In this book, the introduction of new types of electricity production has been treated as a transition starting from the existing power system. A massive introduction of distributed generation (including large-scale renewable energy sources) without any changes in the power system will result in unacceptable levels of quality and reliability. Certain amounts of distributed generation can be integrated, however, in the existing system without any changes. Larger amounts can be integrated by making minor changes (i.e., small amounts of investment). For massive amounts, significant changes (investments) are needed. The hosting capacity has been defined in Chapter 3 as the amount of distributed generation that can be connected (at a specific location, in a certain part of the system, or for the system as a whole) without resulting in unacceptable reliability or voltage quality for other customers. Although it is the primary aims of the power system (reliability, quality, price) that matter, the hosting capacity can also be applied to secondary aims such as protection and overload. The choice of performance index often very much influences the resulting hosting capacity. Such uncertainty is however more common in power system planning and operation than one would get the impression. It should be noted that many of the established rules (e.g., the $(N-1)$ criterion) originated as rules of thumb to make life easier during design and operation of the power system. When such rules become a barrier to the introduction of renewable energy, they need to be reconsidered. We do not in any way suggest that, for example, the reliability of the power system should be reduced to allow more distributed generation to be connected. The primary aim remains (a sufficiently high reliability for the customers), although it might be achieved in different ways.

The impact of distributed generation on overloading and losses is discussed in Chapter 4. In distribution systems, the risk of overload and the losses reduce with limited amounts of distributed generation. The risk of overload becomes larger than that without distributed generation when the maximum production exceeds the sum of maximum and minimum consumption. Losses are less than that without generation as long as the average production is less than twice the average consumption. Rather large amounts of generation can be connected to a distribution feeder before overload or losses become a concern. At subtransmission system, already small amounts of new production can reduce the transport capacity. The operational reserve needed at subtransmission level often sets additional limits. A low hosting capacity for new production at subtransmission level is often due the prioritizing of existing sources of production. Possible methods for increasing the hosting capacity include not only

Integration of Distributed Generation in the Power System, First Edition. Math Bollen and Fainan Hassan.

classical solutions such as building more lines and transformers, but also a wider use of existing methods such as intertrip schemes, power electronics control, and weather-dependent transfer capacity of lines. Possible future methods include risk-based planning and advanced monitoring schemes combined with curtailment of production and consumption.

The risk of overvoltage, discussed in detail in Chapter 5, increases immediately when generation is introduced to the distribution system. Also, reduction in consumption (such as energy saving lamps and lower standby losses of consumer electronics) will increase this risk, just as replacing overhead lines by underground cables. The hosting capacity is very strongly location dependent and also strongly depends on the performance criterion and calculation methods used. The planning levels used by some network operators, together with a worst-case estimation of the minimum voltage drop, could easily result in zero hosting capacity. Reasonable performance limits and more accurate estimations of the voltage drop are easy methods for increasing the hosting capacity. The use of probabilistic methods has also great potential and needs to be further developed. This includes a broad and open discussion among all stakeholders about what the risks are and who carries those risks. Other methods for increasing the hosting capacity include shorter or stronger feeders and more transformers, alternative methods for voltage control or even using distributed generation to control the voltage, and methods for curtailing production and local storage.

Other power quality disturbances are discussed in Chapter 6. Power quality is usually a minor concern with the connection of distributed generation, but some aspects need to be considered. At distribution level, capacitor banks could result in high harmonic voltages and currents due to harmonic resonance. This should be considered routinely during the connection of new installations and also, for example, in the design of collection grids for wind parks. The harmonic emission of modern production units is small at the frequencies that traditionally have been a concern (harmonics 3, 5, 7, 11, 13, etc), but the emission at other frequencies (even harmonics, interharmonics, above 2 kHz) is often higher than that with existing equipment. The consequences of this remain unclear, but further research and development is encouraged so that measures can be taken when needed. The subject should also be taken up in standard-setting organizations. At transmission level, the shift from large conventional production units to distributed generation weakens the system. This results in an increase in disturbance levels in the transmission system. Initial calculations, for voltage dips, indicate that the impact is limited, but this is again a subject that requires further studies. Measures already taken to increase the hosting capacity include setting emission limits to new production. Future methods include a new look at permissable voltage disturbance levels, active and passive harmonic filters, and power electronic converters.

Power system protection, discussed in Chapter 7, is already complicated without the presence of distributed generation. This especially holds for the distribution grid, where it is often difficult to ensure a fast enough operation of the backup protection under all circumstances, while at the same time avoiding any unnecessary trip. The introduction of generation in the distribution system increases only the complexity. For rather small amounts of distributed generation, incorrect operation has already been reported. Changing the protection setting will however solve a lot of those problems, after which hosting capacities of some tens of percent of the transformer capacity can be reached in most cases. A specific problem very difficult to solve is the risk of noncontrolled island operation of the generator together with the feeder load. Very sensitive anti-islanding protection is often used, but the recent concern for transmission system operation implies that the high risk of massive unwanted tripping is no longer acceptable. Increasing the hosting capacity would require more advanced protection schemes, possibly including adaptive settings and an overall monitoring system that ensures the correct operation even during backup. A more widespread use of intertrip schemes is an intermediate step toward such an advanced protection scheme.

Chapter 8 deals with the impact of distributed generation on the operation of the transmission system. The operation of the transmission system concerns to a large extent having sufficient reserves available even in the case when one or more components (such as transmission lines or large production units) unexpectedly fail. The presence of large amounts of distributed generation can interfere with this task in a number of ways: the uncertainty and unpredictability of the production requires additional reserves; a shift from large conventional units to distributed generation implies less units available for providing reserves and maintaining stability; behavior of distributed generation during large disturbances is often unpredictable. The main concerns at this moment are the massive tripping of distributed generation due to the loss of a large production unit or due to a fault at transmission level and the difficulty to predict wind power production. The massive tripping of distributed generation could trigger a blackout or could make the impact of a blackout more severe. Prediction errors could require the transmission system operator to buy expensive balancing power and could in the worst case endanger the operational security of the transmission system. The main solutions used today are setting strict fault ride-through requirements on the production units and prediction of wind power production. Large resources are put on the latter, while also extending the control area by building international connections. Future methods for increasing the hosting capacity include placing the reserves with the consumption instead of the production, overlaying transmission grids (800 kV AC or HVDC) connecting large geographical areas, and allowing or forcing distributed generation to supply ancillary services. A fully open market for ancillary services would allow anybody to supply not only services such as reactive power or operational reserve, but also black start or inertia. Also, at transmission level suitable probabilistic methods are needed for a more optimal planning of the operational reserves.

In the previous chapters, performance limits used by the network operator were often identified as unnecessary barriers to the introduction of renewable energy. Alternative methods were proposed, including risk-based methods, to increase the hosting capacity. Such methods do not require any investment in the power system, so the costs are limited. The objection from many network operators to the use of alternative methods, including less strict limits, should be seen in the light of their task. The tasks of a network operator include ensuring an acceptable quality and reliability for all customers connected to the network. Having strict and fixed planning criteria (including performance criteria) have enabled the network operators to perform this task very well in the past. Many network operators fear that the introduction of distributed generation will imply that they can no longer maintain an acceptable reliability and quality. As a result, they keep very strictly to these limits. Also, they typically do place all costs associated with the connection of a new production unit to the owner of this production unit. This keeps the costs low for other customers but does not necessarily minimize the total costs for society. When introducing less strict limits in the planning stage and risk-based methods, the risk should not be carried only by the network operator. A fair assessment and sharing of the risks, including the financial risks, over all stakeholders is an essential part of enabling an efficient use of the power system for distributed generation.

At the time of writing of this book (2009–2010), a lot of discussion about "smart grids" are going on not only in the research community, by network operators, and by manufacturers of equipment, but also in the popular press and in political circles. In addition to "smart grids," terms such as "intelligent networks" and "active distribution networks" are regularly used, followed by "smart metering" and "advanced metering" that concern the communication between the network and its customers. Such discussions are not included in this book because of the highly deviating interpretations of what these "smart grids" will actually involve. In Chapters 4–8, we however include several solutions that could be classified as "smart grids." There are a number of trends that can be identified in the ongoing discussions and research activities on "smart grids" that will increase the hosting capacity of the power system for distributed generation.

- **Active distribution networks:** The introduction of generation in the distribution system enables a more active operation of the distribution system. The generators can be used, for example, not only to control the voltage, but also to prevent overloads. Using distributed generation for voltage control requires coordination with the existing control devices (automatic transformer tap changers, switched capacitors), which almost certainly involves communication. Communication is also needed when distributed generation is used to limit active power flows.
- **Power electronics:** Many types of distributed generation are equipped with a power electronics interface that could be used actively in the operation of the distribution or transmission system. Voltage control is already mentioned, and reactive power contribution to the transmission grid, electronic inertia, and harmonic filtering are also possible applications. Separate power electronic devices can be used to control the voltage and increase the transfer capacity in the subtransmission and transmission systems. At the highest voltage levels, HVDC links or even HVDC grids can be used to transfer power over large distances. For example, an HVDC grid can be used to connect wind power and cities around the North Sea with hydropower in Norway. The next voltage level, above the existing interconnected grids, is by many envisaged as being an HVDC grid.
- **Advanced protection schemes:** At distribution level, an overall protection scheme could monitor the performance of several feeders, possibly covering two or more voltage levels. Although the existing protection relays would take care of the actual protection, the overall protection scheme would adapt the settings of the individual relays to the actual operational state and intervene in

case of the incorrect operation of an individual relay, for example, resulting in unwanted island operation.

- **Curtailment of small production units:** The production by distributed generation could be limited or the unit could be disconnected whenever its presence would result in unacceptable performance. This could be used to prevent not only overloads and overvoltages but also harmonic distortion in case of high levels. A simple approach, already in use, is to disconnect small generators whenever the voltage at their terminals exceed a certain limit (e.g., 108% of the nominal voltage). More advanced schemes would reduce the production as a function of the voltage or use communication to spread the reduction over different units. Preventing overloads requires a communication infrastructure in any case.
- **Storage:** Storage, for example, in batteries, at low-voltage level can prevent overvoltages and overloads due to solar power or other types of small generation units. Even for medium-voltage installations, battery storage is under development. At transmission level, pumped hydro storage is the only available option at the moment. Apart from compressed air, none of the storage methods under development comes close to the size of pumped storage. Even without pumping, hydropower can be used for storage simply by not using it. The combination of hydropower and renewable sources such as sun and wind is seen as a very good one.
- **Shifting reserves to the customers:** Currently, all operational and planning reserves are in the network or in large production units. Shifting some of the reserves to the consumption side could be a cost-effective way because those reserves would not incur any costs as long as they are not used (which is most of the time). We do not discuss here about the crude methods that are currently used as a method of large resort (disconnecting complete feeders). Instead, sophisticated methods are under development, where the load is somewhat reduced at individual customer level or even at device level. Temperature settings with heating and cooling are examples, as well as reducing the current with any charging device, switching off part of the lighting, or temporarily reducing the speed of electric trains. Having reserves available with the customers will require a complex communication infrastructure but could result in a much more efficient and flexible operation of the power system.
- **Shifting reserves to distributed generation:** In some cases, the distributed generation can contribute to the operational reserves. This will not only improve the operation of the transmission system, but also reduce the amount of energy produced from renewable sources of generation. Distributed generation can be used for downward regulating reserve (where the production is reduced on demand), although it should only occasionally be used for upward regulating reserve because the latter requires a production less than the capacity. A simple method of having downward regulating reserve available is the use of intertrip signals with the connection of large wind parks to the subtransmission system. More advanced schemes can be developed where the maximum production for each wind park is calculated after every (manual or automatic) switching action.

- **New markets:** Several of the above-mentioned developments could be helped by the introduction of appropriate markets. It is, of course, possible to enforce load reduction when needed, but a well-functioning market is generally seen as a better solution. Getting many customers, producers, and consumers involved in different markets will require a complex communication infrastructure. Possible examples are hourly spot market prices being charged directly even to small customers, groups of individual customers offering load reduction to the distribution or transmission system operator via an aggregator, wind turbines offering fault ride-through, and microturbines offering electronic inertia to the system operator. Most of these new markets, with millions of small customers involved, will require large investment in measurement and communication. The possibilities are however very large. Involving all customers in hourly markets and in ancillary services could introduce a flexibility that cannot be achieved with existing methods.
- **Widespread measurements:** Several of the mentioned examples require measurements at many locations, for example, hourly measurements of power consumption. This information can be used to obtain more accurate models of the power system and its load that in turn can be used for improved planning, especially at distribution level. Continuous measurements are also essential when, for example, using load control to prevent overvoltages and overloads in the distribution system.

Next to these advanced solutions, the classical solutions of building more and stronger lines or cables should not be forgotten. However, the introduction of new types of production will require the use of advanced solutions in more cases than in the past. By combining classical and advanced solutions, the power system will not become an unnecessary barrier to the introduction of distributed generation.

BIBLIOGRAPHY

1. *ABB Switchgear Manual*, 9th edition, 1993.
2. BESS, World's largest battery energy storage system, Fairbanks, Alaska, USA, http://www.abb.com, accessed December 30, 2009.
3. T. Aboul-Seoud and J. Jatskevich. Dynamic modeling of induction motor loads for transient voltage stability studies. In *IEEE Canada Electric Power Conference (EPEC)*, October 2008.
4. E. Acha and M. Madrigal. *Power System Harmonics: Computer Modelling and Analysis*, Wiley, Chichester, UK, 2001.
5. T. Ackermann, editor. *Wind Power in Power Systems*. Wiley, Chichester, UK, 2005.
6. T. Ackermann, G. Ancell, L.D. Borup, P.B. Eriksen, B. Ernst, F. Groome, M. Lange, C. Möhrlen, A.G. Orths, J. O'Sullivan, and M. de la Torre. Where the wind blows: power management and forecasting in areas with high wind-energy penetration levels. *IEEE Power and Energy Magazine*, 7(6):65–75, 2009.
7. R.A. Adams, S.W. Middlekauf, E.H. Camm, and J.A. McGee. Solving customer power quality problems due to voltage magnification. *IEEE Transactions on Power Delivery*, 13(4):1515–1520, 1998.
8. V. Akhmatov and P.B. Eriksen. A large wind power system in almost island operation: a Danish case study. *IEEE Transactions on Power Systems*, 22(3):937–943, 2007.
9. Alternative Energy News, Clean energy from flowing waters, http://www.alternative-energy-news.info, accessed March 31, 2010.
10. Alternative Energy News, SunEdison to build Europe's largest solar power plant, http://www.alternative-energy-news.info, accessed March 31, 2010.
11. O. Anaya-Lara, N. Jenkins, J. Ekanayake, P. Cartwright, and M. Hughes. *Wind Energy Generation: Modelling and Control*, Wiley, Chichester, UK, 2009.
12. P.M. Anderson. *Analysis of Faulted Power Systems*, IEEE Press, New York, 1995.
13. P.M. Anderson. *Power System Protection*. IEEE Press, New York, 1999.
14. P.M. Anderson and A.A. Fouad. *Power System Control and Stability*, 2nd edition, IEEE Press, New York, 2003.
15. J. Apt. The spectrum of power from wind turbines. *Journal of Power Sources*, 169:369–374, 2007.
16. J. Arrillaga and N.R. Watson. *Power System Harmonics*, 2nd edition, Wiley, Chichester, UK, 2003.
17. Atlantis Resources Corporation, http://www.atlantisresourcescorporation.com, accessed March 31, 2010.

Integration of Distributed Generation in the Power System, First Edition. Math Bollen and Fainan Hassan.

18. G.P. Azevedo and A.L. Oliviera Filho. Control centers with open architectures [power system EMS]. *IEEE Computer Applications in Power*, (4):27–32, 2001.
19. J. Balcells and D. Gonzalez. Harmonics due to resonance in a wind power plant. In *8th International Conference on Harmonics and Quality of Power*, October 1998.
20. M.E. Baran, S. Teleke, L. Anderson, A. Huang, S. Bhattacharya, and S. Atcitty. STATCOM with energy storage for smoothing intermittent wind farm power. In *IEEE Power Engineering Society General Meeting*, Pittsburgh, PA, July 2008.
21. A. Barin, L.F. Pozzatti, C.G. Carvalho, L.N. Canha, R.Q. Machado, A.R. Abaide, C. Fernandes, and F.A. Farett. Analysis of the impact of distributed generation sources on the operational characteristics of the distribution systems for planning studies. In *International Conference on Electricity Distribution (CIRED)*, Vienna, May 2007.
22. R.E. Barlow and F. Proschan. *Mathematical Theory of Reliability*, classics edition, Society of Industrial and Applied Mathematics (SIAM), Philadelphia, PA, 1996.
23. P. Bartholomeus and H. Overdiep. Energy efficiency and green gas, http://www.netherlands-embassy.org/files/pdf/2.pdf, accessed April 15, 2009.
24. A. Beddoes, M. Gosden, and I. Povey. The performance of an LV network supplying a cluster of 500 houses each with an installed 1 kWe domestic combined heat and power unit. In *International Conference on Electricity Distribution (CIRED)*, Vienna, May 2007.
25. R. Bergeron. An artificial network for emission tests in the frequency range 2–9 kHz. In *18th International Conference on Electricity Distribution (CIRED)*, June 2005.
26. T.R. Betts, R. Gottschalg, and D.G. Infield. ASPIRE: a tool to investigate spectral effects on PV device performance. In *3rd World Conference on Photovoltaic Energy Conversion*, Osaka, Japan, May 2003.
27. A. Bhowmik, A. Maitra, S.M. Halpin, and J.E. Schatz. Determination of allowable penetration levels of distributed generation resources based on harmonic limit considerations. *IEEE Transactions on Power Delivery*, 18(2):619–624, 2003.
28. F. Bignucolo, R. Caldon, and M. Valente. Probabilistic voltage estimation for the active control of distribution networks. In *International Conference on Electricity Distribution (CIRED)*, Vienna, May 2007.
29. R. Billinton and R.N. Allan. *Reliability Evaluation of Power Systems*, 2nd edition, Plenum Press, New York, 1996.
30. G. Blanchet and M. Charbit. *Digital Signal and Image Processing Using MATLAB®*, ISTE, London, 2006.
31. B. Bletterie and H. Brunner. Solar shadows. *Power Engineer*, 20(1):27–29, 2006.
32. B. Bletterie and T. Pfajfar. Impact of photovoltaic generation on voltage variations: how stochastic is PV? In *19th International Conference on Electricity Distribution (CIRED)*, Vienna, May 2007.
33. Z.Q. Bo, X.Z. Dong, R.D. Xu, and A. Klimek. A new integrated protection scheme for distributed generation systems. In *China International Conference on Electricity Distribution (CICED)*, December 2008.
34. M. Bollen. A note on DER and voltage stability. Technical report R08-504, STRI, http://www.stri.se, 2008.
35. M. Bollen, C. Bengtsson, L.H. Nielsen, P.H. Larsen, and S.D. Mikkelsen. Application of regulated shunt reactor for off-shore wind farms. In *CIGRE Symposium*, Brugge, Belgium, October 2007.

36. M. Bollen, M. Häger, and M. Olofsson. Allocation of emission limits for individual emitters at different voltage levels: flicker and harmonics. In *CIGRE Sessions*, Paris, France, August 2010.
37. M. Bollen, M. Häger, and C. Roxenius. Effect of induction motors and other loads on voltage dips: theory and measurements. In *IEEE Bologna PowerTech*, June 2003.
38. M. Bollen, M. Häger, and C. Roxenius. Voltage dips in distribution systems: load effects, measurements and theory. In *17th International Conference on Electricity Distribution (CIRED)*, Barcelona, Spain, May 2003.
39. M. Bollen, A. Holm, Y. He, and P. Owe. A customer-oriented approach towards reliability indices. In *International Conference on Electricity Distribution (CIRED)*, Vienna, May 2007.
40. M. Bollen, J. Zhong, O. Samuelsson, and J. Björnstedt. Performance indicators for microgrids during grid-connected and island operation. In *IEEE Bucharest PowerTech*, June 2009.
41. M.H.J. Bollen. Literature search for reliability data of components in electric distribution networks. Research report, EUT report 93-E-276, Faculty of Electrical Engineering, Eindhoven University of Technology, Eindhoven, The Netherlands, August 1993.
42. M.H.J. Bollen. *Understanding Power Quality Problems: Voltage Sags and Interruptions*, IEEE Press, New York, 2000.
43. M.H.J. Bollen and C. Coujard. Impact of small units for electricity generation. In *1st International Conference on Lifestyle, Health and Technology*, Luleå, Sweden, June 2005.
44. M.H.J. Bollen and I.Y.H. Gu. Characterization of voltage variations in the very-short time scale. *IEEE Transactions on Power Delivery*, 20(2):1198–1199, 2005.
45. M.H.J. Bollen and I.Y.H. Gu. *Signal Processing of Power-Quality Disturbances*, IEEE Press, New York, 2006.
46. M.H.J. Bollen and M. Häger. Impact of increasing penetration of distributed generation on the dip frequency experienced by end-customers. In *International Conference on Electric Distribution Systems (CIRED)*, Turin, Italy, June 2005.
47. M.H.J. Bollen and M. Häger. Power quality: interactions between distributed energy resources, the grid, and other customers. In *1st International Conference on Renewable Energy Sources and Distributed Energy Resources*, Brussels, December 2004.
48. M.H.J. Bollen, M. Häger, and C. Schwaegerl. Quantifying voltage variations on a time scale between 3 seconds and 10 minutes. In *International Conference on Electric Distribution Systems (CIRED)*, Turin, Italy, June 2005.
49. M.H.J. Bollen, F. Hassan, M. Wamundsson, A. Holm, and Y. He. The active use of distributed generation in network planning. In *20th International Conference and Exhibition on Electricity Distribution (CIRED)*, 2009.
50. M.H.J. Bollen and P. Massee. A classification of failures of the protection. In *International Conference on Probabilistic Methods Applied to Power Systems (PMAPS)*, 1991.
51. M.H.J. Bollen, P.F. Ribeiro, E.O.A. Larsson, and C.M. Lundmark. Limits for voltage distortion in the frequency range 2–9 kHz. *IEEE Transactions on Power Delivery*, 23(3):1481–1487, 2008.
52. M.H.J. Bollen, M. Speychal, and K. Lindén. Impact of increasing amounts of wind power on the dip frequency in transmission systems. In *Nordic Wind Power Conference*, Helsinki, Finland, May 2006.
53. M.H.J. Bollen and P. Verde. A framework for regulation of rms voltage and short-duration under and overvoltages. *IEEE Transactions on Power Delivery*, 23(4):2105–2112, 2008.

54. M.H.J. Bollen, L. Wallin, T. Ohnstad, and L. Bertling. On operational risk assessment in transmission systems: weather impact and illustrative example. In *International Conference on Probabilistic Methods Applied to Power Systems (PMAPS 2008)*, Puerto Rico, 2008.

55. M.H.J. Bollen and L.D. Zhang. Different methods for classification of three-phase unbalanced voltage dips due to faults. *Electric Power System Research*, 66(1):59–69, 2003.

56. A.M. Borbely and J.F. Kreider, editors. *Distributed Generation: The Power Paradigm for the New Millennium*, CRC Press, Boca Raton, FL, 2001.

57. P. Bousseau, E. Monnot, G. Malarange, and O. Gonbeau. Distributed generation contribution to voltage control. In *International Conference on Electricity Distribution (CIRED)*, Vienna, May 2007.

58. T.N. Boutsika and S.A. Papathanassiou. Short-circuit calculations in networks with distributed generation. *Electric Power Systems Research*, 78:1181–1191, 2008.

59. G.J. Bowden, P.R. Barker, V.O. Shestopal, and J.W. Twidell. The Weibull distribution function and wind power statistics. *Wind Engineering*, 7(2):85–98, 1983.

60. G. Boyle. *Renewable Energy: Power for a Sustainable Future*, Oxford University Press, 1996.

61. S.M. Brahma and A.A. Girgis. Microprocessor-based reclosing to coordinate fuse and recloser in a system with high penetration of distributed generation. In *IEEE Power Engineering Society Winter Meeting*, 2002.

62. S.M. Brahma and A.A. Girgis. Development of adaptive protection scheme for distribution systems with high penetration of distributed generation. *IEEE Transactions on Power Delivery*, 19(1):56–63, 2004.

63. M. Braun. Reactive power supply by distributed generators. In *IEEE Power Engineering Society General Meeting*, Pittsburgh, PA, July 2008.

64. K. Brekke, H. Seljeseth, and O. Mogstad. Rapid voltage changes: definitions and minimum requirements. In *20th International Conference on Electricity Distribution*, Prague, June 2009.

65. P. Bresesti, W.L. Kling, R.L. Hendriks, and R. Vailati. HVDC connection of offshore wind farms to the transmission system. *IEEE Transaction on Energy Conversion*, 22(1):37–43, 2007.

66. P. Bresesti, W.L. Kling, and R. Vailati. Transmission expansion issues for offshore wind farms integration in Europe. In *IEEE/PES Transmission and Distribution Conference and Exposition*, April 2008.

67. R.W. Brown and R. Paterson. Optimising the scale of a wind farm development at the weak extremity of a rural network. In *International Conference on Large Electric Power Systems (CIGRE)*, Paris, France, August 2008.

68. R. Bründlinger and B. Blêtterie. Unintentional islanding in distribution grids with a high penetration of inverter-based DG: probability for islanding and protection methods. In *IEEE St. Petersburg PowerTech*, July 2005.

69. L. Bryans and A. Kennedy. System role of generation. In *Large-Scale Renewables Energy Seminar*, http://www.soni.ltd.uk/wind.asp, October 2008.

70. J.J. Burke. *Power Distribution Engineering: Fundamentals and Applications*. Marcel Dekker, New York, 1994.

71. T. Burton, D. Sharpe, N. Jenkins, and E. Bossanyi. *Wind Energy Handbook*, Wiley, Chichester, UK, 2001.

72. P.H. Buxton. Technical, standards and control issues of embedded generation. ETSU K/EL/00229/REP, DTI Pub/URN 00/1449, Department of Trade and Industry, UK, 2000.
73. R. Cai. Flicker interaction studies and flickermeter improvement. PhD thesis, Faculty of Electrical Engineering, Eindhoven University of Technology, June 2009.
74. R. Cai, J.F.G. Cobben, J.M.A. Myrzik, and W.L. Kling. Flicker curves of different types of lamps. In *China International Conference on Electricity Distribution*, 2006.
75. Canadian Encyclopedia, Ocean wave energy, http://www.thecanadianencyclopedia.com, accessed March 31, 2010.
76. Canadian Encyclopedia, Tidal energy, http://www.thecanadianencyclopedia.com, accessed March 31, 2010.
77. I.M. Canay, D. Braun, and G.S. Koppl. Delayed current zeros due to out-of-phase synchronizing [of turbogenerators]. *IEEE Transaction on Energy Conversion*, 13(2):124–132, 1998.
78. P. Caramia, G. Carpinelli, and P. Verde. *Power Quality Indices in Liberalized Markets*, Wiley, Chichester, UK, 2009.
79. G. Carannante, C. Fraddanno, M. Pagano, and L. Piegari. Experimental performance of MPPT algorithm for photovoltaic sources subject to inhomogeneous insolation. *IEEE Transactions on Industrial Electronics*, 56(11):4374–4380, 2009.
80. F. Carlsson and V. Neimane. A massive introduction of wind power: changed market conditions? Technical report 08:41, Elforsk, Sweden, http://www.vindenergi.org/Vindforskrapporter/v_132.pdf, 2008.
81. E.S. Cassedy. *Prospects for Sustainable Energy: A Critical Assessment*, Cambridge University Press, Cambridge, UK, 2000.
82. Council of European Energy Regulators, Fourth benchmarking report on quality of electricity supply, Ref. C08-EQS-24-04, http//:www.ceer-eu.org, December 2008.
83. G. Celli, F. Pilo, and F. Pilo. An innovative transient-based protection scheme for MV distribution networks with distributed generation. In *IET 9th International Conference on Developments in Power System Protection (DPSP)*, March 2008.
84. S. Chaitusaney and A. Yokoyama. Impact of protection coordination on sizes of several distributed generation sources. In *7th International Power Engineering Conference (IPEC)*, November 2005.
85. C. Changsong, D. Shanxu, and Y. Jinjun. Research of energy management system of distributed generation based on power forecasting. In *International Conference on Electrical Machines and Systems (ICEMS'08)*, October 2008.
86. S. Chen, C.M. Lo, M.K. Foo, and K.T. How. Testing of fluorescent lamps for its flickering susceptibility towards interharmonic voltages. In *International Power Engineering Conference (IPEC 2007)*, December 2007.
87. S. Cherian and R. Ambrosio. Towards realizing the GridWise^{TM} vision: Integrating the operations and behavior of dispersed energy devices, consumers, and markets. In *IEEE Power Systems Conference and Exposition*, October 2004.
88. H. Cheung, A. Hamlyn, L. Wang, G. Allen, C. Yang, and R. Cheung. Network-integrated adaptive protection for feeders with distributed generations. In *IEEE Power and Energy Society General Meeting*, July 2008.
89. O. Chilard, S. Grenard, O. Devaux, and L. De Alvaro Garcia. Distribution state estimation based on voltage state variables: assessment of results and limitations. In *International Conference on Electricity Distribution (CIRED)*, Prague, June 2009.

90. I.M. Chilvers, N. Jenkins, and P.A. Crossley. The use of 11 kV distance protection to increase generation connected to the distribution network. In *8th IEE International Conference on Developments in Power System Protection*, Vol. 2, April 2004.
91. J.H. Chow, S.H. Javid, J.J. Sanchez-Gasca, C.E.J. Bowler, and J.S. Edmonds. Torsional model identification for turbine-generators. *IEEE Transactions on Energy Conversion*, 1(4):83–91, 1986.
92. S. Chowdhury, S.P. Chowdhury, and P. Crossley. *Microgrids and Active Distribution Networks*, The Institution of Engineering and Technology, London, 2009.
93. S.P. Chowdhury, S. Chowdhury, C.F. Ten, and P.A. Crossley. Islanding protection of distribution systems with distributed generators: a comprehensive survey report. In *IEEE Power and Energy Society General Meeting*, 2008.
94. Modeling new forms of generation and storage. CIGRE Technical Brochure TF 38.01.10, 2000.
95. CIGRE Working Group B4.39, Integration of large scale wind generation using HVDC and power electronics. CIGRE Technical Brochure 370, February 2009.
96. CIGRE/CIRED Joint Working Group C4.107, Economics of power quality. Technical Report, http://www.e-cigre.org, 2011.
97. Technical report.
98. Specification of radio disturbance and immunity measuring apparatus and methods. Part 1. Radio disturbance and immunity measuring apparatus, CISPR 16-1, IEC, October 1999.
99. L. Clifton. Hydro/diesel frequency control. In *Reliable and Affordable Energy for Rural Alaska*, http://www.cliftonlabs.net, September 2002.
100. S. Cole and R. Belmans. Transmission of bulk power: the history and applications of voltage-source converter high-voltage direct current systems. *IEEE Industrial Electronics Magazine*, 3(3):19–24, 2009.
101. S. Cole, D. Van Hertem, and R. Belmans. Connecting Belgium and Germany using HVDC: a preliminary study. In *IEEE Lausanne PowerTech*, July 2007.
102. Cooperative Networks for Renewable Resource Measurements (CONFRRM), Solar energy resource data, http://www.nrel.gov/rredc, accessed May 2, 2009.
103. D. Corbus, D. Lew, G. Jordan, W. Winters, F. Van Hull, J. Manobianco, and B. Zavadil. Up with wind: studying the integration and transmission of higher levels of wind power. *IEEE Power and Energy Magazine*, 7(6):36–46, 2009.
104. S. Corsi and C. Sabelli. General blackout in Italy, Sunday, September 28, 2003, h. 03:28:00. In *IEEE Power Engineering Society General Meeting*, June 2004.
105. M. Crappe. Evolution of European electric power systems in the face of new constraints: impact of decentralized generation. In M. Crappe, editor, *Electric Power Systems*, Wiley, 2008, pp. 37–94.
106. Danish Wind Power Association, Wind power calculator, http://www.windpower.org/en/tour/ wres/pow/index.htm, accessed April 20, 2008.
107. T. Davies. *Protection of Industrial Power Systems*, Pergamon Press, Oxford, 1984.
108. L.V.L. de Abreu, F.A.S. Marques, J. Moran, W. Freitas, and L.C.P. da Silva. Impact of distributed synchronous generators on the dynamic performance of electrical power distribution systems. In *IEEE/PES Transmission and Distribution Conference and Exposition: Latin America*, November 2004, pp. 959–963.

TABLE 8.8 Published Values for the Inertia Constant of Large Generators

Generator type	Size	Inertia constant (s)	Sources
Japanese 60 Hz system	95,000 MW	14–18	223
Thermal unit (2-pole)	Up to 500 MW	3–6	183
Thermal unit (2-pole)		2.5–6	256
Thermal unit (2-pole)		4–7	179, 428
Thermal unit (4-pole)	Up to 100 MW	6–10	183
Thermal unit (4-pole)		4–10	256
Thermal unit (4-pole)		6–9	179, 428
Large thermal unit		3.5	72
Large turbogenerator		4	440
Turbogenerators	50 MW class	2–5	351
Turbogenerator		2.8	91
Turbogenerator (2-pole)		3–5	14
Turbogenerator (4-pole)		5–8	14
Noncondensing unit		3–4	179, 428
Thermal unit, forced cooling		7–10	194
Thermal unit, natural cooling		10–15	194
Combined cycle gas turbine		7–8	72
Hydro unit		2–4	179, 183, 256, 428
Hydro unit		6–8	194
Large hydro unit		4	72
Large hydro unit		3.0–5.5	14
Synchronous condenser		1–1.25	179, 428

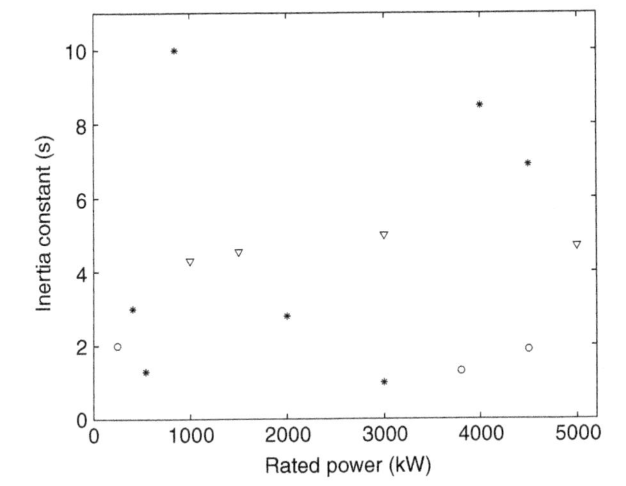

Figure 8.26 Inertia constant of small hydropower units (stars), synchronous generators (circles), and wind turbines (triangles).

TABLE 8.9 Published Values for the Inertia Constant of Motors and Small Generators

Generator type	Size	Inertia constant (s)	Sources
Synchronous motor		2	440
Synchronous motor		3–5	194
Synchronous motor		1–5	179
Induction motor		3	3
Induction motor	4.3 MVA	2.0	388
Induction motor	94 kVA–4.7 MVA	0.45–1.98	388
Synchronous generators	1.5–5 MVA	0.5–2	108
Small diesel generators		0.8–1.5	247
Induction generators		3–5	164
Landfill unit		1.0	72
Waste unit		3	72
Biomass unit		3	72
Small hydro unit		1.5	72
Small hydro unit		1.5–4.0	14
Wind turbine	2 MW	2.4–6.8	249
Wind turbine		2.2	313
Wind turbine		3	72
Wind turbine		4.3	69
CHP with combined cycle gas turbine		4–6	72
CHP with heat recovery steam turbine		3.5	72
CHP with gas turbine		6	72
CHP with internal combustion engine		1.7	72
CHP with steam turbine		3	72

[226, 388]. The figure shows a wide range in inertia constants for the hydropower units, without any correlation with the generator size.

The inertia constant of small generators is, as far as the limited amount of data available shows, somewhat smaller than the inertia constant of large conventional generators (Table 8.9). The inertia constant of wind turbines is of the same range as that of large conventional generator units. The inertia constant of small thermal generators, such as the ones used in combined heat-and-power installations is only about half the value for large units. During periods of low wind, the amount of connected units per MW consumption will be higher for wind power than for conventional generation. As a result, there is also more kinetic energy available per MW consumption. For high wind speed throughout the system, a situation could occur in which the amount of kinetic energy becomes somewhat less. The values in the above tables and figures however indicate that the reduction will not be spectacular. The inertia on the consumption side also contributes to the system inertia and this one is not impacted by the introduction of distributed generation.

Distributed generators connected through a power electronics interface do not contribute at all to the inertia constant of the system. When conventional generators are replaced by such distributed generation, the amount of kinetic energy in the system

Using electronic inertia, alternative solutions are possible, some of which might be better than the classical solution. The amount of energy made available (i.e., the reduction in rotational speed) could, for example, depend on the rate of change of frequency. In that case, more energy would be made available when the system stability is endangered more. Control schemes could also be developed where a drop in frequency results in a higher effective inertia constant than a rise in frequency. The same generator could help in preventing a drop in frequency without slowing down the frequency recovery. With microturbines, a combination of underfrequency protection and electronic inertia could be installed where the turbine speed is reduced very quickly once the frequency drops below a certain threshold. The reader may notice that such schemes introduce a gray zone including both system inertia and primary reserve. In fact, some authors consider the kinetic energy as a form of primary reserve.

8.8 FREQUENCY STABILITY

The frequency stability in a power system is determined by the balance between all production and consumption in the whole synchronous system. Any unbalance will result in a change in frequency, according to

$$\frac{\mathrm{d}f}{\mathrm{d}t} = \frac{1}{2} f_0 \frac{P_{\mathrm{gen}} - P_{\mathrm{load}}}{SH} \tag{8.31}$$

where f_0 is the nominal power system frequency, P_{gen} the total production, P_{load} the total consumption, and SH the total kinetic energy of the rotating mass connected to the system. See the discussions on inertia and kinetic energy in Section 8.7.

The reaction of the production units to a sudden loss of production is shown schematically in Figure 8.27. The change in frequency is determined, according to (8.31), by the difference between production and consumption. Note that the figure is not to scale. For the frequency to return to its original, pre-event, value, the area below the dashed line should be equal to the area above the curve. The frequency disturbance starts with the sudden loss of a large production unit or an HVDC

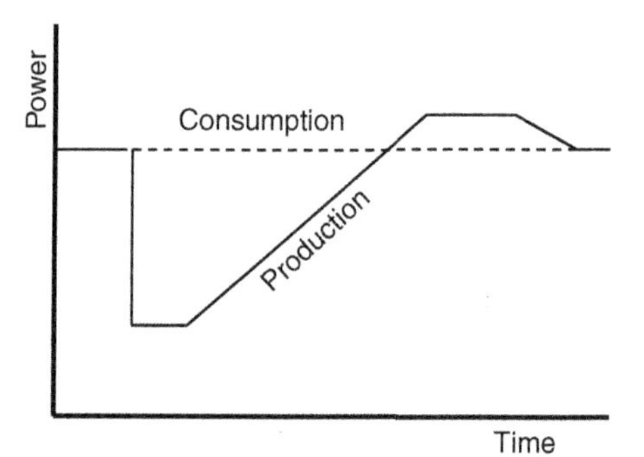

Figure 8.27 Recovery of the total production in a synchronous area due to a sudden shortage of production.

will decrease. It is this reduction that most concerns the system operators, as it could result in frequency as well as angular instability.

Example 8.15 Consider a region with 10,000 MVA installed generation in the form of high-inertia thermal units. Assume that the average inertia constant equals

$$\frac{\sum S_i H_i}{S_b} = 7\,\mathrm{s}$$

with $S_b = 10,000$ MVA. The total kinetic energy in the system is 70,000 MWs. Connecting 5000 MW of wind power would allow, on a windy day, to reduce the amount of thermal generation by up to 6000 MVA. If we neglect the inertia of the load and assume that wind power is connected by a power electronics interface not contributing to the system inertia, the total kinetic energy goes down to 28,000 MWs, a reduction by a factor of 2.5. Here, we assumed that all large thermal units have the same inertia constant.

Exercise 8.9 What will be the percentage reduction in kinetic energy when the amount of kinetic energy in the load is half the amount in the generation?

The issue of inertia of modern wind turbines is discussed clearly in Section 5.3.7 of Ref. 164. It is explained that what matters is not the actual amount of kinetic energy available in the rotating mass, but the speed at which this energy can be made available to the power system. For wind turbines based on induction generators, the energy is made available when the generator speed follows the power system frequency. In fact, during a fast drop in frequency, the generator slip will increase (compared to the actual frequency) and it will quickly produce more electrical power at the expense of its kinetic energy, thus slowing down in speed. The effective inertia constant is between 3 and 5 s according to theoretical studies and measurements [207, 278] as quoted by Ref. 164.

No such natural transfer of energy from the rotating mass to the grid takes place when a power electronics converter is used, as is the case for double-fed induction generators, full power converters, and HVDC connections to wind parks. These types of generation would not contribute to the kinetic energy available to the system. It is however possible to build "electronic inertia," for example, by using the rate of change of frequency (ROCOF) as an input to the torque control of the actual turbine [164, p. 190; 305]. With all implementations of electronic inertia, it is important that there is some kind of additional energy transfer from the generation side to the grid side of the converter once the system frequency drops. A simple implementation, for example for microturbines, would be to let the turbine speed be proportional to the system frequency. A reduction in system frequency would result in a reduction in turbine speed and the difference in kinetic energy would become available.

With nonrotating distributed generation (such as solar panels), such an implementation would require some energy storage or some reserve. The latter would result in a reduction in the production of renewable energy. The loss of renewable energy may be limited by having the electronic inertia only activated when there is not enough kinetic energy connected to the system. Such a reasoning could naturally result in a "kinetic energy market."

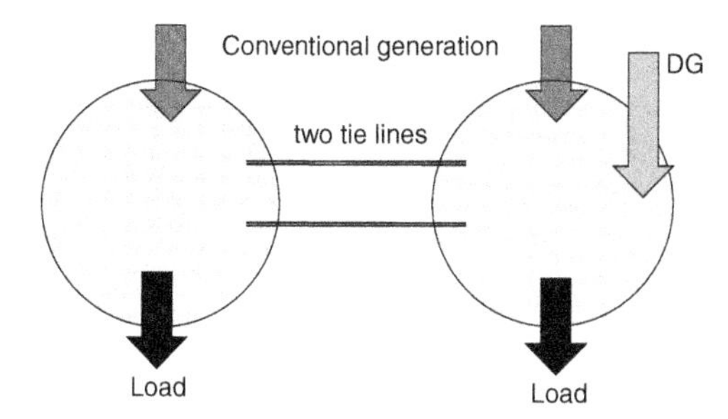

Figure 8.23 Example system with two main load-generation areas.

a longer time close to the voltage stability limit. In that case, the transfer should be reduced. The concern is what happens when one of the lines of the corridor is lost. The flow of power through the corridor should be less than the transfer capacity of the corridor. The flow of power is equal to the unbalance between the two regions. Using the notation in Figure 8.24, for the power flow through the corridor we get:

$$P_{\text{corr}} = (P_1 - L_1) = (P_2 + P_{\text{DG}} - L_2) \tag{8.22}$$

According to the $(N - 1)$ criterion, the system should be stable for all individual events ("contingencies") that might occur in the system. The loss of the largest generator is often the worst-case event. The amount of power transport through the corridor should therefore always be less than the transfer capacity minus the amount of primary reserve that has to be imported upon loss of the largest generator on the receiving size. The shift from large conventional stations to distributed generation may result in the largest units being in operation less often. If the loss of the largest unit is the worst case, the shift to smaller units will increase the amount of power that can be transported. The secure transfer capacity should also consider any transfer of primary reserve after the loss of a large production unit.

Let P_{VSL} be the voltage stability limit for the transfer through the corridor and $P_{\text{gen,large}}$ the import upon loss of the largest generator on the receiving end. The secure transfer capacity is in that case equal to

$$P_{\text{corr,max}} = P_{\text{VSL}} - P_{\text{gen,large}} \tag{8.23}$$

For an important corridor, the worst case may no longer be the loss of the largest power station, but the loss of one of the lines through the corridor. Assuming that the

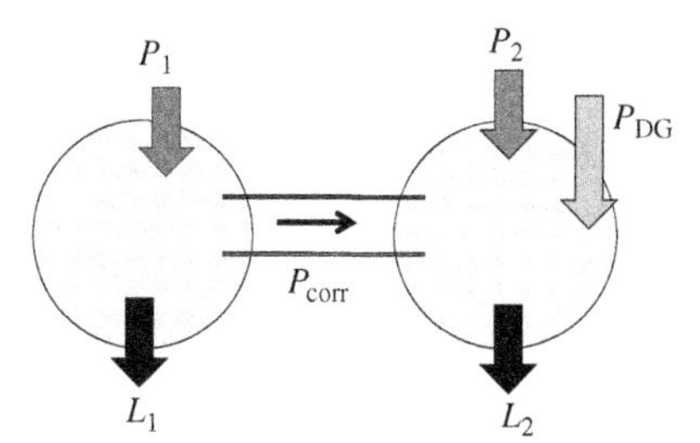

Figure 8.24 Power flows in the example system from Figure 8.23.

TABLE 8.7 Effect of Reactive Power Load on the Transport Capacity of a Transmission Line

Q/P	$\cos\phi$	P_{max}
0%	1.000	$0.5\frac{U^2}{X}$
10%	0.995	$0.45\frac{U^2}{X}$
20%	0.98	$0.41\frac{U^2}{X}$
30%	0.96	$0.37\frac{U^2}{X}$
40%	0.93	$0.34\frac{U^2}{X}$

causing the voltage to decrease further. Some network operators block the operation of tap changers as a last resort when the system moves toward voltage collapse.

From Figure 8.22 and Table 8.7, it is evident that the power factor is a critical parameter for voltage stability. For low power factors, voltage collapse is more probable. That is why, it is justified for long transmission lines to keep the power factor close to unity with the use of capacitors, synchronous compensators, and so on.

Example 8.13 Consider a transmission system where the load power factor is 0.93 or higher. The transport capacity of a transmission line, according to Table 8.7, is then

$$P_{max} = 0.34\frac{U^2}{X}$$

Consider a 400 kV transmission line with an impedance of 300 mΩ/km. The transfer capacity according to (8.22) is 1800 MW for a 100 km line, 360 MW for a 500 km line, and only 180 MW for a 1000 km line. In most cases, the power factor would be higher (0.93 was the lower limit after all), so the transport capacity is typically higher. However, the load power factor is not predictable, so the lowest value (worst case) has to be considered for design of the system.

Suppose that the system contains a corridor consisting of eight such transmission lines in parallel, each with a length of 500 km. The total amount of power that can be transported over this transmission corridor is $8 \times 360\,\text{MW} = 2880\,\text{MW}$.

However, the loss of one line would lead to voltage collapse in such a case. Operating under the $(N-1)$ criterion gives a secure transfer capacity of $7 \times 360\,\text{MW} = 2520\,\text{MW}$.

Using perfect reactive power compensation (e.g., with a number of static var compensators) would increase the secure transfer capacity to 3720 MW.

If, next to perfect reactive power compensation, the operating voltage is increased to 430 kV, the secure transfer capacity increases to 4300 MW. Without building extra lines, the secure transfer capacity has increased from 2520 to 4300 MW, that is, an increase by 70%.

A very simplified example system, to illustrate the impact of distributed generation on voltage stability, is shown in Figure 8.23. The system consists of two geographically separated areas, connected through two transmission lines ("tie lines" in the figure). Within each of the two areas, the system is strong enough; the limitation in transfer capacity is the voltage stability limit of the corridor between the two areas.

The concern with voltage stability is never when all lines are in operation. The $(N-1)$ criterion implies that it is not allowed to operate a transmission system for

Note that the short-circuit capacity required is the one available after the fault. This is important when considering redundancy or compliance with the $(N-1)$ criterion. For example, when a group of induction generators is connected to the grid via two cables, so at to improve the reliability, the short-circuit capacity should be 20 times the rating of the machines, with one of the cables not available. The same holds for industrial installations with large amounts of induction motor load. One may of course decide not to have any redundancy. In that case, the factor of 20 should hold for both cables in operation, but a fault in one of the cables could result in short-term voltage collapse and loss of the group of induction generators.

The calculations performed in Example 8.11 give an indication of the limitations posed by the postfault inrush of induction generators. This is obviously a worst-case situation, where peak consumption and maximum production are assumed to occur together, and exactly at this moment a three-phase fault occurs close to the substation. This may seem a high set of coincidences, but it is this worst-case approach that has resulted in the high level of reliability that we have come to expect from the transmission system. A consequence of this approach is obviously that it will make it more expensive to connect large amounts of distributed generation or large wind parks.

The operation of the transmission system should be such that no single fault results in instability. When the fault level varies, the system should be stable, even for the lowest fault level. Curtailment of production could be used during operational states with low fault levels. The curtailment of production should consist of taking induction generators out of operation, not just of reducing the production of individual generators. In this case, there is no need for an automatic disconnection after, for example, the loss of a transmission line (such as in an "intertrip" or "special protection" scheme). Under the rules for the operation of the transmission system, the system is allowed to be insecure for up to 10–30 min. This allows manual curtailment of production.

It may however be possible in some case to improve the stability by using intertrip schemes. This will strongly depend on the specific properties of the system. When short-term voltage stability is expected for a limited number of fault locations, the wind park causing the voltage collapse can be disconnected quickly whenever any of these faults occur. This can be achieved through intertrips from the breakers clearing those faults to the breakers that will disconnect the wind park. More advanced methods might be possible where only some of the turbines are disconnected and where the amount of turbines disconnected depends on the amount of production immediately before the fault.

When the voltage collapse would impact only one customer and the customer would accept being disconnected occasionally, it is allowed to operate without reserve. Such a situation could occur not only with the connection of a large wind park but also with the connection of an industrial installation. The customer would in that case accept a deterioration of the reliability against a reduction in the connection fee.

For a more accurate estimation of the hosting capacity, a detailed dynamic model of both the consumption and the generation is needed. Such a model needs to include the consumption and generation connected to nearby substations as well. Detailed models of wind parks (the main source of induction generators) are typically available.

current, with mainly a reactive component. This high reactive current results in a reduced voltage at the terminals of the motor. The reduced voltage gives a reduction in electrical torque: the electrical torque is proportional to the square of the voltage for the same rotational speed. When the electrical torque is less than the mechanical torque (the load torque), the motor will not be able to accelerate back to its normal speed: the motor will stall. The operational torque of a motor is typically around 50% of the maximum torque (the pull-out torque), so a voltage drop down to 70% will prevent the motor from coming back.

Consider a motor load of rated power S_{M} connected to a system with short-circuit capacity S_k. Immediately after the fault, the motor load consumes reactive power $6 \times S_{\mathrm{M}}$ resulting in a voltage drop

$$\Delta U = \frac{6 S_{\mathrm{M}}}{S_k} \tag{8.11}$$

The voltage drop should at most be 0.3 pu for the system to be stable, so that

$$\frac{6 S_{\mathrm{M}}}{S_k} < 0.3 \tag{8.12}$$

Rewriting this gives the following requirement for the short-circuit power, which is a rule of thumb in the design of industrial distribution systems with large amounts of induction motor load:

$$S_k > 20 \times S_{\mathrm{M}} \tag{8.13}$$

When the short-circuit capacity is more than 20 times the rated power of the motor load, the system is stable.

The behavior of induction generators is very similar to that of induction motors, as discussed, for example, in Refs. 11, 268, and 293. Both induction motors and induction generators require a large amount of reactive power after a fault. The same reasoning as given before for induction motors also applies to induction generators, so the rule of thumb for the minimum system strength will hold for both. Whenever the total amount of induction machines (generators and motors) is more than about 5% of the short-circuit capacity, short-term voltage collapse could become an issue.

Example 8.11 Consider a 132 kV substation with 2316 MVA fault level. (This is the median value for all 132 kV buses in the United Kingdom; see Table 6.9.) The peak load is 150 MW. The amount of induction motor load is estimated at 25% with 0.85 power factor.

The total rating of induction motor load supplied from this bus is

$$\frac{0.25 \times 150}{0.85} = 44.1\,\mathrm{MVA}$$

The rule of thumb gives us a maximum amount of induction motor load equal to 5% of the fault level, which is 115.8 MVA in this case. Beyond the 44.1 MVA already present on consumption side, at most 71 MVA can be added as induction generators.

Exercise 8.5 How sensitive is the result on the assumptions made for the induction motor load?

109. H.S. de Bronzeado, P.A.C. Rosas, E.A.N. Feitosa, and M.S. Miranda. Behavior of wind turbines under Brazilian wind conditions and their interaction with the grid. In *International Conference on Harmonics and Quality of Power*, Athens, Greece, October 1998.

110. P. Delarue, P. Bartholomeus, F. Minne, and E. de Jaeger. Study of harmonic currents introduced by three-phase PWM-converters connected to the grid. In *17th International Conference on Electricity Distribution (CIRED)*, Barcelona, Spain, May 2003.

111. M. Delfanti, M. Merlo, M. Pozzi, V. Olivieri, and M. Gallanti. Power flows in the Italian distribution electric system with dispersed generation. In *International Conference on Electricity Distribution (CIRED)*, Prague, June 2009.

112. G. Delille, B. François, G. Malarange, and J.L. Fraisse. Energy storage systems in distribution grids: new assets to upgrade distribution networks abilities. In *International Conference on Electricity Distribution (CIRED)*, Prague, June 2009.

113. J. Deuse and G. Bourgain, editors. *EU-DEEP Results: Integrating Distributed Energy Resources into Today's Electrical System*, ExpandDER, www.expandDER.com, 2009.

114. J. Deuse, S. Grenard, and M.H.J. Bollen. EU-DEEP integrated project: technical implications of the "hosting-capacity" of the system for DER. *International Journal of Distributed Energy Resources*, 4(1):17–34, 2008.

115. J. Deuse, S. Grenard, M.H.J. Bollen, M. Häger, and F. Sollerkvist. Effective impact of DER on distribution system protection. In *International Conference on Electricity Distribution (CIRED)*, Vienna, May 2007.

116. J. Deuse, K. Karoui, H. Crisciu, L. Gertmar, O. Samuelsson, P. Karlsson, V. Chuvychin, L. Ribickis, and M.H.J. Bollen. Interactions of dispersed energy resources with power system in normal and emergency conditions. In *International Conference on Large Electric Systems (CIGRE)*, Paris, France, August 2006.

117. E.P. Dick, D. Fisher, R. Marttila, and C. Mulkins. Point-on-wave capacitor switching and adjustable-speed drives. *IEEE Transactions on Power Delivery*, 11(3):1367–1372, 1996.

118. R. Doherty and M. O'Malley. New approach to quantify reserve demand in systems with significant installed wind capacity. *IEEE Transactions on Power Systems*, 20(2):587–595, 2005.

119. D.S. Dorr. Point of utilization power quality study results. *IEEE Transactions on Industry Applications*, 31(4):658–666, 1995.

120. D.S. Dorr, M.B. Hughes, T.M. Gruzs, R.E. Juewicz, and J.L. McClaine. Interpreting recent power quality surveys to define the electrical environment. *IEEE Transactions on Industry Applications*, 33(5):1480–1487, 1997.

121. J. Drapela and P. Toman. Interharmonic-flicker curves of lamps and compatibility level for interharmonic voltages. In *IEEE Lausanne PowerTech*, July 2007.

122. J. Du, Y. Lu, G. Zhu, and X. Lin. An asymmetrical fault location method based on communication system in distribution network with DGs. In *IEEE/PES Power Systems Conference and Exposition*, March 2009.

123. R.C. Dugan and T.E. McDermott. Operating conflicts for distributed generation on distribution systems. In *Rural Electric Power Conference*, April 2001.

124. R.C. Dugan and T.E. McDermott. Distributed generation. *IEEE Industry Applications Magazine*, 8(2):19–25, 2002.

125. R.C. Dugan, M.F. McGranaghan, S. Santosa, and H.W. Beaty. *Electric Power Systems Quality*, 2nd edition, McGraw-Hill, New York, 2003.

126. A. Dysko, C. Booth, O. Anaya-Lara, and G.M. Burt. Reducing unnecessary disconnection of renewable generation from the power system. *IET Renewable Power Generation*, 1(1):41–48, 2007.

127. J. Ehnberg. Generation reliability for isolated power systems with solar, wind and hydro generation. Licentiate thesis, Department of Electric Power Engineering, Chalmers University of Technology, Gothenburg, Sweden, 2003.

128. S.G.J. Ehnberg and M.H.J. Bollen. Simulations of global solar radiation based on cloud observations. *Solar Energy*, 78(2):157–162, 2005.

129. W. El-Khattam and T.S. Sidhu. Restoration of directional overcurrent relay coordination in distributed generation systems utilizing fault current limiter. *IEEE Transactions on Power Delivery*, 23(3):576–585, 2008.

130. Electricity Storage Association, Technologies: pumped hydro, http://www.electricitystorage.org/site/technologies/pumped_hydro/, last updated April 2009.

131. Electricity Training Association, *Power System Protection*, Vol. 3: *Application*, Institution of Electrical Engineers, London, 1995.

132. Driftuppföljning av vindkraftverk årsrapport 2008. Elforsk rapport 09:37, http://www.vindenergi. org/driftuppfolj.htm, 2009.

133. H. Emanuel, M. Schellschmidt, S. Wachtel, and S. Adloff. Power quality measurements of wind energy converters with full-scale converter according to IEC 61400-21. In *10th International Conference on Electrical Power Quality and Utilisation (EPQU)*, 2009.

134. EN 50160.

135. EN 50438, Requirements for the connection of micro-generators in parallel with public low-voltage distribution networks.

136. The Encyclopedia of Earth. http://www.eoearth.org/, accessed April 3, 2010.

137. Technical regulation for electricity generating facilities of 11 kW or lower. Technical regulations tf 3.2.1, Energinet, http://www.energinet.dk, 2008.

138. Technical regulation for thermal power station units of 1.5 MW and higher. Technical regulations tf 3.2.3, Energinet, http://www.energinet.dk, 2008.

139. Technical regulation for thermal power station units larger than 11 kW and smaller than 1.5 MW. Technical regulations tf 3.2.4, Energinet, http://www.energinet.dk, 2008.

140. Wind turbines connected to grids with voltages above 100 kV: technical regulations for the properties and the control of wind turbines. Technical regulations tf 3.2.5, Energinet, http://www.energinet.dk, 2004.

141. Wind turbines connected to grids with voltages below 100 kV: technical regulations for the properties and the control of wind turbines. Technical regulations tf 3.2.6, Energinet, http://www.energinet.dk, 2003.

142. energinet.dk, Download of market data. http://http://www.energinet.dk, 2009.

143. Energy Information Administration, Inga hydroelectric facility, November 2002. http://www.eia.doe. gov, accessed April 3, 2010.

144. Energy Information Administration, Annual energy outlook 2009, March 2009, http://www.eia.doe. gov, accessed April 3, 2010.

145. J.H.R. Enslin and P.J.M. Heskes. Harmonic interaction between a large number of distributed power inverters and the distribution network. *IEEE Transactions on Power Electronics*, 19(6):1586–1593, 2004.

146. J.H.R. Enslin, W.T.J. Hulshorst, A.M.S. Atmadji, P.J.M. Heskes, A. Kotsopoulos, J.F.G. Cobben, and P. van der Sluijs. Harmonic interaction between large numbers of photovoltaic inverters and the distribution network. In *IEEE Bologna PowerTech*, June 2003.

147. K. Eriksson. Operational experience of HVDC light. In *IEE AC–DC Power Transmission*, November 2001.

148. B. Ernst, B. Oakleaf, M.L. Ahlstrom, M. Lange, C. Moehrlen, B. Lange, U. Focken, and K. Rohrig. Predicting the wind: models and methods for wind forecasting for utility operations planning. *IEEE Power and Energy Magazine*, 5(6):78–89, 2007.

149. G. Esposito, D. Zaninelli, G.C. Lazaroiu, and N. Golovanov. Impact of embedded generation on the voltage quality of distribution networks. In *9th International Conference on Electrical Power Quality and Utilisation (EPQU)*, October 2007.

150. Balance management harmonisation and integration: 4th report. European Transmission System Operators, January 2007.

151. Experimental data of 5 experiments single site tests: Grenoble and Athens aggregation tests: United Kingdom, Germany and Greece. EU-DEEP Deliverable D8, http://www.eu-deep.com, February 2009.

152. Towards smart power networks: lessons learned from European research FP5 projects. European Communities, 2005.

153. European SmartGrids Technology Platform: vision and strategy for Europe's electricity networks of the future. European Communities, 2006.

154. European Wind Integration Study (EWIS): towards a successful integration of wind power into European electricity grids. Final report.

155. P. Fairley. Europe backs supergrids: recent efforts show hope for regional transmission planning for renewable power. *Technology Review*, http://www.technologyreview.com/energy/21742/, December 2, 2008.

156. P. Fairley. China's potent wind potential: forecasters see no need for new coal and nuclear power plants. *Technology Review*, http://www.technologyreview.com/energy/23460/, September 14, 2009.

157. F.A. Farret and M.G. Simoes. *Integration of Alternative Sources of Energy*. Wiley, Hoboken, NJ, 2006.

158. N. Femia, G. Petrone, G. Spagnuolo, and M. Vitelli. Optimizing duty-cycle perturbation of P&O MPPT technique. In *IEEE 35th Annual Power Electronics Specialists Conference (PESC'04)*, Aachen, Germany, 2004.

159. M.A. Figueroa and E. Orduna. Ultra-high-speed protection for medium voltage distribution networks with distributed generation. In *IEEE/PES Transmission and Distribution Conference and Exposition: Latin America*, August 2008.

160. M. Fila, J. Hiscock, D. Reid, P. Lang, and G.A. Taylor. Flexible voltage control in distribution networks with distributed generation: modelling analysis and field trial comparison. In *International Conference on Electricity Distribution (CIRED)*, Prague, June 2009.

161. R. Fioravanti, K. Vu, and W. Stadlin. Large-scale solutions: storage, renewables and wholesale markets. *IEEE Power and Energy Magazine*, 7(4):48–57, 2009.

162. N. Flourentzou, V.G. Agelidis, and G.D. Demetriades. VSC-based HVDC power transmission systems: an overview. *IEEE Transactions on Power Electronics*, 24(3):592–602, 2009.

163. Desertec Foundation, http://www.desertec.org/, 2009.

164. B. Fox, D. Flynn, L. Bryans, N. Jenkins, D. Milborrow, M. O'Malley, R. Watson, and O. Anaya-Lara. *Wind Power Integration: Connection and System Operational Aspects*, The Institution of Engineering and Technology, Stevenage, UK, 2008.

165. FRCC automatic underfrequency load shedding program. Regional Reliability Standard PRC-006-FRCC-01, Florida Reliability Coordination Council, http://www.frcc.com, 2009.

166. W. Freitas, W. Xu, C.A. Affonso, and Z. Huang. Comparative analysis of ROCOF and vector surge relays for distributed generation applications. *IEEE Transactions on Power Delivery*, 20(2):1315–1324, 2005.

167. L. Freris and D. Infield. *Renewable Energy in Power Systems*, Wiley, Chichester, UK, 2008.

168. J. Frunt, W.L. Kling, J.M.A. Myrzik, F.A. Nobel, and D.A.M. Klaar. Effects of further integration of distributed generation on the electricity market. In *Proceedings of the 41st International Universities Power Engineering Conference (UPEC)*, September 2006.

169. D.E. Frye, S. Pond, and W.P Elliott. Note on the kinetic energy spectrum of coastal winds. *Monthly Weather Review*, 100(9):671–673, 1972.

170. H.B. Funmilayo and K.L. Butler-Purry. An approach to mitigate the impact of distributed generation on the overcurrent protection scheme for radial feeders. In *IEEE/PES Power Systems Conference and Exposition*, March 2009.

171. GE 1.5 MW, 60 Hz wind turbine general data. Technical report, GE Wind Energy, http://www.pes-psrc.org, 2003.

172. M. Geidl. Protection of power systems with distributed generation: state of the art. Technical report, Power Systems Laboratory, Swiss Federal Institute of Technology, Zürich, July 2005.

173. H. Geng, D. Xu, B. Wu, and G. Yang. Active damping for PMSG based WECS with DC link current estimation. *IEEE Transactions on Industrial Electronics*, 2010.

174. J.M. Gers and E.J. Holmes. *Protection of Electricity Distribution Networks*, The Institution of Electrical Engineers, Stevenage, UK, 1998.

175. P. Gipe. *Wind Power: Renewable Energy for Home, Farm, and Business*, Chelsea Green Publishing Company, White River Junction, VT, 2004.

176. A. Girgis and S. Brahma. Effect of distributed generation on protective device coordination in distribution system. In *Large Engineering Systems Conference on Power Engineering (LESCOPE)*, July 2001.

177. 50Hertz Transmission GmbH. Wind power production data 2009, www.vattenfall.de, accessed January 16, 2010.

178. T. Gönen. *Electric Power Distribution System Engineering*, McGraw-Hill, New York, 1986.

179. J.J. Grainger and W.D. Stevenson. *Power System Analysis*, McGraw-Hill, New York, 1994.

180. W. Grant, D. Edelson, J. Dumas, J. Zack, M. Ahlstrom, J. Kehler, P. Storck, J. Lerner, K. Parks, and C. Finley. Change in the air: operational challenges in wind-power production and prediction. *IEEE Power and Energy Magazine*, 7(6):47–58, 2009.

181. A. Greenwood. *Electrical Transients in Power Systems*, 2nd edition, Wiley, New York, 1991.

182. A.F. Gubina, A. Keane, P. Meibom, J. O'Sullivan, O. Goulding, T. McCartan, and M. O'Malley. New tool for integration of wind power forecasting into power system operation. In *IEEE Bucharest PowerTech*, June 2009.

183. A.E. Guile and W. Paterson. *Electrical Power Systems*, Vol. 2, Pergamon Press, Oxford, 1977.

184. E.W. Gunther and H. Mehta. A survey of distribution system power quality: preliminary results. *IEEE Transactions on Power Delivery*, 10(1):322–329, 1995.

185. M. Häger, F. Sollerkvist, and M.H.J. Bollen. The impact of distributed energy resources on distribution-system protection. In *Nordic Distribution Automation Conference (NORDAC)*, Stockholm, August 2006.

186. S. Häggmark, V. Neimane, U. Axelsson, P. Holmberg, G. Karlsson, K. Kauhaniemi, M. Olsson, and C. Liljegren. Aspects of different distributed generation technologies: CODGUNET WP3, March 2003.

187. M. Halpin and E. De Jaeger. Suggestions for overall EMC coordination with regard to rapid voltage changes. In *20th International Conference on Electricity Distribution (CIRED)*, June 2009.

188. M. Halpin, R. Cai, E. De Jaeger, I. Papac, S. Perera, and X. Yang. A review of flicker objectives related to complaints, measurements and analysis techniques. In *20th International Conference on Electricity Distribution (CIRED)*, June 2009.

189. R. Hara, H. Kita, T. Tanabe, H. Sugihara, A. Kuwayama, and S. Miwa. Testing the technologies: demonstration grid-connected photovoltaic projects in Japan. *IEEE Power and Energy Magazine*, 7(3):77–85, 2009.

190. H. Haraguchi, S. Morimoto, and M. Sanada. Suitable design of a PMSG for a small-scale wind power generator. In *International Conference on Electrical Machines and Systems (ICEMS'09)*, 2009.

191. H. Haraguchi, S. Morimoto, and M. Sanada. Suitable design of a PMSG for large-scale wind power generator. In *IEEE Energy Conversion Congress and Exposition (ECCE'09)*, 2009.

192. J. Harrison. Micro cogeneration in Britain. In M. Pehnt, editor, *Microcogeneration: Towards Decentralized Energy Systems*, Springer, 2005.

193. W.G. Hartmann. How not to nuisance-trip distributed generation. In *IEEE Industrial and Commercial Power Systems Technical Conference*, May 2005.

194. Y. Hase. *Handbook of Power System Engineering*, Wiley, Chichester, UK, 2008.

195. F. Hassan. Converter-interfaced distributed generation-grid interconnection issues. PhD thesis, Chalmers University of Technology, Gothenburg, Sweden, 2007.

196. F. Hassan and M. Bollen. Active islanding of a current-controlled converter-interfaced DG. In *IEEE Bucharest PowerTech*, July 2009.

197. F. Hassan, M. Wämundson, and M. Bollen. Combined active and reactive power control with converter interfaced energy sources. In *IEEE International Conference on Sustainable Energy Technologies (ICSET)*, 2008.

198. J.E. Hay. Study of shortwave radiation on non-horizontal surfaces. Report 79-12, Atmospheric Environmental Services, Ontario, 1979.

199. J. Van Hecke, N. Janssens, J. Deuse, and F. Promel. Coordinated voltage control experience in Belgium. In *CIGRE Session*, Paris, France, Report 38-111, 2000.

200. S. Heier. *Grid Integration of Wind Energy Conversion Systems*, 2nd edition, Wiley, Chichester, UK, 2006.

201. S.D. Henry, D.T. Rizy, and T.L. Baldwin. The application of droop-control in distributed energy resources to extend the voltage collapse margin. In *IEEE Industry Application Society Industrial and Commercial Power Systems Technical Conference*, 2008.

202. C.A. Hernandez-Aramburo, T.C. Green, and N. Mugniot. Fuel consumption minimization of a microgrid. *IEEE Transactions on Industry Applications*, 41(3):673–681, 2005.

203. P.J.M. Heskes, J.F.G. Cobben, and H.H.C. de Moor. Harmonic distortion in residential areas due to large scale PV implementation is predictable. *International Journal of Distributed Energy Resources*, 1(1):17–32, 2005.

204. M. Hird, N. Jenkins, and P. Taylor. 11 kV voltage controller: practical considerations. In *International Conference on Electricity Distribution (CIRED)*, Barcelona, 2003.

205. J. Hiscock, N. Hiscock, and A. Kennedy. Advanced voltage control for networks with distributed generation. In *International Conference on Electricity Distribution (CIRED)*, Vienna, May 2007.

206. M. Hojo, H. Hatano, and Y. Fuwa. Voltage rise suppression by reactive power control with cooperating photovoltaic generation systems. In *International Conference on Electricity Distribution (CIRED)*, Prague, June 2009.

207. L. Holdsworth, J.B. Ekanayake, and N. Jenkins. Power system frequency response from fixed-speed and double-fed induction generator based wind turbines. *Wind Energy*, 7:21–35, 2004.

208. A. Holm, P. Owe, M.H.J. Bollen, J. Pylvänäinen, and H. Paananen. CDI service level index: pilot use in network planning. In *International Conference on Electricity Distribution (CIRED)*, Prague, June 2009.

209. S.H. Horowith and A.G. Phadke. *Power System Relaying*, 2nd edition, Research Studies Press, Taunton, UK, 1995.

210. IEC 61000-2-2, Compatibility levels for low-frequency conducted disturbances and signalling in public low-voltage power supply systems.

211. IEC 61000-3-6, Assessment of emission limits for the connection of distorting installations to MV, HV and EHV power systems.

212. IEC 61000-3-7, Assessment of emission limits for the connection of fluctuating installations to MV, HV and EHV power systems.

213. IEC 61000-3-13, Assessment of emission limits for the connection of unbalanced installations to MV, HV and EHV power systems.

214. IEC 61000-3-15, Assessment of low-frequency electromagnetic immunity and emission requirements for dispersed generation systems in LV network.

215. IEC 61000-4-7, General guide on harmonics and interharmonics measurements and instrumentation, for power supply systems and equipment connected thereto.

216. IEC 61850, Communication networks and systems in substations. Part 1. Introduction and overview.

217. IEEEE Standard 519-1992, Recommended practices and requirements for harmonic control in electrical power systems.

218. IEEEE Standard 1346-1998, Recommended practice for evaluating electric power system compatibility with electronic process equipment.

219. IEEEE Standard 1366-2003, Guide for electric power distribution reliability indices,.

220. IEEEE Standard 1547-2003, Standard for interconnecting distributed resources with electric power systems.

221. D. Infield and F. Li. Integrating micro-generation into distribution systems: a review of recent research. In *IEEE Power Engineering Society General Meeting*, July 2008.

222. D.G. Infield, P. Onions, A.D. Simmons, and G.A. Smith. Power quality from multiple grid-connected single-phase inverters. *IEEE Transactions on Power Delivery*, 19(4):1983–1989, 2004.

223. T. Inoue, H. Taniguchi, Y. Ikeguchi, and K. Yoshida. Estimation of power system inertia constant and capacity of spinning-reserve support generators using measured frequency transients. *IEEE Transactions on Power Systems*, 12(1):136–143, 1997.

224. International Energy Agency, Key world energy statistics 2009, http://www.iea.org, 2009.

225. International Power First Hydro Company, http://www.fhc.co.uk/index.asp, accessed December 29, 2009.

226. A. Ishchenko, J.M.A. Myrzik, and W.L. Kling. Transient stability analysis of distribution network with dispersed generation. In *Proceedings of the 41st International Universities Power Engineering Conference*, September 2006.

227. IVGFT Task Force, Accommodating high levels of variable generation: to ensure the reliability of the bulk power system. Technical report, NERC, http://www.nerc.com, April 2009.

228. A. Jager. Modélisation et simulation d'une microturbine à gaz. Internal technical report, restricted distribution, Laborelec, Linkebeek, Belgium, November 2003.

229. J. Jäger, T. Keil, L. Shang, and R. Krebs. New protection co-ordination methods in the presence of distributed generation. In *8th IEE International Conference on Developments in Power System Protection*, Vol. 1, April 2004.

230. J. Jäger and L. Shang. High-impedance protection applications for tripping acceleration in networks with DG. In *IEEE/PES Transmission and Distribution Conference and Exhibition: Asia and Pacific*, 2005.

231. N. Janssens and J. Trecat. General aspects of the control, regulation and security of the energy network in alternating current. In M. Crappe, editor, *Electric Power Systems*, Wiley, 2008, pp. 5–35.

232. N. Jenkins, R. Allan, P. Crossley, D. Kirschen, and G. Strbac. *Embedded Generation*, Institution of Electrical Engineers, London, 2000.

233. N. Jenkins, J. Peças Lopes, and N. Hatziargyriou. One day short course on integration of dispersed generation in distribution networks. In *Oporto Power Tech*, September 2001.

234. W. Jewell and R. Ramakumar. The effects of moving clouds on electric utilities with dispersed photovoltaic generation. *IEEE Transactions on Energy Conversion*, 2(4):570–576, 1987.

235. K. Jiarong and X. Shaojun. Research on the power sharing of the parallel inverters without control interconnection basing on droop characteristics. In *CES/IEEE 5th International Electronics and Motion Control Conference, IPEMC*, 2006.

236. V. John, Z. Ye, and A. Kolwalkar. Investigation of anti-islanding protection of power converter based distributed generators using frequency domain analysis. *IEEE Transactions on Power Electronics*, 19(5):1177–1183, 2004.

237. D. Jovcic. Interconnecting offshore wind farms using multiterminal VSC-based HVDC. In *IEEE Power Engineering Society General Meeting*, 2006.

238. K. Kauhaniemi and L. Kumpulainen. Impact of distributed generation on the protection of distribution networks. In *8th IEE International Conference on Developments in Power System Protection*, April 2004.

239. G. Kaur and M.Y. Vaziri. Effects of distributed generation (DG) interconnections on protection of distribution feeders. In *IEEE Power Engineering Society General Meeting*, 2006.

240. T. Keppler, N.R. Watson, J. Arrillaga, and S. Chen. Theoretical assessment of light flicker caused by sub- and interharmonic frequencies. *IEEE Transactions on Power Delivery*, 18(1):329–333, 2003.

241. E.C. Kern, E.M. Gulachenski, and G.A. Kern. Cloud effects on distributed photovoltaic generation: slow transients at the Gardner, Massachusetts photovoltaic experiment. *IEEE Transactions on Energy Conversion*, 4(2):184–190, 1989.

242. E.C. Kern and M.C. Russel. Spatial and temporal irradiance variations over large array fields. In *IEEE Photovoltaic Specialists Conference*, Las Vegas, NV, 1988.

243. E.W. Kimbark. *Power System Stability*, IEEE Press, New York, 1995.

244. B. Kirby, J. Kueck, T. Laughner, and K. Morris. Spinning reserve from hotel load response: initial progress. Technical report ORNL/TM-2008/217, Oak Ridge National Laboratory, October 2008.

245. B.J. Kirby. Spinning reserve from responsive loads. Technical report ORNL/TM-2003/19, Oak Ridge National Laboratory, March 2003.

246. S.A. Klein. Accommodating wind power variability by real-time adjustment of controllable loads. In *IEEE Power Engineering Society General Meeting*, Pittsburgh, PA, Paper 1430, July 2008.

247. J. Klimstra. Fault ride-through capability of engine-driven power plants. In *PowerGen Europe*, Cologne, http://www.wartsila.com, May 2009.

248. W.L. Kling, D.A.M. Klaar, J.H. Schuld, A.J.L.M. Kanters, C.G.A. Koreman, H.F. Reijnders, and C.J.G. Spoorenberg. Phase shifting transformers installed in the Netherlands in order to increase available international transmission capacity. In *CIGRE Sessions*, Paris, France, 2004.

249. H. Knudsen and J.N. Nielsen. Introduction to the modelling of wind turbines. In T. Ackermann, editor, *Wind Power in Power Systems*, Wiley, Chichester, UK, 2005, pp. 525–554.

250. A. Kotsopoulos, P.J.M. Heskes, and M.J. Jansen. Zero-crossing distortion in grid-connected PV inverters. *IEEE Transactions on Industry Electronics*, 52(2):558–565, 2005.

251. X.I. Koutiva, T.D. Vrionis, N.A. Vovos, and G.B. Giannakopoulos. Optimal integration of an offshore wind farm to a weak AC grid. *IEEE Transactions on Power Delivery*, 21(2):987–994, 2006.

252. P.C. Krause, O. Wasynczuk, and S.D. Sudhoff. *Analysis of Electric Machinery*, IEEE Press, New York, 1995.

253. E. Kreyszig. *Advanced Engineering Mathematics*, 7th edition, Wiley, New York, 1993.

254. L.K. Kumpulainen and K.T. Kauhaniemi. Aspects of the effects of distributed generation in single-line-to-earth faults. In *International Conference on Future Power Systems*, November 2005.

255. L.K. Kumpulainen and K.T. Kauhaniemi. Analysis of the impact of distributed generation on automatic reclosing. In *IEEE, PES Power Systems Conference and Exposition*, Vol. 1, October 2004, pp. 603–608.

256. P. Kundur. *Power System Stability and Control*, McGraw-Hill, New York, 1994.

257. P.D. Ladakakos, M.G. Ioannides, and M.I. Koulouvari. Assessment of wind turbines impact on the power quality of autonomous weak grids. In *International Conference on Harmonics and Quality of Power (ICHQP)*, Athens, Greece, October 1998.

258. E. Lakervi and E.J. Holmes. *Electricity Distribution Network Design*, Institution of Electrical Engineers, London, 1995.

259. Å. Larsson. Flicker emission of wind turbines during continuous operation. *IEEE Transactions on Energy Conversion*, 17(1):114–118, 2002.

260. E.O.A. Larsson, M.H.J. Bollen, M.G. Wahlberg, C.M. Lundmark, and S.K. Rönnberg. Measurements of high-frequency (2–150 kHz) distortion in low-voltage networks. *IEEE Transactions on Power Delivery*, 25(3):1749–1757, 2010.

261. E.O.A. Larsson, C.M. Lundmark, and M.H.J. Bollen. Distortion of fluorescent lamps in the frequency range 2–150 kHz. In *International Conference on Harmonics and Quality of Power (ICHQP)*, October 2006.
262. M. Larsson. Coordinated voltage control in electric power systems. Doctoral dissertation, Department of Industrial Electrical Engineering and Automation, Lund University, 2000.
263. R.H. Lasseter. Microgrids. In *IEEE Power Engineering Society Winter Meeting*, January 2002.
264. M. A. Laughton and D. J. Warne. *Electrical Engineers Reference Book*, 16th edition, Elsevier Science, 2003.
265. D.M. Laverty, D.J. Morrow, T. Littler, and P.A. Crossley. Loss-of-mains detection by Internet based differential ROCOF. In *International Conference on Developments on Power System Protection (DPSP)*, March 2008.
266. J. Lawhorn, D. Osborn, J. Caspary, B. Nickell, D. Larson, W. Lasher, and M.E. Rahman. The view from the top: perspective on value-based planning processes for delivering renewable energy resources. *IEEE Power and Energy Magazine*, 7(6):76–88, 2009.
267. C.D. Le. Fault ride-through of wind parks with induction generators. Technical report R09-552, STRI AB, Ludvika, Sweden, 2009.
268. C.D. Le and M.H.J. Bollen. Ride-through of induction generator based wind park with switched capacitor, SVC, or STATCOM. In *IEEE/PES General Meeting*, 2010.
269. W.-J. Lee and R. Kenarangui. Energy management for motors, system, and electrical equipment. *IEEE Transactions on Industry Applications*, 38(2):602–607, 2002.
270. E.E. Lewis. *Introduction to Reliability Engineering*, Wiley, New York, 1987.
271. H. Li and Z. Chen. Transient stability analysis of wind turbines with induction generators considering blades and shaft flexibility. In *33rd Annual Conference of the IEEE Industrial Electronics Society (IECON)*, November 2007.
272. H.Y. Li and H. Leite. Increasing distributed generation using automatic voltage reference setting technique. In *IEEE Power Engineering Society General Meeting*, Pittsburgh, PA, July 2008.
273. J. Li, N. Samaan, and S. Williams. Modeling of large wind farm systems for dynamic and harmonics analysis. In *IEEE/PES Transmission and Distribution Conference and Exposition*, April 2008.
274. X. Lin, Y.-P. Lu, and L.-H. Wang. The study of the protection revision method based on the DG effect to protection sensitivity. In *IEEE International Conference on Sustainable Energy Technologies (ICSET)*, November 2008.
275. B. Lindgren. A power converter for photovoltaic applications. Licentiate thesis, Chalmers University of Technology, Gothenburg, Sweden, February 2000.
276. B. Lindgren. Power-generation, power-electronics and power-systems issues of power converters for photovoltaic applications. PhD thesis, Chalmers University of Technology, Gothenburg, Sweden, December 2002.
277. M. Lindholm and T.W. Rasmussen. Harmonic analysis of doubly fed induction generators. In *Power Electronics and Drive Systems*, November 2003.
278. T.L. Littler, B. Fox, and D. Flynn. Measurement-based estimation of wind farm inertia. In *IEEE St. Petersburg Power Tech*, June 2005.
279. K.-H. Liu and L. Wang. Analysis of measured harmonic currents and voltages contributed by a commercial wind power system. In *IEEE Power Engineering Society General Meeting*, June 2007.

280. P. Loevenbruck and A. Sacre. Impact of distributed generation on losses, draw off costs from transmission network and investments of the French distribution network operator ERDF. In *International Conference on Electricity Distribution (CIRED)*, Prague, June 2009.

281. L.A.C. Lopes and Y.Z. Zhang. Islanding detection assessment of multi-inverter systems with active frequency drifting methods. *IEEE Transactions in Power Delivery*, 23(1):480–486, 2008.

282. P. Lund. The Danish cell project. Part 1. Background and general approach. In *IEEE Power Engineering Society General Meeting*, June 2007.

283. P. Lund. Cell controller pilot project: intelligent mobilization to distributed power generation. In *3rd International Conference on Integration of Renewable and Distributed Energy Resources*, Nice, France, December 2008.

284. C.M. Lundmark, S. K. Rönnberg, M. Wahlberg, E.O.A. Larsson, and M.H.J. Bollen. EMC filter common mode resonance. In *IEEE Bucharest PowerTech*, June 2009.

285. C.C.Y. Ma and M. Iqbal. Statistical comparison of models for estimating solar radiation on incident surfaces. *Solar Energy*, 31(3):313–317, 1983.

286. D.J.C. MacKay. *Sustainable Energy: Without the Hot Air*, UIT Cambridge, Cambridge, UK, 2009.

287. K.J.P. Macken, M.H.J. Bollen, and R.J.M. Belmans. Mitigation of voltage dips through distributed generation systems. *IEEE Transactions on Industry Applications*, 40(6):1686–1693, 2004.

288. L. Magnell. Måttliga balanskostnader vid storskalig introduktion av vindkraft. *ERA*, (10):54, 2008.

289. S.T. Mak and F. de la Rosa. Waveform distortion issues based on field data recordings in MV and LV facilities. In *Power Engineering Society General Meeting*, June 2005.

290. K. Maki, S. Repo, and P. Jarventausta. General procedure of protection planning for installation of distributed generation in distribution network. *International Journal of Distributed Energy Resources*, 2(1):47–64, 2006.

291. K. Maki, S. Repo, and P. Jarventausta. Methods for assessing the protection impacts of distributed generation in network planning activities. In *IET 9th International Conference on Developments in Power System Protection*, March 2008.

292. J.F. Manwell, J.G. McGowan, and A.L. Rogers. *Wind Energy Explained: Theory, Design and Application*, Wiley, Chichester, UK, 2002.

293. M. Martins, Y. Sun, and M.H.J. Bollen. Voltage stability of wind parks and similarities with large industrial systems. In *Nordic Wind Power Conference*, Gothenburg, Sweden, March 2004.

294. J.I. Marvik, A. Petterteig, and H.K. Hoidalen. Analysis of fault detection and location in medium voltage radial networks with distributed generation. In *IEEE Lausanne PowerTech*, July 2007.

295. M.N. Marwali and A. Keyhani. Control of distributed generation systems. Part I. Voltages and currents control. *IEEE Transactions on Power Electronics*, 19(6):1541–1550, 2004.

296. G.M. Masters. *Renewable and Efficient Electric Power Systems*, Wiley–Interscience, Hoboken, NJ, 2004.

297. R.M Mathur and R.K. Varma. *Thyristor-Based FACTS Controllers for Electrical Transmission Systems*, IEEE Press, New York, 2002.

298. L. Max. Design and control of a DC collection grid for a wind farm. PhD thesis, Chalmers University of Technology, Gothenburg, Sweden, 2009.

299. M.F. McGranaghan, T.E. Grebe, G. Hensley, M. Samotyj, and T. Singh. Impact of utility-switched capacitors on customer systems. II. Adjustable-speed drive concerns. *IEEE Transactions on Power Delivery*, 6(4):1623–1628, 1991.

300. J. Meeus. *Astronomical Algorithms*, 2nd edition, Willmann-Bell, Richmond, VA, 1999.

301. L. Mihalache. Voltage amplitude regulation method for paralleling 400 Hz GPU inverters without intercommunication. In *IEEE Applied Power Electronics Conference and Exposition (APEC)*, 2009.

302. F. Minne. Analysis of the measurements done on the wind turbine of Schelle. Internal technical report, restricted distribution, Laborelec, Linkebeek, Belgium, September 2002.

303. C.V. Moreno, H.A. Duarte, and J.U. Garcia. Propagation of flicker in electric power networks due to wind energy conversions systems. *IEEE Transactions on Energy Conversion*, 17(2):267–272, 2002.

304. J. Morren and S.W.H. de Haan. Short-circuit current of wind turbines with double fed induction generator. *IEEE Transactions on Energy Conversion*, 22(1):174–180, 2007.

305. J. Morrena, J. Pierik, and S.W.H. de Haan. Inertial response of variable speed wind turbines. *Electric Power Systems Research*, 76(11):980–987, 2006.

306. A.B. Morton, S. Cowdroy, J.R.A. Hill, M. Halliday, and G.D. Nicholson. AC or DC? Economics of grid connection design for offshore wind farms. In *IEE International Conference on AC and DC Power Transmission*, March 2006.

307. M. Moyer. Fusions false dawn. *Scientific American*, March 2010.

308. C. Mozina. A tutorial on the impact of distributed generation (DG) on distribution systems. In *Texas A&M Relay Conference*, http://www.scribd.com/doc/8913190/DISTRIBUTED-GENERATION-DG, 2008.

309. C.J. Mozina. Distributed generator interconnect protection practices. In *IEEE/PES Transmission and Distribution Conference and Exhibition*, May 2006.

310. C.J. Mozina. The impact of distributed generation. *PAC World (Power, Automation and Control Magazine)*, 2008, pp. 18–25.

311. Index Mundi, Chili hydroelectric power production by year, http://www.indexmundi.com. accessed March 3, 2010.

312. Index Mundi, Spain hydroelectric power production by year, http://www .indexmundi.com, accessed March 3, 2010.

313. S.M. Muyeen, M.H. Ali, R. Takahashi, T. Murata, J. Tamura, Y. Tomaki, A. Sakahara, and E. Sasano. Comparative study on transient stability analysis of wind turbine generator system using different drive train models. *IET Renewable Power Generation*, 1(2):131–141, 2007.

314. P. Myrda. A call to action: realizing the U.S. smart grid. *PAC World (Protection, Automation and Control Magazine)*, 2008, pp. 47–51.

315. A.D. Rajapakse, N. Perera, and T.E. Buchholzer. Isolation of faults in distribution networks with distributed generators. *IEEE Transactions on Power Delivery*, 23(4):2347–2355, 2008.

316. J. Nacsa, G.L. Kovács, and G. Haidegger. Intelligent, open architecture controller using knowledge server. *SPIE Proceedings*, 4563:25–33, 2001.

317. M. Nagpal, F. Plumptre, R. Fulton, and T.G. Martinich. Dispersed generation interconnection utility perspective. *IEEE Transactions on Industry Applications*, 42(3):864–872, 2006.

318. R. Nair. Hydro-turbine and generator design relationships. Part II. Bulb and rim units. *IEEE Transactions on Energy Conversion*, 1(2):164–171, 1986.

319. Grid Code Review Panel National Grid. Annual summary report for ROCOF tripping incidents (1 August 2003 to 31 July 2004). Technical Report Paper GCRP 04/24, National Grid, September 2004.
320. National Grid PLC, GB Seven-year statement, http://www.nationalgrid.com, May 2008.
321. New grid for 90 percent renewables. New Energy 6/2009.
322. L. Nielsen, L. Prahm, R. Berkowicz, and K. Conradsen. Net incoming radiation estimated from hourly global radiation and/or cloud observations. *Journal of Climatology*, 1(3):255–272, 1981.
323. Nordic grid code 2007 (Nordic collection of rules). Technical report, Nordel, http://www.entsoe.eu, 2007.
324. A. Notholt-Vergara, R. Bründlinger, E. de Jong, and D. Giebel. Protection, dynamic grid support and fault ride through. International White Book on the Grid Integration of Static Converters, http://www.der-lab.net/, September 2009.
325. G. Notton, C. Cristofari, M. Muselli, P. Poggi, and N. Heraud. Hourly solar irradiations estimation: from horizontal measurements to inclined data. In *1st International Symposium on Environment Identities and Mediterranean Area*, July 2006.
326. A. Nourai and C. Schafer. Changing the electricity game: storage for transmission and distribution. *IEEE Power and Energy Magazine*, 7(4):42–47, 2009.
327. L.F. Ochoa, A. Keane, C. Dent, and G.P. Harrison. Applying active network management schemes to an Irish distribution network for wind power maximisation. In *International Conference on Electricity Distribution (CIRED)*, Prague, June 2009.
328. M. Odenberg, Y. Fredriksson, and T. Kåberger. Tillfälliga pristoppar visar att elmarknaden fungerar (Occasional price peaks proof that the electricity market is functioning, in Swedish), Dagens Nyheter, http://www.dn.se, January 25, 2010.
329. Queensland Department of National Resources and Mining. Historical Meteorological data surfaces, http://www.longpaddock.qld.gov.au/silo/, accessed April 19, 2009.
330. A.R. Oliva and J.C. Balda. A PV dispersed generator: a power quality analysis within the IEEE 519. *IEEE Transactions on Power Delivery*, 18(2):525–530, 2003.
331. G.A. St. Onge and W.A. Stewart. Rock island's second powerhouse uses bulb-type hydro units. *IEEE Transactions on Power Apparatus and Systems*, 96(5):1690–1696, 1977.
332. Svenska Kraftnät (Swedish Transmission System Operator). Sveriges produktion, förbrukning, import/export, http://www.svk.se/Energimarknaden/El/Statistik/, accessed March 28, 2009.
333. J.A. Orr and A.E. Emanuel. On the need for strict second harmonic limits. *IEEE Transactions on Power Delivery*, 15(3):967–971, 2000.
334. A.J. Pansini. *Power Systems Stability*. PenWell Publishing, Tulsa, OK, 1992.
335. M.P. Papadopoulos, S.A. Papathanassiou, S.T. Tentzerakis, and N.G. Boulaxis. Investigation of the flicker emission by grid connected wind turbines. In *International Conference on Harmonics and Quality of Power*, 1998.
336. T.M. Papazoglou and A. Gigandidou. Impact and benefits of distributed wind generation on quality and security in the case of the Cretan EPS. In *CIGRE/IEEE International Symposium, Quality and Security of Electric Power Delivery Systems*, Montreal, Canada, October 2003.
337. M.R. Patel. *Wind and Solar Power Systems*, CRC Press, Boca Rotan, FL, 2002.
338. M. Pavella, D. Ernst, and D. Ruiz-Vega. *Transient Stability of Power Systems: A Unified Approach to Assessment and Control*, Kluwer Academic Publishers, Norwell, MA, 2000.

339. M. Pavella and P.G. Murthy. *Transient Stability of Power Systems: Theory and Practice*, Wiley, Chichester, UK, 1994.

340. A. Payman, S. Pierfederici, and F.M. Tabar. Energy management in a fuel cell/supercapacitor multisource/multiload electrical hybrid systems. *IEEE Transactions on Power Electronics*, 24(12):2681–2691, 2009.

341. F.F. Pazos. Failure of anti-islanding protections in large PV plants. In *International Conference on Electricity Distribution (CIRED)*, Prague, June 2009.

342. D. Peftitsis, G. Adamidis, and A. Balouktsis. A new MPPT method for photovoltaic generation systems based on hill climbing algorithm. In *Proceedings of the International Conference on Electrical Machines*, 2008.

343. D. Peftitsis, G. Adamidis, and A. Fyntanakis. Modulation of three phase rectifier in connection with PMSG for maximum energy extraction. In *European Conference on Power Electronics and Applications (EPE'09)*, 2009.

344. R. Perez, P. Ineichen, R. Seals, J. Michalsky, and R. Stewart. Modeling daylight availability and irradiance components from direct and global irradiance. *Solar Energy*, 44(5):271–289, 1990.

345. R. Perez, R. Seals, P. Ineichen, R. Stewart, and D. Menicucci. A new simplified version of the Perez diffuse irradiance model for tilted surfaces. *Solar Energy*, 39(3):221–231, 1987.

346. R. Perez, R. Stewart, R. Arbogast, J. Seals, and J. Scott. An anisotropic hourly diffuse radiation model for surfaces: description, performance validation, site dependency evaluation. *Solar Energy*, 36(6):221–231, 1986.

347. S. Phuttapatimok, A. Sangswang, M. Seapan, D. Chenvidhya, and K. Kirtikara. Evaluation of fault contribution in the presence of PV grid-connected systems. In *33rd IEEE Photovoltic Specialists Conference*, May 2008.

348. R. Piwko, E. Camm, A. Ellis, E. Muljadi, R. Zavadil, R. Walling, M. O'Malley, G. Irwin, and S. Saylors. A whirl of activity. *IEEE Power and Energy Magazine*, 7(6):26–35, 2009.

349. M. Prodanovic, K. De Brabandere, J. Van Den Keybus, T. Green, and J. Driesen. Harmonic and reactive power compensation as ancillary services in inverter-based distributed generation. *IET Generation, Transmission and Distribution*, 1(3):432–438, 2007.

350. M. Prodanovic and T.C. Green. Control and filter design of three-phase inverters for high power quality grid connection. *IEEE Transactions on Power Electronics*, 18(1):373–380, 2003.

351. Generator interconnection facility study report. Technical report, Transmission Department, Progress Energy Carolinas, http://www.oatioasis.com, 2004.

352. G.C. Pyo, H.W. Kang, and S.I. Moon. A new operation method for grid-connected PV system considering voltage regulation in distribution system. In *IEEE Power Engineering Society General Meeting*, Pittsburgh, PA, July 2008.

353. M.R. Qader, M.H.J. Bollen, and R.N. Allan. Stochastic prediction of voltage sags in a large transmission system. *IEEE Transactions on Industry Applications*, 35(1):152–162, 1999.

354. V. Quaschning. *Erneuerbare Energier und Klimaschutz*. Carl Hanser Verlag, München, 2008.

355. A.H. Rafa, O. Anaya-Lara, and J.R. McDonald. Flexible control of converter-interfaced microgeneration. In *IEEE Power Engineering Society General Meeting*, Pittsburgh, PA, July 2007.

356. C.J. Ramos, A.P. Martins, and A.S. Carvalho. Rotor current controller with voltage harmonics compensation for a DFIG operating under unbalanced and distorted stator voltage. In *33rd Annual Conference of the IEEE Industrial Electronics Society (IECON)*, November 2007.

357. A. Reidy and R. Watson. Comparison of VSC based HVDC and HVAC interconnections to a large offshore wind farm. In *IEEE Power Engineering Society General Meeting*, June 2005.

358. B. Renders. Converter-gekoppelde decentrale generatoren en netkwaliteit in laagspanningsnetten (in Dutch). PhD thesis, Vakgroep Elektrische Energie, Systemen en Automatisering, Universiteit Gent, Ghent, Belgium, 2009.

359. B. Renders, L. Degroote, B. Meersman, and L. Vandevelde. Re-adding damping to the distribution network: harmonics and voltage dips. In *13th International Conference on Harmonics and Quality of Power (ICHQP)*, 2008.

360. B. Renders, K. De Gusseme, W.R. Ryckaert, K. Stockman, L. Vandevelde, and M.H.J. Bollen. Distributed generation for mitigating voltage dips in low-voltage distribution grids. *IEEE Transactions on Power Delivery*, 23(3):1581–1588, 2008.

361. B. Renders, K.D. Gusseme, W.R. Ryckaert, and L. Vandevelde. Input impedance of grid-connected converters with programmable harmonic resistance. *IET Electric Power Applications*, 1(3):355–361, 2007.

362. Renewable Energy Development, Tidal power: Kaipara Harbour, March 11, 2008, http://renewableenergydev.com, accessed March 31, 2010.

363. Renewable Energy Development, Tidal power: Wando Hoenggan Waterway. March 17, 2008, http://renewableenergydev.com, accessed March 31, 2010.

364. La Rance Tidal Power Plant. Renewable energy in the UK, http://www.reuk.co.uk, accessed March 31, 2010.

365. P. Richardson and A. Keane. Impact of high penetrations of micro-generation on low voltage distribution networks. In *International Conference on Electricity Distribution (CIRED)*, Prague, June 2009.

366. Data-ICT-dienst Rijkswaterstaat, Delft, The Netherlands, http://www.waterbase.nl/, accessed March 14, 2010.

367. B. Roberts. Capturing grid power: performance, purpose and promise of different storage technologies. *IEEE Power and Energy Magazine*, 7(4):32–41, 2009.

368. D. Roberts, J. Hiscock, A. Creighton, N. Johnson, T. Moore, and A.J. Beddoes. Increasing voltage headroom for the connection of distributed generation. In *International Conference on Electricity Distribution (CIRED)*, Prague, Paper 0468, June 2009.

369. A.G. González Rodríguez, A. González Rodríguez, and M. Burgos Payán. Estimating wind turbines mechanical constants. In *International Conference on Renewable Energy and Power Quality*, www.icrepq.com, 2007.

370. S. Rönnberg, M. Wahlberg, M. Bollen, A. Larsson, and M. Lundmark. Measurements of interaction between equipment in the frequency range 9 to 95 kHz. In *International Conference on Electricity Distribution (CIRED)*, Prague, June 2009.

371. S. K. Rönnberg, M. Wahlberg, E.O.A. Larsson, M.H.J. Bollen, and C.M. Lundmark. Interaction between equipment and power line communication: 9–95 kHz. In *IEEE Bucharest PowerTech*, June 2009.

372. Z. Saad-Saoud and N. Jenkins. Models of predicting flicker induced by large wind turbines. *IEEE Transactions on Energy Conversion*, 14(3):743–748, 1999.

373. A.M. Salamah, S.J.. Finney, and B.W. Williams. Single-phase voltage source inverter with a bidirectional buckboost stage for harmonic injection and distributed generation. *IEEE Transactions on Power Electronics*, 24(2):376–387, 2009.

374. S. Schostan, K.-D. Dettmann, D. Schulz, and J. Plotkin. Investigation of an atypical sixth harmonic current level of a 5 MW wind turbine configuration. In *International Conference on "Computer as a Tool" (EUROCON)*, September 2007.

375. Seabased AB, http://www.seabased.com, accessed March 31, 2010.

376. C. Seife. *Sun in a Bottle*, Viking, 2008.

377. H. Seljeseth. Voltage variations averaging interval: 10 minutes or 1 minute? In *Technical Workshop on Voltage Quality Standards*, Milan, Italy, September 29, 2006.

378. H. Seljeseth, K. Sand, and K.E. Fossen. Laboratory tests of electrical appliances immunity against voltage swells. In *International Conference on Electricity Distribution (CIRED)*, Paper 0781, June 2009.

379. T. Senjyu, Y. Miyazato, A. Yona, N. Urasaki, and T. Funabashi. Optimal distribution voltage control and coordination with distributed generation. *IEEE Transactions on Power Delivery*, 23(2):1236–1242, 2008.

380. A. Shafiu, T. Bopp, I. Chilvers, G. Strbac, N. Jenkins, and H. Li. Active management and protection of distribution networks with distributed generation. In *IEEE Power Engineering Society General Meeting*. Vol. 1, June 2004.

381. A. Shafiu, V. Thornley, N. Jenkins, and A. Maloyd. Control of active networks. In *International Conference on Electricity Distribution (CIRED)*, Turin, 2005.

382. J.A. Silva, H.B. Funmilayo, and K.L. Butler-Purry. Impact of distributed generation on the IEEE 34 node radial test feeder with overcurrent protection. In *39th North American Power Symposium*, September 2007, pp. 49–57.

383. R. Simanjorang, Y. Miura, T. Ise, S. Sugimoto, and H. Fujita. Application of series type BTB converter for minimizing circulating current and balancing power transformers in loop distribution lines. In *Power Conversion Conference (PCC'07)*, April 2–5, 2007, pp. 997–1004.

384. G. Sinden. *Wind Power and the UK Wind Resource*. Environmental Change Institute, University of Oxford, 2005.

385. C.L. Smallwood and J. Wennermark. Benefits of distribution automation: improving the reliability, availability, and operation of United's electric distribution system. *IEEE Industry Applications Magazine*, January/February 2010, pp. 65–73.

386. J.C. Smith and B. Parsons. What does 20 percent look like? Developments on wind technology and systems. *IEEE Power and Energy Magazine*, 5(6):22–33, 2007.

387. J.C. Smith and B. Parsons. It's in the air: increased wind integration around the globe. *IEEE Power and Energy Magazine*, 7(6):14–24, 2009.

388. J.R. Smith and M.-J. Chen. *Three-Phase Electrical Machine Systems: Computer Simulations*, Research Studies Press, Taunton, UK, 1993.

389. B. Sørensen. *Renewable Energy: Its Physics, Engineering, Environmental Impacts, Economics and Planning*, 2nd edition, Academic Press, San Diego, CA, 2000.

390. South Africa Info, SA's hydro power potential. July 2004, http://www.southafrica.info, accessed April 3, 2010.

391. M. Speychal, M.H.J. Bollen, and K. Lindén. Impact of increasing levels of DER on the dip frequency in transmission systems. In *Probabilistic Methods Applied to Power Systems (PMAPS)*, Stockholm, Sweden, June 2006.

392. M. Steibler. *Wind Energy Systems for Electric Power Generation*, Springer, Heidelberg, 2008.

393. N. Stern. *Stern Review Report on the Economics of Climate Change*. Cambridge University Press, 2006.

394. J. Stevens, R. Bonn, J. Ginn, S. Gonzalez, and G. Kern. Development and testing of an approach to anti-islanding in utility-interconnected photovoltaic systems. Sandia report SAND 2000-1939, Sandia National Laboratories, August 2000.

395. C. Strauss. *Practical Electrical Network Automation and Communication Systems*, Elsevier, 2003.

396. G. Strbac, J. Mutale, and T. Bopp. Business models in a world characterised by distributed generation. D.1.2. Analysis of distributed generation characteristics. Technical report, EESD Project NNE5/2001/256 BUSMOD, 2004.

397. T. Sun. Power quality of grid-connected wind turbines with DFIG and their interaction with the grid. PhD thesis, Institute of Energy Technology, Aalborg University of Technology, Aalborg, Denmark, March 2004.

398. Kriterier vid spänningsgodhet vid leveransspänning över 1000 V (Voltage quality criteria for voltage levels above 1000 V, in Swedish). Svensk Energi, no. 30391, November 2004.

399. J.O.G. Tande. Impact of wind turbines on voltage quality. In *8th International Conference on Harmonics and Quality of Power*, October 1998, pp. 1158–1161.

400. C. Taylor. *Power System Voltage Stability*, McGraw-Hill, New York, 1994.

401. C.F. Ten and P.A. Crossley. Evaluation of ROCOF relay performance on networks with distributed generation. In *International Conference on Developments in Power System Protection (DPSP 2008)*, March 2008.

402. Tenessee Valley Authority, Raccoon Mountain Pumped-Storage Plant, http://www.tva.gov/sites/ raccoonmt.htm, accessed December 29, 2009.

403. S. Tentzerakis, N. Paraskevopoulou, S. Papathanassiou, and P. Papadopoulos. Measurement of wind farm harmonic emissions. In *IEEE Power Electronics Specialists Conference (PESC 2008)*, 2008.

404. S.T. Tentzerakis and S.A. Papathanassiou. An investigation of the harmonic emissions of wind turbines. *IEEE Transactions on Energy Conversion*, 22(1):150–158, 2007.

405. T. Thiringer. Power quality measurements performed on a low-voltage grid equipped with two wind turbines. *IEEE Transactions on Energy Conversion*, 11(3):601–606, 1996.

406. T. Thiringer and J. A. Dahlberg. Periodic pulsations from a three-bladed wind turbine. *IEEE Transactions on Energy Conversion*, 16(3):128–133, 2001.

407. T. Thiringer, T. Petru, and S. Lundberg. Flicker contribution from wind turbine installations. *IEEE Transactions on Energy Conversion*, 19(1):157–163, 2004.

408. M. Thomson and D.G. Infield. Impact of widespread photovoltaics generation on distribution systems. *IET Renewable Power Generation*, 1(1):33–40, 2007.

409. V. Thornley, N. Jenkins, P. Reay, J. Hill, and C. Barbier. Field experience with active network management of distribution networks with distributed generation. In *International Conference on Electricity Distribution (CIRED)*, Vienna, May 2007.

410. A. Timbus, M. Larsson, and C. Yuen. Active management of distributed energy resources using standardized communications and modern information technologies. *IEEE Transaction on Industrial Electronics*, 56(10):4029–4037, 2009.

411. S. Toma, T. Senjyu, Y. Miyazato, A. Yona, T. Funabashi, A.Y Saber, and C.-H. Kim. Optimal coordinated voltage control in distribution system. In *IEEE Power Engineering Society General Meeting*, Pittsburgh, PA, July 2007.

412. T. Tran-Quoc, T.M.C Le, C. Kieny, N. Hadjsaid, S. Bacha, C. Duvauchelle, and A. Almeida. Local voltage control of PVs in distribution networks. In *International Conference on Electricity Distribution (CIRED)*, Prague, June 2009.

413. T. Tran-Quoc, E. Monnot, G. Rami, A. Almeida, Ch. Kieny, and N. Hadjsaid. Intelligent voltage control in distribution network with distributed generation. In *International Conference on Electricity Distribution (CIRED)*, Vienna, May 2007.

414. Final report of the investigation committee on the 28 September 2003 blackout in Italy. Technical report, UCTE, April 2004, http://www.ucte.org/_library/otherreports/20040427_UCTE_IC_Final_report.pdf, accessed February 22, 2009.

415. Final report system disturbance on 4 November 2006. Technical report, UCTE, 2006.

416. UTCE Operations Handbook, http://www.entsoe.eu, accessed December 2009.

417. T. van Cutsem and C. Vournas. *Voltage Stability of Electric Power Systems*, Kluwer Academic Publishers, Dordrecht, The Netherlands, 1998.

418. I. van der Hoven. Power spectrum of horizontal wind speed in the frequency range from 0.0007 to 900 cycles per hour. *Journal of Meteorology*, 14(2):160–164, 1957.

419. F. Vandenberghe. Lessons and conclusions from the 28 September 2003 blackout in Italy. In *IEA Workshop*, March 29, 2004, http://www.iea.org/textbase/work/2004/transmission/vandenberghe.pdf accessed February 22 , 2009.

420. Vattenfall, Our project on CCS: carbon capture and storage, http://www.vattenfall.com, accessed April 3, 2010.

421. J. Verboomen, D. Van Hertem, P. Schavemaker, W. Kling, and R. Belmans. Border-flow control by means of phase shifting transformers, In *IEEE Lausanne PowerTech*, July 2007.

422. J. Vesterlund. Wind power plants—investigate conditions for measurements: current and voltage (in Swedish). MSc thesis, Department of Applied Physics and Electronics, Umeå University, January 8, 2009.

423. F.A. Viawan and D. Karlsson. Voltage and reactive power control in systems with synchronous machine-based distributed generation. *IEEE Transactions on Power Delivery*, 23(2):1079–1087, 2008.

424. F.A. Viawan, F. Vuinovich, and A. Sannino. Probabilistic approach to the design of photovoltaic distributed generation in low voltage feeder. In *International Conference on Probabilistic Methods Applied to Power Systems*, Stockholm, June 2006.

425. A. Viehweider, B. Bletterie, and D. Burnier De Castro. Advanced coordinated voltage control strategies for active distribution network operation. In *International Conference on Electricity Distribution (CIRED)*, Prague, June 2009.

426. C. Vilar, J. Usaola, and H. Amaris. A frequency domain approach to wind turbines for flicker analysis. *IEEE Transactions on Energy Conversion*, 18(2):335–341, 2003.

427. N. Wade, P. Taylor, P. Lang, and J. Svensson. Energy storage for power flow management and voltage control on an 11 kV UK distribution network. In *International Conference on Electricity Distribution (CIRED)*, Prague, June 2009.

428. C.L. Wadhwa. *Electrical Power Systems*, 2nd edition, Wiley Eastern Limited, New Delhi, 1991.

429. V.E. Wagner, J.P. Staniak, and T.L. Orloff. Utility capacitor switching and adjustable-speed drives. *IEEE Transactions on Industry Applications*, 27(4):645–651, 1991.

430. G.J. Wakileh. Harmonics in rotating machines. *Electric Power Systems Research*, 66(1):31–37, 2003.

431. G.R. Walker and P.C. Sernia. Cascaded DC–DC converter connection of photovoltaic modules. *IEEE Transaction on Power Electronics*, 19(4):1130–1139, 2004.

432. R.A. Walling, R. Saint, R.C. Dugan, J. Burke, and L.A. Kojovic. Summary of distributed resources impact on power delivery systems. *IEEE Transactions on Power Delivery*, 23(3):1636–1644, 2008.

433. M. Wämundson, M.H.J. Bollen, L. Bertling, and J. Svensson. On using operational risk assessment in transmission system operation. In *IEEE/PES General Meeting*, 2010.

434. J. Wang, D. Xu, B. Wu, and Z. Luo. A low-cost rectifier topology with variable-speed control capability for high-power PMSG wind turbines. In *IEEE Energy Conversion Congress and Exposition (ECCE'09)*, 2009.

435. S. Wang. Distributed generation and its effect on distribution network system. In *International Conference on Electricity Distribution (CIRED)*, Prague, June 2009.

436. W. Wang, Z.-C. Pan, W. Cong, C.-G. Yu, and F. Gu. Impact of distributed generation on relay protection and its improved measures. In *China International Conference on Electricity Distribution (CICED)*, December 2008.

437. I. Wangensteen. *Power System Economics: The Nordic Electricity Market*. Tapir Academic Press, Trondheim, 2007.

438. I. Wasiak, M.C. Thoma, C.E.T. Foote, R. Mienski, R. Pawelek, P. Gburczyk, and G.M. Burt. A power-quality management algorithm for low-voltage grids with distributed resources. *IEEE Transactions on Power Delivery*, 23(2):1055–1062, 2008.

439. Wavebob, http://www.wavebob.com, accessed March 31, 2010.

440. B.M. Weedy and B.J. Cory. *Electric Power Systems*, 4th edition Wiley, Chichester, UK, 1998.

441. D. Westermann, S. Nicolai, and P. Bretschneider. Energy management for distribution networks with storage systems – a hierarchical approach. In *IEEE Power Engineering Society General Meeting*, Pittsburgh, PA, July 2008.

442. Whispergen heat and power system, http:www.whispergen.com, accessed April 4, 2010.

443. W. Wiechowski and P.B. Eriksen. Selected studies on offshore wind farm cable connections: challenges and experience of the Danish TSO. In *IEEE Power and Energy Society General Meeting*, July 2008.

444. Annapolis Royal Generating Station, http://en.wikipedia.org/, accessed March 31, 2010.

445. Carbon capture and storage, http://en.wikipedia.org/, accessed April 3, 2010.

446. Compressed-air energy storage, http://en.wikipedia.org/, accessed December 29, 2009.

447. High-voltage direct current, http://en.wikipedia.org/, accessed February 23, 2010.

448. Pacific DC intertie, http://en.wikipedia.org/, accessed February 23, 2010.

449. Pumped-storage hydroelectricity, http://en.wikipedia.org/, accessed December 29, 2009.

450. Rance Tidal Power Station, http://en.wikipedia.org/, accessed March 31, 2010.

451. Solar energy generating systems, http://en.wikipedia.org/, accessed March 14, 2010.

452. Solar power, http://en.wikipedia.org/, accessed March 14, 2010.

453. Wave power, http://en.wikipedia.org/, accessed March 31, 2010.

454. H.L. Willis. *Power Distribution Planning Reference Book*, Marcel Dekker, New York, 1997.

455. S. Wilson, M. Gosden, and I. Povey. Experiences gained from the monitoring of an LV network supplying a micro-CHP cluster. In *International Conference on Electricity Distribution (CIRED)*, Prague, June 2009.

456. T. Wizelius. *Vindkraft i teori och praktik*, Studentlitteratur, Lund, 2002.

457. J. Wong, P. Baroutis, R. Chadha, R. Iravani, M. Graovac, and X. Wang. A methodology for evaluation of permissible depth of penetration of distributed generation in urban distribution systems. In *IEEE Power Engineering Society General Meeting*, Pittsburgh, PA, July 2008.

458. A.J. Wood. *Local Energy: Distributed Generation of Heat and Power*, The Institution of Engineering and Technology, London, UK, 2008.

459. World Nuclear Association, World nuclear power reactors and uranium requirements, http://www.world-nuclear.org, accessed April 3, 2010.

460. A. Woyte, V.V. Thong, and R. Belmans. Voltage fluctuations on distribution level introduced by photovoltaic systems. *IEEE Transactions on Energy Conversion*, 21(1):202–209, 2006.

461. World Wind Energy Report 2008, World Wind Energy Association (WWEA), Bonn, Germany, http://www.wwindea.org, February 2009.

462. S. Yanagawa, T. Kato, W. Kai, A. Tabata, Y. Yokomizu, T. Okamoto, and Y. Suzuoki. Evaluation of LFC capacity for output fluctuation of photovoltaic power generation systems based on multi-point observation of insolation. In *Power Engineering Society Summer Meeting*, Vancouver, BC, Canada, July 2001.

463. Y. Yang and K.-E. Högberg. Wind power and its impact on operational risk assessment – tripping of generation due to voltage dips. Report R08-522, STRI AB, Ludvika, Sweden, http://www.statnett.no, 2008.

464. Z. Yu and K. Ogboenyira. Renewable energy through micro-inverters. *Power Electronics Technology*, April 2009.

465. S. Yuvarajan and J. Shoeb. A fast and accurate maximum power point tracker for PV systems. In *23rd IEEE Applied Power Electronics Conference and Exposition (APEC'08)*, 2008.

466. R. Zavadil, N. Miller, A. Ellis, E. Muljadi, E. Camm, and B. Kirby. Interconnecting wind generation into the power system. *IEEE Power and Energy Magazine*, 5(6):47–58, 2007.

467. X. Zeng, K.K. Li, W.L. Chan, and S. Sheng. Multi-agents based protection for distributed generation systems. In *IEEE International Conference on Electric Utility Deregulation, Restructuring and Power Technologies (DRPT)*, April 2004.

468. L.D. Zhang and M.H.J. Bollen. Characteristic of voltage dips (sags) in power systems. *IEEE Transactions on Power Delivery*, 15(2):827–832, 2000.

469. Q. Zhang, L. Qian, C. Zhang, and D. Cartes. Study on grid connected inverter used in high power wind generation system. In *41st IAS Annual Meeting*, October 2006.

470. X. Zhu, D. Xu, and P. Chen. Energy management analysis and design for a 5 kW PEMFC distributed power system. In *IEEE Power Electronics Specialists Conference (PESC'08)*, June 2008.

INDEX

Integration of Distributed Generation in the Power System, First Edition. Math Bollen and Fainan Hassan.

1. *Principles of Electric Machines with Power Electronic Applications, Second Edition*
M. E. El-Hawary

2. *Pulse Width Modulation for Power Converters: Principles and Practice*
D. Grahame Holmes and Thomas Lipo

3. *Analysis of Electric Machinery and Drive Systems, Second Edition*
Paul C. Krause, Oleg Wasynczuk, and Scott D. Sudhoff

4. *Risk Assessment of Power Systems: Models, Methods, and Applications*
Wenyuan Li

5. *Optimization Principles: Practical Applications to the Operations of Markets of the Electric Power Industry*
Narayan S. Rau

6. *Electric Economics: Regulation and Deregulation*
Geoffrey Rothwell and Tomas Gomez

7. *Electric Power Systems: Analysis and Control*
Fabio Saccomanno

8. *Electrical Insulation for Rotating Machines: Design, Evaluation, Aging, Testing, and Repair*
Greg Stone, Edward A. Boulter, Ian Culbert, and Hussein Dhirani

9. *Signal Processing of Power Quality Disturbances*
Math H. J. Bollen and Irene Y. H. Gu

10. *Instantaneous Power Theory and Applications to Power Conditioning*
Hirofumi Akagi, Edson H. Watanabe, and Mauricio Aredes

11. *Maintaining Mission Critical Systems in a 24/7 Environment, Second Edition*
Peter M. Curtis

12. *Elements of Tidal-Electric Engineering*
Robert H. Clark

13. *Handbook of Large Turbo-Generator Operation Maintenance, Second Edition*
Geoff Klempner and Isidor Kerszenbaum

14. *Introduction to Electrical Power Systems*
Mohamed E. El-Hawary

15. *Modeling and Control of Fuel Cells: Disturbed Generation Applications*
M. Hashem Nehrir and Caisheng Wang

16. *Power Distribution System Reliability: Practical Methods and Applications*
Ali A. Chowdhury and Don O. Koval

17. *Introduction to FACTS Controllers: Theory, Modeling, and Applications*
Kalyan K. Sen and Mey Ling Sen

18. *Economic Market Design and Planning for Electric Power Systems*
James Momoh and Lamine Mili

19. *Operation and Control of Electric Energy Processing Systems*
James Momoh and Lamine Mili

20. *Restructured Electric Power Systems: Analysis of Electricity Markets with Equilibrium Models*
Xiao-Ping Zhang

21. *An Introduction to Wavelet Modulated Inverters*
S.A. Saleh and M. Azizur Rahman

22. *Probabilistic Transmission System Planning*
Wenyuan Li

23. *Control of Electric Machine Drive Systems*
Seung-Ki Sul

24. *High Voltage and Electrical Insulation Engineering*
Ravindra Arora and Wolfgang Mosch

25. *Practical Lighting Design with LEDs*
Ron Lenk and Carol Lenk

26. *Electricity Power Generation: The Changing Dimensions*
Digambar M. Tagare

27. *Electric Distribution Systems*
Abdelhay A. Sallam and Om P. Malik

28. *Maintaining Mission Critical Systems in a 24/7 Environment, Second Edition*
Peter M. Curtis

29. *Power Conversion and Control of Wind Energy Systems*
Bin Wu, Yongqiang Lang, Navid Zargan, and Samir Kouro

30. *Integration of Distributed Generation in the Power System*
Math Bollen and Fainan Hassan

Forthcoming Titles

Doubly Fed Induction Machine: Modeling and Control for Wind Energy Generation Applications
Gonzalo Abad, Jesus Lopez, Miguel Rodriguez, Luis Marroyo, and Grzegorz Iwanski

High Voltage Protection for Telecommunications
Steven W. Blume

CPSIA information can be obtained
at www.ICGtesting.com
Printed in the USA
BVOW06*0741060817
490848BV00003B/1/P